Neuroscience Discoveries

Neuroscience Discoveries

Edmund T. Rolls

The MIT Press
Cambridge, Massachusetts
London, England

The MIT Press
Massachusetts Institute of Technology
77 Massachusetts Avenue
Cambridge, MA 02139
mitpress.mit.edu

The MIT Press would like to thank the anonymous peer reviewers who provided comments on drafts of this book. The generous work of academic experts is essential for establishing the authority and quality of our publications. We acknowledge with gratitude the contributions of these otherwise uncredited readers.

This book was typeset by the author in LaTeX using the WinEdt editor. Printed and bound in the United States of America.

Library of Congress Cataloging-in-Publication Data is available.

ISBN: 978-0-262-05772-1

10 9 8 7 6 5 4 3 2 1

EU Authorised Representative: Easy Access System Europe, Mustamäe tee 50, 10621 Tallinn, Estonia | Email: gpsr.requests@easproject.com

Contents

Preface

Enormous progress has been made in understanding how the brain works in the last 50 years. This book has the aim of showing how discoveries have been and are being made in understanding the brain. That is, the book aims to show the thought processes, collaborations, and the developments of this area of science in this golden era for making advances in understanding the brain processes that underlie behavior.

This is highlighted by the fact that when I was an undergraduate at the University of Cambridge reading medicine, almost always whenever I asked the distinguished members of the academic staff there: "What do we know about this part of the brain and how it functions?" the answer was usually and honestly "very little". An aim of this book is to show the thinking that was behind many discoveries that have been made since then about how our brains work.

I hope that this book will help to show not only some of the thinking behind research discoveries, but will also be valuable to others, by encouraging others to see how some ideas in neuroscience have been generated, and to try the same processes themselves.

When providing insight to others on how discoveries in at least this area of science are made, it is important to be able to provide a first-person account of how some of the discoveries have been and are being made, for that will illuminate some of the thought processes involved, and is not the type of material that appears in scientific papers or in textbooks. The value of first-hand information about how scientists have made discoveries in neuroscience was highlighted in 2025 in an issue of the journal *Hippocampus* devoted to discoveries about the hippocampus (Hasselmo and Nadel, 2025).

For this reason, many of the discoveries highlighted in this book by a first-hand description are the discoveries made by Edmund Rolls, often in collaboration with colleagues from other disciplines such as theoretical physics and mathematics. Indeed, an aim of this book is to highlight the importance of multidisciplinary collaboration when making discoveries about the brain, for to understand the brain and how it computes, evidence from a very wide range of disciplines is needed, including from neuronal neurophysiology during behavior, functional neuroimaging, neuroanatomy, neuropharmacology, psychology, clinical neuropsychology, neuropsychiatry, theoretical physics, mathematics, and computer science, all of which feature in the discoveries described here.

The overall plan of the book is as follows.

In Chapter 1, background information is provided about the approach being taken to understanding *What* each part of the brain does, and about *How* it implements these functions. Chapter 1 also shows how relatively little was known about how the brain works until some of the discoveries made by David Hubel and Torsten Wiesel in the 1960s, which I built on by performing many investigations of neuronal activity in many brain regions during behavior, and complementing this with neuroimaging in humans, with clinical neuropsychology in patients with brain damage, and with computational neuroscience to provide an understanding of how the brain might work. Chapter 1 also provides an introduction to some of the different cortical regions and their functions.

Chapter 2 describes how some of my early discoveries on reward systems in the brain were made, and of the role of serendipity in them.

That research led to the research described in Chapter 3, in which brain reward systems such as the orbitofrontal cortex are being analyzed during behavior, and how this is leading to theories of emotion, and motivation, and to some of their disorders such as depression.

Chapter 4 describes how we wished to find how visual stimuli reach some of these brain regions involved in emotion, so we recorded in the inferior temporal visual cortex, and discovered not only face cells that encode face identity, but also face expression cells. Those discoveries led on to discoveries about the key properties of what is represented in these cortical regions: how the neuronal representations are invariant with respect to transforms such as different views or sizes of the stimuli. We went on to show how those invariant representations are fundamental for understanding the operation of memory and emotion brain system that receive from the temporal lobe visual cortical areas, for they enable generalization to other views etc of a person when they are shown later. This generalization enables our brains to automatically recall the appropriate memory and emotion-related associations of the person or object, even when it has transformed to a different view of the person or object. But to show how all of these representations and their properties that we discovered are computed, we needed to develop a theory and model of how these visual and cortical regions compute transform-invariant representations of objects and faces, etc. And we did go on to develop a biologically plausible theory of how the cortex builds these transform-invariant object, face and person representations for visual perception.

Chapter 5 describes how a revolution is taking place in our understanding of the hippocampal episodic memory system in our brains, in that the spatial representation in the brains of humans and other primates is typically the location in visual space at which we are looking. That discovery is in contrast to what is found in rodents (rats and mice), in which hippocampal neurons represent the place where the individual is located. Our discovery has major implication for what humans and other primates store and can later recall in our episodic memory systems that can associate what happened with where it happened as encoded by where we are looking. Our discovery also has major implications for how we navigate in the world using visual stimuli such as landmarks, instead of relying on place-to-place self-motion in the dark, which was the model of navigation developed for rodents. These discoveries on memory systems are complemented by discoveries about how these systems work using analytic computational models that show how the memory systems in our brains could work computationally, and by discoveries of the cortical connectivity of our memory systems in humans using tractography, fMRI and magnetoencephalography.

Chapter 6 describes how a foundation is being built for how the human brain functions. The foundation involves measuring the connectivity between the smallest cortical regions that can be anatomically and functionally delineated, and analyzing the computations performed in each of these cortical regions.

Chapter 7 describes how computational neuroscience is helping us to understand how the brain works probabilistically with 'noise' or randomness generated by the close to random spike firing times of individual neurons for a given mean firing rate; and how this stochastic cortical neurodynamics has major applications to understanding human decision-making, psychiatric disorders, and even creativity.

Chapter 8 describes how we have made progress in understanding the computations of many brain systems for sensory processing, memory, and emotion using biologically plausible computations with the synaptic connection strengths between neurons being learned by rules that rely on the presynaptic and postsynaptic firing rates. A key issue here is that these brain computations as presently understood are very different indeed from the deep learning error backpropagation methods that are applied in artificial intelligence algorithms. In order to understand the more complex processing involved in human language, will it be necessary and possible to develop deep learning methods that are biologically plausible; or does the human brain solve language in very different ways to those used in current AI machines? The book thus finishes by pointing to key questions that need to be addressed in future about how the human brain computes for language and reasoning, and what the relations are between the algorithms used for computations in the brain, and the algorithms used for AI, which may well be quite different. These are key issues for the future in neuroscience.

Part of the material described in the book reflects research performed over many years in collaboration with many colleagues, whose tremendous contributions are warmly appreciated. The contributions of many will be evident from the references cited in the text. In addition, I have benefited enormously from the discussions I have had with a large number of colleagues and friends, many of whom I hope will see areas of neuroscience in the text that they have been able to illuminate. Much of the work described would not have been possible without financial support from a number of sources, particularly the Medical Research Council of the UK, the Human Frontier Science Program, the Wellcome Trust, the McDonnell-Pew Foundation, and the Commission of the European Communities.

The book was typeset by the author in LaTeXusing the WinEdt editor. The .pdf version of the book has clickable links.

Updates to the publications cited in this book and .pdf files of many papers are available at https://www.oxcns.org. A summary of some of my discoveries in neuroscience with links to the published papers is available at https://www.oxcns.org/Discoveries.html

Edmund T. Rolls dedicates this work to the overlapping group: his family, friends, and colleagues: *in salutem praesentium, in memoriam absentium.*

Edmund T Rolls, MA, DPhil, DSc, Hon DSc
Oxford Centre for Computational Neuroscience
Oxford, England
https://www.oxcns.org
ORCID ID: https://orcid.org/0000-0003-3025-1292

1 Introduction

1.1 Introduction

The overall aim of this book is to provide insight into how scientific discoveries have been and are being made about how our brains function.

This insight can come best from those who have made the discoveries, for then they can describe how one research idea provided the foundation for the next insight and step forward.

I am in the privileged position of having made many neuroscience discoveries from 1970 until the present, which has been a golden era of advance in understanding our brains. The aim of this book is to show what was known at the time of each discovery, and how the thought processes led to each new discovery. That is the type of information that is not typically provided in research papers or textbooks, where instead one is encouraged to provide a general overview of the field as an introduction to a research paper, rather than to how the new ideas for the research were actually generated. That is one reason why I think that books such as the present book thus may be not only interesting, but also valuable to others, by encouraging others to see how some ideas have been generated, and to try the same processes themselves.

Another aim of this book is to show how by being guided by one's previous research discoveries, one can build whole areas of research by moving steadily forward by allowing each discovery to build on the previous discovery. At least in neuroscience this is very important, because we are dealing with brain systems that need to be followed from their start in the brain forward through the system, in order to understand how each stage contributes to the overall function or computation being performed. Each system might be involved in vision, or taste, or smell, or touch, and an aim of this book is to show how by analyzing what occurs at each successive stage of each system, it is possible to generate clear hypotheses and eventually theories about how whole brain systems operate.

It is then possible to go beyond the individual brain system, and understand how the different systems combine to enable for example flavor to be built from the taste, smell, texture and even sight of food.

But it is possible to go beyond this, by moving ahead to how memory systems work based on the sensory inputs; and how reward and pleasure systems and emotional systems in our brains operate, again with inputs from sensory systems.

Now, all of this might provide an approach to **what** is being performed (or computed) by each brain region, but it is then necessary for a true understanding to move forward to **how** it is computed, that is to the mechanism level. This is what the title of my book **Brain Computations: What and How** (Rolls, 2021a) was aiming at. Understanding brain computation at the mechanistic level is very important, for it is typically at that level that we can understand some mental disorders, and how we can sometimes treat brain dysfunction, for example with drugs that operate on key brain systems.

Indeed, key aspects of my research in neuroscience are to combine evidence from neuronal activity in the brain with computational neuroscience approaches; to complement that by investigations of the connectivity between each computational region in the human brain to help understand what each brain region computes given its inputs and outputs; and to combine this with neuroimaging and neuropsychology in humans to help understand how brain systems work (compute) in humans in health and disease.

An aim of this book is thus to show that these are some of the ways in which a coherent body of research, about in this case brain function, can be generated. Indeed, what is described here and elsewhere (Rolls, 2023c, 2026d) shows how whole series of discoveries can be important in order to accumulate evidence of what is represented in different brain systems and in different parts of brain systems. This approach is essential for understanding the computational functions performed by different brain regions and by the brain as a whole (Rolls, 2023c, 2026d). I show how this integrative strategy with sets of related discoveries is important to understand how each function such as vision, memory, and emotion, is performed by our brains. However, neuroscientists who understand this integrative approach should also be aware that awards for scientific research are often given for single discoveries, rather than a set of integrated discoveries of the type necessary to understand how our brains work, that is, perform their computations.

Another aim of this book is to emphasize that conceptual advances, and theories about how a brain system works or computes, are an essential part of how neuroscience has been developing. Making conceptual advances about how our brains work is more important than just using the latest technique to perform a detailed analysis of what is essentially already known conceptually, although these technical advances are often what is reported in high impact factor journals. An aim is accordingly to provide insight into how some of these conceptual advances, and empirical advances that are far beyond anything known at the time, have been and are being made, using first-person knowledge about some of these discoveries made by me with my colleagues.

The aim to show how conceptual advances are made in understanding brain function happens is greatly facilitated by the fact that this book covers a period from about 1970 onwards when there was very little understanding of what was computed by each brain region and how it was computed, through the period up to the present day. This is a period in which major conceptual advances have been made in understanding our brains. In a sense, this is the amazing period for neuroscience when many conceptual advances were made, about for example the brain systems in primates including humans involved in emotion, reward, decision-making, vision, memory, and psychiatric disorders such as depression, that are described in this book.

It is important to emphasize that discoveries in the neurosciences typically involve a research group or team, and often benefit enormously from interdisciplinary collaboration. I therefore wish to acknowledge all of my collaborators who have been so important in the discoveries of my research group, on which much of this book is based. The book is based on the discoveries in which I have been involved, because a first-person account of how discoveries are being made in neuroscience is the aim here. The names of my collaborators in these discoveries will be evident from the authors on each research paper. In addition, an aim of this book is to highlight some of my interdisciplinary collaborations, as they are so important for making substantial advances in understanding systems as complex as our brains. Most of my research has been on humans or non-human primates (rather than rodents whose brain systems are somewhat different), because I read pre-clinical medicine at Cambridge, have a vocation for medicine, and want my research to be relevant to advancing understanding of the human brain in health and disease, including mental disorders.

The present book does not aim to be a comprehensive guide to our current understanding of the computations performed by many brain systems. Instead, this book focusses on how discoveries in neuroscience have been and are being made, and some of the thought processes involved. It is hoped that at least some parts of the book will be of interest to the non-specialist, and it does not matter given the aims of this book if not all the neuroscience can be understood, for the thinking that leads to discoveries is of interest in its own right. In fact, a wider guide to the computations performed by many brain systems is made available as an Open Access book **Brain Computations and Principles; and AI** (Rolls, 2026d), and that may be consulted for more details of the actual neuroscience research. A .pdf of that book (and many of my other books) with clickable links can be downloaded free from my website https://www.oxcns.org.

1.2 A start in neuroscience

Here I describe the early pathway that led me towards making discoveries in neuroscience, which is I think of interest for it illustrates how decisions have to be made along the pathway to discoveries. I made a start in neuroscience at school, and then at Cambridge University by reading preclinical medicine, and then I was offered the opportunity to perform research for a DPhil at Oxford University, as described next.

While at Hardye's School, Dorchester, Dorset, close to the College where my father lectured, I developed a vocation to train in medicine, and was offered a place at the London Hospital Medical School. I applied there because I had an interest in the brain and neurology, and knew that Lord Brain of Brain's **Clinical Neurology** (Brain, 1960) was there. My interest in how the brain works had been stimulated by reading **Nerves, Brain and Man** (Grayson, 1961), and while at school I built an oscilloscope from a 3.5 inch cathode ray tube (requiring 4,000 volts on its anode) and valves (vacuum tubes) that were easily available at that time, and a low noise valve preamplifier, and was able to record from earthworm giant axons. This experience in electronics was invaluable for my later research, for part of what helped my research was being able to perform every part of it, including the electronics required, and programming computers, and even building small computational devices to measure neuronal firing rates before computers became generally available.

The University of Cambridge offered a 3-year course in preclinical medicine, and I opted for that, and graduated with a preclinical medicine qualification from Cambridge (exemption from the 2nd MB), and a BA in Natural Sciences (which later became an MA). The Physiological Laboratory at Cambridge was outstanding in neurophysiology, with Lord Adrian who was the first person to record neuronal action potentials there, and also Alan Hodgkin, and a host of others including Horace Barlow, Fergus Campbell, John Robson, Giles Brindley (with whom David Marr was working), Pat Merton, and later Colin Blakemore. All taught me how to carefully evaluate experimental evidence. But the great disappointment was that when I asked: "What do we know about this or that part of the brain and how it functions?" the answer was usually and honestly "very little".

That point emphasizes how little was known about neuroscience when I was an undergraduate. An aim of this book is to show the thinking that was behind many of the discoveries that have been made since then about how our brains work.

However, while I was an undergraduate and doing some study in a vacation in the University of Birmingham Medical School Library, I came across some of the early discoveries of David Hubel and Torsten Wiesel on the responses of visual cortex neurons to bars and edges (Hubel and Wiesel, 1962), and realized that for me, that was the way forward to start to understand better what each part of the brain was involved in, and the best way to find out how the brain worked. That early paper was a revelation to me, and I decided at that time that I would try to do some neurophysiological research, before coming back as I then intended to complete the clinical part of my medical training, at the London Hospital Medical School.

I was very fortunate as an undergraduate at Cambridge to be reading a Natural Sciences degree that also covered preclinical medicine, and was able to learn a great deal more about the brain by also reading Experimental Psychology. In Experimental Psychology I received lectures from Larry

Weiskrantz, Alan Cowey, and Richard Gregory, especially about what was known about visual and memory systems in the brain. In fact, I and David Marr (who had recently graduated in mathematics) were both present at Larry Weiskrantz's lectures to undergraduates on the hippocampus and memory, and Larry's lectures inspired both of us.

As an undergraduate, I was able to perform research on the relation between retinal cone and ganglion cell density (which I counted in 6 micron sections of the retina that I prepared with wonderful help from the histology chief technical member of staff in the Physiological Laboratory at Cambridge), and cortical magnification factor in the primary visual cortex V1, which were published in due course with Alan Cowey (Rolls and Cowey, 1970; Cowey and Rolls, 1975).

But for graduate research, I really wanted to do neurophysiology as I felt that provided a new way forward for scientific discoveries in the brain. Larry Weiskrantz suggested that I might apply to the University of Oxford, where David Vowles had a neurophysiology lab. I decided to take up neurophysiology for a graduate degree. When I arrived in Oxford to start my DPhil on neurophysiology research in the then Institute of Experimental Psychology, I found that Larry Weiskrantz was just joining at the same time, as the new (and first) Professor of Experimental Psychology at Oxford, and that Alan Cowey and Nick Humphrey were coming from Cambridge to Oxford at the same time.

The neurophysiology lab at Oxford that I set up was for a project on hormonal effects on single neuron activity in the anaesthetized brain of doves, to investigate how hormones affect behavior. But the hormones took approximately 20 minutes to influence neuronal activity, so it was slow work, with hours spent seeking to obtain repeatable evidence for significant effects on neuronal activity. During my first few months, David Vowles decided to leave Oxford and take up a Chair at Edinburgh, and I decided to stay in Oxford to continue with neurophysiology. After my first 8 months as a graduate student, I had the opportunity to travel to the University of Pennsylvania for the summer with Barbara Simons, where she could do research on angiotensin and thirst with Alan Epstein her former mentor from the University of Pennsylvania, and James Fitzsimons from Cambridge, where she was the first female PhD student in Physiology.

While at the University of Pennsylvania for 4 months of the summer, I was very fortunate to be invited by Randy Gallistel of Psychology to set up a neurophysiology lab at Penn to work on the neurophysiology of brain-stimulation reward. Because we were using electrical brain-stimulation reward, I soon found that I could investigate which neurons were activated by the reward, for they typically responded within 2–10 ms, allowing rapid replication of effects, which was a contrast to the slow hormone effects I had been investigating previously in Oxford. In that 4 months at the University of Pennsylvania, I was able to show that in rats two separate pathways were being activated by the brain-stimulation reward, distinguishable by their refractory periods, and by the different brain regions in which they were found. Being able to trace different pathways in the brain involved in different aspects of reward was the subject of my first publication, in the journal Science (Gallistel, Rolls and Greene, 1969).

When I returned to Oxford in October to start my second year of DPhil research, I decided to work on the neural bases of brain-stimulation reward, as that area of research allowed new discoveries to be made more rapidly and easily. There was no supervisor at Oxford for this area of research, so I supervised myself, and in fact very soon while I was a graduate student, I had a DPhil student of my own to supervise, Peter Kelly, who wanted to join my lab after obtaining the top degree in Physiology at Oxford.

The development of this new research on brain systems involved in reward, appetite, and emotion, and how discoveries were made, is taken up in Chapters 2 and 3.

1.3 An introduction to some of the brain systems described in this book

Much introductory information useful for understanding brain function, and the background to many of the discoveries described in this book, is available in Chapter 1 of the Open Access book **Brain Computations and Principles; and AI** (Rolls, 2026d) available for free download at https://www.oxcns.org. In addition, many of the research papers for the discoveries described in this book are available at https://www.oxcns.org. Here I provide an introduction to some of the cortical regions and systems present in the human brain that are described in this book.

A key part of my research is to understand what is computed in different brain regions and systems, and how it is computed (Rolls, 2021a, 2023c, 2026d). To understand the computations in the brain, it is essential to understand the systems-level organization of the brain, in order to understand how the networks in each region provide a particular computational function as part of an overall computational scheme. Discoveries on several of these brain systems-level processing streams are described in this book. These discoveries are based primarily on studies in non-human primates and humans so that they are relevant to understanding human brain function in health and disease, for the systems-level organization of many brain systems is very different between rodents, and primates including humans, as described in Section 29.4 of **Brain Computations and Principles; and AI** (Rolls, 2026d).

Here I provide an overview of some of the cortical regions described in this book, using as a basis the cortical regions and divisions described in the Human Connectome Project Multimodal Parcellation (HCP-MMP) atlas (Glasser, Coalson, Robinson, Hacker, Harwell, Yacoub, Ugurbil, Andersson, Beckmann, Jenkinson, Smith and Van Essen, 2016a), which is becoming increasingly used (Rolls, 2023c, 2026d). This atlas provides a very useful parcellation of human cerebral cortical regions (Glasser et al., 2016a,b). The atlas is multimodal in that each region is defined by a combination of criteria: architecture (myelin content and cortical thickness maps); resting-state functional connectivity; and task-based fMRI activation in seven tasks (Glasser et al., 2016a). The HCP-MMP v1.0 atlas includes 180 cortical regions in each hemisphere, based on analyses in 210 participants. This approach and its use of a surface-based analysis to iron out problems associated with particular regions being found at different locations with respect to sulci and 3D coordinates potentially makes this atlas able to discriminate more reliably between different cortical regions with potentially different functions and connectivity than any other atlas based on evidence from a single modality such as only functional connectivity.

Given the usefulness of the parcellation provided in the HCP-MMP atlas (Glasser et al., 2016a), we have provided an extended version of the atlas, HCPex, that extends it with 66 subcortical areas (Huang, Rolls, Feng and Lin, 2022). HCPex is implemented in a volumetric form useful for many types of neuroimaging software (e.g. SPM (Friston et al., 2006) https://www.fil.ion.ucl.ac.uk/spm/ and FSL (Jenkinson et al., 2012; Smith et al., 2004) https://fsl.fmrib.ox.ac.uk), and provides a reordered version of the cortical areas with an option to revert to the original order (Huang, Rolls, Feng and Lin, 2022). The order of cortical regions defined in HCPex and shown in Tables A.1–A.6 is used throughout this book. This atlas helps in the analysis of brain structure and function, for many of the cortical regions defined in the HCP-MMP / HCPex atlas are known about functionally, many can be related to corresponding regions in macaques in which much detailed neurophysiology is available, and the atlas helps new investigations of the brain to be related to the known functions of those anatomically and functionally defined brain regions.

The cortical regions defined in the HCP-MMP / HCPex atlas are shown in Figs. 1.1 and 1.2, with a list of the cortical regions in Appendix A in Tables A.1–A.6 and of the subcortical regions in HCPex in Table A.7. Coronal slices to further help visualization of the locations in the brain of

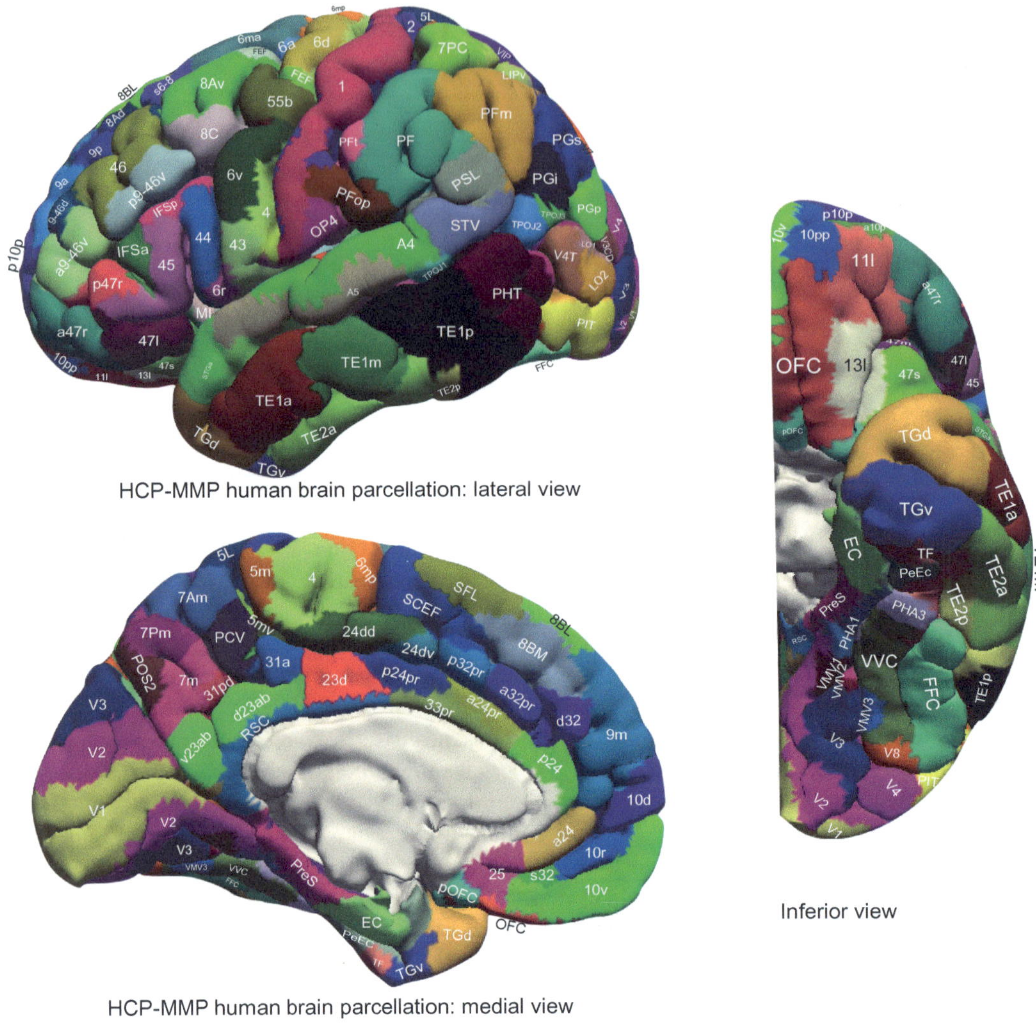

Fig. 1.1 Anatomical regions in the Human Connectome Project Multimodal Parcellation atlas (HCP-MMP) of the human cortical regions. Cortical regions are shown as defined in the HCP-MMP atlas (Glasser et al., 2016a), and in its extended version HCPex (Huang, Rolls et al. (2022)). The regions are shown on images of the human brain without the sulci expanded to show which cortical HCP-MMP regions are normally visible, for comparison with Fig. 1.2. Abbreviations are provided in Tables A.1–A.6.

these cortical regions are provided by Huang, Rolls, Feng and Lin (2022). Descriptions of many of these cortical regions including their connectivity and functions are provided in the different Chapters of this book. Short summaries of some of the key Divisions in the HCP-MMP and HCPex atlases (Glasser et al., 2016a; Huang et al., 2022) are provided next with reference to Tables A.1–A.6.

The main **visual processing cortical regions** are illustrated in Figs. 4.1, 6.1, 5.8 and 6.3. Their connectivity and functions are described in Chapters 4, 5 and 6 of this book, and in Chapters 2 and 3 of Rolls (2026d), and by Rolls et al. (2023b); Rolls (2024c); Rolls et al. (2023e, 2024c). In the HCP-MMP atlas, these visual cortical regions are in the Primary Visual, Early Visual, Dorsal Visual Stream, Ventral Visual Stream, and MT+ Complex Divisions in Table A.1, and in the Lateral Temporal Division in Table A.3.

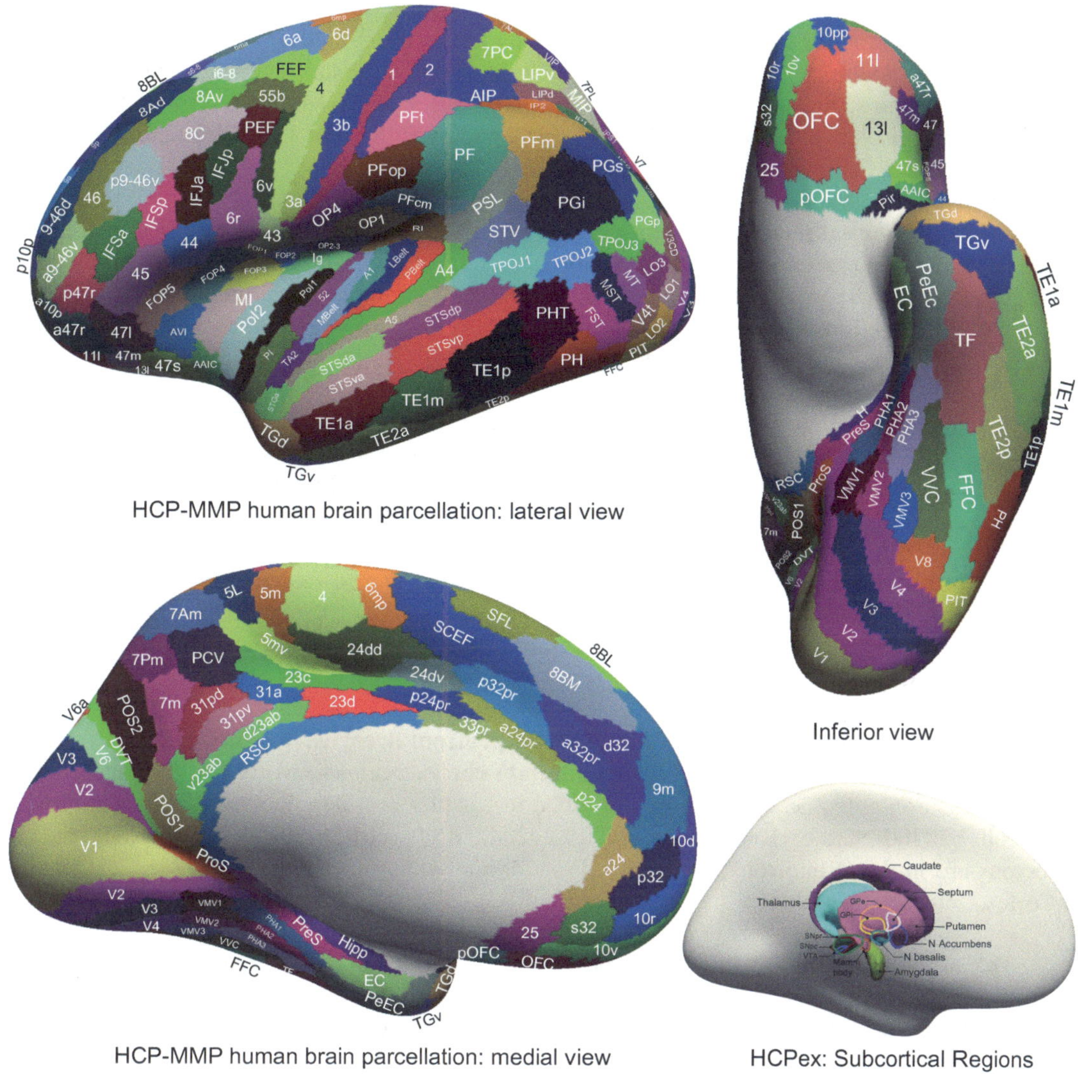

Fig. 1.2 Anatomical regions in the Human Connectome Project Multimodal Parcellation atlas HCP-MMP of the human cortical regions. Cortical Regions are shown as defined in the HCP-MMP atlas (Glasser et al., 2016a), and in its extended version HCPex which includes the thalamus and the other subcortical areas shown (Huang, Rolls et al. (2022)). The regions are shown on images of the human brain with the sulci expanded to show all cortical HCP-MMP regions, for comparison with Fig. 1.1. Abbreviations are provided in Tables A.1–A.6.

The main **somatosensory and motor cortical regions** are illustrated in Fig. 6.4 and described in Section 6.6 in this book, and in Chapters 6 and 15 of Rolls (2026d), and by Rolls et al. (2023c). The somatosensory and motor cortical regions are in the SomatoSensory, ParaCentral and MidCingulate, Premotor, and Posterior Opercular Divisions in Table A.2, and in the Insula / Frontal Opercular Division in Table A.3. Regions 23c, 24dd and 24dv are the midcingulate premotor cortex described by Vogt (2016) and Rolls et al. (2023g).

The main **auditory cortical regions** are illustrated in Fig. 6.6 and described in Section 6.7 in this book and in Chapter 7 of Rolls (2026d), and by Rolls et al. (2023f). The auditory cortical regions are in the Early Auditory, and Auditory Association Divisions in Table A.2. However some of the regions in the 'Auditory Association' Division are in fact multimodal, include visual and auditory responsiveness to social stimuli, and are also involved in semantic representations for language, as described in Chapters 8, 2 and 14 of Rolls (2026d), and by Rolls et al. (2022c, 2023b); Pitcher and Ungerleider (2021).

The **hippocampal memory system regions** are described in Chapter 5 and Section 6.10.5 of this book, in Chapters 9 and 10 of Rolls (2026d), and by Rolls et al. (2022a, 2023g,a); Rolls (2023a, 2022a); Rolls and Treves (2024); Rolls (2025a, 2026c). The hippocampal memory system regions are in the Medial Temporal Division in Table A.3, with its closely related regions in the Posterior Cingulate Division shown in Table A.4.

The human **language-related cortical regions** are illustrated in Figs. 6.7, 6.8, 6.9 and 6.11 and described in Section 6.10 of this book, in Chapter 14 of Rolls (2026d), and by Rolls et al. (2022c). The semantic language-related cortical regions are in the Temporo-Parietal-Occipital TPO Division in Table A.3, and regions 44, 45 and 47l (Broca's area) are in Table A.6.

The **parietal cortex regions** involved in visuo-spatial and related multimodal functions are illustrated in Fig. 6.5 and described in Section 6.5 of this book, in Chapter 10 of Rolls (2026d), and by Rolls et al. (2023a). The parietal cortex regions are in the Superior Parietal and Inferior Parietal Divisions in Table A.4.

The **orbitofrontal cortex and ventromedial prefrontal cortex regions** involved in reward valuation and emotion, and the **anterior cingulate cortex regions** involved in linking reward to action in action-outcome learning, are described in Chapter 3 of this book, in Chapters 11 and 12 of Rolls (2026d), and by Rolls et al. (2022b, 2023d); Rolls (2014a, 2019c, 2023b, 2025e). These regions are in the Orbitofrontal and Polar Frontal, and Anterior Cingulate and Medial Prefrontal Cortex Divisions in Table A.5.

The **subcortical regions** that are in the HCPex atlas (Huang, Rolls, Feng and Lin, 2022) and involved in many functions are listed in Table A.7, and their connectivity and functions are described in Chapter 16 and elsewhere in Rolls (2026d), and elsewhere (Rolls et al., 2023d, 2022b; Rolls, 2022a).

1.4 Introduction to three neuronal network architectures

A key part of my approach to neuroscience is to go beyond the phenomenological, empirical, investigation of hypotheses, and to combine this with theories and models of the computations performed by each brain region. I find it very productive to think about the brain computationally, for the brain is a computational system, and understanding it as such is helping to guide the empirical research towards addressing fundamental conceptual issues about how our brains work. At about the time that we were discovering cells that responded to the sight of food (Chapter 3), and face cells (Chapter 4), I was inspired by some of the books by Teuvo Kohonen (Kohonen, 1977, 1984) and by the book *Parallel Models of Associative Memory* (Hinton and Anderson, 1981), which helped to introduce models of how networks of neurons might perform well-defined computations.

In this section I introduce some key neuronal network architectures found in the brain, and show what they compute. These networks are important in our current understanding of how computations are performed by the brain (Rolls, 2026d). Reference to these networks is made in the Chapters in this book. These networks are biologically plausible in that the information required to change a synaptic weight by learning is available at the synapse, from the activity in the presynaptic terminal onto a neuron, and the activity of the postsynaptic neuron. Deep learning using error backpropagation as commonly used in Artificial Intelligence does not, at least so far, seem biologically plausible because it does not use a local learning rule of the type just described (Rolls, 2026d), so for now is not used in the biologically plausible models that I have developed for understanding brain function (see Chapter 8 and Rolls (2026d)). Details of the neuronal network architectures introduced next are provided in **Brain Computations and Principles; and AI** (Rolls, 2026d).

In the first architecture (see Fig. 1.3a and b), pattern associations can be learned. The output neurons are driven by an unconditioned stimulus (for example the taste of food). A conditioned stimulus (for example the sight of food) reaches the output neurons by associatively modifiable synapses w_{ij}. These associatively modifiable synapses increase in strength if there is high presynaptic firing (e.g. of the Conditioned Stimulus Inputs in Fig. 1.3a) and high post-synaptic firing or activation (e.g. firing of the output neuron shown in Fig. 1.3a). This is sometimes termed Hebbian synaptic modification, after Donald Hebb who proposed that a learning rule such as this might be present in the brain (Hebb, 1949), which it is (Takeuchi, Duszkiewicz and Morris, 2014; Rolls, 2023c; Morris, Amaral, Bliss, Duff and O'Keefe, 2025). If the conditioned stimulus is paired during learning with activation of the output neurons produced by the unconditioned stimulus, then later, after learning, due to the associative synaptic modification, the conditioned stimulus alone will produce the same output as the unconditioned stimulus. The networks can be used for many types of associative learning between two stimuli, including associating a visual stimulus with a reward to generate the expected value of the visual stimulus. **Pattern association networks** are described in Appendix B.2 of Rolls (2026d).

In the second architecture, an **autoassociation or attractor network**, the output neurons have recurrent associatively modifiable synaptic connections w_{ij} to other neurons in the network (see Fig. 1.3c). When an external input causes the output neurons to fire, then associative links are formed through the modifiable synapses that connect the set of neurons that is active. Later, if only a fraction of the original input pattern[1] is presented, then the associative synaptic connections or weights allow the whole of the memory to be retrieved. This is called completion. Because the components of the pattern are associated with each other as a result of the associatively modifiable recurrent connections, this is called an autoassociative memory. It is believed to be used in the cortex for many purposes, including short-term memory; episodic memory, in which the parts of a memory of an episode are associated together; and helping to define the response properties of cortical neurons, which have collaterals between themselves within a limited region. Autoassociation or attractor networks are described in Appendix B.3 of Rolls (2026d).

In the third architecture, a **competitive network**, the main input to the output neurons is received through associatively modifiable synapses w_{ij} (see Fig. 1.3d). Because of the initial values of the synaptic strengths, or because every axon does not contact every output neuron in what is termed 'diluted connectivity', different patterns tend to activate different output neurons. When one pattern is being presented, the most strongly activated neurons tend via lateral inhibition to inhibit the other neurons. For this reason the network is called competitive. During the presentation of that pattern, associative modification of the active axons onto the active postsynaptic neuron takes place. Later, that or similar patterns will have a greater chance of activating that neuron or set of neurons. Other neurons learn to respond to other input patterns. In this way, a network is built that can categorize

[1] Pattern refers here to which neurons are active in a vector of neuronal activity. Each element of the vector is the activity of a neuron, indicated in the brain by the neuron's firing rate.

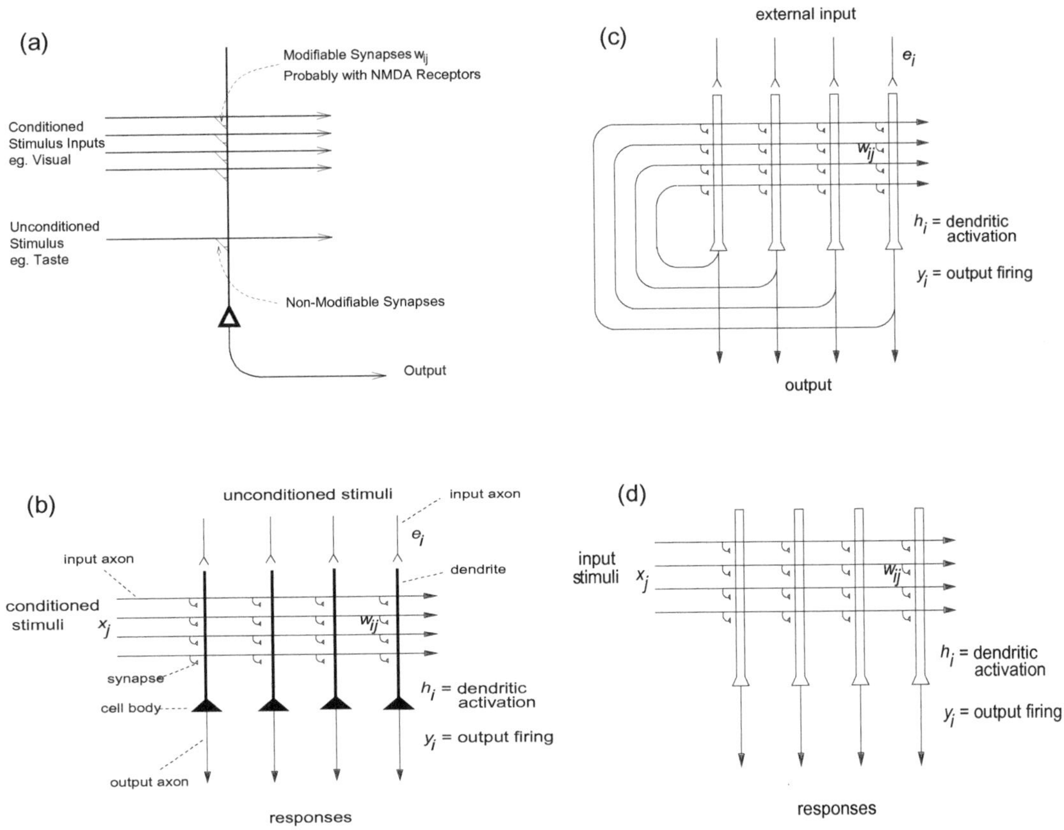

Fig. 1.3 Three network architectures that use local learning rules: (a) pattern association network introduced with a single output neuron; (b) pattern association network; (c) autoassociation network; (d) competitive network. Cortical pyramidal cell bodies are represented by triangular shapes, and the dendrites of each cell are shown above, with the axons shown below. The negative feedback inhibitory neurons are not shown.

patterns, placing similar patterns into the same category. In this way, somewhat similar flavors of a food could be placed in the same category, and different flavors into different categories. This is useful as a preprocessor for sensory information, self-organizes to produce feature analyzers, and finds uses in many other parts of the brain too. Competitive networks are described in Appendix B.4 of Rolls (2026d), and, importantly, perform unsupervised learning in that there is no teacher specifying how each neuron should respond. Instead, the categories are learned from the statistics present in the inputs. In contrast, in deep learning trained by error backpropagation, there is an external target or teacher for how each output neuron should respond to every input stimulus, and any error for each output neuron must be corrected by training the whole deep network (LeCun, Bengio and Hinton, 2015; Rolls, 2023c, 2026d).

Competitive networks enable new representations in the brain to be formed, by learning to respond to different combinations of inputs. The inputs are sometimes referred to as features. An example in vision might be that a common co-occurrence of two input features, such as a horizontal and vertical line, might enable a 'T' junction to be learned. Combinations of just a few features (termed low order combinations) need to be learned at any one stage, otherwise there would be a combinatorial explosion in the number of feature analyzers needed (Feldman, 1985). The building of feature analyzers using competitive learning is a key aspect of the operation of hierarchical networks, as described in Section 4.7.

All three architectures require inhibitory interneurons, which receive inputs from the principal excitatory neurons in the network (usually the pyramidal cells shown in Fig. 1.3) and implement feedback inhibition by connections to the pyramidal cells. The inhibition is usually implemented by GABA neurons, and maintains a small proportion of the pyramidal cells active.

These three architectures illustrate some of the important types of learning that can be implemented in Neural Networks, as described in more detail in Appendix B of Rolls (2026d). First, competitive learning is an example of **unsupervised learning**, in that the system self-organizes based on the statistics of the inputs being received, without an external teacher to guide its learning. This is used in the model of invariant visual object recognition, VisNet, described in Section 4.7. This type of learning must be very common in the brain, in that there is often no external teacher to help the system self-organise, let alone an external teacher for every neuron. Unsupervised learning is relatively little used in the machine learning community at present, and there are likely to be important lessons learned from how unsupervised learning is used in the brain.

Pattern association learning is an example of **supervised learning**, in that each output neuron has an external input that specifies the output that each neuron should produce. This external input is e_i in Fig. 1.3b. In the pattern association networks shown in Fig. 1.3b the inputs x_j learn to produce the required outputs just by associative synaptic modification, and this type of learning is common in the brain. In machine learning, the learning rule may be much more powerful, and may involve calculating an error between the output that is required for each neuron, and the actual output that is being produced by the inputs x_j. A one-layer network with such error correction learning is referred to as a one-layer error correction network or one-layer perceptron, and is described in Section B.11 of Rolls (2026d). The brain does in general not have the architecture required for this type of learning, as there is no separate teacher for each neuron to specify what its firing should be for every input, and no way to compare the actual firing with the target firing to compute an error to correct the appropriate synaptic weight. The only part of the brain where the architecture might be suitable is the cerebellum, where there is a single climbing fibre for each Purkinje cell, as described in Chapter 17 of Rolls (2023c).

The error correction type of learning can be implemented in machine learning in a multilayer network, in which inputs are applied to the bottom layer of the network, and every neuron in the output layer has a teacher. In the backpropagation of error algorithm (LeCun, Bengio and Hinton, 2015) (see also Appendix B of Rolls (2026d)), the error is backpropagated though all the layers of the network to calculate the error that should be used to correct the firing of every neuron in all the hidden layers not connected to the input or output. This is a non-local type of synaptic modification, and could not easily be implemented in the cortex, partly because there is no teacher for each neuron, and partly because of the implausibility of computing the backpropagated errors through every layer or level in the hierarchy, although attempts have been made (O'Reilly and Munakata, 2000; Guerguiev, Lillicrap and Richards, 2017; Lillicrap, Santoro, Marris, Akerman and Hinton, 2020; Song, Lukasiewicz, Xu and Bogacz, 2020). Although deep learning trained by error backpropagation is used widely for classifying inputs in machine learning, there is no clear evidence yet that it is or could be implemented in the cortex (Lillicrap et al., 2020; Rolls, 2026d; Song et al., 2024). The brain may solve similar problems, but in other ways, with one example being invariant visual object recognition as described in Section 4.7.

Another type of learning is **reinforcement learning**, in which there is a single reward signal for the whole network (Sutton and Barto, 2018; Rolls, 2023c, 2026d). This type of learning may be implemented in the brain for the slow learning of stimulus-response habits in the basal ganglia, using dopamine to provide a reward prediction error signal (Schultz, 2024, 2006, 2016a; Rolls, 2023c, 2026d).

2 Reward produced by Brain Stimulation

2.1 Introduction

One of my first interests in the neurosciences was brain systems involved in reward and emotion (Rolls, 2014b). This interest started in the last part of my first year as a DPhil graduate student at Oxford, as described in Section 1.2.

Olds and Milner (1954) had discovered that electrical stimulation at some sites in the brain is rewarding, in that animals including humans will work to obtain the stimulation. James Olds told me about how the discovery was made. James had noticed that a research seminar had been advertised at McGill University to be given by Neal Miller, and remembered that the word 'reinforcement' was in the seminar title. James did not go to the seminar, but if he had, would have found that the seminar was about negative reinforcement, and systems involved in avoidance learning. But perhaps seeing the word 'reinforcement' had remained in James Olds' mind. What happened then was that James was performing an experiment that involved electrical stimulation of the brain that was not to do with reward. James noticed that when the weak electrical stimulation was being delivered via an electrode, the rat seemed to have an 'excited happy look'. The next time that the rat went to the same corner of the open field environment, James delivered the stimulation again. Within about 3 such trials of electrical stimulation in that place, James could see that the rat was returning to that place for more. Thus the discovery was made that electrical stimulation of the brain could be rewarding in some brain regions, that is, that the stimulation could be positively reinforcing. Perhaps the word 'reinforcement' in the seminar title had helped? In any case, this is an example of where a scientific discovery was made because an experimenter was doing an experiment with a different hypothesis, but noticed something interesting. Chance may have played a role in this discovery, but being open to seeing something unusual happening in an experiment, and being able to follow it up, was important in this discovery (Olds and Milner, 1954; Olds, 1977).

I have found the same about many of the discoveries that I have made. In many cases, as described in this book, we were among the first to record neuronal activity in many brain regions of awake behaving primates, and frequently it would take us 6 months of research in which we performed what I termed 'clinical tests' (like the clinical testing one might perform in neurology to reach a diagnosis) before we had clear hypotheses about what the neurons were responding to and encoding, that could then be tested in hypothesis-driven research. Interactive testing of neuronal activity to develop new hypotheses is not possible with methods in which hundreds of neurons are recorded simultaneously in fixed paradigms, and then the neuronal activity is analyzed afterwards offline. That style of research does not provide a possibility to refine and test hypotheses about exactly what individual neurons may be responding to. I do think that there is advantage to performing some testing of neurons with natural stimuli in natural environments. Using natural stimuli in the real world is how we discovered sight of food neurons, the brain pathways involved in natural rewards such as food and social and emotional stimuli (Chapter 3), face identity and face expression neurons (Chapter 4), and hippocampal spatial view neurons (Chapter 5).

As a matter of fact, the discovery that neurons in the primary visual cortex can be driven by bars and edges also involved acute observation that was followed up, at a time when other hypotheses were being tested (Hubel and Wiesel, 1962). Hubel and Wiesel were testing their visual cortex neurons with slides in a projector that showed the conventional circular visual stimuli; and the discovery that bars or edges were much more effective was made when they noticed a neuronal response when the slide dropped into position in the projector, and an edge stimulus was unintentionally delivered, which made a cortical neuron respond (see e.g. Hubel and Wiesel (1962) which was referred to in Chapter 1).

2.2 Neuronal activity produced by brain-stimulation reward in rodents

To understand the neural mechanisms of brain-stimulation reward it is first necessary to know where brain self-stimulation sites are located. Therefore I decided to extend earlier research (Olds and Olds, 1965) by showing that one group of brain-stimulation reward sites is located along the general course of the medial forebrain bundle, which passes lateral to the midline from the ventral tegmental area of the midbrain posteriorly, through the lateral hypothalamus, preoptic area and nucleus accumbens, to reach the prefrontal cortex (orbitofrontal cortex in the monkey) anteriorly (Rolls, 1971, 1974, 1975, 1976; Rolls and Cooper, 1974). Many cell groups and neural pathways follow this path or much of this general course. For example, there is the medial forebrain bundle itself, interconnecting forebrain and brainstem regions with hypothalamic systems. There are fibres connected to neurons in prefrontal self-stimulation sites, which pass many self-stimulation sites in their course through the brain (Rolls and Cooper, 1973, 1974).

We showed that a second group of brain-stimulation reward sites is in limbic and related areas such as the amygdala, nucleus accumbens, and prefrontal cortex (orbitofrontal cortex in the monkey) (Rolls, 1974, 1975).

The discovery that reward sites in the brain extend into the orbitofrontal cortex was a seminal discovery, which stimulated my interest in the orbitofrontal cortex and reward. That started me on the journey of all the discoveries about the orbitofrontal cortex and its connected regions in reward, emotion, and motivation described in Chapter 3 and elsewhere (Rolls, 2005, 2014a, 2023c, 2026d, 2023b).

A key new discovery then came from the neuronal recording that I performed next, to analyze which neuronal systems are activated by reward. I discovered that this second group of brain-stimulation reward sites is highly interconnected neurophysiologically with the first group lying along the general course of the medial forebrain bundle, in that stimulation in any one of these reward sites activates neurons in the others, not only in rodents, but also importantly in primates (Rolls, 1974, 1975, 2005) as described next, allowing the research to become directly relevant to understanding the functions of the orbitofrontal cortex in humans. That is important, because the orbitofrontal cortex is greatly developed in primates including humans from what is present in rodents (Rolls, 2023c, 2026d, 2023b; Passingham, 2021).

2.3 Neuronal activity produced by brain-stimulation reward in primates

I extended my research on neuronal activity elicited by brain-stimulation reward in rodents to primates (squirrel monkeys and macaques), because my real interest and continuing interest is to understand the human brain better in health and disease.

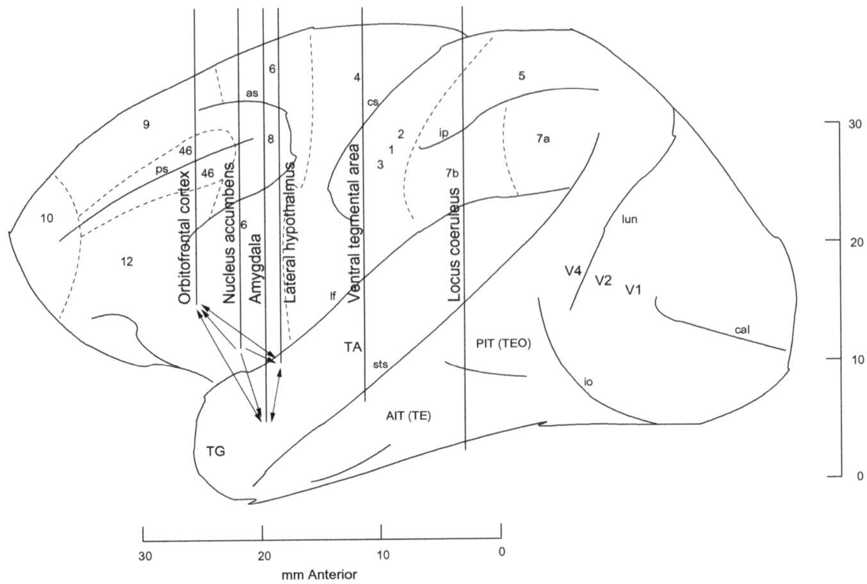

Fig. 2.1 Some brain sites in primates at which electrical stimulation of the brain can produce reward. The arrows indicate that stimulation at any one of these sites activates neurons in all of the other sites connected by arrows, as shown by Rolls, Burton and Mora (1980). The scales show the stereotaxic base planes. Abbreviations: as, arcuate sulcus; cal, calcarine sulcus; cs, central sulcus; lf, lateral (or Sylvian) fissure; lun, lunate sulcus; ps, principal sulcus; io, inferior occipital sulcus; ip, intraparietal sulcus (which has been opened to reveal some of the areas it contains); sts, superior temporal sulcus (which has been opened to reveal some of the areas it contains). AIT, anterior inferior temporal cortex; FST, visual motion processing area; LIP, lateral intraparietal area; MST, visual motion processing area; MT, visual motion processing area (also called V5); PIT, posterior inferior temporal cortex; STP, superior temporal plane; TA, architectonic area including auditory association cortex; TE, architectonic area including high order visual association cortex, and some of its subareas TEa and TEm; TG, architectonic area in the temporal pole; V1–V4, visual areas V1–V4; VIP, ventral intraparietal area; TEO, architectonic area including posterior visual association cortex. The numerals refer to architectonic areas, and have the following approximate functional equivalence: 1,2,3, somatosensory cortex (posterior to the central sulcus); 4, motor cortex; 5, superior parietal lobule; 7a, inferior parietal lobule, visual part; 7b, inferior parietal lobule, somatosensory part; 6, lateral premotor cortex; 8, frontal eye field; 12, part of orbitofrontal cortex; 46, dorsolateral prefrontal cortex.

In the monkey we discovered that neurons in the lateral hypothalamus, orbitofrontal cortex, amygdala, nucleus accumbens, and ventral tegmental area are activated during self-stimulation of any one of these sites or of the nucleus accumbens (see Fig. 2.1) (Rolls, 1974, 1975, 1976; Rolls, Burton and Mora, 1980c; Mora, Avrith and Rolls, 1980). Thus in the monkey, there is a highly interconnected set of structures, stimulation in any one of which will support self-stimulation and will activate neurons in the other structures.

Following a review paper on this research on brain-stimulation reward (Rolls, 1974), I was invited to write a book about reward systems in the brain, and that was my first book, **The Brain and Reward** (Rolls, 1975), which started to open up some of my later ideas (Rolls, 2023b) about reward systems in the brain, and how they relate to the brain systems involved in emotion, appetite, and motivation.

The discovery of the role of the orbitofrontal cortex in brain-stimulation reward in primates was foundational for later discoveries on the functions of the orbitofrontal cortex in reward, non-reward, and emotion in primates including humans, and their disorders including depression in humans (Rolls, 2014a, 2019c; Rolls et al., 2020b; Rolls, 2023b; Zhang et al., 2024a).

2.4 Neuronal activity produced by brain-stimulation reward and natural reward

But then I realized that to advance understanding it would be important to investigate whether these neurons activated from brain-stimulation reward sites are involved in natural rewards, such as the taste of food, for that offered an avenue to advance our understanding of brain systems involved in food reward, obesity, and potentially other types of rewarded / emotional behavior (Rolls, 2016f). So I set up to record from single neurons in the brain of behaving primates, which led to the research described in Chapter 3, with the initial findings relating to brain-stimulation reward and natural rewards described next.

We discovered when recording in the lateral hypothalamus and substantia innominata (which is lateral and rostral to the lateral hypothalamus) from neurons activated by brain-stimulation reward, that some of the neurons activated by brain-stimulation reward (approximately 13% in one sample of 764 neurons) altered their activity in relation to feeding (Rolls, 1974, 1975, 1976, 1979; Rolls, Burton and Mora, 1980c). In fact, we discovered the neurons with feeding-related responses when we sought to determine what functions are performed by the hypothalamic neurons activated by brain-stimulation reward.

In the original sample of 764 neurons, we discovered that some neurons altered their activity only during the eating of some substances, so that their activity appeared to be associated with the taste of food. But we were then amazed to discover that many more of these hypothalamic neurons (approximately 11% of the total sample) altered their activity before food entered the mouth, while the monkey was looking at the food (Rolls, Burton and Mora, 1976; Rolls, Sanghera and Roper-Hall, 1979). We then discovered that the activity of this second set of neurons only occurred to the sight of food if the monkey was hungry (Burton, Rolls and Mora, 1976), and becomes associated during learning with the sight of an object when it is a food (Mora, Rolls and Burton, 1976). These discoveries showed that these neurons were responding to food reward value, and to stimuli learned by association to be related to expected reward value, that is, to expected value.

An example of one of these investigations is illustrated in Fig. 2.2 (Rolls, Burton and Mora, 1980c). This shows that some of the hypothalamic neurons activated from other brain-stimulation reward sites were also activated by the sight of food; and that self-stimulation occurred through the recording microelectrode when it was close to these neurons activated by the natural reward of food.

The implication of these discoveries is that brain-stimulation reward can occur because it activates neurons activated by natural rewards, such as in this case food reward. Consistent with this discovery, the brain-stimulation reward at these sites was dependent on hunger, and was attenuated after feeding to satiety, providing further evidence that the brain-stimulation reward at these sites was acting like the natural reward of food.

The discovery of the importance of the orbitofrontal cortex in brain-stimulation reward (Rolls, Burton and Mora, 1980c; Rolls, 2005) is highly relevant to the search more recently for brain sites where electrical stimulation in humans may relieve depression, with one brain site the subcallosal anterior cingulate cortex (Mayberg et al., 2005; Lozano et al., 2012; Lujan et al., 2013; Mayberg et al., 2016; Holtzheimer et al., 2017; Riva-Posse et al., 2018, 2020; Siddiqi et al., 2021; Zhu et al., 2021). Our results described here and in Chapter 3 provide evidence that the orbitofrontal cortex is a key brain area that deserves further investigation in humans as a reward site. I have proposed that the human orbitofrontal cortex may be a key brain region in which electrical stimulation may relieve depression (Rolls, 2018b, 2019a,e,c; Rolls, Cheng and Feng, 2020b; Rolls, 2023b,c, 2026d; Zhang, Rolls, Wang, Xie, Cheng and Feng, 2024a).

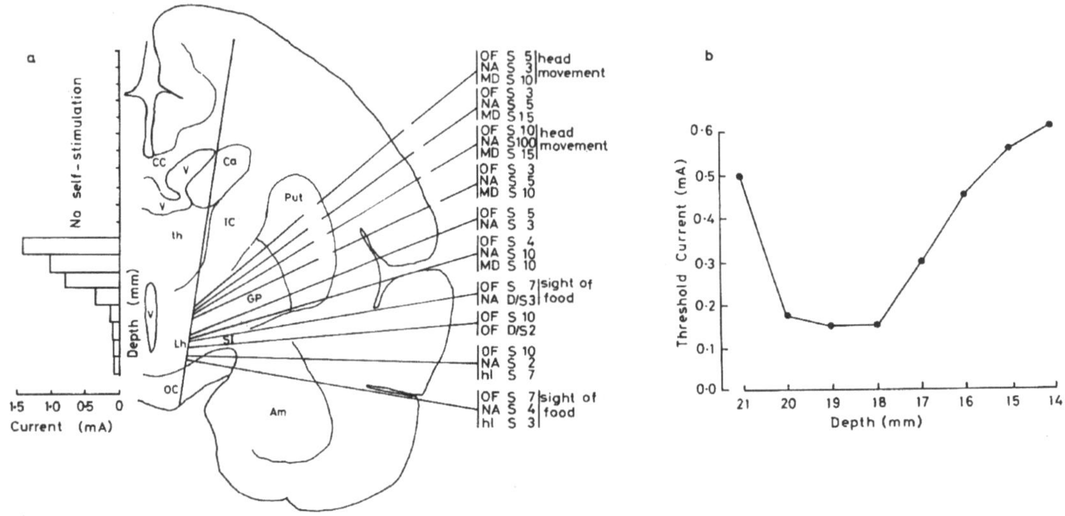

Fig. 2.2 a. Neurons activated by both brain-stimulation reward and the sight of food were found at the lower end of this microelectrode track in the macaque lateral hypothalamus. Neurons higher up the track in the globus pallidus were activated by brain-stimulation reward and also by head movements. The neurons were trans-synaptically (S) or possibly in some cases directly (D/S) activated with the latencies shown in ms from self-stimulation sites in the orbitofrontal cortex (OF), nucleus accumbens (NA), mediodorsal nucleus of the thalamus (MD), or lateral hypothalamus (hl). b. Self-stimulation through the recording microelectrode occurred with low currents if the microelectrode was in the hypothalamus close to the neurons activated by the sight of food. (Data from Rolls 1975 The Brain and Emotion: Pergamon Press; and Rolls, Burton and Mora 1980 Neurophysiological analysis of brain-stimulation reward in the monkey. Brain Research 194: 339-357.)

Extending the research on reward systems and emotion to primates from rodents was important, because I had been trained in preclinical medicine at the University of Cambridge, and had and have a vocation to understand brain function in humans not only in health, but also in disease. It turns out that this was important foresight, for we now know that there are many systems-level differences from rodents in the organization of the primate including human brain, some of which are set out in Section 29.4 of **Brain Computations and Principles; and AI** (Rolls, 2026d).

Part of the importance of the discoveries described in this Chapter is that we made fundamental discoveries about the sites of reward systems in the primate brain. One of the reward sites that we discovered was the orbitofrontal cortex, and that discovery provided a foundation for future discoveries on the functions of the orbitofrontal cortex in reward value, emotion, and motivation in primates including humans (Rolls, 2014a, 2023b). Another part of the importance of the discoveries was the identification of neurons in primates, in for example the lateral hypothalamus, that respond to natural rewards such as the taste or sight of food. These were the first or among the first discoveries of reward neurons, that is neurons that respond to the reward value of a stimulus such as food, in any species, and as far as I know the first in primates.

3 Natural rewards: brain systems involved in reward value, emotion, and motivation, and their disorders including depression

3.1 Introduction

As described in Chapter 2, I discovered that brain systems involved in brain-stimulation reward in primates as well as rodents include the lateral hypothalamus and structures along the medial forebrain bundle, and in addition the orbitofrontal cortex and the amygdala (Rolls, 1974; Rolls et al., 1980c; Rolls, 2005). But then I realized that to advance understanding it would be important to investigate whether these neurons activated from brain-stimulation reward sites are involved in natural rewards, such as the taste of food, for that offered an avenue to advance our understanding of brain systems involved in food reward, obesity, and potentially other types of rewarded / emotional behavior. So, as introduced in Chapter 2, I set up to record from single neurons in the brain of behaving primates, and some of the discoveries that followed are described in this Chapter.

3.2 The discovery of lateral hypothalamic neurons responding to the sight and taste of food reward

We soon discovered that neurons in the primate lateral hypothalamus activated from brain-stimulation reward sites could also in many cases be activated by the taste of food, a specific natural reward, providing an explanation for why brain-stimulation reward of some sites occurs (Rolls, Burton and Mora, 1980c; Burton, Rolls and Mora, 1976), as described in Chapter 2.

But on one occasion during such an experiment on taste reward I pulled a peanut out of my lab coat, and offered it to the primate for being so cooperative in the investigations. I was amazed to find that the taste reward neuron had a very large increase in firing rate to the sight of the peanut. This occurred even before the peanut was close enough for the monkey to reach out for the peanut, place the peanut in its mouth, and eat it. I then repeated this three times, first showing the peanut and then letting the monkey eat it, and each time the neuron responded to the sight of the peanut with a large increase of firing rate. I checked the firing rate values being printed out on the ASR33 teletype by the PDP11 computer which I had programmed in Fortran, and found that the lateral hypothalamic neuron was indeed responding to the sight of food (Rolls, Burton and Mora, 1976; Rolls, Sanghera and Roper-Hall, 1979).

We performed many investigations that showed that these were food reward neurons. We showed that they gradually stopped responding when the food reward was devalued by feeding to satiety (Burton, Rolls and Mora, 1976), as shown for example in Fig. 3.1. We also showed that these neurons would learn to respond to a visual stimulus such as the sight of a triangle if it signaled taste reward in a visual discrimination task. Moreover, we discovered that these neurons would reverse their response in one trial if another visual stimulus such as the sight of a square was associated with food reward, and would stop responding to the triangle if it was now in the reversal associated with very mildly aversive salt solution (Mora, Rolls and Burton, 1976). We had discovered neurons that responded to the sight and usually also the taste of food, and that represented the reward value of the food (Rolls, 2023c, 2026d, 2014a).

3.3 The discovery of reward-specific satiety: sensory-specific satiety

While performing these experiments in which the macaques were fed to satiety and a lateral hypothalamic food reward neuron stopped responding to the sight of food such as fruit juice, one day, after the monkey was satiated with fruit juice, I offered the monkey a peanut, and was amazed to see a large response from the lateral hypothalamic neuron to the different food (Rolls et al., 1986). I fed the macaque to satiety with the peanut, and the neuron gradually stopped responding to the sight of the peanut. But then I offered another food, a piece of banana, and discovered that the neuron, still unresponsive to the fruit juice and peanut, now responded to the sight of the banana, which the monkey readily accepted and ate avidly (see Fig. 3.1) (Rolls et al., 1986; Rolls, 1981b).

I had discovered sensory-specific satiety in these neurophysiological investigations, and the effects that a variety of food could have in increasing food intake, at least in a single meal, and the results were very replicable (Rolls et al., 1986).

I then organised some practical classes at the University of Oxford for undergraduate students, to test whether I could extend my discovery of sensory-specific satiety to humans at the behavioral level, and the experiments worked beautifully. I invited my former wife, Dr Barbara Rolls, who was working on thirst at the time, to participate in these practical classes on the appetite for food, and she became interested, and collaborated with me in a number of studies on sensory-specific satiety in humans that we published together (Rolls et al., 1983; Rolls and Rolls, 1982; Rolls et al., 1982a, 1981b, 1982a, 1981a)[2].

To illustrate sensory-specific satiety in action, in one experiment that I designed we fed Oxford schoolboys either one color of Smarties to eat, or four different colors. (Smarties, similar to M&Ms, are small discs of chocolate covered with a thin, colored sugar.) We found that the decrease in the pleasantness of the color of Smartie eaten was significantly greater for the color eaten than for the other colors, even though the flavors and nutritional content of the particular Smarties used in this experiment were identical (Rolls et al., 1982b). This experiment thus shows that just the color of a food is sufficient to produce some reward-specific satiety. The effect was nothing to do with

[2]Barbara Rolls later told me that she was subsequently awarded a prize for the discovery of sensory-specific satiety, which she accepted. I mention this, so that those who may take up research in the neurosciences should know in advance that they will sometimes not be recognized for discoveries that they have made. I suspect that this is quite common (and indeed Candace Pert felt the same about her discovery of opiate receptors, for which she was not included in the Nobel prize (Pert, 1997)). However, I was very pleased to hear about the award of the prize for the discovery of sensory-specific satiety (to Barbara Rolls), for sensory-specific satiety and its associated brain mechanisms are probably the most important factors in how much food is eaten in a single meal (Rolls, 2016f; Rolls et al., 2025b; Rolls, 2026b). Sensory-specific satiety is thus very relevant to over-eating and obesity (Rolls, 2016f, 2026d). Not only did I discover sensory-specific satiety and its brain mechanisms and how relevant it is to human behavior, but I also discovered that there is a long-term form of sensory-specific satiety that can build up over months if the same foods are eaten. I discovered this with one of my Oxford undergraduate students in an Ethiopian refugee camp (Rolls and de Waal, 1985), and we recommended that the foods should be changed from time to time so that people will want to keep eating the nutritionally good food that is being provided. I also went on to discover that even only smelling or tasting a food for as long as it would be eaten in a meal can produce some sensory-specific satiety, in an investigation with Juliet Rolls (Rolls and Rolls, 1997), whom I was bringing up in Oxford. This was Juliet's undergraduate research project, for which she was awarded a prize. This discovery is important, for it helps to elucidate the mechanisms of sensory-specific satiety (Rolls and Rolls, 1997; Rolls et al., 2025b).

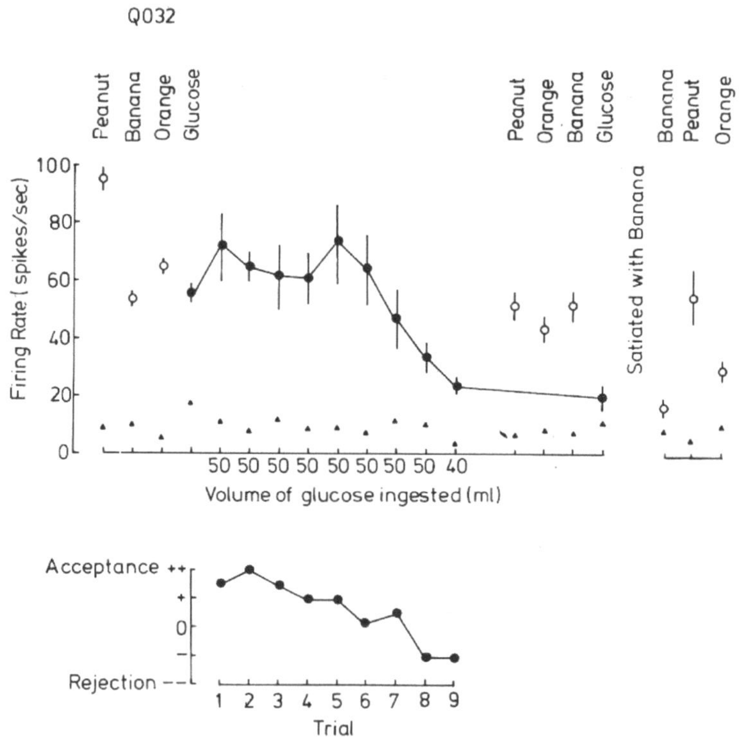

Fig. 3.1 The effect of feeding the monkey to satiety with 20% glucose solution on the responses of a lateral hypothalamic neuron to the sight of the glucose (filled circles) and to the sight of other foods (open circles). After the monkey had fed to satiety with glucose, the neuron responded much less to the sight of glucose, but still responded to the other foods. The mean responses of the neuron ($\pm$ s.e.m.) are shown to the stimuli at different stages of the experiment. The satiety of the monkey, shown below, was measured by whether he accepted or rejected the glucose. After satiety with glucose, a second satiety experiment was performed in which banana was fed to satiety, and after this, the responses of the neuron decreased to the sight of the banana, but remained to the sight of peanut and orange, for which he still had an appetite. (Data from Rolls E T, Murzi E, Yaxley S, Thorpe S J, and Simpson S J. 1986 Sensory-specific satiety: food-specific reduction in responsiveness of ventral forebrain neurons after feeding in the monkey. Brain Res 368: 79-86; and Rolls E T. 1981. Central nervous mechanisms related to feeding and appetite. British Medical Bulletin 37: 131-134.)

the nutrient content, which was identical. That is why I named this reduction in the reward value and subjective pleasantness of a food eaten to satiety 'sensory-specific satiety'. Consistently, some sensory-specific satiety for the odor of a food was produced over the typical timecourse of a meal even when the food was not eaten or even placed in the mouth (Rolls and Rolls, 1997). Importantly, sensory-specific satiety is not a sensory adaptation, for the intensity of the food eaten to satiety hardly changes, whereas its pleasantness decreases to zero (Rolls and Rolls, 1997). Consistently, we found that effects of satiety including sensory-specific satiety are not found in sensory regions such as the primate primary taste cortex and inferior temporal visual cortex (Rolls, 2023c, 2026d, 2015a, 2016f,c; Rolls et al., 1977). Because the change found is in the pleasantness and reward value of the stimulus, sensory-specific satiety can also be termed 'reward-specific satiety' (Rolls et al., 2025b).

3.4 The discovery of orbitofrontal cortex neurons responding to reward, implementing sensory-specific satiety, and leading towards a theory of the functions of the orbitofrontal cortex in emotion and motivation

I next thought it important to discover where the representation of reward value and its modulation by appetite (hunger in the case of food reward) is first evident in the visual and taste cortical pathways, in order to understand the information processing for rewards in the brain. Some of the brain regions and pathways described next are shown in Fig. 3.2.

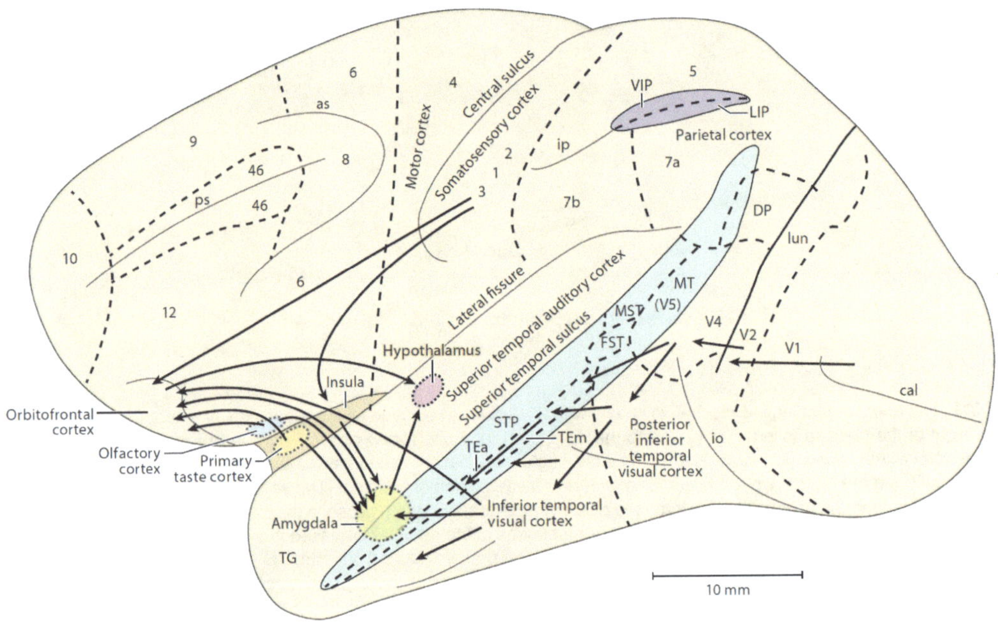

Fig. 3.2 Some of the connections of the primate orbitofrontal cortex and amygdala shown on this lateral view of the brain of the macaque monkey, showing the connections from the primary taste and olfactory cortices. Connections are also shown in the 'ventral visual system' from V1 to V2, V4, the inferior temporal visual cortex, etc., with some connections reaching the amygdala and orbitofrontal cortex. In addition, connections from the somatosensory cortical areas 1, 2, and 3 that reach the orbitofrontal cortex directly and via the insular cortex, and that reach the amygdala via the insular cortex, are shown. as, arcuate sulcus; cal, calcarine sulcus; cs, central sulcus; lf, lateral (or Sylvian) fissure; lun, lunate sulcus; ps, principal sulcus; io, inferior occipital sulcus; ip, intraparietal sulcus (which has been opened to reveal some of the areas it contains); sts, superior temporal sulcus (which has been opened to reveal some of the areas it contains). AIT, anterior inferior temporal cortex; FST, visual motion processing area; LIP, lateral intraparietal area; MST, visual motion processing area; MT, visual motion processing area (also called V5); PIT, posterior inferior temporal cortex; STP, superior temporal plane; TA, architectonic area including auditory association cortex; TE, architectonic area including high order visual association cortex, and some of its subareas TEa and TEm; TG, architectonic area in the temporal pole; V1–V4, visual areas V1–V4; VIP, ventral intraparietal area; TEO, architectonic area including posterior visual association cortex. The numerals refer to architectonic areas, and have the following approximate functional equivalence: 1,2,3, somatosensory cortex (posterior to the central sulcus); 4, motor cortex; 5, superior parietal lobule; 7a, inferior parietal lobule, visual part; 7b, inferior parietal lobule, somatosensory part; 6, lateral premotor cortex; 8, frontal eye field; 12, part of orbitofrontal cortex; 46, dorsolateral prefrontal cortex. (Data from Rolls E T 2016 Reward systems in the brain and nutrition. Annual Review of Nutrition 36: 435–470.

I knew that neurons in the primate orbitofrontal cortex were projecting to lateral hypothalamic reward sites from the studies of brain-stimulation reward described in Chapter 2 (Rolls et al., 1980d). I also knew that lesions of the orbitofrontal cortex could impair reward-related learning in primates

(Butter, 1969; Iversen and Mishkin, 1970). So we set up what were some of the first experiments ever to investigate neuronal activity in the primate orbitofrontal cortex (Thorpe, Rolls and Maddison, 1983).

We discovered neurons in the orbitofrontal cortex that responded to the taste and/or sight and/or smell of food, and showed that the visual neurons encoded the reward value, in that devaluation by feeding to satiety abolished the neuronal responses to the sight of food (Thorpe, Rolls and Maddison, 1983). These **orbitofrontal cortex neurons encode reward value** in that they reversed the visual stimuli to which they responded in a visual discrimination reversal task (Thorpe et al., 1983; Rolls et al., 1996b,a; Critchley and Rolls, 1996b,a; Rolls and Baylis, 1994; Kringelbach et al., 2003). In reward reversal, these neurons could reverse the visual stimulus to which they responded in a rule-based or model-based way, in that if in reversal a reward to one stimulus was not obtained, the monkey would choose the other stimulus on the very next trial and would obtain reward, even though the previous reward association of that stimulus had been with an aversive taste (Thorpe et al., 1983). I have been told by those who work with rodents that no single-trial rule-based reward reversal has been demonstrated in rodents (Trevor Robbins, personal communication), yet I argue that this very rapid change of behavior when a reward is no longer obtained is fundamental to much primate including human social behavior (Rolls, 2023b).

Developing the argument further, the firing of these orbitofrontal cortex neurons often clearly reflects the preferences of the monkey (Thorpe et al., 1983). Moreover, we discovered that these neurons encode reward value and are related to appetite control in that devaluation by feeding to satiety was found, and, indeed, we discovered that sensory-specific satiety was implemented in the primate orbitofrontal cortex as illustrated in Fig. 3.3. This figure shows sensory-specific satiety for an orbitofrontal cortex neuron with taste responses (Rolls et al., 1989c). These orbitofrontal cortex reward value neurons probably project to the hypothalamic neurons that we discovered with similar responses, as indicated in Figs. 3.2 and 3.12 (Rolls, 2014a).

The orbitofrontal cortex seemed special in these types of neuronal response, for when we recorded in the primate amygdala, we discovered neurons that could respond to the sight and taste of food, but their responses did not often reflect the exact preferences of the monkey, the neurons were relatively slow to reverse in the reversal of a visual discrimination task, and the effects of satiety were not as clear as in the hypothalamus and orbitofrontal cortex (Kadohisa et al., 2005; Rolls, 1984; Sanghera et al., 1979).

I decided that the next important question to understand reward value processing in the brain was to find out at what stage of taste processing the reward value of the taste was represented by the neuronal firing. To address this fundamental question, we then performed a series of studies to analyze the systems-level functional architecture of taste processing in the primate brain. We discovered that the primate orbitofrontal cortex includes a secondary taste cortical region in that it receives inputs anatomically very strongly from the primary taste cortex in the anterior insula and adjacent frontal operculum (see Fig. 3.2) (Baylis, Rolls and Baylis, 1995), and has taste-responsive neurons (Baylis et al., 1995; Rolls et al., 1989c, 1990). Extending understanding of the reward system in the orbitofrontal cortex further, we discovered that olfactory and taste stimuli converge onto single neurons to implement the flavor of food, and its reward value (Rolls et al., 1997a; Rolls and Baylis, 1994).

We then discovered that the primate **anterior insula primary taste cortex encodes the identity of the 5 tastes** sweet, salt, bitter, sour, and umami (glutamate), **and does not encode the reward value of the taste** in that the neurons continued to respond to a taste after it had been devalued by feeding to satiety (Kadohisa et al., 2005; Rolls, 2016c; Yaxley et al., 1990, 1988; Scott et al., 1986). These discoveries were helped by a wonderful collaboration with Tom Scott, an expert on

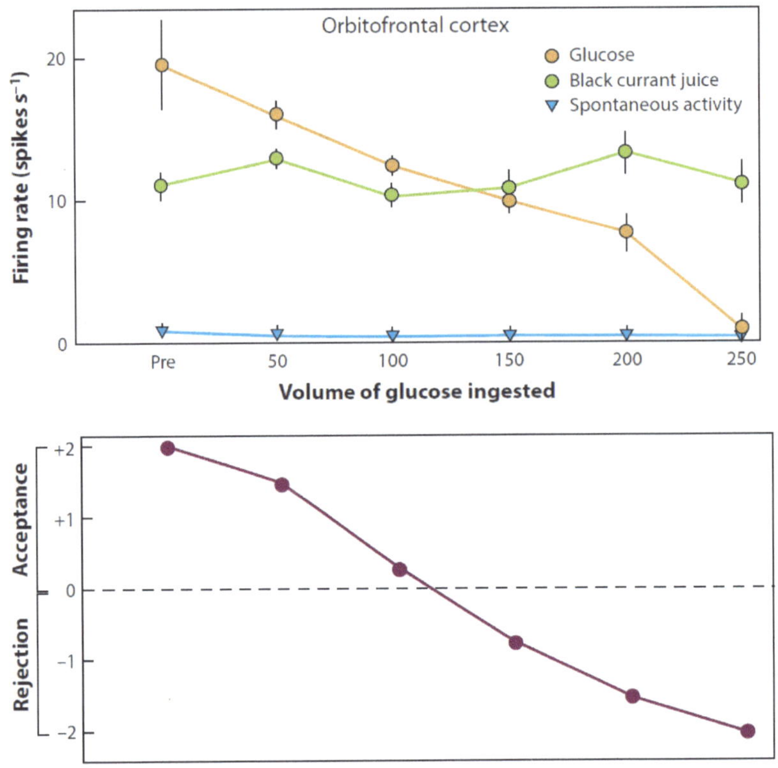

Fig. 3.3 The effect of feeding to satiety with glucose solution on the responses (rate $\pm$ s.e.m.) of a neuron in the secondary taste cortex to the taste of glucose (open circles) and of blackcurrant juice (BJ). The spontaneous firing rate is also indicated (SA). Below the neuronal response data, the behavioral measure of the acceptance or rejection of the solution on a scale from +2 (strong acceptance) to –2 (strong rejection) is shown. The solution used to feed to satiety was 20% glucose. The monkey was fed 50 ml of the solution at each stage of the experiment as indicated along the abscissa, until he was satiated as shown by whether he accepted or rejected the solution. Pre is the firing rate of the neuron before the satiety experiment started. (Data from Rolls E T, Sienkiewicz Z J and Yaxley S. 1989 Hunger modulates the responses to gustatory stimuli of single neurons in the caudolateral orbitofrontal cortex of the macaque monkey. European Journal of Neuroscience 1: 53–60.)

taste physiology in rodents, who came to Oxford to work collaboratively on processing in the taste pathways of primates. This research highlights the frequent importance of interdisciplinary collaboration when studying a system as complex as the brain.

We also investigated the activity of neurons in the inferior temporal visual cortex, to investigate the functional architecture of visual inputs into the orbitofrontal cortex reward system. The research question I asked was: Where is reward represented by neuronal firing in the visual pathways?

We discovered that **neurons in the inferior temporal visual cortex** respond beautifully to objects (see Chapter 4), but **do not encode the reward value of the objects** in that devaluation did not influence the responses of inferior temporal cortex neurons, and inferior temporal cortex neurons do not reverse their visual responses during visual discrimination reversal (Rolls et al., 1977, 2003a) (Fig. 4.2). Because the inferior temporal visual cortex projects to the orbitofrontal cortex (Rolls et al., 2022b), this showed that the first stage in the visual pathways in primates in which reward value was made explicit by neuronal firing was in the orbitofrontal cortex.

The importance of these discoveries is that they show that in primates, first the sensory perceptual qualities of stimuli that are independent of reward are built in cortical regions such as the insular taste cortex and the inferior temporal visual cortex; and then in the orbitofrontal cortex the stimuli are represented in terms of their reward value. Thus perception is kept separate from reward value and emotion in terms of neural processing in primates, including humans. These discoveries revealed a fundamental principle of operation of the primate including human brain, that sensory analysis and perception is kept separate from reward value and emotional processing (Rolls, 2014a, 2016b, 2026d). This principle has great evolutionary adaptive value, in that we can identify potential rewards such as the sight of food and can learn about their location, even when they are not currently rewarding. These concepts are incorporated into the neural architecture shown in Fig. 3.12.

Extending the reward system further, we discovered that the **reward value and pleasantness of fat** in food in the mouth is implemented by a texture channel that responds to low values of the coefficient of sliding friction (i.e. are very slippery or lubricating) (Rolls et al., 2018b; Rolls, 2010b; Grabenhorst et al., 2010b; Verhagen et al., 2004; De Araujo and Rolls, 2004; Verhagen et al., 2003; Rolls et al., 1999a). This discovery has many implications for the development of foods that will have the pleasant mouth feel of fat (as in for example ice cream or chocolate), but which need contain no fat. An example of an orbitofrontal cortex neuron that encodes fat in the mouth by the coefficient of sliding friction is shown in Fig. 3.4. The coefficient of sliding friction measures how easily two surfaces will slide across each other, and a general term to describe it is slipperiness. The point is clear in Fig. 3.4: slippery oils that are non-fat such as silicone oil are effective stimuli for this fat-encoding neural system in the mouth. The importance of interdisciplinary collaboration in neuroscience is emphasized again by this research (Rolls et al., 2018b), which was collaborative with Ian Norton and his colleagues with expertise in tribology and rheology at the University of Birmingham, UK.

The responses of these orbitofrontal cortex neurons are related to the reward value of fat in the mouth, in that the responses decrease to zero during feeding to satiety with a fatty food such as cream (Rolls, Critchley, Browning, Hernadi and Lenard, 1999a). The fat-sensitive neurons did not have responses to lauric or linoleic acid, which are fatty acids present in some fats, so the responses of these fat-sensitive neurons were not related to fatty acid sensing. We discovered that a separate population of neurons do have responses to linoleic or lauric fatty acids, but not to fat in the mouth. Their responses may be related to "off tastes" of food produced for example by butyric acid, which food manufacturers aim to minimize in foods (Rolls, 2020a).

While performing these investigations, I wanted to be sure that the fat-sensing was not by just the thickness of a food. So we included a set of stimuli of increasing viscosity (thickness) in the investigations. This led us to the next discovery, that another population of neurons has responses related to the log of viscosity (which correlates with subjective ratings of food thickness (Kadohisa et al., 2005)), but not with the coefficient of sliding friction (Rolls et al., 2018b), showing that a number of different oral texture signals are represented in the orbitofrontal cortex and also insular taste cortex (Verhagen et al., 2004; Kadohisa et al., 2005; Rolls, 2020a).

This discovery leading to a new understanding of the representation of oral fat by the coefficient of sliding friction in the cerebral cortex opens the way for the systematic development of foods with the pleasant mouth-feel of fat, together with ideal nutritional content, and has great potential to contribute to healthy eating and a healthy body weight (Rolls et al., 2018b; Rolls, 2020a, 2026d).

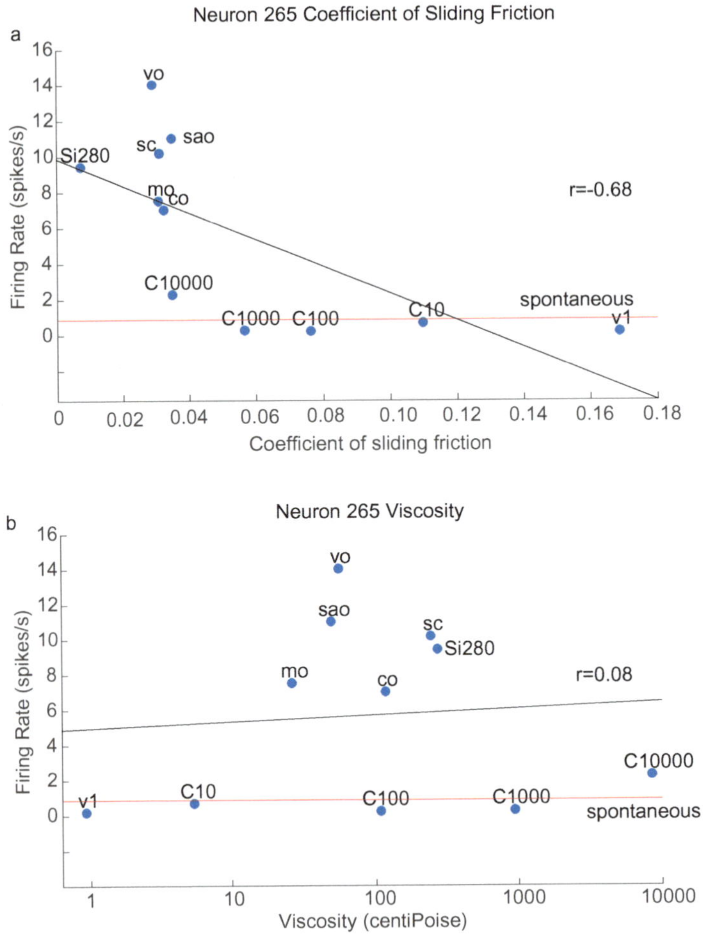

Fig. 3.4 An orbitofrontal cortex neuron with responses non-linearly correlated with decreases in the coefficient of sliding friction (a). The neuron responds almost not at all until the coefficient of sliding friction falls below 0.04. The neuron is thus very selective for fat texture, because of its non-linear response in relation to the coefficient of sliding friction. The linear regression line has a correlation of $r = -0.68$ ($p=0.02$). (b): There is a much weaker relation to viscosity ($r = 0.08$, $p=0.82$), with the oils producing a larger response than predicted linearly. Further, a regression line through the non-oil stimuli would have a lower slope. C10-C10000: carboxymethyl cellulose with the nominal viscosity of 10, 100, 1000 and 10,000 cP. v1: water (1 cP). co: coconut oil; mo: mineral oil; sao: safflower oil; vo: vegetable oil; sc: single cream. Si280: silicone oil with a nominal viscosity of 280 cP. Li: linoleic acid; La; lauric acid. The horizontal red line indicates the spontaneous firing rate. The Pearson correlation between the firing rate of each neuron and (a) the coefficient of sliding friction, and (b) the viscosity, was calculated to show to what extent the firing of a neuron reflected one or other of these measures. Linear regression lines are shown in the Figure for how the firing rates were related to the coefficient of sliding friction, or to the log of the viscosity. (Data from Rolls E T, Mills T, Norton A B, Lazidis A, and Norton I T. 2018 The neuronal encoding of oral fat by the coefficient of sliding friction in the cerebral cortex and amygdala. Cerebral Cortex 28: 4080–4089.)

3.5 Pleasure and reward value in the human brain

3.5.1 Representations of reward value and pleasure in the human brain

I then wanted to test whether these discoveries at the neuronal level (in primates) about reward systems in the brain apply to humans. I therefore undertook a collaboration with the Imaging Centre at the University of Nottingham where Sir Peter Mansfield had made discoveries that made functional magnetic resonance imaging of the brain (fMRI) feasible. I worked with Richard Bowtell and Sue Francis on a physics-built 3T fMRI machine in which heating was measured by placing one's hands

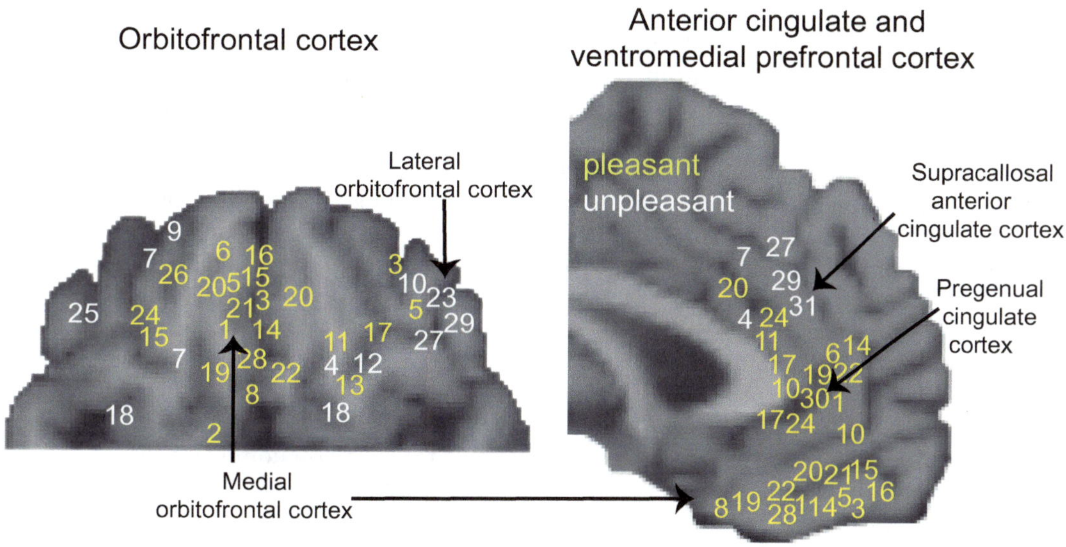

Fig. 3.5 Subjective pleasure is correlated with activations (yellow) in the human medial orbitofrontal cortex and pregenual anterior cingulate and ventromedial prefrontal cortex. Subjective unpleasantness is correlated with activations (white) in the human lateral orbitofrontal cortex and supracallosal anterior cingulate cortex. Yellow: sites where activations correlate with subjective pleasantness. White: sites where activations correlate with subjective unpleasantness. The numbers refer to effects found in specific studies cited in the paper shown below. Taste: 1, 2; odor: 3-10; flavor: 11-16; oral texture: 17, 18; chocolate: 19; water: 20; wine: 21; oral temperature: 22, 23; somatosensory temperature: 24, 25; the sight of touch: 26, 27; facial attractiveness: 28, 29; erotic pictures: 30; laser-induced pain: 31. (Data from Grabenhorst F and Rolls E T 2011 Value, pleasure and choice in the ventral prefrontal cortex. Trends in Cognitive Sciences 15: 56–67.)

on the system. I believe that we were one of the pioneers of fMRI neuroimaging at 3T. Whenever we tried a new experiment, I always went into the magnet first. In those early investigations at Nottingham, we were able to show that orbitofrontal cortex activations are related to olfactory sensory-specific satiety (O'Doherty, Rolls, Francis, Bowtell, McGlone, Kobal, Renner and Ahne, 2000); that pleasant and unpleasant tastes are represented separately in the human orbitofrontal cortex (O'Doherty, Rolls, Francis, Bowtell and McGlone, 2001b); and that pleasant touch is represented in the medial orbitofrontal cortex, and unpleasant / mildly painful touch is represented in the lateral orbitofrontal cortex (Francis, Rolls, Bowtell, McGlone, O'Doherty, Browning, Clare and Smith, 1999; Rolls, O'Doherty, Kringelbach, Francis, Bowtell and McGlone, 2003d).

After that pioneering start, we continued with fMRI at the University of Oxford, where the physicist Peter Jezzard provided us with a neuroimaging protocol that enabled us to shim (tune) the 3T magnet for every individual to obtain signal from the orbitofrontal cortex, which helped us to perform investigations on neuroimaging of the orbitofrontal cortex before most other neuroimaging centers could. We were able to perform a whole series of fMRI investigations in humans to test whether the same neural processing that we had discovered in the primate orbitofrontal cortex and related regions is present in humans.

We discovered that in humans too, the activation of the orbitofrontal cortex by food is related to reward value and sensory-specific satiety for food. A notable advance in these investigations was that we could relate the activations to the subjectively experienced pleasure produced by the stimuli when they are rewarding (Kringelbach, O'Doherty, Rolls and Andrews, 2003). Studies such as this pioneered the neural basis of subjective pleasure in humans for a wide range of stimuli, including not only food flavor (Kringelbach, O'Doherty, Rolls and Andrews, 2003; De Araujo, Rolls, Kringelbach,

McGlone and Phillips, 2003b), but also taste (Grabenhorst and Rolls, 2008; De Araujo, Kringelbach, Rolls and McGlone, 2003a; Grabenhorst, D'Souza, Parris, Rolls and Passingham, 2010a), smell (Grabenhorst, Rolls, Margot, da Silva and Velazco, 2007; Rolls, Grabenhorst, Margot, da Silva and Velazco, 2008a), warm touch (Rolls, Grabenhorst and Parris, 2008b), and pleasant vs painful touch (Rolls, O'Doherty, Kringelbach, Francis, Bowtell and McGlone, 2003d; Rolls, Grabenhorst and Franco, 2009; McCabe, Rolls, Bilderbeck and McGlone, 2008), all of which we discovered are related to orbitofrontal cortex activation. We also showed that there is a representation of the reward value and pleasantness of fat in the mouth in the human orbitofrontal cortex and a region to which it is connected, the anterior cingulate cortex (Grabenhorst, Rolls, Parris and D'Souza, 2010b).

One of the key discoveries in these fMRI investigations was about where human subjective, reportable, pleasantness (as well as reward value), is represented in the brain. In many of these studies, we discovered that there is a linear relationship between the subjective pleasantness of the stimuli, and the activation of the orbitofrontal cortex, with an example shown in Fig. 3.8.

Another key discovery was that monetary (or points) reward is represented in the medial orbitofrontal cortex, and monetary loss is represented in the lateral orbitofrontal cortex (O'Doherty, Kringelbach, Rolls, Hornak and Andrews, 2001a; Rolls, Vatansever, Li, Cheng and Feng, 2020e) (see Fig. 3.11), which we confirmed in a large cohort of almost 2,000 adolescents (Xie et al., 2021b) as illustrated in Fig. 3.6. This was an extension beyond what was possible in primates, for here we have an abstract type of reward, money, or points being won or lost in a task, that is also represented in the human orbitofrontal cortex.

Another key discovery is that in humans there is some topology present in the orbitofrontal cortex and a region to which it projects, the anterior cingulate cortex, with many rewards represented in the human medial orbitofrontal cortex and pregenual anterior cingulate cortex, and many punishers and also non-reward represented in the lateral orbitofrontal cortex and supracallosal anterior cingulate cortex, as illustrated in Fig. 3.5 (Kringelbach and Rolls, 2004; Grabenhorst and Rolls, 2011) (see examples in Figs. 3.6, 3.8, 3.11, 3.28). Part of the importance of this is in relation to understanding the roles of different parts of the orbitofrontal cortex in depression (Section 3.13).

3.5.2 Social rewards, and abstract, monetary, rewards

Given the discoveries on the orbitofrontal cortex about reward information processing described above, I thought it important to see whether this could be extended to other types of reward, and more generally to emotion.

One extension was to social types of reward: we discovered orbitofrontal cortex visual neurons in macaques that respond to the sight of face expression, or to face identity (Rolls, Critchley, Browning and Inoue, 2006a), both of which are important in producing appropriate social behavior to another individual. This implicates the orbitofrontal cortex in social reward. This research was extended by the finding that some orbitofrontal cortex neurons are tuned to socially relevant categories of faces, such as juveniles (Barat, Wirth and Duhamel, 2018). Socially relevant stimuli such as the sight of face expression and face expression change also activate the human orbitofrontal cortex (Kringelbach and Rolls, 2003, 2004).

Extending the analysis further, as introduced above we made the discovery that an abstract type of reward, monetary reward, is represented in the human orbitofrontal cortex (O'Doherty et al., 2001a). This discovery helped to seed interest in the orbitofrontal cortex and neuroeconomics, and enormous interest followed (Padoa-Schioppa and Assad, 2006; Padoa-Schioppa, 2007; Padoa-Schioppa and Assad, 2008; Padoa-Schioppa and Conen, 2017; Glimcher and Fehr, 2013; Ballesta et al., 2020, 2022; Zhang et al., 2024b; Battista et al., 2025). One interesting advance is that we confirmed activation of the medial orbitofrontal cortex when winning points in a deterministic Go/NoGo task (like that used

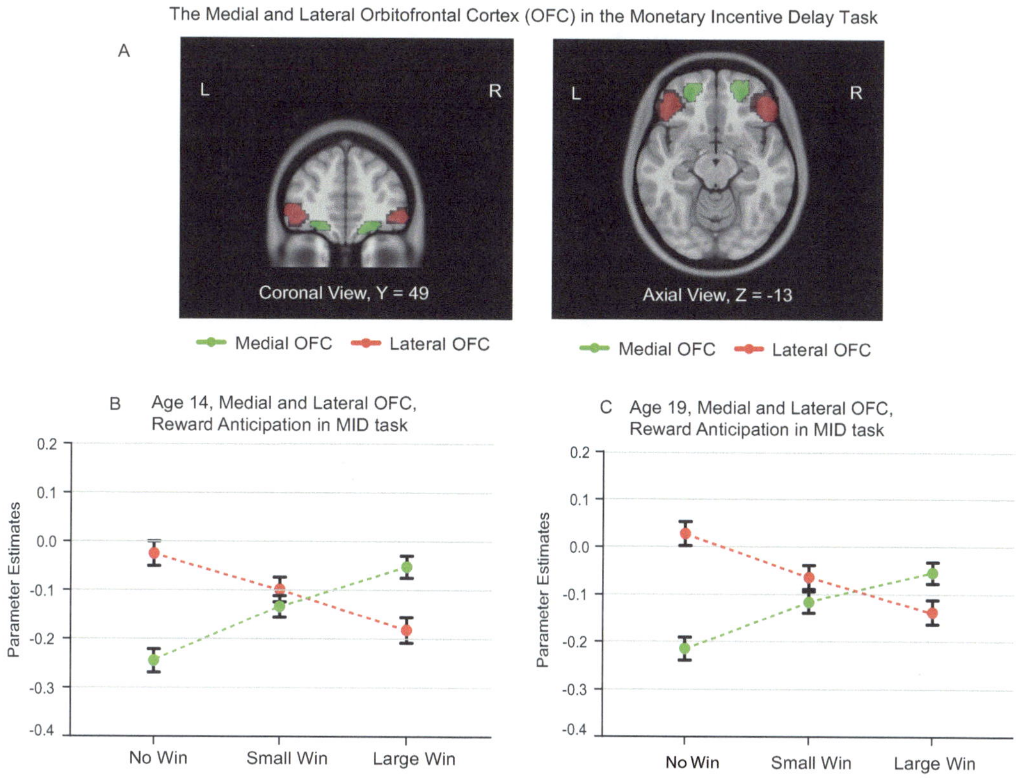

Fig. 3.6 The lateral orbitofrontal cortex is activated by not winning, and the medial orbitofrontal cortex by winning, in the monetary incentive delay task. The lateral orbitofrontal cortex region in which activations increased towards no reward (No Win) in the monetary incentive delay task are shown in red in 1140 participants at age 19 and in 1877 overlapping participants at age 14. The conditions were Large Win (10 points) to Small Win (2 points) to No Win (0 points) (at 19; sweets were used at 14). The medial orbitofrontal cortex region in which activations increased with increasing reward from No Win to Small Win to High Win) is shown in green. The parameter estimates are shown from the activations for the participants (mean $\pm$ sem) with the lateral orbitofrontal cortex in red and medial orbitofrontal cortex in green. The interaction term showing the sensitivity of the medial orbitofrontal cortex to reward and the lateral orbitofrontal cortex to non-reward was significant at $p=10^{-50}$ at age 19 and $p< 10^{-72}$ at age 14. (Data from Xie C, Jia T, Rolls E T et al. 2021 Reward vs non-reward sensitivity of the medial vs lateral orbitofrontal cortex related to the severity of depressive symptoms. Biological Psychiatry: Cognitive Neuroscience and Neuroimaging 6: 259–269.)

in macaques (Thorpe et al., 1983; Rolls et al., 1996a; Critchley and Rolls, 1996b)), and activation of the lateral orbitofrontal cortex when losing points (Rolls, Vatansever, Li, Cheng and Feng, 2020e) (Fig. 3.11B). Another advance was the demonstration in a group of almost 2000 adolescents that winning points in a monetary incentive delay task activated the medial orbitofrontal cortex, and that losing points activated the lateral orbitofrontal cortex (Xie, Jia, Rolls and al., 2021a) (see Fig. 3.6).

3.5.3 Effects of cognition and attention on reward value and pleasure in the brain

But we were able to make another set of key discoveries about how pleasure systems are organised in the human brain. We asked the question: 'How do cognition and attention influence emotion and reward processing in the brain?'. One possibility was that reward and pleasure signals in the brain might be sent to brain regions involved in cognition and attention, where interactions would occur to enable cognition and attention to influence reward and emotion processing. Another possibility

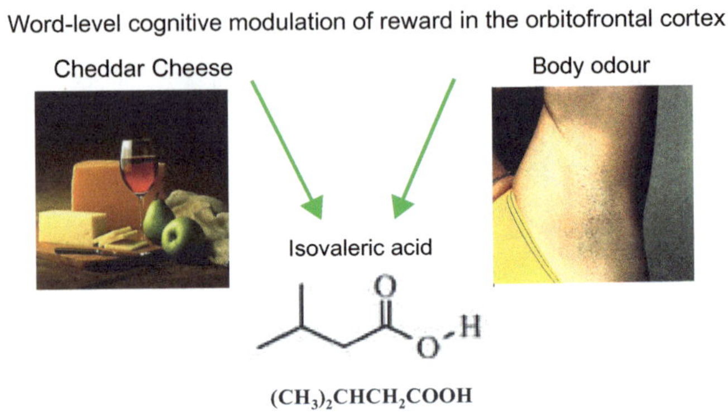

Fig. 3.7 Top-down effects of cognition on the pleasantness of reward representations in the orbitofrontal cortex and anterior cingulate cortex. In this experiment, we discovered that word-level descriptors can influence reward value / pleasantness representations in the orbitofrontal cortex and regions connected to it (De Araujo et al., 2005). In the experiment, the same odor, isovaleric acid, was always presented, but on some trials a word label indicated that it was the odor of cheddar cheese, and on other trials that it was body odor. The word labels influenced the activations in the orbitofrontal and anterior cingulate cortex; and the pleasantness ratings of the odor. (No visual images of food were used in the experiment: just the visual word labels.)

was that the cognition and attentional signals might be sent into the pleasure, reward, and emotion systems such as the orbitofrontal cortex to influence the actual representations of reward value and pleasure in the emotion-related parts of the brain such as the orbitofrontal cortex.

We investigated this first by measuring the effects of cognitive inputs on the representations of the subjective pleasantness and reward value for odors. We discovered that word level descriptors can influence the subjective pleasantness of olfactory stimuli, and the representation of olfactory stimuli in the human orbitofrontal cortex (De Araujo, Rolls, Velazco, Margot and Cayeux, 2005) (Fig. 3.7). The modulation is rather like the top-down effects of attention on perception (Desimone and Duncan, 1995; Rolls and Deco, 2002; Deco and Rolls, 2003, 2005d), not only phenomenologically, but also probably computationally using biased competition or biased activation as described in Section 4.3.5 of Rolls (2023c).

In the discovery of top-down effects of cognition on olfactory processing in the brain, a standard test odor, isovaleric acid (with a small amount of cheddar cheese odor added to make it more pleasant), was used as the test olfactory stimulus delivered with an olfactometer during functional neuroimaging with fMRI (De Araujo, Rolls, Velazco, Margot and Cayeux, 2005). This odor is somewhat ambiguous, and might be interpreted as the odor emitted by a cheese-like odor (rather like brie), or might be interpreted as a rather pungent and unpleasant body odor. A word was shown during the 8 s odor delivery. On some trials, the test odor was accompanied by the visually presented word ‘Cheddar cheese’. On other trials, the test odor was accompanied by the visually presented word ‘Body odor’. A word label was used rather than a picture label to make the modulating input very abstract and cognitive. First, we showed that the word labels influenced the pleasantness ratings of the test odor.

We went on to show that the word label modulated the activation to the odor in brain regions activated by odors such as the orbitofrontal cortex (secondary olfactory cortex), anterior cingulate cortex, and amygdala (De Araujo et al., 2005). For example, in the medial orbitofrontal cortex the word label ‘Cheddar cheese’ caused a larger activation to be produced to the test odor than when the word label ‘Body odor’ was being presented. In these medial orbitofrontal cortex regions and the amygdala, the activations were correlated with the pleasantness ratings, as shown in Fig. 3.8. This is consistent with the finding that the pleasantness of odors is represented in the medial orbitofrontal cortex (Rolls, Kringelbach and De Araujo, 2003c).

These findings show that cognition can influence and indeed modulate reward-related (including

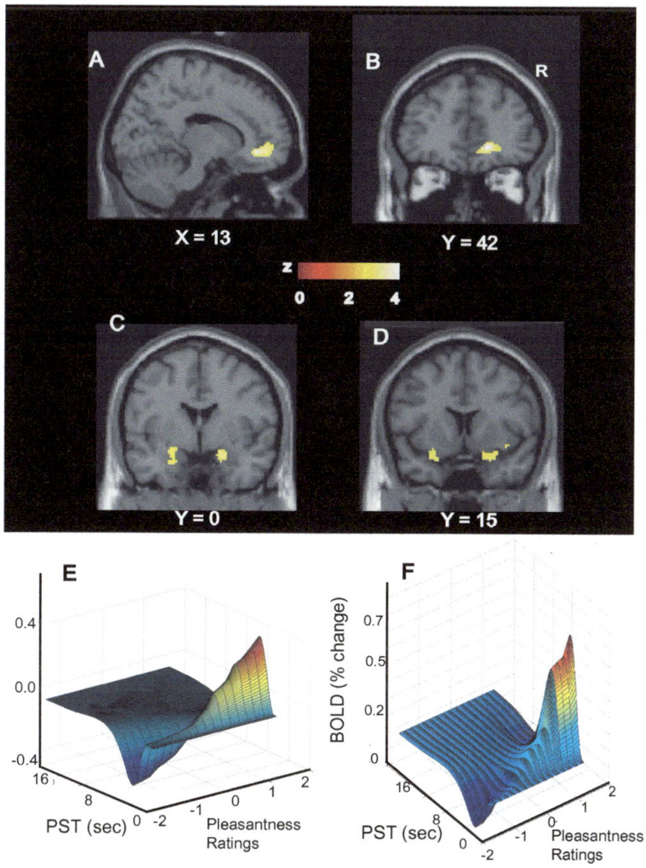

Fig. 3.8 Top-down effects of cognition on olfactory processing. Group (random) effects analysis showing the brain regions where the BOLD signal was correlated with pleasantness ratings given to the test odor. The pleasantness ratings were being modulated by the word labels. (A) Activations in the rostral anterior cingulate cortex, in the region adjoining the medial OFC, shown in a sagittal slice. (B) The same activation shown coronally. (C) Bilateral activations in the amygdala. (D) These activations extended anteriorly to the primary olfactory cortex. The image was thresholded at $p<0.0001$ uncorrected in order to show the extent of the activation. (E) Parametric plots of the data averaged across all subjects showing that the percentage BOLD change (fitted) correlates with the pleasantness ratings in the region shown in A and B. The parametric plots were very similar for the primary olfactory region shown in D. PST - Post-stimulus time (s). (F) Parametric plots for the amygdala region shown in C. (Data from De Araujo I E, Rolls E T, Velazco M I, Margot C, and Cayeux I. 2005 Cognitive modulation of olfactory processing. Neuron 46: 671–679.)

subjective pleasantness) processing in the orbitofrontal cortex, and in the amygdala (De Araujo, Rolls, Velazco, Margot and Cayeux, 2005).

We next extended this area of research, and discovered that this word-level cognitive modulation of reward value in the orbitofrontal cortex is quite a general property, for it can be extended to taste and flavor (Grabenhorst, Rolls and Bilderbeck, 2008a), and to the pleasantness of the sight and feel of touch (McCabe, Rolls, Bilderbeck and McGlone, 2008).

We also discovered that another type of top-down influence, top-down attention, well known to influence sensory representations in the visual system (Desimone and Duncan, 1995), can modulate reward representations in the orbitofrontal cortex. In one example, when participants were instructed to remember and rate the *pleasantness* of a taste stimulus, 0.1 M monosodium glutamate (umami taste), activations were greater in the medial orbitofrontal and pregenual anterior cingulate cortex than when participants were instructed to remember and rate the *intensity* of the taste (Grabenhorst and Rolls, 2008). When the subjects were instructed to remember and rate the intensity, activations were greater in the insular taste cortex (Grabenhorst and Rolls, 2008).

The concept here is that humans can selectively pay attention to the reward value / pleasantness of stimuli, by biasing up activations in the orbitofrontal cortex where reward value and subjective pleasantness are represented; or can selectively pay attention to the physical properties of stimuli such as their identity and intensity by biasing activations upwards in the primary taste cortex, where the physical properties of the stimuli such as their identity and intensity are represented.

In another example, when participants were instructed to remember and rate the pleasantness of a jasmin odor, fMRI activations were greater in the medial orbitofrontal and pregenual anterior cingulate cortex than when subjects were instructed to remember and rate the intensity of the odor (Rolls et al., 2008a).

These discoveries do have important implications for humans. One implication is that humans can use top-down attention to influence how their brains respond to emotional stimuli: whether to enhance or minimize the emotional impact of ongoing events, and of remembered events. Another implication is that the human reward system right down to the fundamental level of the orbitofrontal cortex can be influenced, in terms of how pleasant stimuli or products are, by even word level descriptions, and this will apply to advertising and many types of cognitive level inputs about affective stimuli and events.

I then felt it important to take these empirical discoveries forward to provide a computational understanding of the underlying mechanisms for how cognition and attention influence reward and subjective pleasantness systems in the human brain. I therefore developed our earlier models of top-down biasing of activity in sensory systems (Deco and Rolls, 2002a, 2003, 2004; Deco et al., 2004; Deco and Rolls, 2005a,d) (see also Desimone and Duncan (1995)) to provide a biased competition understanding of how these top-down effects of cognition and attention influence reward and pleasantness representations in the orbitofrontal cortex (Grabenhorst and Rolls, 2010; Ge et al., 2012; Luo et al., 2013; Rolls, 2013b) (see Rolls (2023c, 2026d)).

These investigations provide a good example of how interdisciplinary research is often important in making advances in understanding brain function. In primates it was possible to investigate at the neuronal level very precisely what is represented and how it is represented at different stages of each of the processing systems. In humans, it is possible to complement this by making discoveries about where human subjective affective feelings are related to brain activity. With computational neuroscience, it is possible to gain understanding of what the underlying mechanisms are of the operation of these brain systems (Rolls, 2023c, 2026d).

3.6 A mechanism for reward-specific satiety and for reward-specific motivation

We discovered that after sensory-specific satiety on one reward, an orbitofrontal cortex reward neuron can still be activated by a different reward (see Sections 3.3 and 3.4, and Fig. 3.3). This computation can also be termed 'reward-specific satiety' (Pastor-Bernier et al., 2021), though the term 'sensory-specific satiety' emphasizes the point that this type of satiety reflects the sensory properties of the reward. In addition, we discovered that the neurons in the insular primary taste cortex and inferior temporal visual cortex that provide inputs to the orbitofrontal cortex are not influenced by satiety, and remain responsive to stimuli after satiety is reached, in primates (see Section 3.4).

These discoveries led me to generate the hypothesis that the neural mechanism of sensory-specific satiety is a decrease in the release of neurotransmitter from the presynaptic terminals of the selective

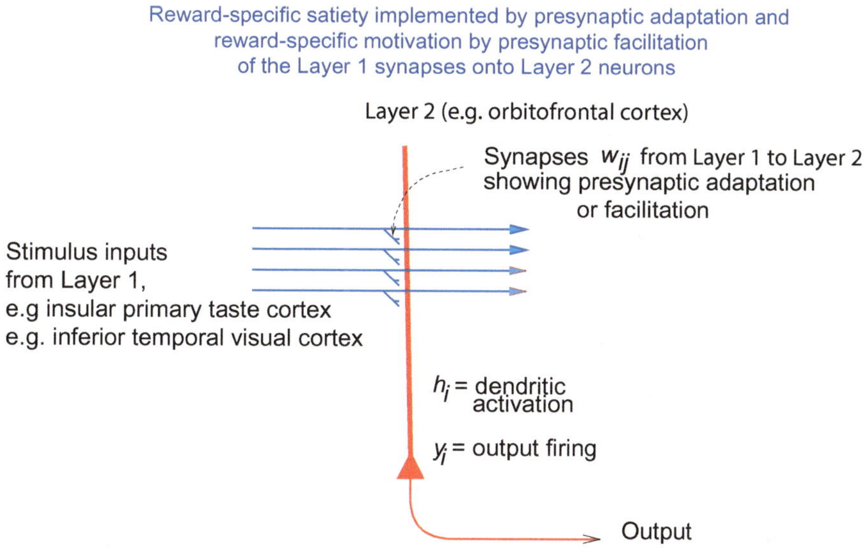

Fig. 3.9 Architecture for reward-specific satiety (also known as sensory-specific satiety) implemented by slow presynaptic adaptation in the synaptic terminals from Layer 1 to Layer 2. Layer 2 in primates including humans is the orbitofrontal cortex. In the same architecture, reward-specific motivation (incentive motivation) can be implemented by shorter term synaptic facilitation. (Data from Rolls E T, Zhang C and Feng J. 2025 Reward-specific satiety and reward-specific motivation: neural bases and significance. Cerebral Cortex 35: bhaf248.)

taste and visual neurons onto orbitofrontal cortex neurons. The synaptic mechanism involved is termed presynaptic adaptation, and with that theory, we set up a model to demonstrate and test my theory using an integrate-and-fire neuronal network, with the architecture illustrated in Fig. 3.9 (Rolls et al., 2025b). This process is important to understand, because sensory-specific satiety is a property of almost all reward systems, but of no punishment system (Rolls, 2014a, 2023b), and has clear adaptive value, by promoting choice of a wide range of nutrients and other behaviors that are important for reproductive fitness (Rolls, 2014a, 2023b). I propose that this mechanism applies to most reward systems, with potentially differences in the time course of the adaptation for different reward systems, and probably differences between different individuals.

In an example of how one discovery can lead to another, I then had the idea that pre-synaptic facilitation in the same terminals might account for the finding that when a reward is first offered, its reward value and the 'wanting' for the reward may increase for a short initial period. This is called incentive motivation (Rolls, 2014a, 2023b), or perhaps better reward-specific motivation (Rolls et al., 2025b) in that a small amount of reward may increase its reward value and make one want it more. (Donald Hebb (1949) called this the 'salted nut phenomenon', for if a street vendor offers a free sample of some nuts, one may want to purchase some. Another example is that the smell of bread cooking may encourage people to want to go into the shop to buy some bread.) I have proposed that incentive motivation has biological adaptive value by locking an individual onto a reward for at least some time, which is a much more efficient foraging strategy than changing behavior to find a different reward when the reward value drops a little, as the two rewards might be far apart (Rolls, 2014a, 2023b). An alternative way of computing efficient foraging could be to evaluate how much energy and other costs might be incurred in travelling between two distant types of reward just to obtain what had now become a slightly better reward as a result of having consumed some of the other reward. The incentive motivation influence on foraging seems much simpler and more efficient.

The theory I propose for the implementation of incentive motivation (which I propose can more clearly be termed 'reward-specific motivation') is that it occurs because the synaptic terminals onto reward neurons, in for example the orbitofrontal cortex, release more transmitter early on when they start to respond to a stimulus, thereby producing by presynaptic facilitation increased firing of orbitofrontal cortex reward neurons when a reward is first made available (Rolls et al., 2025b). The proposal is that presynaptic facilitation in the architecture shown in Fig. 3.9 produces reward-specific motivation, and we modelled this (Rolls et al., 2025b).

I have also proposed that reward-specific motivation, and reward-specific satiety, may be a key part of the neural mechanism involved in 'exploit vs explore' (i.e. stay with the current reward, or explore for a reward) in foraging (Rolls, 2023b; Rolls et al., 2025b), in which the frontal pole cortex is also implicated (Rolls et al., 2024a).

These discoveries are not only fundamental to understanding appetite control, but also to most rewarded and emotional behavior, for reward-specific motivation and reward-specific satiety / sensory-specific satiety are properties of almost all of our reward systems, and none of our punishment systems (Rolls, 2014a, 2023b; Rolls et al., 2025b).

3.7 The discovery that the orbitofrontal cortex encodes non-reward when reward is removed, and in reward reversal and extinction

While testing orbitofrontal cortex neurons for whether they code for reward value, we discovered another population of orbitofrontal cortex neurons that respond to non-reward. These macaque orbitofrontal cortex neurons respond when an expected reward is removed and not obtained, and in reward reversal or extinction (Thorpe, Rolls and Maddison, 1983), and these have been termed non-reward neurons (Rolls, 2014a, 2019c,a, 2023c, 2026d) (see example in Fig. 3.10). Most of these neurons respond when a reward is seen but then removed, and they can be selective for non-reward in reversal and extinction tasks (Thorpe et al., 1983). They can be described as negative reward prediction error neurons, in that they respond when a reward outcome is less than was expected (Rolls, 2019c). These neurons do not respond to expected punishers (e.g. the discriminative stimulus for saline in Fig. 3.10 (Thorpe et al., 1983)), but other neurons do respond to expected punishers (Rolls, Critchley, Mason and Wakeman, 1996a), showing that non-reward and punishment are represented by different neurons in the orbitofrontal cortex. The finding of non-reward neurons is robust, in that 18/494 (3.6%) of the neurons in the original study responded to non-reward (Thorpe, Rolls and Maddison, 1983); consistent results were found in different tasks in a complementary study (10/140 non-reward neurons in the orbitofrontal cortex or 7.1%) (Rosenkilde, Bauer and Fuster, 1981); and an fMRI study has shown that the macaque lateral orbitofrontal cortex is activated when an expected reward is not obtained during reversal (Chau et al., 2015).

The hypothesis is that the non-reward responsiveness of these neurons is computed in the orbitofrontal cortex. It can be computed here, because this is the first cortical region in primates in which expected value and outcome value are represented, as summarized in Fig. 3.12. Expected value refers to the reward value of a stimulus that provides evidence about how much reward is expected, for example neurons that respond to the sight of food. Outcome value refers to how much reward is actually obtained as a result of a choice, for example what taste is received in a visual discrimination task. The evidence is set out fully by Rolls (2019c; 2021a; 2023c; 2026d), and these two signals are those required to compute non-reward, that is, that reward outcome is less than the expected value.

My next aim was to investigate whether the discoveries on non-reward systems in primates can be extended to non-reward in the human orbitofrontal cortex. We tested this, and discovered that

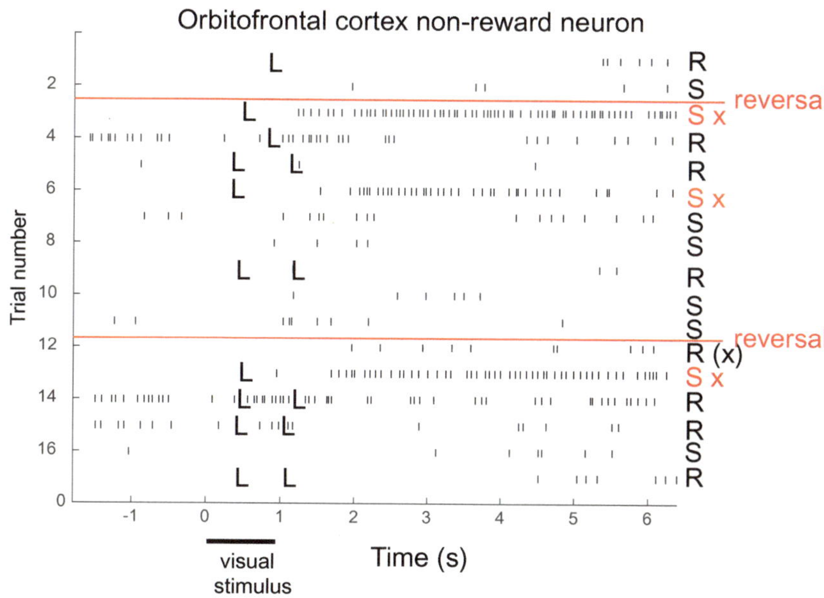

Fig. 3.10 Evidence that the lateral orbitofrontal cortex is activated by non-reward. Non-reward error-related neurons maintain their firing after non-reward is obtained. Responses of an orbitofrontal cortex neuron that responded only when the macaque licked to a visual stimulus during reversal, expecting to obtain fruit juice reward, but actually obtained the taste of aversive saline because it was the first trial of reversal (trials 3, 6, and 13). Each vertical line represents an action potential; each L indicates a lick response in the Go-NoGo visual discrimination task. The visual stimulus was shown at time 0 for 1 s. The neuron did not respond on most reward (R) or saline (S) trials, but did respond on the trials marked S x, which were the first or second trials after a reversal of the visual discrimination on which the monkey licked to obtain reward, but actually obtained saline because the task had been reversed. The two times at which the reward contingencies were reversed are indicated. After responding to non-reward, when the expected reward was not obtained, the neuron fired for many seconds, and was sometimes still firing at the start of the next trial. It is notable that after an expected reward was not obtained due to a reversal contingency being applied, on the very next trial the macaque selected the previously non-rewarded stimulus. This shows that rapid reversal can be performed by a non-associative process, and must be rule-based. (Data from Thorpe S J , Rolls E T and Maddison S. 1983 The orbitofrontal cortex: neuronal activity in the behaving monkey. Experimental Brain Research 49: 93–115.)

the human lateral orbitofrontal cortex is activated when an expected reward is not obtained in a visual discrimination reversal task, that is when the expected social reward of a smile face expression was not obtained (Kringelbach and Rolls, 2003) (Fig. 3.11A), and when money is not received in a monetary reward task (O'Doherty et al., 2001a; Xie et al., 2021b) (Fig. 3.6), and in a rule-based deterministic one-trial reward reversal task for points when expected reward is not obtained, as illustrated in Fig. 3.11B (Rolls, Vatansever, Li, Cheng and Feng, 2020e).

We also discovered that the human lateral orbitofrontal cortex is activated by punishing, subjectively unpleasant, stimuli (Grabenhorst and Rolls, 2011; Rolls, 2019c, 2021a, 2019a) (see Fig. 3.5). Examples include unpleasant odors (Rolls et al., 2003c), pain (Rolls et al., 2003d), losing money (O'Doherty et al., 2001a), and receiving an angry face expression indicating that behavior should change in a reward reversal (Kringelbach and Rolls, 2003). The human right lateral orbitofrontal cortex / inferior frontal gyrus is also activated when behavioral correction is required in the stop-signal task (Aron et al., 2014; Deng et al., 2017).

These discoveries show that one way in which the human orbitofrontal cortex is involved in decision-making and emotion is by representing rewards, punishers, and errors made during decision-making. This is supported by the problems that orbitofrontal cortex damage produces in decision-making, which including failing to respond correctly to non-reward, as described next.

A Social Reversal

B Rule-based Reward Reversal for points

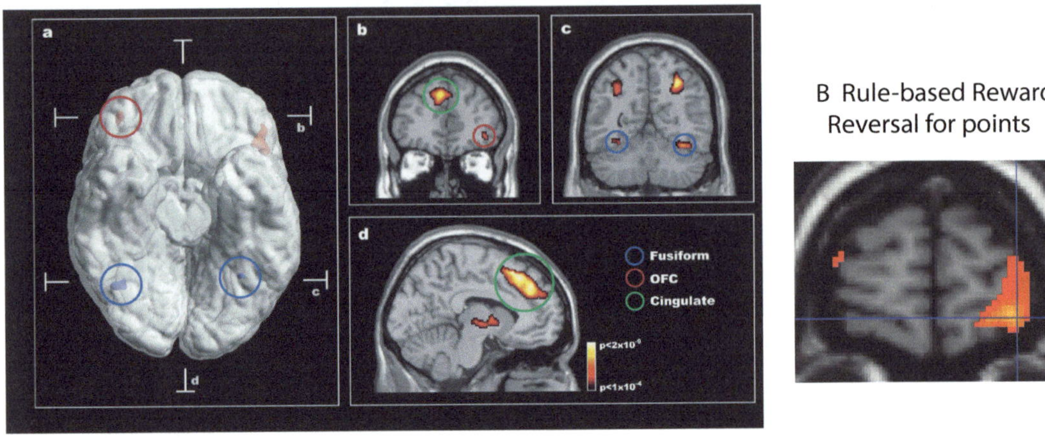

Fig. 3.11 Evidence that the lateral orbitofrontal cortex is activated by non-reward. A. Activation of the lateral orbitofrontal cortex in a visual discrimination reversal task on reversal trials, when a face was selected but the expected reward was not obtained, indicating that the subject should select the other face in future to obtain the reward. a) A ventral view of the human brain with indication of the location of the two coronal slices (b,c) and the transverse slice (d). The activations with the red circle in the lateral orbitofrontal cortex (OFC, peaks at [42 42 -8] and [-46 30 -8]) show the activation on reversal trials compared to the non-reversal trials. For comparison, the activations with the blue circle show the fusiform face area produced just by face expressions, not by reversal, which are also indicated in the coronal slice in (c). b) A coronal slice showing the activation in the right orbitofrontal cortex on reversal trials. Activation is also shown in the supracallosal anterior cingulate region (Cingulate, green circle) that is also known to be activated by many punishing, unpleasant, stimuli (see Grabenhorst and Rolls (2011)). (Data from Kringelbach M L and Rolls ET 2003 Neural correlates of rapid reversal learning in a simple model of human social interaction. NeuroImage 20: 1371–83.) B. Activations in the human right lateral orbitofrontal cortex on the reversal trials of a Go/NoGo visual discrimination reversal task in humans. This was a deterministic task in which points could be won or lost and the reversal was rule-based in one trial, analogous to the task in which non-reward neurons were discovered in macaques. (Data from Rolls E T, Vatansever D et al, 2020 Rapid rule-based reward reversal and the lateral orbitofrontal cortex. Cerebral Cortex Communications 1: tgaa087).

3.8 Discoveries on the effects of damage to the human orbitofrontal cortex

Based on the neurophysiological evidence that we were accumulating about the orbitofrontal cortex, I was keen given my training in preclinical medicine to investigate whether the functions revealed in these studies had applications to understanding and potentially treating the effects of damage to the orbitofrontal in humans. Accordingly I initiated a research programme on this (Rolls, Hornak, Wade and McGrath, 1994a; Hornak, Rolls and Wade, 1996; Rolls, 2018b, 2019c; Hornak, O'Doherty, Bramham, Rolls, Morris, Bullock and Polkey, 2004; Berlin, Rolls and Kischka, 2004; Berlin and Rolls, 2004; Rolls, 2019a; Rolls, Cheng and Feng, 2020b; Rolls, 2021d, 2023c, 2026d). We worked initially with patients with closed head injuries that damaged the orbitofrontal cortex, which can occur in for example motor bicycle accidents. Later we further tested the hypotheses in patients with restricted surgical excisions of parts of the orbitofrontal cortex to treat tumours, epilepsy, etc.

I started with visual discrimination reversal learning, for that had the potential to be impaired if the reward and non-reward systems we had identified in the orbitofrontal cortex were malfunctioning, and to account for the behavior of these patients which can be disinhibited and without full regard for the consequences of their decisions. Such an impairment might account I surmised for the observation by the neurologists with whom I worked that some of these patients would use funds from an insurance claim to buy another motorbike. We found that damage to the human orbitofrontal cortex can impair reward reversal learning during decision-making in humans (Rolls et al., 1994a; Hornak et al., 2004) (cf Fellows (2011)), who continue responding to the previously rewarded, now non-rewarded, stimulus. The change in contingency between the stimulus and reward vs non-reward is not processed correctly.

This type of flexibility to change behavior rapidly when reward is no longer being received is important in primate including human social interactions. I have shown that many of the effects of damage to the human orbitofrontal cortex, including the difficulty in responding appropriately to the changed circumstances of the patient, and the changed personality including impulsivity, can be related to these impairments in responding to non-reward and punishers (Rolls, 2018b, 2019c; Hornak, O'Doherty, Bramham, Rolls, Morris, Bullock and Polkey, 2004; Rolls, Hornak, Wade and McGrath, 1994a; Berlin, Rolls and Kischka, 2004; Berlin and Rolls, 2004; Rolls, 2019a; Rolls, Cheng and Feng, 2020b; Rolls, 2021d,a).

We also discovered that patients with orbitofrontal cortex damage can be impaired on identifying the emotional expression seen on a face or heard in a voice (Hornak, Rolls and Wade, 1996; Hornak, Bramham, Rolls, Morris, O'Doherty, Bullock and Polkey, 2003), consistent with our discovery of neurons in the macaque orbitofrontal cortex that respond to face expression, and to vocal stimuli (Rolls, Critchley, Browning and Inoue, 2006a). Part of the importance of this discovery is that it shows some of the ways in which altered function of the orbitofrontal cortex could alter and impair social behavior, in which the ability to correctly decode and respond to social signals conveyed by face and voice expression is very important.

Together, these effects of damage to the human orbitofrontal cortex help to show that causally the orbitofrontal cortex is involved in reward-related behavior in ways that are consistent with what we discovered neurophysiologically, and moreover have implications for how to help people with damage to the orbitofrontal cortex (Rolls, 2018b, 2019c; Hornak, O'Doherty, Bramham, Rolls, Morris, Bullock and Polkey, 2004; Rolls, Hornak, Wade and McGrath, 1994a; Berlin, Rolls and Kischka, 2004; Berlin and Rolls, 2004; Rolls, 2019a; Rolls, Cheng and Feng, 2020b; Rolls, 2021d, 2023c, 2026d).

Applying this approach to another mental condition, we discovered that in autism there is disconnectivity (measured by reduced functional connectivity) between the face expression representation area we discovered in the cortex in the superior temporal sulcus (Hasselmo, Rolls, Baylis and Nalwa, 1989b; Hasselmo, Rolls and Baylis, 1989a; Rolls, 2024c) and the orbitofrontal cortex (Cheng, Rolls, Gu, Zhang and Feng, 2015; Rolls, Zhou, Cheng, Gilson, Deco and Feng, 2020f). An implication is that in autism there are impairments in recognizing the social significance of face including facial expression stimuli, which in turn has implications for understanding the differences in behavior and for treatment by training.

3.9 The systems-level processing of reward value in primates including humans

I have developed a framework in the context of the discoveries described here for the systems-level organization of the brain for reward value, emotion, and motivation in primates including humans, and this is shown in Fig. 3.12 (Rolls, 2023b, 2025e). This synthesis is built on discoveries from neuronal recordings, the effects of brain damage, and fMRI in humans and macaques some of which is summarized above and elsewhere (Rolls, 2018b, 2021a, 2014a, 2019e; Rolls et al., 2020b; Rolls, 2023c, 2026d, 2023b, 2025e,d). Part of the evidence for what is shown in Fig. 3.12 comes from reward devaluation, in which when the reward value is changed, for example by feeding to satiety, neural responses to stimuli are little affected in Tier 1, but decrease to zero in Tier 2. Part of the evidence comes from the learning of associations between stimuli and reward value, which occurs mainly in Tier 2. Part of the evidence comes from the effects of brain damage on emotion, which

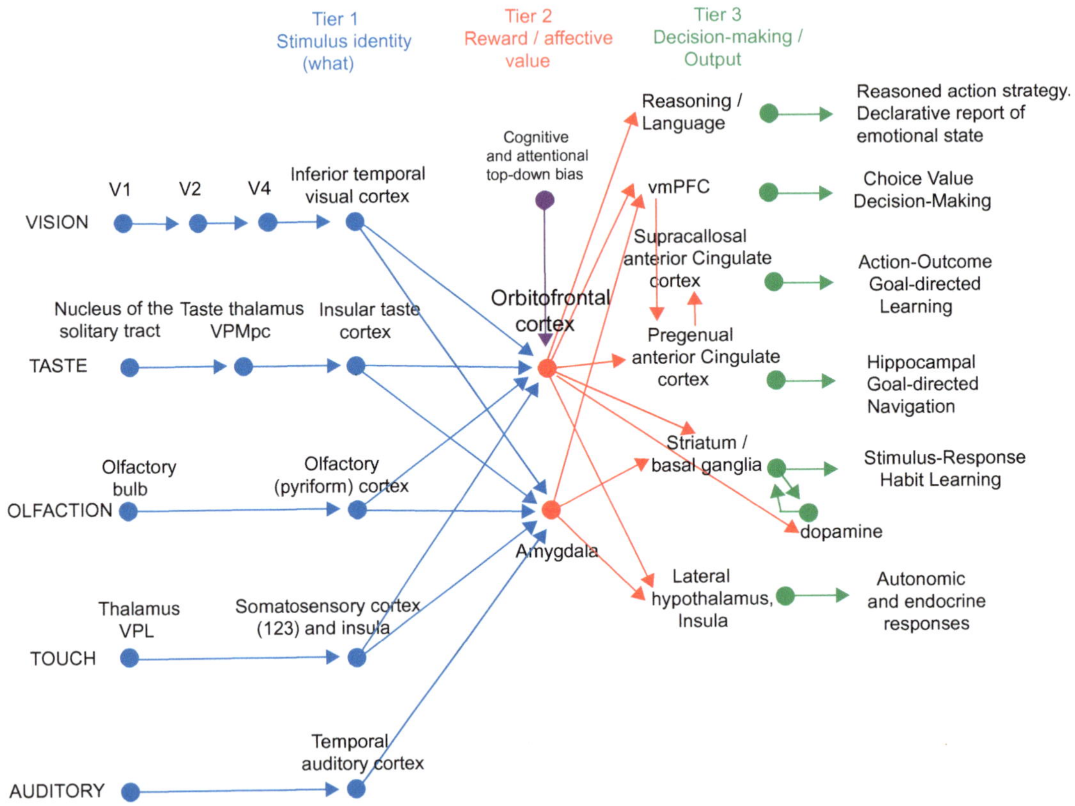

Fig. 3.12 The systems level organization of the brain for reward value, emotion, and motivation in primates including humans. In Tier 1, representations are built of visual, taste, olfactory and tactile stimuli that are independent of reward value and therefore of emotion. In Tier 2, reward value and emotion are represented. A pathway for top-down attentional and cognitive modulation of emotion is shown in purple. In Tier 3 actions are learned in the supracallosal (or dorsal) anterior cingulate cortex to obtain the reward values signaled by the orbitofrontal cortex and amygdala that are relayed in part via the pregenual anterior cingulate cortex and vmPFC. Decisions between stimuli of different reward value can be taken in the ventromedial prefrontal cortex, vmPFC. In Tier 3, orbitofrontal cortex inputs to the reasoning / language systems enable affective value to be incorporated and reported. In Tier 3, stimulus-response habits can also be produced using reinforcement learning. In Tier 3 autonomic responses can also be produced to emotion-provoking stimuli. Auditory inputs also reach the amygdala. V1 - primary visual (striate) cortex; V2 and V4 - further cortical visual areas. PFC - prefrontal cortex. The Medial PFC area 10 is part of the ventromedial prefrontal cortex (vmPFC). VPL - ventro-postero-lateral nucleus of the thalamus, which conveys somatosensory information to the primary somatosensory cortex (areas 1, 2 and 3). VPMpc - ventro-postero-medial nucleus pars parvocellularis of the thalamus, which conveys taste information to the primary taste cortex. (Data from Rolls E T (2025) Emotion, Motivation, Reasoning, and how their brain systems are related. Brain Sciences 15: 507.)

occur primarily after damage to the orbitofrontal cortex and amygdala in Tier 2, and the anterior cingulate cortex in Tier 3 (Rolls, 2021d).

A key point emphasized in Fig. 3.12 is that there are multiple routes to output from the key brain regions involved in reward value, emotion, and motivation, the orbitofrontal cortex and the amygdala, in Tier 2 (Rolls, 2023b, 2026d). The multiple routes are organized in a set of hierarchies, with each level in the system added later in evolution, but with all levels left in operation over the course of evolution (Rolls, 2016b). The result of this is that a response such as an autonomic response to a stimulus that happens also to be rewarding might be produced by only the lower levels of the system operating, without necessarily the highest e.g. explicit (declarative, conscious) levels being involved.

The lowest levels in the hierarchy are involved in reflexes, including for example reflex withdrawal of a limb to a nociceptive stimulus, and autonomic responses.

The first level in the hierarchy illustrated in Fig. 3.12 can produce learned autonomic and some other behavioral responses to for example a previously neutral visual or auditory stimulus after it has been paired with a nociceptive stimulus or with a good taste stimulus. This route involves stimulus-reinforcer learning in the amygdala and orbitofrontal cortex.

A second level in the hierarchy shown in Fig. 3.12 is the route from the orbitofrontal cortex and amygdala via the basal ganglia including the ventral striatum to produce implicit stimulus-response habits. The responses are not under the control of the reward value of the stimulus, in that after devaluing the stimulus outsidethe task, the stimulus will still elicit the response for several trials (Dickinson, 2016; Rolls, 2023b).

A third level in the hierarchy is to the hippocampus to provide the goals for navigation.

A fourth level in the hierarchy that is important in emotion is from especially the orbitofrontal cortex to the anterior cingulate cortex for goal-directed action. In this route to action, the evidence is that the orbitofrontal cortex encodes the value of stimuli (e.g. the reward value of the taste and sight of food), and the anterior cingulate cortex is the route to action, and encodes the value of actions taking into account the cost of the action (Rolls, 2023b; Rushworth et al., 2012). The emotional states implemented at this level may not necessarily be conscious (Rolls, 2025e).

A fifth level, in the ventromedial prefrontal cortex, is implicated in choice value decision-making.

A sixth level in the hierarchy shown in Fig. 3.12 is from the orbitofrontal cortex (and much less the amygdala (Rolls et al., 2023d)) via multiple step reasoning systems involving syntax and language, which can be associated with explicit conscious states. A higher order syntactic thought system for correcting first order thoughts is proposed to help correct this reasoning, and to be associated with the explicit conscious states (Rolls, 2020c, 2008b, 2014a, 2023c, 2026d) (see Section 3.11.6). Connectivity from the lateral orbitofrontal cortex to language regions provides a route for emotional states to enter the reasoning system (Rolls et al., 2023d; Rolls, 2025e).

It is emphasized that each of these types of output have adaptive value in preparing individuals to deal physiologically and behaviorally with what may generally be described as emotion-provoking events (Rolls, 2023b).

Part of the importance of these discoveries in primates and humans is that they apply to humans and other primates, whereas the organization of reward value processing and therefore emotion in the rodent brain is very different (Rolls, 2019c, 2021a, 2023c, 2026d, 2023b), and the rodent provides a poor model of the functional organization of the primate including human brain (Section 29.4 of Rolls (2026d)).

3.10 Discoveries on the connectivity of the human orbitofrontal cortex, and its relation to function

Discoveries are now described on the connectivity of the human orbitofrontal cortex, and its implications for functions (Rolls et al., 2022b, 2023d). The evidence is based on use of the Human Connectome Project - Multimodal Cortical Parcellation (HCP-MMP, Section 1.3 and Appendix A (Glasser et al., 2016a)), and measuring effective connectivity, complemented by functional connectivity and diffusion tractography with data from the Human Connectome Project. The measures used, effective and functional connectivity, and tractography, are described in Section 6.1, and Chapter 6 shows how performing these investigations is helping to provide a connectional framework for helping to understand how cortical regions communicate the computations that they perform to each other. This is essential for understanding the operation and functions of the cerebral cortex as a whole.

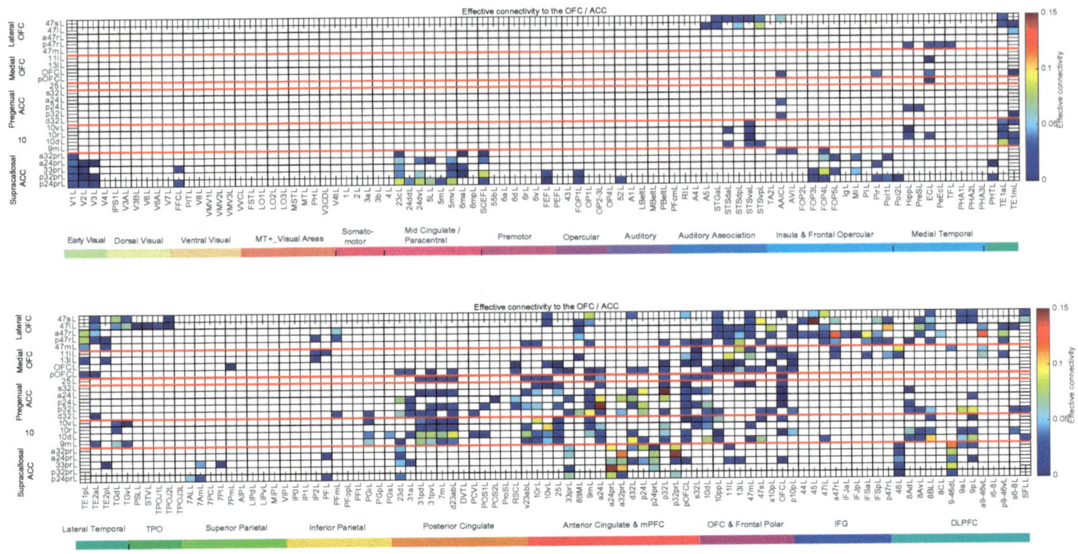

Fig. 3.13 Effective connectivity to the orbitofrontal cortex (OFC) and anterior cingulate cortex (ACC) from all 180 cortical regions in the left hemisphere of the human brain (see text for description). The area 10 group consists of 10v, 10r, 10d and 9m, and are included here because they are close to and have somewhat similar connectivity to the pregenual ACC group, and are part of what is sometimes termed the ventromedial prefrontal cortex. The effective connectivity is read from column to row. All effective connectivities greater than 0 are shown, and effective connectivities of 0 are shown as a blank. The connectivities from the first set of cortical regions are shown above, and from the second set below. All effective connectivity maps are scaled to show 0.15 as the maximum, as this is the highest effective connectivity found between this set of brain regions. The effective connectivity algorithm for the whole brain is set to have a maximum of 0.2, and this was for connectivity between V1 and V2. The abbreviations are listed in Appendix A and the cortical parcellation is introduced in Section 1.3. (Data from Rolls E T, Deco G, Huang C-C and Feng J 2022 The human orbitofrontal cortex, vmPFC, and anterior cingulate cortex effective connectome: emotion, memory, and action. Cerebral Cortex 33: 350-356.)

An example of a map for the effective connectivity to the medial and lateral orbitofrontal cortex, to pregenual anterior cingulate cortex, to ventromedial prefrontal cortex area 10 regions, and to the supracallosal anterior cingulate cortex, from all cortical regions in the left hemisphere, to indicate the underlying measurements, is provided in Fig. 3.13 from Rolls et al. (2022b). This shows for example connectivity to some of the orbitofrontal cortex and ventromedial prefrontal cortex (area 10) regions, but much less to anterior cingulate cortex regions, from inferior temporal cortex regions TE1a, TE1m, TE1p, TE2a and TE2p, and also from the temporal pole regions TGd and TGv (Rolls et al., 2022b).

In Sections 3.10.1 and 3.10.2 the connectivities of the medial and lateral orbitofrontal cortex are presented at a detailed level, and syntheses and implications of the connectivities are highlighted in Sections 3.10.3–3.10.6. The analyses that we have provided for these connectivities (Rolls et al., 2022b) are important for understanding how the orbitofrontal cortex and its connected regions operate in emotion, motivation, and reward-related decision-making (Rolls, 2023b,c, 2026d).

3.10.1 Human medial orbitofrontal cortex connectivity

The medial orbitofrontal cortex regions defined in the HCP-MMP atlas are 11l, 13l, OFC, and pOFC. A summary of the effective connectivity of the human medial orbitofrontal cortex is shown in Fig. 3.14 (Rolls et al., 2022b, 2023d).

Parts of the medial orbitofrontal cortex (11l, 13l, OFC and pOFC, which are interconnected) have effective connectivity with the taste/olfactory/visceral anterior agranular insular complex (AAIC); the piriform (olfactory) cortex; the entorhinal cortex (EC); the inferior temporal visual cortex (TE1p,

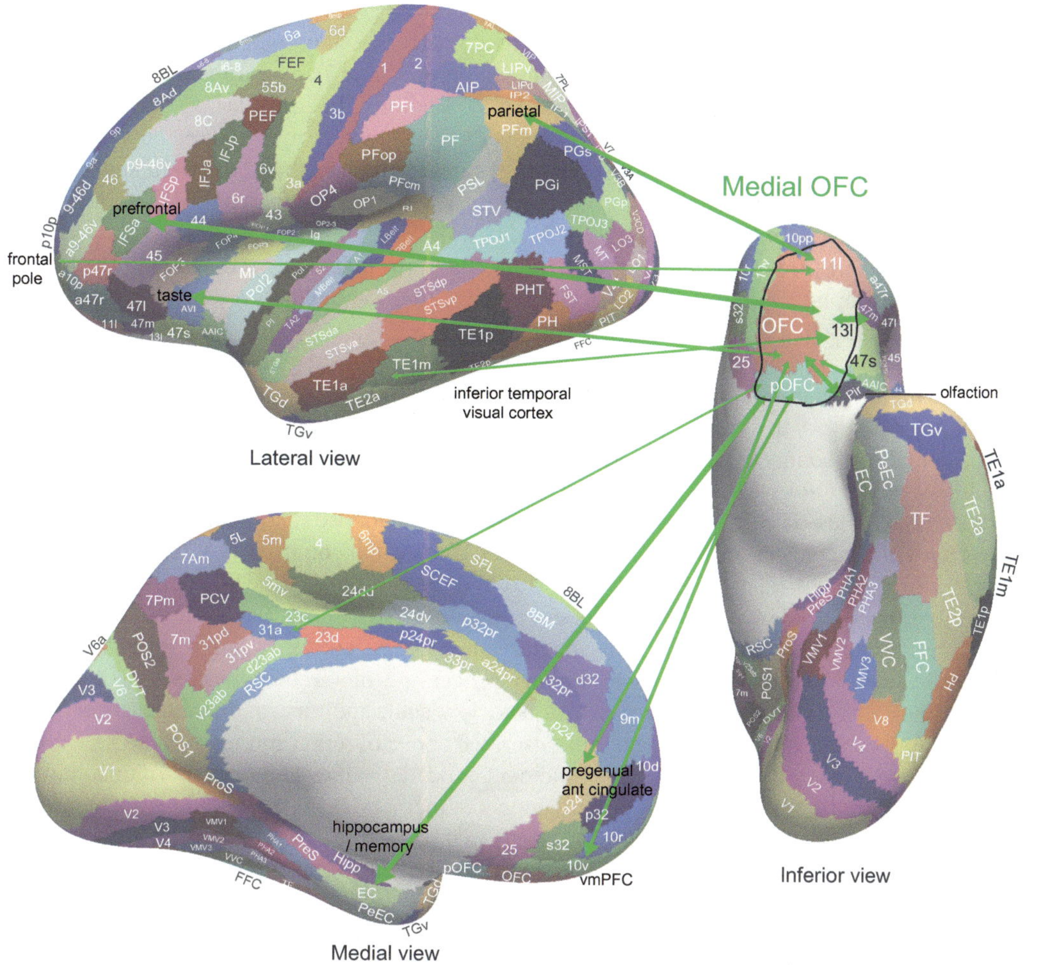

Fig. 3.14 Summary of the effective connectivity of the human medial orbitofrontal cortex. The medial orbitofrontal cortex has taste, olfactory and inferior temporal visual cortex inputs, and connectivity with the hippocampus, pregenual anterior cingulate cortex, ventromedial prefrontal cortex (vmPFC), posterior cingulate cortex (e.g. 31), parietal cortex, inferior prefrontal cortex, and frontal pole. The main regions with which the medial OFC has connectivity are indicated by names with the words in black font. The width of the arrows and the size of the arrow heads in each direction reflects the strength of the effective connectivity. The abbreviations for the cortical regions are listed in Appendix A. (Data from Rolls E T 2023 Emotion, motivation, decision-making, the orbitofrontal cortex, anterior cingulate cortex, and the amygdala. Brain Structure and Function 228: 1201-1257.)

TE2a, TE2p); superior medial parietal 7Pm; inferior parietal PF which is somatosensory (Rolls et al., 2023a); with parts of the posterior cingulate cortex (31pv, 7m, d23ab) related to memory (Rolls et al., 2023g); with the pregenual anterior cingulate cortex (s32, a24, p24, p32, d32) and much less with the supracallosal anterior cingulate cortex (only 33pr); with ventromedial prefrontal 10r, 10d and 9m; with the frontal pole (10pp, p10p, a10p); with lateral orbitofrontal cortex (47m, 47s, a47r); and dorsolateral prefrontal cortex (46 and a9-46v) (Rolls et al., 2022b). The connectivities are stronger towards the medial orbitofrontal cortex for the inferior temporal visual cortex and frontal pole regions. Medial orbitofrontal cortex regions also have effective connectivity directed towards the caudate nucleus and nucleus accumbens (Rolls et al., 2022b).

Region OFC is remarkable for effective connectivities directed towards more cortical regions than other parts of the medial orbitofrontal cortex, including somatosensory cortex regions 5L and 5m;

the fusiform face area (FFC) and some other relatively early visual cortical areas; and some parietal areas including PGp, PGs, and some superior parietal parts of area 7 and intraparietal areas described elsewhere (Rolls et al., 2023a; Rolls, 2023c, 2026d). Region OFC also has effective connectivity directed towards the substantia nigra pars compacta (SNpc) and ventral tegmental area (VTA) (Rolls et al., 2022b), which contain dopaminergic neurons. I have proposed that the reward and non-reward signals needed by the dopamine reward prediction error neurons (Schultz, 2017, 2024) are provided by the orbitofrontal cortex (Rolls, 2017a, 2023c, 2026d).

Regions 11l and 13l have outputs directed to inferior prefrontal areas (IFS and IFJ regions) and to dorsolateral prefrontal areas 46 and a9-46v (Rolls et al., 2023c) both involved in short-term memory (Rolls, 2023c, 2026d)

pOFC is the only cortical area with effective connectivity directed to the nucleus basalis of Meynert which includes cholinergic neurons that project to the neocortex (Section 3.10.4) (Huang et al., 2022; Zaborszky et al., 2018, 2008).

The functional connectivity is generally consistent, adding some evidence for interactions with frontal opercular FOP4 which is probably taste-related, and the somatosensory insula (MI) (Rolls et al., 2022b, 2023c). The diffusion tractography provides evidence that medial orbitofrontal cortex regions have direct connections with the anterior agranular insular complex (AAIC) which is probably taste/olfactory/visceral-related, and with the piriform cortex (Pir); with the hippocampal system; and with temporal pole TGd involved in semantic representations (Rolls et al., 2022c).

To synthesize, with more below, the medial orbitofrontal cortex is found in humans to have connectivity with regions at the ends of sensory processing hierarchies that provide evidence for 'what' stimulus is present, including taste, olfactory, visual, auditory, and somatosensory cortical systems. This is consistent with the functions described above of representing unlearned rewards; and learning stimulus-reward associations to implement expected value. The medial orbitofrontal cortex also has connectivity that enables it to provide inputs to memory systems, including the hippocampal memory system; and the pregenual cingulate cortex that has onward connectivity to the hippocampal memory system; and parts of the posterior inferior parietal cortex that are related to memory. The medial orbitofrontal cortex also has some connectivity with the supracallosal anterior cingulate cortex implicated in guiding actions to obtain rewards and avoid punishers in action-outcome learning (Rushworth et al., 2012; Rolls, 2019e, 2023c, 2026d); and with the lateral orbitofrontal cortex.

3.10.2 Human lateral orbitofrontal cortex connectivity

The lateral orbitofrontal cortex regions defined in the HCP-MMP atlas are 47s, 47l, a47r, p47r and 47m. A summary of the effective connectivity of the human lateral orbitofrontal cortex is shown in Fig. 3.15 (Rolls et al., 2022b, 2023d).

The lateral orbitofrontal cortex regions a47r, p47r and 47m share generally similar effective connectivities from the visual inferior temporal cortex (TE areas); from parts of the parietal cortex (PFm which receives visual and auditory object-level information and IP2 which is visuomotor (Rolls et al., 2023a)); from the medial orbitofrontal cortex (11l, 13l, pOFC); from the inferior frontal gyrus regions including IFJ, IFS and BA45; from the dorsolateral prefrontal cortex (8Av, 8BL, a9-46v and p9-46v) implicated in short-term memory (Rolls, 2023c, 2026d); and from the frontal pole (a10p, p10p, 10pp) (Rolls et al., 2022b, 2023d, 2024a). Region 47m (which is relatively medial in this group) also has effective connectivity with the hippocampal memory system (Hipp, EC, perirhinal, and TF); and with ventromedial prefrontal region 10r; and with the frontal pole (10d, and 9m (Rolls et al., 2024a)). Although in most cases there are effective connectivities from a47r, p47r and 47m to these other cortical regions, the effective connectivities are in most cases stronger towards the lateral orbitofrontal cortex, except that the connectivities are stronger from the lateral orbitofrontal cortex

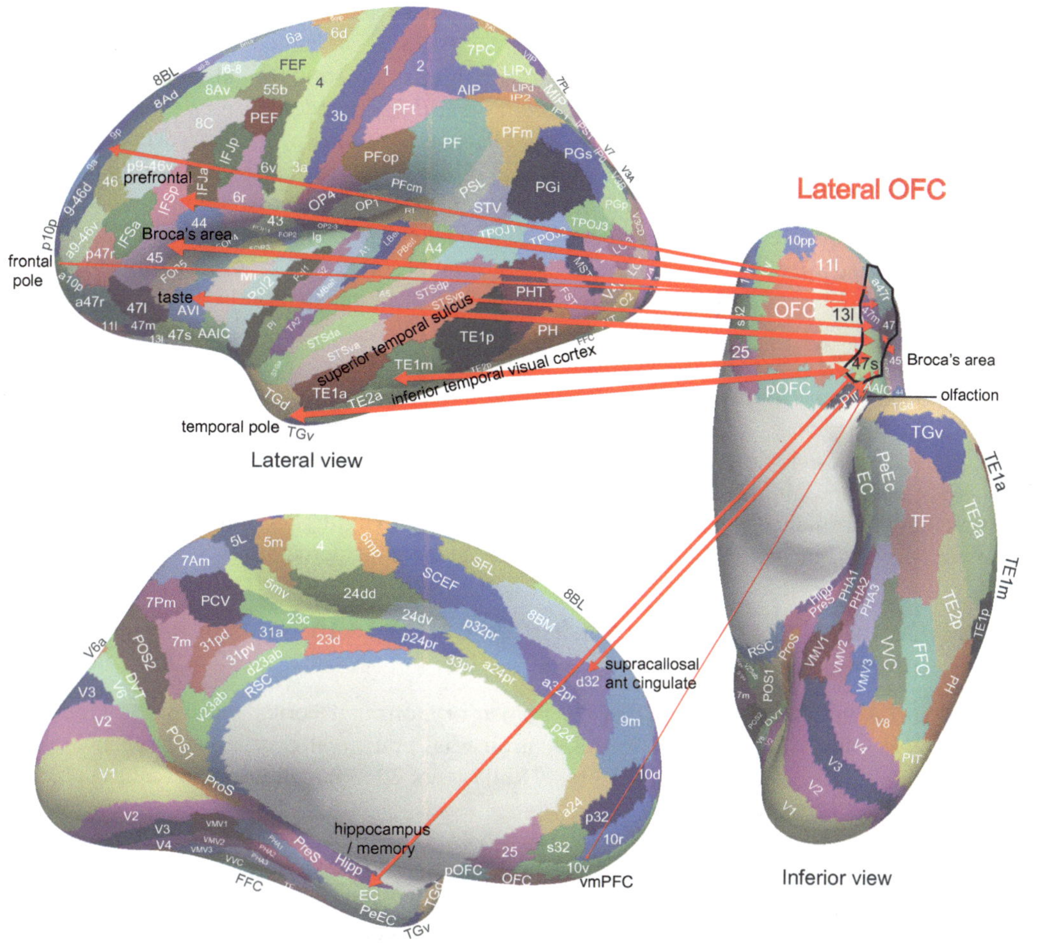

Fig. 3.15 Summary of the effective connectivity of the human lateral orbitofrontal cortex. The lateral orbitofrontal cortex has taste, olfactory and inferior temporal visual cortex inputs, and connectivity with the hippocampus, supracallosal (dorsal) anterior cingulate cortex, inferior and dorsolateral prefrontal cortex, and frontal pole. However, the lateral OFC also has connectivity with language regions (the cortex in the superior temporal sulcus and Broca's area). The main regions with which the lateral OFC has connectivity are indicated by names with the words in black font. The width of the arrows and the size of the arrow heads in each direction reflects the strength of the effective connectivity. The abbreviations are listed in Appendix A. (Data from Rolls E T 2023 Emotion, motivation, decision-making, the orbitofrontal cortex, anterior cingulate cortex, and the amygdala. Brain Structure and Function 228: 1201-1257.)

towards the set of inferior prefrontal cortex language-related regions including 45, 44, IFJ and IFS. The diffusion tractography provides in addition evidence for connections of these parts of the lateral orbitofrontal cortex with the anterior ventral insular region (AVI) and the frontal opercular areas FOP4 and FOP5 which include the insular taste cortex (Rolls, 2015a, 2016c; Rolls et al., 2022b, 2023d); with the anterior agranular insular complex (AAIC) which may be visceral (Rolls, 2016c) and also has taste-olfactory convergence (De Araujo et al., 2003b); with the middle insular region (MI) which is somatosensory; and with the piriform (olfactory) cortex.

Lateral orbitofrontal cortex regions 47s and 47l (which tend to be more posterior, and are close to Broca's region 45) have effective connectivity (Rolls et al., 2022b, 2023d) with regions involved in language (Rolls et al., 2022c). For example, 47s and 47l have effective connectivity with superior temporal (STS and STG) auditory association / semantic cortical areas; with the temporal pole TG areas implicated in semantic representations; with the PeriSylvian language (PSL), STV and TPOJ regions involved in language (Rolls et al., 2022c); with the superior frontal language area (SFL); and directed to inferior prefrontal regions including IFJ, IFS, 45 and 44 (Broca's area). Apart from these language-related connectivities, 47s and 47l have connections with other cortical regions that are similar to those of the other parts of the lateral orbitofrontal cortex (a47r, p47r and 47m) with which they also have connectivity, and accordingly I propose that 47s and 47l provide access from the orbitofrontal cortex reward / punishment system to language regions for reports of pleasantness, unpleasantness, and affective value. The lateral orbitofrontal cortex thus provides a route for emotion-related stimuli to enter the language system for conscious declarative reports (Rolls, 2023b).

A synthesis of the effective connectivity of the human orbitofrontal cortex showing how it receives inputs from all sensory systems (taste, olfactory, visual, somatosensory, and auditory), is provided in Fig. 3.16 (Rolls et al., 2022b). The sensory inputs are important in enabling the orbitofrontal cortex to form representations of the reward value of stimuli in all of these sensory modalities. Fig. 3.16 also shows how the orbitofrontal cortex has onward connectivity to further regions via the ventromedial prefrontal cortex and anterior cingulate cortex. Highlights of this human orbitofrontal cortex connectivity, and their implications for function, are considered next.

3.10.3 Connectivity of the orbitofrontal cortex to the hippocampal memory system, and memory consolidation

Fig. 3.16 shows how a major output of the human orbitofrontal cortex reaches the hippocampal memory system, both directly, and via the ventromedial prefrontal cortex, and pregenual anterior cingulate cortex and posterior cingulate cortex (Rolls et al., 2022b; Rolls, 2023b; Rolls et al., 2023g; Rolls, 2022a, 2023c, 2026d).

It is proposed in Sections 5.3.7.3, 5.3.7.4 and 6.10.5 that this connectivity enables reward and punishment information that relates to emotion to become a key component of episodic memory, together with 'where' or 'when' and 'what' information (Rolls, 2022a, 2023a). During recall of an episodic memory, the reward / aversive / emotion component can then be recalled back to the orbitofrontal cortex, as described in Chapter 5.

However, even beyond that, it is proposed that the reward / aversive input to the hippocampus provides a 'value' component to each episodic memory, with episodic memories with a 'value' component more likely to be frequently recalled and thought about in humans, and thereby making an important contribution to what becomes consolidated into long-term, typically semantic, memory. This is described further in Section 6.10.5 and by Rolls (2022a). This proposal also helps to provide an account of how damage to the orbitofrontal cortex / ventromedial prefrontal cortex / anterior cingulate cortex impairs episodic memory (Section 6.10.5, and Rolls (2022a); McCormick et al. (2018); De Luca et al. (2019)).

3.10.4 Connectivity of the orbitofrontal cortex to the cholinergic basal forebrain nucleus of Meynert, and memory consolidation

Fig. 3.16 shows (in yellow) how another key region of the human orbitofrontal cortex, pOFC, has connectivity to the cholinergic basal forebrain nucleus of Meynert. This nucleus provides the cholinergic innervation of the neocortex, where acetylcholine is involved in memory consolidation. Accordingly it is proposed that rewarding or punishing stimuli and events, which produce emotion, are decoded by the orbitofrontal cortex, and activate the basal nucleus cholinergic projection to the neocortex. Thus the orbitofrontal cortex plays an important role in this way too in memory

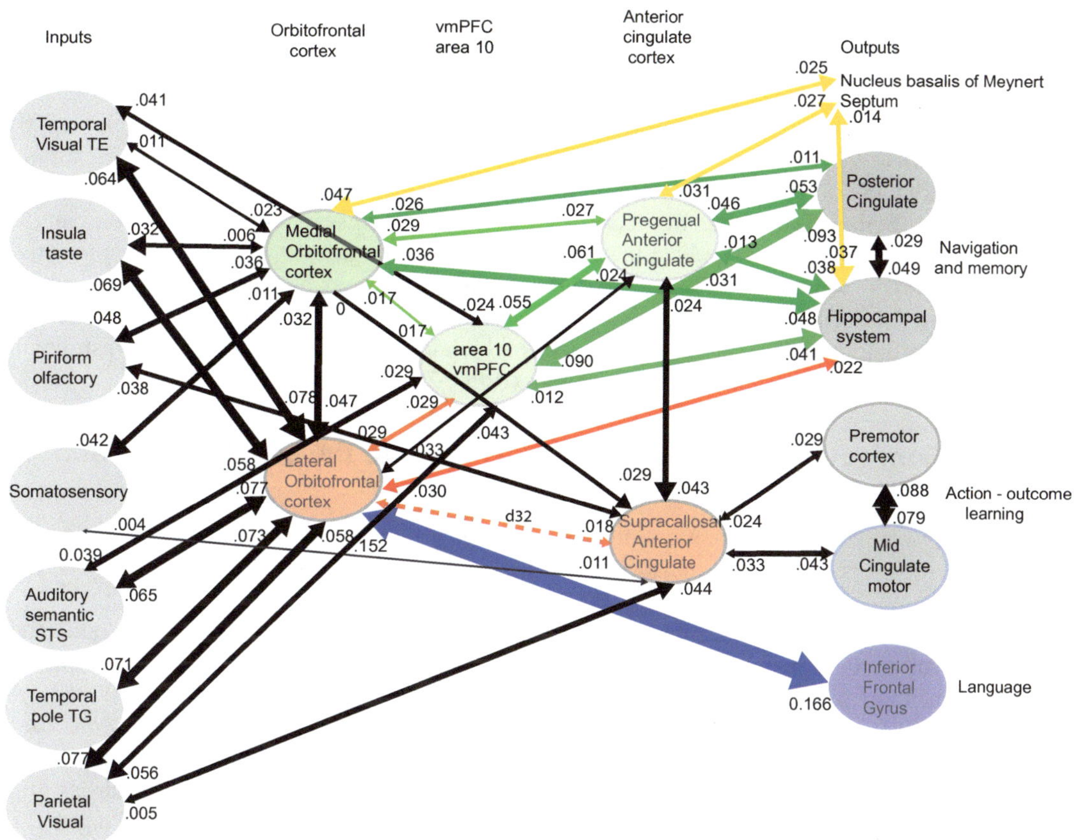

Fig. 3.16 Synthesis of the effective connectivity of the orbitofrontal cortex, vmPFC, and anterior cingulate cortex shown in the middle, with inputs on the left and outputs on the right. The regions included for each of the 5 central ellipses are as defined in the paper cited below. The green ellipses are areas that are activated mainly by rewards, and the red ellipses are areas activated mainly by punishers or non-reward (not receiving an expected reward). The lateral orbitofrontal cortex has connectivity with the supracallosal anterior cingulate cortex via region d32 as shown. The width of the arrows is proportional to the effective connectivity in the highest direction, and the size of the arrows reflects the strength of the effective connectivity in each direction. The effective connectivities shown are for the strongest link where more than one link between regions applies for a group of brain regions. Effective connectivities with hippocampal memory system regions are shown in green; with premotor / mid-cingulate regions in red; with inferior prefrontal language system in blue; and in yellow to the basal forebrain nuclei of Meynert which contains cholinergic neurons that project to the neocortex and to the septal nuclei which contain cholinergic neurons that project to the hippocampus. The Somatosensory regions include 5 and parietal PF and PFop, which also connect to the pregenual anterior cingulate but are not shown for clarity; the Parietal regions include visual parietal regions 7, PGi and PFm. The diffusion tractography showed connections between the lateral orbitofrontal cortex and the supracallosal anterior cingulate cortex. The connectivity with dorsolateral prefrontal cortex is not included here for clarity. Connectivity is shown for the five groups in the centre of the Figure, and does not include for example connectivity between somatosensory and premotor cortical regions. (Data from Rolls E T, Deco G, Huang C-C. and Feng J. 2022 The human orbitofrontal cortex, vmPFC, and anterior cingulate cortex effective connectome: emotion, memory, and action. Cerebral Cortex 33: 350-356.)

consolidation. The evidence for this is presented in Section 6.10.5, and by Rolls (2022a). This proposal also helps to provide an account of how damage to the orbitofrontal cortex / ventromedial prefrontal cortex / anterior cingulate cortex impairs episodic memory (Section 6.10.5, and Rolls (2022a)).

3.10.5 Connectivity of the orbitofrontal cortex to the supracallosal anterior cingulate cortex for action-outcome learning

Fig. 3.16 shows how the human orbitofrontal cortex also has connectivity to the supracallosal anterior cingulate cortex, which is implicated in action-outcome learning. For this type of learning, it is proposed that the orbitofrontal cortex provides the reward or non-reward or punishment outcome that follows an action. The supracallosal anterior cingulate cortex is the part of the anterior cingulate cortex that is proposed to be involved in action-outcome learning, for it has extensive connectivity with premotor cortex regions including the mid-cingulate motor area that are involved in actions, as shown in Fig. 3.16. The evidence for this is described by Rolls et al. (2022b); Rolls (2023c, 2026d). This is the key output of the orbitofrontal cortex by which it is involved in goal-directed instrumental learning, which is an important aspect of emotion (Rolls, 2014a, 2023b,c, 2026d).

3.10.6 Connectivity of the orbitofrontal cortex to the basal ganglia and dopaminergic system

The orbitofrontal cortex outputs to the head of the caudate nucleus and ventral striatum provide a route for reinforcers to guide stimulus-response habit learning, as described in Chapter 16 of Rolls (2023c).

Very interestingly, in humans, medial orbitofrontal cortex region OFC has effective connectivity to the substantia nigra pars compacta and ventral tegmental area (Rolls et al., 2022b), where the dopamine neurons are located (Rockland et al., 1994). As described by Rolls (2023c, 2026d), this is the most likely source of the reward-related information needed if dopamine neurons are to reflect reward prediction error (Schultz, 2016c, 2017; Rolls, 2023c, 2026d). Part of the interest and importance of this is that Wolfram Schultz has not emphasized how dopamine neurons might receive their reward prediction error inputs (Schultz, 2017, 2016c,a), and I am proposing that the orbitofrontal cortex is at least part of the pathway (Rolls, 2017a), for as I discovered and described earlier in this Chapter, the orbitofrontal cortex represents at least reward outcome, expected reward value, and negative reward prediction error (i.e. receiving less reward as the outcome than expected).

3.11 A theory of emotion and motivation, and their brain systems

3.11.1 A theory of emotion

The discoveries that we made described here provide the basis for a theory of emotion and motivation in primates including humans that I have developed (Rolls, 2014a, 2023b, 2025e). In Rolls' theory of emotion, emotions are states elicited by rewards or punishers or not receiving a reward, and motivational states are states in which we are trying to obtain the reward, and avoid punishers or non-reward (Rolls, 2014a, 2023b). More specifically, in this approach, which had precursors (Weiskrantz, 1968; Gray, 1975), emotions can be defined as states elicited by rewards and punishers, that is, by instrumental reinforcers (Rolls, 2014a, 2023b, 2022b, 2019c, 2018b, 2013a, 2005, 1999b). When this theory was being developed (Rolls, 1999b, 2000a, 2005), I think that it had a major effect on research worldwide on emotion. Scientists now had an operational approach to analyze what had before this seemed the difficult issue of emotion (for example, how to define, measure, and study it), for now the approach could be to investigate the operation of reward and punishment systems in the brain. An outline of this approach to emotion (Rolls, 2014a, 2023b, 2025e) is described next.

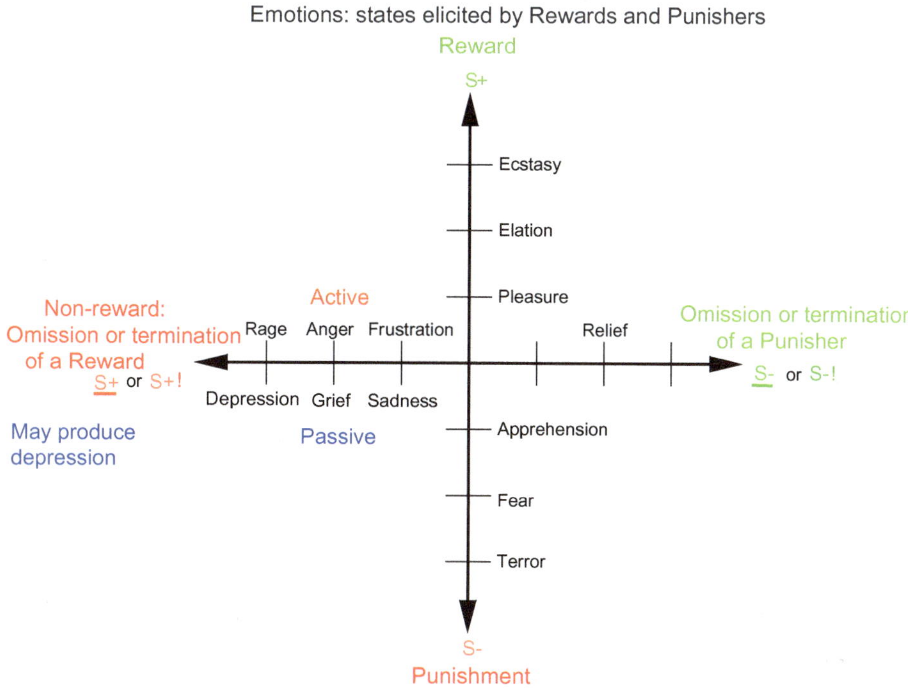

Fig. 3.17 Some of the emotions associated with different reinforcement contingencies are indicated. Intensity increases away from the center of the diagram, on a continuous scale. The classification scheme created by the different reinforcement contingencies consists with respect to the action of (1) the delivery of a reward (S+), (2) the delivery of a punisher (S–), (3) the omission of a reward (S+) (extinction) or the termination of a reward (S+!) (time out), and (4) the omission of a punisher (S–) (avoidance) or the termination of a punisher (S–!) (escape). Note that the vertical axis describes emotions associated with the delivery of a reward (up) or punisher (down). The horizontal axis describes emotions associated with the non-delivery of an expected reward (left) or the non-delivery of an expected punisher (right). For the contingency of non-reward (horizontal axis, left) different emotions can arise depending on whether an active action is possible to respond to the non-reward, or whether no action is possible, which is labelled as the passive condition. In the passive condition, non-reward may produce depression. The diagram summarizes emotions that might result for one reinforcer as a result of different contingencies. Every separate reinforcer has the potential to operate according to contingencies such as these. This diagram does not imply a dimensional theory of emotion, but shows the types of emotional state that might be produced by a specific reinforcer. Each different reinforcer will produce different emotional states, but the contingencies will operate as shown to produce different specific emotional states for each different reinforcer. (Data from Rolls E T 2025 Emotion, Motivation, Reasoning, and how their brain systems are related. Brain Sciences 15: 507.)

A reward is anything for which an animal will work. A punisher is anything that an animal will work to escape or avoid, or that will suppress actions on which it is contingent (see Rolls (2014a) and Rolls (2023b) for details and explanation of these definitions, and of how these processes relate to emotion). The force of 'instrumental' in this definition is that the emotional states are seen as defining the goals for arbitrary behavioral actions, made to obtain the instrumental reinforcer. This is very different from classical conditioning, in which a response, typically autonomic, may be elicited to a stimulus, without any need for an intervening state. An important part of the function of emotion is to provide an intervening state, after for example the delivery of a non-reward, in which behavior can be reorganized, with perhaps a new action substituted for what was performed previously, in order to obtain the goal / reward (Rolls, 2014a).

An outline of a classification scheme for how different reinforcement contingencies (delivery of reward, omission or termination of reward, delivery of punishment, omission or termination of punishment) are related to different emotions is shown in Fig. 3.17 (Rolls, 2014a, 2023b). Movement away from the center of the diagram represents increasing intensity of emotion, on a continuous scale. The diagram shows that emotions associated with the delivery of a reward (S+) include pleasure, elation and ecstasy. Of course, other emotional labels can be included along the same axis. Emotions associated with the delivery of a punisher (S–) include apprehension, fear, and terror (see Fig. 3.17). Emotions associated with the omission of a reward ($\underline{\text{S+}}$) or the termination of a reward (S+!) include frustration, anger and rage. Emotions associated with the omission of a punisher ($\underline{\text{S–}}$) or the termination of a punisher (S–!) include relief.

An important point about Fig. 3.17 is that there are a large number of different primary (unlearned, gene-specified) reinforcers, and that for example the reward label S+ shows states that might be elicited by just one type of reward, such as a pleasant touch. There will be a different reward axis (S+) and non-reward axis ($\underline{\text{S+}}$ and S+!) for each type of reward (e.g. pleasant touch vs sweet taste); and, correspondingly, a different punisher axis (S–) and non-punisher axis ($\underline{\text{S–}}$ and S–!) for each type of punisher (e.g. pain vs bitter taste). Examples of the reinforcers include food reward, water reward, pleasant touch, pain, a potential mate, kin including children, and altruism, with many examples provided in Table 2.1 of **Emotion and Decision-Making Explained** (Rolls, 2014a) (available on line at https://www.oxcns.org) and by Rolls (2023b).

Previously neutral stimuli can become associated by learning with these primary reinforcers, and become secondary reinforcers, which thereby become goals for action. An example is the sight of a new food, which may become a secondary reinforcer by association between its sight and the primary reinforcer of its taste (Rolls, 2014a).

I argue that brains are designed around reward and punishment value systems, because this is the way that genes can build a complex system that will produce appropriate but flexible behavior to increase their fitness (Rolls, 2014a). The way that evolution by natural selection does this is to build us with reward and punishment systems that will direct our behavior towards goals in such a way that survival and in particular reproductive fitness are achieved. By specifying goals, rather than particular responses, genes leave much more open the possible behavioral strategies that might be required to increase their fitness. Specifying particular responses would be inefficient in terms of behavioral flexibility as environments change during evolution, and also would be more genetically costly to specify (in terms of the information to be encoded and the possibility of error). This view of the evolutionary adaptive value for genes to build organisms using reward- and punishment-decoding and action systems in the brain places me squarely in line as a scientist developing ideas from Darwin (1859) onwards (Hamilton, 1964, 1996; Dawkins, 1976), and is a key part of Rolls' theory of emotion and motivation (Rolls, 2014a, 2013a, 2018b, 2019c, 2023b, 2025e). The theory goes beyond what was described in **The Selfish Gene** (Dawkins, 1976) in that rewards and punishers are introduced as key variables that genes vary, rather than behavioral responses, providing a much more efficient way to direct behavior towards reproductive fitness (Section 3.11.5.3).

The theory helps in understanding much of sensory information processing in the cortex, followed by reward and punishment value encoding, followed by decision-making and action selection to obtain the goals identified by the sensory/reinforcer decoding systems. Value coding systems must be separate from purely sensory or motor systems, and while a goal is being sought, or if a goal is not obtained, the value-related representation must remain to direct further goal-directed behavior, and it is these continuing goal-related states to which emotion is related (Rolls, 2023b, 2025e).

The theory is strongly supported by the effects of damage to the orbitofrontal cortex, which impairs emotional feelings and behavior, face and voice expression recognition, and socially appropriate behavior, as well as reward reversal learning (Section 3.8) (Rolls, Hornak, Wade and McGrath, 1994a; Hornak, Rolls and Wade, 1996; Rolls, 1999c; Hornak, Bramham, Rolls, Morris, O'Doherty, Bullock and Polkey, 2003; Hornak, O'Doherty, Bramham, Rolls, Morris, Bullock and Polkey, 2004; Berlin, Rolls and Kischka, 2004; Berlin, Rolls and Iversen, 2005; Fellows and Farah, 2003; Camille, Tsuchida and Fellows, 2011; Rolls, 2021d).

3.11.2 A wide range of emotions

I have developed this theory of emotion (Rolls, 2014a, 2023b, 2025e) to show how a very wide range of emotions can be accounted for, as a result of the operation of a number of factors, including the following:
1. The reinforcement contingency (e.g., whether reward or punishment is given, or withheld) (see Fig. 3.17).
2. The intensity of the reinforcer (see Fig. 3.17).
3. Any environmental stimulus might have a number of different reinforcement associations. (For example, a stimulus might be associated both with the presentation of a reward and of a punisher, allowing states such as conflict and guilt to arise.)
4. Emotions elicited by stimuli associated with different primary reinforcers will be different, as described above, with different primary reinforcers each producing different affective states.
5. Emotions elicited by different secondary reinforcing stimuli will be different from each other (even if the primary reinforcer is similar). For example, the same touch to the arm but by different people might give rise to very different emotions. Cognitive states and semantic knowledge can contribute to emotion in these ways, as well as in other ways that might arise because for example of reasoning in the rational brain system.
6. The emotion elicited can depend on whether an active or passive behavioral response is possible. (For example, if an active behavioral response can occur to the omission of a positive reinforcer, then anger might be produced, but if only passive behavior is possible, then sadness, depression or grief might occur: see Fig. 3.17.)

By combining these six factors, I have shown how it is possible to account for a very wide range of emotions (Rolls, 2014a).

3.11.3 The functions of emotions

The most important function of emotion is as part of the processes of learning goal-directed actions to obtain rewards or avoid punishers.

The first process involved is stimulus-reinforcer association learning; emotional states are produced as a result (Rolls, 2014a). An example might be learning that the sight of a person is associated with rewards, which might produce the emotion of happiness. This process is implemented in structures such as the orbitofrontal cortex and amygdala (Figs. 3.2 and 3.12) (Grabenhorst and Rolls, 2011; Rolls and Grabenhorst, 2008; Rolls, 2014a, 2023b, 2025e).

The second process involved is instrumental learning of an action made to approach and obtain the reward (an outcome of the action) or to avoid or escape from the punisher (an outcome). This is action-outcome learning when the actions are being guided by the goals, and involves brain regions such as the anterior cingulate cortex (Rushworth et al., 2011, 2012; Rolls, 2018b, 2021a, 2014a, 2019e, 2023c, 2026d, 2023b, 2025e). Emotion is an integral part of this, for it is the state elicited in the first stage, by stimuli that are decoded as rewards or punishers (Rolls, 2014a). The emotional state intervenes between the stimulus which is decoded as the goal, and the selection of an action to obtain the goal. The decoded reward value of the goal in fact provides motivation to perform an action to obtain the goal (Rolls, 2025e), which may take some time. The behavior is under control of the reward value of the goal, in that if the reward is devalued, for example by feeding a food until satiety is reached, then on the very next occasion that the stimulus (the food) is offered, no action will be performed to try to obtain it (Rolls, 2014a, 2025e).

The striatum, rest of the basal ganglia, and dopamine system can become involved when the behavior becomes automatic, and habit-based, that is, uses stimulus-response connections (Fig. 3.12). In this situation, very little emotion may be elicited by the stimulus, as the behavior has now become automated as a stimulus-response habit. For this type of learning, if the reward is devalued outside the situation, then the very next time that the stimulus is offered, the automated response is likely to be performed, providing evidence that the behavior is no longer being guided by the reward value of the stimulus (Rolls, 2014a, 2025e). The dopamine system is involved in this type of rather slow habit-based learning, it is thought by providing an error signal to the striatum which implements this type of habit learning (Schultz, 2017, 2016b,a). I have proposed that the dopamine system probably receives its inputs from the orbitofrontal cortex (Rolls et al., 2022b; Rolls, 2017a; Rolls et al., 2022b; Rolls, 2023c, 2026d).

Other functions of emotion include the elicitation of autonomic responses, via pathways for example from the amygdala, and from the orbitofrontal cortex, to the anteroventral visceral/autonomic insula and to the subgenual cingulate cortex (Rolls, 2014a, 2019c,e, 2013a; Critchley and Harrison, 2013; Quadt et al., 2022).

3.11.4 The functions of the orbitofrontal cortex vs. amygdala in emotion

The amygdala in rodents, in which the orbitofrontal cortex is much less developed than in primates (Passingham, 2021; Passingham and Wise, 2012), has been implicated in emotion-related responses such as conditioned autonomic responses, conditioned freezing behavior, cortical arousal, and learned incentive effects in fear conditioning in which an auditory tone is associated with foot shock (Quirk et al., 1996; LeDoux, 1996, 1995; Davis, 2011; LeDoux and Pine, 2016).

Although the primate amygdala has some of the same connections as the orbitofrontal cortex in monkeys (see Fig. 3.12) (Rolls, 2014a, 2023c, 2026d), we have discovered that in humans the amygdala has much less connectivity back to the neocortex than the orbitofrontal cortex (Rolls et al., 2023d). Given the paucity of amygdalo-neocortical effective connectivity in humans, and the richness of its subcortical outputs in rodents (Quirk et al., 1996) and in humans (Klein-Flugge et al., 2022), it is proposed that the human amygdala is involved primarily in autonomic and conditioned responses via brainstem connectivity, rather than in reported (declarative) emotion (Rolls et al., 2023d). The contrasting strong connectivity of the orbitofrontal cortex with language regions including Broca's area is illustrated in Figs. 3.15 and 3.16.

This new evidence about the connectivity of the human amygdala is consistent with the evidence that the effects of damage to the human amygdala on emotion and emotional experience are limited (LeDoux et al., 2018; LeDoux and Pine, 2016; Delgado et al., 2011; Whalen and Phelps, 2009; Rolls et al., 2023d; Feinstein et al., 2011; Damasio et al., 2013; Adolphs et al., 2005; Kennedy and Adolphs, 2011) and much less than the effects of damage to the orbitofrontal cortex (Hornak et al., 2004, 2003, 1996; Rolls et al., 1994a; Fellows, 2011; Camille et al., 2011; Rolls, 2019c, 2021d). Indeed, LeDoux and colleagues have now emphasized the evidence that the human amygdala has limited involvement in subjective emotional experience (LeDoux et al., 2018; LeDoux and Brown, 2017; LeDoux and Pine, 2016; LeDoux, 2012; Taschereau-Dumouchel et al., 2022; LeDoux, 2020). In this context, LeDoux's conundrum is: if not the amygdala for subjective emotional experience, then what (LeDoux, 2020)? My answer based on the research described here and elsewhere (Rolls et al., 2023d; Rolls, 2023c, 2026d, 2023b) is: the human orbitofrontal cortex is the key brain region involved in subjective emotion (Rolls et al., 2023d; Rolls, 2023c, 2026d, 2014a, 2019c, 2023b, 2025e).

The problem of over-interpreting the role of the amygdala in emotion was that rodent studies showed that some responses such as classically conditioned autonomic responses and freezing are elicited by the amygdala with its outputs to brainstem systems, and it was inferred that therefore the amygdala is involved in emotion in the way that it is experienced by humans (Quirk et al., 1996; LeDoux, 2000, 1996, 1995). It turned out later that humans with amygdala damage had similar response-related changes, but little impairment in subjectively experienced and reported emotions (LeDoux et al., 2018; LeDoux and Pine, 2016; Delgado et al., 2011; Whalen and Phelps, 2009; Rolls et al., 2023d; Damasio et al., 2013). Further, pharmacological blockade of these peripheral responses in humans did not impair the conscious emotional subjective states (LeDoux and Pine, 2016; LeDoux and Hofmann, 2018). It is important therefore not to infer subjective reported emotional states in humans from responses such as conditioned autonomic and freezing responses (Rolls et al., 2023d; Rolls, 2023b,c, 2026d, 2025e; Dawkins, 2026).

3.11.5 A reasoning, rational, route to action, vs. the emotional route to action

3.11.5.1 The reasoning, rational, route to action

Routes to action that are related to the emotional system are described above, and are indicated in Fig. 3.12. But I became interested in an alternative route to action, the route to action by reasoning, and describe it here. I suggest that having these two routes to action, the emotional and the rational, makes humans especially interesting and complex in their decision-making, because the interests of the two routes to action are somewhat different (Rolls, 2014a, 2023b, 2025e).

The reasoning route to action in humans, and perhaps some other species, involves a computation with many "if ... then" statements, to implement a multi-step plan to obtain a reward. In this case, the reward may actually be deferred as part of the plan, which might involve working first to obtain one reward, and only then to work for a second more highly valued reward, if this was thought to be overall an optimal strategy in terms of resource usage (e.g., time). In this case, syntax is required, because the many symbols (e.g., names of people) that are part of the plan must be correctly linked or bound. Such linking might be of the form: "if A does this, then B is likely to do this, and this will cause C to do this ..". The requirement of syntax for this type of planning implies that involvement of language systems in the brain is required for this type of planning (Rolls, 2014a, 2018b, 2021a, 2023c, 2026d). Thus the explicit language system in humans may allow working for deferred rewards by enabling use of a one-off, individual, plan appropriate for each situation.

3.11.5.2 Decisions between the emotional and reasoning systems

The question then arises of how decisions are made in animals such as humans that have both the implicit, direct reward-based, and the explicit, rational, planning systems (Rolls, 2011a, 2014a, 2018b, 2026a, 2021a, 2019d, 2023b,c, 2026d, 2025e). One particular situation in which the first, implicit, system may be especially important is when rapid reactions to stimuli with reward or punishment value must be made, for then the direct connections from structures such as the orbitofrontal cortex to the basal ganglia may allow rapid responses (Rolls, 2014a, 2025e). Another is when there may be too many factors to be taken into account easily by the explicit, rational, planning, system, when the implicit system may be used to guide responses. In contrast, when the implicit system continually makes errors, it would then be beneficial for the individual to switch from direct action based on obtaining what the orbitofrontal cortex system decodes as being the most positively rewarding choice currently available, to the explicit conscious control system that can evaluate with its long-term planning algorithms what action should be performed next. Indeed, it would be adaptive for the explicit system to regularly be assessing performance by (i.e. monitoring) the goal-based and habit-based systems, and to switch itself in to control behavior quite frequently, as otherwise the adaptive value of having the explicit reasoning system would be less than optimal (Rolls, 2025e).

I have described how there is often a conflict between these systems, in that the first, implicit, emotional, system is able to guide behavior particularly to obtain the greatest immediate reinforcement, whereas the explicit system can potentially enable immediate rewards to be deferred, and longer-term, multi-step, plans to be formed (Rolls, 2026a, 2019d, 2025e). One example is that a delicious but fattening food such as ice cream might be offered, and the implicit emotional system might make that rewarding, but the rational system might reason that in the interests of long-term health and reducing the risk of obesity, it might be wise to not accept the offer of that delicious food. This type of conflict will occur in animals with a syntactic planning ability, that is in humans and any other animals that have the ability to process a series of "if ... then" stages of planning. This is a property of the human language system, and the extent to which it is a property of non-human primates is not yet fully clear. In any case, such conflict may be an important aspect of the operation of at least the human mind, because it is so essential for humans to correctly decide, at every moment, whether for example to invest in a relationship or a group that may offer long-term benefits, or whether to directly pursue immediate benefits (Rolls, 2014a, 2018b, 2026a, 2019d, 2025e).

3.11.5.3 The Selfish Gene vs. The Selfish Phenotype, and evolution

I have provided evidence above that there are two main routes to decision-making and goal-directed action, the emotion, goal-related route, and the reasoning route. I now move to the next concept, that the 'interests' of the first and second routes to decision-making and action are different. As argued very convincingly by Richard Dawkins in **The Selfish Gene** (Dawkins, 1989), and by others (Hamilton, 1964, 1996; Ridley, 1993), many behaviors occur in the interests of the survival (reproductive fitness) of the genes, not of the individual (nor of the group), and much behavior can be understood in this way.

My thinking has extended this approach by arguing that an important role for some genes in evolution is to define the goals (i.e. the rewards and punishers) for actions that will lead to better survival of those genes; that emotions are the states associated with these gene-defined goals; and that the defining of goals (stimuli with reward value) for actions rather that actions themselves is an efficient way for genes to operate, as it leaves flexibility of choice of action open until the animal is alive (Rolls, 2014a, 2018b, 2023b,c, 2026d). This provides great simplification of the genotype as action details do not need to be specified, just rewarding and punishing stimuli, and also provides flexibility of action in the face of changing environments faced by the genes. This is a useful and interesting advance beyond what was considered in **The Selfish Gene** and later books (Dawkins, 1986, 1982, 1976, 1989). In any case, the interests that are implied when the first route to action is chosen are those of the "selfish genes", not those of the individual, the phenotype.

However, I have developed the argument that the second route to action allows, by reasoning, decisions to be taken that might not be in the interests of the genes, might be longer term decisions, and might be in the interests of the individual, the phenotype (Rolls, 2011a, 2014a, 2023b, 2025e). An example might be a choice not to have children, but instead to devote oneself to science, medicine, music, and/or literature. The reasoning, rational, system presumably evolved because taking longer-term decisions involving planning rather than choosing a gene-defined goal might be advantageous at least sometimes for genes. But an unforeseen consequence of the evolution of the rational system might be that the decisions would, sometimes, not be to the advantage of any genes in the organism. After all, evolution by natural selection operates utilizing genetic variation like a Blind Watchmaker (Dawkins, 1986). In this sense, the interests when the second route to decision-making is used are at least sometimes those of the "selfish phenotype". Hence the decision-making referred to above is between a first system where the goals are gene-defined, and a second rational system in which the decisions may be made in the interests of the genes, or in the interests of the phenotype (the individual) and not in the interests of the genes (Rolls, 2025e).

If the multistep syntactic reasoning / planning system fails, there may be a credit assignment problem, in that the faulty step in the series of steps needs to be identified. I have argued that this computation requires a higher order thought system, that can think about the first order thoughts (the plan), and can correct the plan. This higher order thought system needs syntax, as it has to perform computations on the first order syntactic thoughts, the multistep plan. I have proposed that it is a property of this Higher Order Syntactic Thought system that when it operates, it would feel like something, to be thinking about one's own first order thoughts, and that is a basis for my HOST theory of consciousness described in Section 3.11.6 (Rolls, 2020c, 2011a, 2008b, 2007a, 2025e). It is suggested that anything that is being dealt with by the HOST computational system becomes conscious, and that is also how qualia, raw sensory feels, are produced (Rolls, 2020c).

3.11.5.4 The relation between the mind and the brain, causality, and levels of explanation

What is described above about emotional feelings, and about explicit thought and reasoning, raises the issue of what the relation is between the mind and the brain (Descartes, 1644; Dennett, 1991; Block, 2005; Carruthers, 2019). That led me to propose a computational neuroscience theory of the relation between the mind and the brain (Rolls, 2021i).

In the neuroscience-based approach that I propose for the relation between the mind and the brain, the proposal is that events at the sub-neuronal, neuronal, neuronal network, and mental, levels take place simultaneously to perform a computation that can be described at a high level as a mental state, with content about the world (Rolls, 2021g,i). I argue that as the processes at the different levels of explanation take place at the same time, they are linked by a non-causal relationship: causality can best be described in brains as operating within levels across time, but not between levels. This mind-brain theory allows mental events to be different in kind from the mechanistic events that underlie them; but does not lead one to argue that mental events cause brain events, or vice versa: they are different levels of explanation of the operation of the computational system (Rolls, 2021g,i). The approach was developed in part by considering the relations between the hardware in a computer and the software being run by the computer, and about how we might think about causality in this simpler system (Rolls, 2021g,i). This computational neuroscience levels of explanation approach to causality (Rolls, 2021i, 2023c, 2026d) provides an opportunity to proceed beyond Cartesian dualism (Descartes, 1644) and physical reductionism (Carruthers, 2019) in considering the relations between the mind and the brain (Rolls, 2021g,i, 2020c, 2026d).

3.11.6 Consciousness and emotional feelings: a Higher Order Syntactic Thought theory

An important aspect of emotion that many wish to understand better is why emotional states feel like something. That is part of the much larger problem of consciousness about which there are many diverse views, and no clear way forward of deciding which are the most promising directions (Bayne et al., 2024; Mudrik et al., 2025; Dawkins, 2026). Because consciousness is a large issue in understanding brain function, I have thought about it, and have developed an approach (Rolls, 2020c, 2004a, 2025e, 2026d).

One popular approach, the global workspace hypothesis of consciousness, holds that many different brain systems need to interact to account for consciousness (Dehaene, 2014; Dehaene et al., 2017; van Vugt et al., 2018; Lau, 2022). However, there are problems with this approach, for example about exactly how global the workspace needs to be for consciousness to suddenly be present (Rolls, 2020c; Lau, 2022). Humans can perform many complex tasks, including it appears driving a car for a short period, without being able to report about the driving if they were thinking about something else, which they can report about. Yet driving a car might be thought to require a rather large workspace.

Instead, I have proposed a different theory of phenomenal consciousness, in which there is a particular computational system involved in which Higher Order Syntactic Thoughts (HOSTs) are used to perform credit assignment on first order thoughts of multiple step plans to correct them by manipulating symbols in a syntactic type of working memory (Rolls, 2020c). This provides a good adaptive reason for the evolution of this kind of computational module, with which, it is proposed, phenomenal consciousness is associated (Rolls, 2020c).

Some advantages of this HOST approach to phenomenal consciousness compared to other Higher Order Thought (HOT) theories (Rosenthal, 2012; Lau and Rosenthal, 2011) have been described (Rolls, 2020c). For example, a multi-step thought or planning system has a potential credit assignment problem (which step in the multistep process caused the process to fail), and the incorrect step needs to be identified, for which a natural method is to think about the plan, and work out which step may have produced the error (Rolls, 2020c, 2008b). I hypothesize that the HOST system which requires the ability to manipulate first (or lower) order symbols in working memory might utilize parts of the prefrontal cortex implicated in working memory, and especially the left inferior frontal gyrus, which is involved in language and probably syntactical processing (Rolls, 2020c; Rolls et al., 2022c; Rolls and Deco, 2015b; Rolls, 2023c, 2026d). Overall, the approach that I advocate is to identify the computations that are linked to consciousness, and to analyze the neural bases of those computations (Rolls, 2020c).

Rolls' Higher Order Syntactic Thought theory of consciousness (Rolls, 2020c) differs from that of Lau (2022), who proposes that consciousness involves a higher order monitoring of perceptual reality that could be used to correct problems such as an incorrect threshold for perceptual input. However, that type of problem could include a signal-to-noise ratio problem, which need have nothing to do with consciousness. A simple algorithm can separate neural noise (Lau's source of interference) from a signal. And it is not clear how a Higher Order Thought could correct the engineering issue of a signal to noise ratio problem by adjusting the threshold by just thinking about it. But if the theory (Lau, 2022) is that one needs to reason about the sensory (or any other processing) to see whether it needs to be corrected, then it becomes a theory very much like my HOST theory of consciousness (Rolls, 2020c), except that I argue that the monitoring that is needed and that would be computationally useful is that of thoughts, in which a credit assignment problem could arise, and which could be corrected by a higher order syntactic thought about a lower order syntactic thought, and this would be very useful for any multistep planning (Rolls, 2020c).

I then argue that it is plausible to think that a system that was performing this type of computation, thinking about its own thoughts, with the lower order thoughts grounded in the world, would feel conscious. That is of course a plausibility argument, but a plausible one (Rolls, 2020c, 1997a,b, 2004a, 2007a,c, 2008b, 2010a, 2011a, 2014a, 2025e). And the HOST theory is testable, for example by intervening by temporary inactivation of this type of computation in the brain, to test whether consciousness is affected by this intervention.

To finish this section on emotion, a more full description of Rolls' theory of emotion is available in **Emotion and Decision-Making Explained** (Rolls, 2014a), in **The Brain, Emotion, and Depression** (Rolls, 2018b), and in Rolls (2023b, 2025e), all available for free download from https://www.oxcns.org.

In this theory of emotion, the primate including human orbitofrontal cortex plays a key role by receiving inputs from the ends of sensory processing systems for visual objects and faces, taste, olfaction, touch and auditory stimuli in which reward is not represented, and then converting these sensory representations to new representations in which value (reward or punisher or non-reward) is the key part of the representation (Rolls, 2023b, 2025e).

I show that outputs from this system for emotion in the human brain reach language cortical

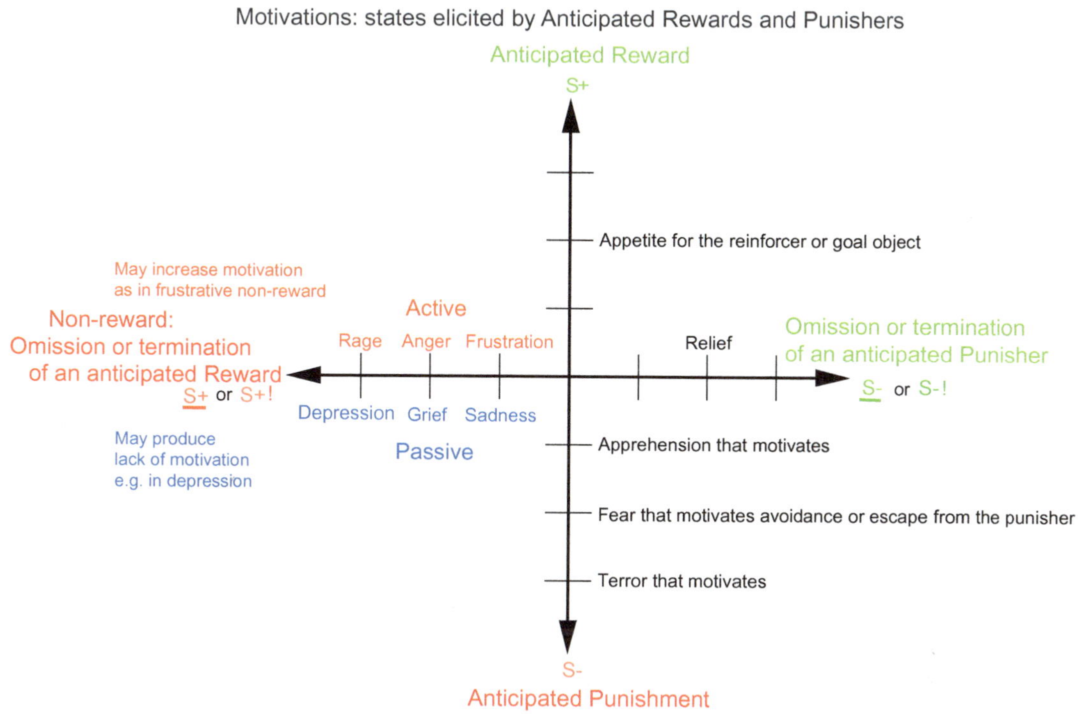

Fig. 3.18 Some of the motivational states associated with different reinforcement contingencies for anticipated rewards and punishers are indicated. An anticipated reward motivates behavior to obtain the reward which is the goal of an instrumental action; and an anticipated punisher motivates behavior to avoid or escape from the punisher which is the goal of an instrumental action. Intensity increases away from the center of the diagram, on a continuous scale. The classification scheme shows how different motivational states created by the different reinforcement contingencies consist of being in a state that leads to (1) wanting to perform an action to obtain an anticipated reward (S+), (2) wanting to perform an action to escape from or avoid an anticipated punisher (S-), (3) wanting to perform an action, or not, because of the omission of an anticipated reward (S+) (extinction) or the termination of a reward (S+!) (time out), and (4) wanting to perform an action, or not, because of the omission of a punisher (S–) (avoidance) or the termination of a punisher (S-!) (escape). Note that the vertical axis describes motivational states associated with the expected delivery of a reward (up) or punisher (down). The horizontal axis describes motivational states associated with the non-delivery of an expected reward (left) or the non-delivery of an expected punisher (right), and any action depends on what is possible in the environment. The diagram summarizes motivational states that might arise for one reinforcer as a result of different contingencies. Every separate reinforcer has the potential to operate according to contingencies such as these. Each different reinforcer will produce different motivational states, but the contingencies will operate as shown to produce different specific motivational states for each different reinforcer. Emotional states elicited by expected reinforcers can produce motivational states to work for an expected reward or avoid an expected punisher. (Data from Rolls E T 2025 Emotion, Motivation, Reasoning, and how their brain systems are related. Brain Sciences 15: 507.)

regions for declarative statements about subjective pleasantness; the anterior cingulate cortex for action - goal outcome learning; the striatum for habit responses; and the ventral striatum as a route that with the habenula may allow reward to reach dopamine neurons in the midbrain (Rolls, 2023b, 2017a) (see Fig. 3.12).

3.11.7 A theory of motivation

In the context of the discoveries I have made about the brain systems involved in reward processing, I have developed a theory of motivation (Rolls, 2023b, 2025e), which complements and utilizes many of the same brain systems as my theory of emotion. My definition of motivation is that motivational states are states that are present when rewards and punishers, that is, instrumental reinforcers, are the goals for action (Rolls, 2014a, 2016g, 2023b, 2025e).

The force of 'instrumental' in this definition is that the motivational states are seen as defining the goals for arbitrary behavioral actions, made to obtain the instrumental reinforcer, the goal for the action. This is very different from classical conditioning, in which a response, typically autonomic, may be elicited to a stimulus without any need for an intervening state (Rolls, 2014a) (see above). This is also very different from instrumental, habit, stimulus-response, behavior in which the behavior is not under control of the reward value of the goal, based on the criterion of devaluation (see above). For stimulus-response habit instrumental behavior, and unlearned responses, the term drive can be used to describe the state that influences whether a response will be made (Rolls, 2025e). The motivational states (such as hunger) modulate the reward value / goal value of instrumental reinforcers that have particular functions (Rolls, 2014a, 2016g, 2023b, 2025e). It is important in this approach to motivation that the reward values of potential goals are regulated appropriately, with for example factors such as plasma glucose, gastric distension and absorbed food acting to control the reward value of food (Rolls, 2016f, 2005); and cellular and extracellular dehydration modulating the reward value of water (Rolls and Rolls, 1981; Rolls et al., 1980b,a; Rolls, 2005).

An example of a motivational state might thus be a hunger state in which the animal will perform goal-directed actions to obtain the reinforcer or goal. Another example is that the omission or termination of a reward ('extinction' and 'time out' respectively) can produce a motivational state of frustration, in which the probability of the action may become reduced if no action is possible to regain the reward, or may increase if further motivated attempts are likely to lead to the reward (Rolls, 2014a, 2016g, 2023b, 2025e).

These examples show that the reinforcement contingency as well as the particular reinforcer or goal object (e.g. food, water, aversive stimulation) lead to particular motivational states. The types of motivational state related to different reinforcement contingencies such as anticipated reward or anticipated punishment are illustrated in Fig. 3.18. The diagram summarizes motivational states that might relate to one reinforcer as a result of different contingencies. Every separate reinforcer has the potential to operate according to contingencies such as these, in an analogous way to that described above for emotional states. Each different reinforcer will produce different motivational states, but the contingencies will operate as shown to produce different specific motivational states for each different reinforcer. Thus hunger might be present when the appetite is for the goal object of food, and thirst when the appetite is for the goal object of water. Definitions of reinforcers, and of the contingencies with which they operate, are elaborated by Rolls (2014a).

We must be clear about the difference between motivation and emotion. According to Rolls' theory of emotion described above, emotion is the state that results from having received, or not having received, the (conditioned or unconditioned) instrumental reinforcer, the goal object (Rolls, 2014a, 2018b, 2023b, 2025e). In contrast, motivation is the state when the instrumental reinforcer is anticipated (Fig. 3.18), is wanted, is being worked for, before the outcome stage, where the outcome is the delivery or not of the reinforcer, the reward or punisher.

A key and unifying property of this theory of motivation and emotion is that the goal objects can be the same for motivation and emotion, simplifying the biological specification used to specify the unconditioned reinforcers for both emotion and motivation, with the difference being that motivation is the phase before the outcome when the outcome is anticipated, and emotion is the phase after the conditioned reinforcer or unconditioned outcome, the reward or punisher, has been received. The biological specification of emotion and motivation is simplified in that the same set of genes set out in Table 2.1 of **Emotion and Decision-Making Explained** (Rolls, 2014a) (available on line at https://www.oxcns.org) and in Rolls (2023b, 2025e, 2026d) is used to define the rewards and the punishers, the goals, related to both emotion and motivation.

An additional property is that emotions, states occurring after the delivery or not of the reinforcer, can be motivating (Rolls, 2014a, 2025e). A good example is that if an expected reward is not obtained, then the frustrative non-reward can be motivating, and make the animal (including humans) work harder to obtain the goal object (Rolls, 2014a, 2025e).

Two-stages of learning may be involved in such motivational goal-controlled instrumental learning. In the first stage, associations are learned between conditioned stimuli (such as the sight of food), and unconditioned (gene-defined primary reinforcing) stimuli (such as the taste of food). This produces a motivational state to obtain the goal object, the conditioned stimulus. In the second stage, actions are learned to obtain the goal object. The motivational state thus intervenes between the stimulus that defines a goal object, and the action that may then be learned and performed to obtain the goal object. This again makes this a unifying approach to understanding both motivation as well as emotion (Rolls, 2025e).

Because Rolls' theories of motivation and emotion are based on reward value systems, the concept is that normally 'wanting' a goal object (motivation), and the liking when the goal object is obtained (the emotion), are closely related to each other. Wanting and liking are normally very closely related, with the same modifiers of the reward value such as feeding to satiety affecting both the wanting and the liking (Rolls, 2025e). An exception, which is I think an exception as it applies mainly to effects produced by drugs that bypass much of the brain's processing systems for rewards (Rolls, 2025e), is that drugs that release dopamine such as cocaine and amphetamine, the psychomotor stimulants, may have large effects on wanting (drug craving), with much less effect to produce liking (Berridge et al., 2009; Robinson and Berridge, 2025; Rolls, 2025e).

Another issue is that normally the sight and taste of food provide the reward value for eating, and placing food in the stomach or duodenal tract reduces the motivational state of hunger and produces satiety (Rolls, 2005, 2014a, 2026d). This is shown by sham feeding experiments in which reward (and subjective pleasure in humans) is produced when food is only tasted and then drains out of the stomach; and by experiments in which food is placed directly into the stomach or duodenum, which produces little reward or subjective pleasure in humans, but does decrease hunger (motivation) (Gibbs et al., 1981; Rolls, 2005, 2014a, 2026d). Confusion about this can arise when it is sometimes said that gut signals produce reward (De Araujo, Schatzker and Small, 2020), but this is just because there is some rather slow conditioning between taste and gut signals to produce conditioned appetite and conditioned satiety (Booth, 1985; Rolls, 2016f, 2023b, 2025e, 2026d), and a key role of food directly sensed in the gut is to produce satiety (Gibbs et al., 1981; Rolls, 2005, 2016f, 2026d).

These theories of motivation and emotion are I believe conceptually important (as well as elegant), for they link much of the empirical research on brain systems involved in reward processing, in motivation, and in emotion, together, with a large unifying framework (Rolls, 2005, 2014a, 2023c, 2026d, 2023b, 2025e, 2026d).

3.12 Reward-related decision-making

From fMRI studies, we have shown that the orbitofrontal cortex represents value on a continuous scale, with the fMRI BOLD signal linearly related to subjective, conscious, pleasantness ratings (Rolls, Grabenhorst and Parris, 2008b; Grabenhorst, Rolls and Bilderbeck, 2008a; Grabenhorst, Rolls, Margot, da Silva and Velazco, 2007; Grabenhorst, Rolls, Parris and D'Souza, 2010b; Grabenhorst, D'Souza, Parris, Rolls and Passingham, 2010a; Rolls, Grabenhorst and Parris, 2010c). The linear correlation of the orbitofrontal cortex BOLD signal with the simultaneously rated pleasantness of stimuli is further evidence we discovered that the orbitofrontal cortex in involved in subjectively experienced emotion (see example in Fig. 3.8).

In contrast, a more anterior and medial cortical region, which includes part of medial area 10 and is sometimes termed the 'ventromedial prefrontal cortex' (vmPFC), is activated when choices must be made between rewards (Grabenhorst et al., 2008b; Rolls et al., 2010c; Rolls and Grabenhorst, 2008). This was shown in tasks in which decisions were made about which stimulus of two was more pleasant (including olfactory (Rolls, Grabenhorst and Parris, 2010c) stimuli, and thermal (Grabenhorst, Rolls and Parris, 2008b) stimuli, compared to making ratings of pleasantness about the same stimuli.

By this time, I had developed an interest in how attractor networks could be used not only for memory, but also for decision-making, as part of my quest for *how* computations are performed in the brain (Rolls and Deco, 2010). I decided to put the attractor network theory of decision-making to empirical test, as described next. The research is I believe very interesting, for it reveals many of the properties of how we believe that reward-related decisions are made in the brain; and shows how we have further developed and then tested computational models of decision-making.

The model of decision-making that we used was developed from an integrate-and-fire model described by Wang (2002; 2008), but modified so that the spontaneous firing rate state was stable even when the decision cues were applied, as this meant that decisions were only taken if noise from the stochastic firing of neurons was present, and this modelled reaction time distributions better (Rolls and Deco, 2010). We have applied the model successfully to several experimental paradigms including attention and short-term memory as well as decision-making (Rolls and Deco, 2002; Deco and Rolls, 2002b, 2003, 2004; Deco, Rolls and Horwitz, 2004; Szabo, Almeida, Deco and Stetter, 2004; Deco and Rolls, 2005d, 2006; Rolls, Grabenhorst and Deco, 2010a,b; Rolls and Deco, 2010; Smerieri, Rolls and Feng, 2010; Deco, Rolls and Romo, 2010; Rolls and Deco, 2011c; Rolls, 2012a; Martinez-Garcia, Rolls, Deco and Romo, 2011; Rolls, Dempere-Marco and Deco, 2013; Deco, Rolls, Albantakis and Romo, 2013; Rolls and Deco, 2015b,a, 2016; Rolls and Mills, 2019).

3.12.1 An attractor decision-making network

Consider the attractor network architecture shown in Fig. 3.19a (with more details provided by Rolls (2023c; 2026d)). A set of cortical neurons has recurrent collateral excitatory synaptic connections w_{ij} from the other neurons. The evidence for decision 1 is applied via the λ_1 inputs, and for decision 2 via the λ_2 inputs. The synaptic weights w_{ij} have been associatively modified during training in the presence of λ_1 and at a different time of λ_2. The Hebbian or associative synaptic modification is such that if the presynaptic terminal and the postsynaptic neuron are simultaneously active, the synaptic connections become stronger. There are inhibitory neurons (not shown in Fig. 3.19a) that keep the total firing in the network within bounds, and in fact implement competition between the neuronal populations. As a result of the associative synaptic modification, there are strong connections within the set of neurons activated by λ_1, and strong connections within the set of neurons activated by λ_2. These strengthened synapses provide positive feedback, so that if the whole or part of λ_1 is applied, that set of neurons becomes active, and maintains its activity for a long period even when the input λ_1 is removed. The neurons activated by λ_1 if firing inhibit the neurons activated by λ_2 through the inhibitory interneurons, so that just one population wins the competition and maintains its activity. This thus provides a model not only of memory, and its retrieval, but also of decision-making. This is called an attractor network, because a subset of the neurons within either population is sufficient to attract the system into a state in which all the neurons in that population are active, by using the strengthened recurrent collateral connections. The properties of attractor or autoassociation networks are described elsewhere (Rolls, 2023c, 2026d, 2016b; Deco et al., 2013; Rolls and Deco, 2010; Hertz et al., 1991; Hopfield, 1982).

For decision-making, when λ_1 and λ_2 are applied simultaneously, each attractor competes through the inhibitory interneurons (not shown), until one attractor population of neurons wins the competition, and the network falls into one of the high firing rate attractor states that represents

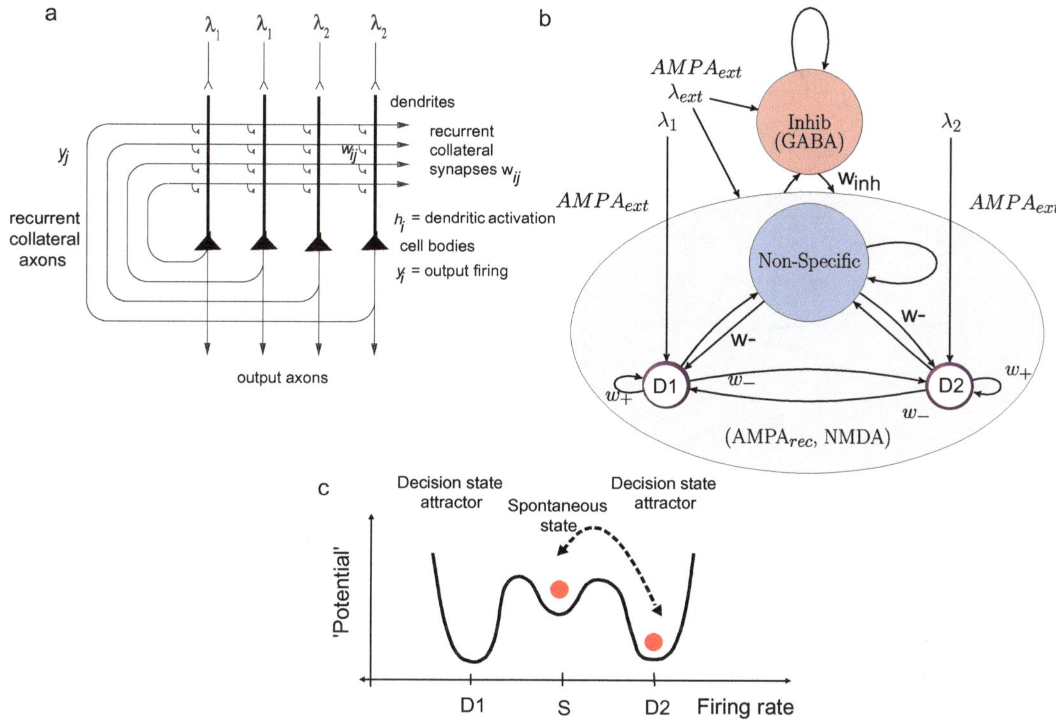

Fig. 3.19 (a) Attractor or autoassociation single network architecture for decision-making. The cell body of each neuron is shown as a triangle (like a cortical pyramidal cell), the dendrite is vertical, and receives recurrent collateral synaptic connections w_{ij} from the other neurons. The evidence for decision 1 is applied via the λ_1 inputs, and for decision 2 via the λ_2 inputs. The synaptic weights w_{ij} have been associatively modified during training in the presence of λ_1 and at a different time of λ_2. When λ_1 and λ_2 are applied, each attractor competes through the inhibitory interneurons (not shown), until one wins the competition, and the network falls into one of the high firing rate attractors that represents the decision. The noise in the network caused by the random spiking times of the neurons (for a given mean rate) means that on some trials, for given inputs, the neurons in the decision 1 (D1) attractor are more likely to win, and on other trials the neurons in the decision 2 (D2) attractor are more likely to win. This makes the decision-making probabilistic, for, as shown in (c), the noise influences when the system will jump out of the spontaneous firing stable (low energy) state S, and whether it jumps into the high firing state for decision 1 (D1) or decision 2 (D2). (b) The architecture of the integrate-and-fire network used to model decision-making (see text). The synaptic weights between the neural populations (decision pools D1 and D2, the non-specific pool, and the inhibitory pool) are 1 except where indicated. In particular, the recurrent weights, indicated by a recurrent arrow, between the neurons within an attractor decision-making pool have strong weights w_+, and between the different pools have the weak strength w_-. (c) A multistable 'effective energy landscape' for decision-making with stable states shown as low 'potential' basins. Even when the inputs are being applied to the network, the spontaneous firing rate state is stable, and noise provokes transitions from the low firing rate spontaneous state S into the high firing rate decision attractor state D1 or D2. If the noise is greater, the escaping time to a decision state, and thus the decision or reaction time, will be shorter. (Data from Rolls E T and Deco G 2010 The Noisy Brain: Stochastic Dynamics as a Principle of Brain Function. Oxford University Press: Oxford.)

the decision. When the network starts from a state of spontaneous firing, the biasing inputs encourage one of the attractors to gradually win the competition, but this process is influenced by the Poisson-like firing (spiking) of the neurons, so that which attractor wins is probabilistic. (Poisson-like indicates that the firing times are random for a given mean firing rate.) If the evidence in favor of the two decisions is equal, the network chooses each decision probabilistically on 50% of the trials. The model shows how probabilistic decision-making could be implemented in the brain. The model also shows how the evidence can be accumulated over long periods of time because of the integrating action of the attractor short-term memory network, with the recurrent collaterals feeding back information to be combined with the continuing inputs λ_1 and λ_2. The model produces shorter reaction or decision times as a function of the magnitude of the difference between the evidence for the two decisions:

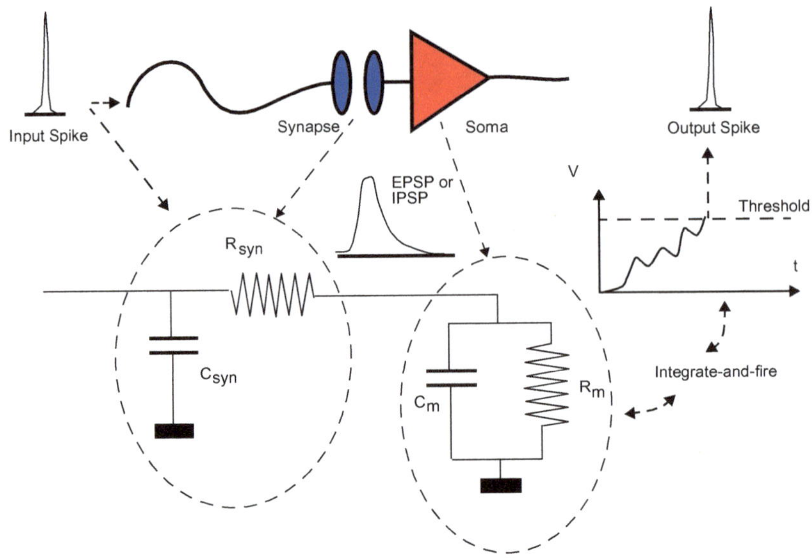

Fig. 3.20 Integrate-and-fire neuron. The basic circuit of an integrate-and-fire model consists of the neuron's membrane capacitance C_{m} in parallel with the membrane's resistance R_{m} (the reciprocal of the membrane conductance g_{m}) driven by a synaptic current with a conductance and time constant determined by the synaptic resistance R_{syn} (the reciprocal of the synaptic conductance g_j) and capacitance C_{syn} shown in the figure. These effects produce excitatory or inhibitory postsynaptic potentials, EPSPs or IPSPs. These potentials are integrated by the cell, and if a threshold V_{thr} is reached a δ-pulse (spike) is fired and transmitted to other neurons, and the membrane potential is reset. (Data from Deco G and Rolls E T 2003 Attention and working memory: a dynamical model of neuronal activity in the prefrontal cortex. European Journal of Neuroscience 18: 2374–2390.)

difficult decisions take longer, partly because the firing rates take longer to reach a decision threshold if the difference between the inputs is small.

3.12.2 A model of reward-related decision-making mechanisms in the orbitofrontal cortex

The aim that we set ourselves next was to make a reward-related decision-making model for the orbitofrontal cortex. We could then test predictions of the model in an fMRI investigation (Rolls, 2023c, 2026d; Rolls et al., 2010a,b). The analyses performed with the model relate to the ways in which probabilistic decision-making is influenced by the easiness vs the difficulty of the decision, and how confidence in a decision emerges as a property of the neuronal attractor network decision-making process.

We first simulated integrate-and-fire models of attractor-based choice decision-making, and predicted from them the neuronal firing rates in the decision attractor populations and the fMRI BOLD signals on easy vs difficult trials (Rolls, Grabenhorst and Deco, 2010a) as described next. We then performed two fMRI investigations of decision-making about the reward value and subjective pleasantness of thermal and olfactory stimuli, and showed that brain regions implicated by other analyses in decision-making (Rolls and Grabenhorst, 2008; Grabenhorst, Rolls and Parris, 2008b; Rolls, Grabenhorst and Parris, 2010c) do show the predicted difference between activations on easy vs difficult trials in the orbitofrontal cortex (Rolls, Grabenhorst and Deco, 2010a).

The attractor network model for probabilistic decision-making that we used (Fig. 3.19) is implemented with integrate-and-fire neurons of the type illustrated in Fig. 3.20. The evidence λ_1 for decision 1 is applied to pool D1 through 800 synapses onto each neuron, and the evidence λ_2 for decision 2 is applied to pool D2 in the same way (Fig. 3.19b). The same sets of 800 synapses on every neuron in the network also receive external inputs at a rate λ_{ext}=3 spikes/s, a typical firing rate for the spontaneous activity of cortical neurons, to reflect background activity from other cortical areas. The distribution of these inputs onto each neuron over time is what would be produced by Poisson spikes trains at 3 spikes/s on each of the 800 externally connected synapses onto each neuron, and this is one of the sources of noise in the network.

Figure 3.21a and e show for the decision-making model the mean firing rates of the two neuronal populations D1 and D2 for two trial types, easy trials (ΔI=160 Hz) and difficult trials (ΔI=0). (ΔI is the difference in spikes/s summed across all synapses to each neuron between the two inputs, λ_1 to population D1, and λ_2 to population D2), with a large ΔI producing an easy decision in which confidence might be high. The results are shown for correct trials, that is, trials on which the D1 population won the competition and fired with a rate of $>$ 10 spikes/s more than the rate of D2 for the last 1000 ms of the simulation runs. Figure 3.21b shows the mean firing rates of the four populations of neurons on a difficult trial, and Fig. 3.21c shows the rastergrams for the same trial, for which the energy landscape is also shown in Fig. 3.19c. Figure 3.21d shows the firing rates on another difficult trial (ΔI=0) to illustrate the variability shown from trial to trial, with on this trial prolonged competition between the D1 and D2 attractors until the D1 attractor finally won after approximately 1100 ms. Figure 3.21f shows firing rate plots for the four neuronal populations in an example of a single easy trial (ΔI=160), Fig. 3.21g shows the synaptic currents in the four neuronal populations on the same trial, and Fig. 3.21h shows rastergrams for the same trial (Rolls, Grabenhorst and Deco, 2010a).

Two important points are made by the results shown in Fig. 3.21. First, the network falls into its decision attractor faster on easy trials than on difficult trials. We would accordingly expect reaction times to be shorter on easy than on difficult trials. We might also expect the BOLD signal related to the activity of the network to be higher on easy than on difficult trials because it starts sooner on easy trials.

Second, the mean firing rate after the network has settled into the correct decision attractor is higher on easy than on difficult trials. The reason for this is that on easy trials, the firing rate of the winning attractor neuronal population depends not only on the recurrent collateral connectivity, but also on the relatively larger continuing input λ_1 to the population that wins. We might therefore expect the BOLD signal related to the activity of the network to be higher on easy than on difficult trials because the maintained activity in the winning attractor is higher on easy trials.

The effects evident in Fig. 3.21 are quantified, and elucidated over a range of values for ΔI, next.

Figure 3.22a shows the firing rates (mean $\pm$ sd) on correct trials when in the D1 attractor as a function of ΔI. ΔI=0 corresponds to the most difficult decision, and ΔI=160 corresponds to easy. The firing rates for both the winning population D1 and for the losing population D2 are shown. The firing rates were measured in the last 1 s of firing, i.e. between t=3 and t=4 s. It is clear that the mean firing rate of the winning population increases monotonically as ΔI increases, and interestingly, the increase is approximately linear (Pearson $r = 0.995$, p$< 10^{-6}$).

As shown in Fig. 3.22a, the firing rates of the losing population decrease as ΔI increases. The decrease of firing rate of the losing population is due in part to feedback inhibition through the inhibitory neurons by the winning population. Thus the difference of firing rates between the winning and losing populations, as well as the firing rate of the winning population D1, both clearly reflect ΔI, and in a sense the confidence in the decision.

The time for the network to reach the correct D1 attractor, i.e. the reaction or decision time of the network, is shown as a function of ΔI in Fig. 3.22b (mean $\pm$ sd). Interestingly, the reaction time continues to decrease ($r = -0.95, p < 10^{-4}$) over a wide range of ΔI, even when as shown

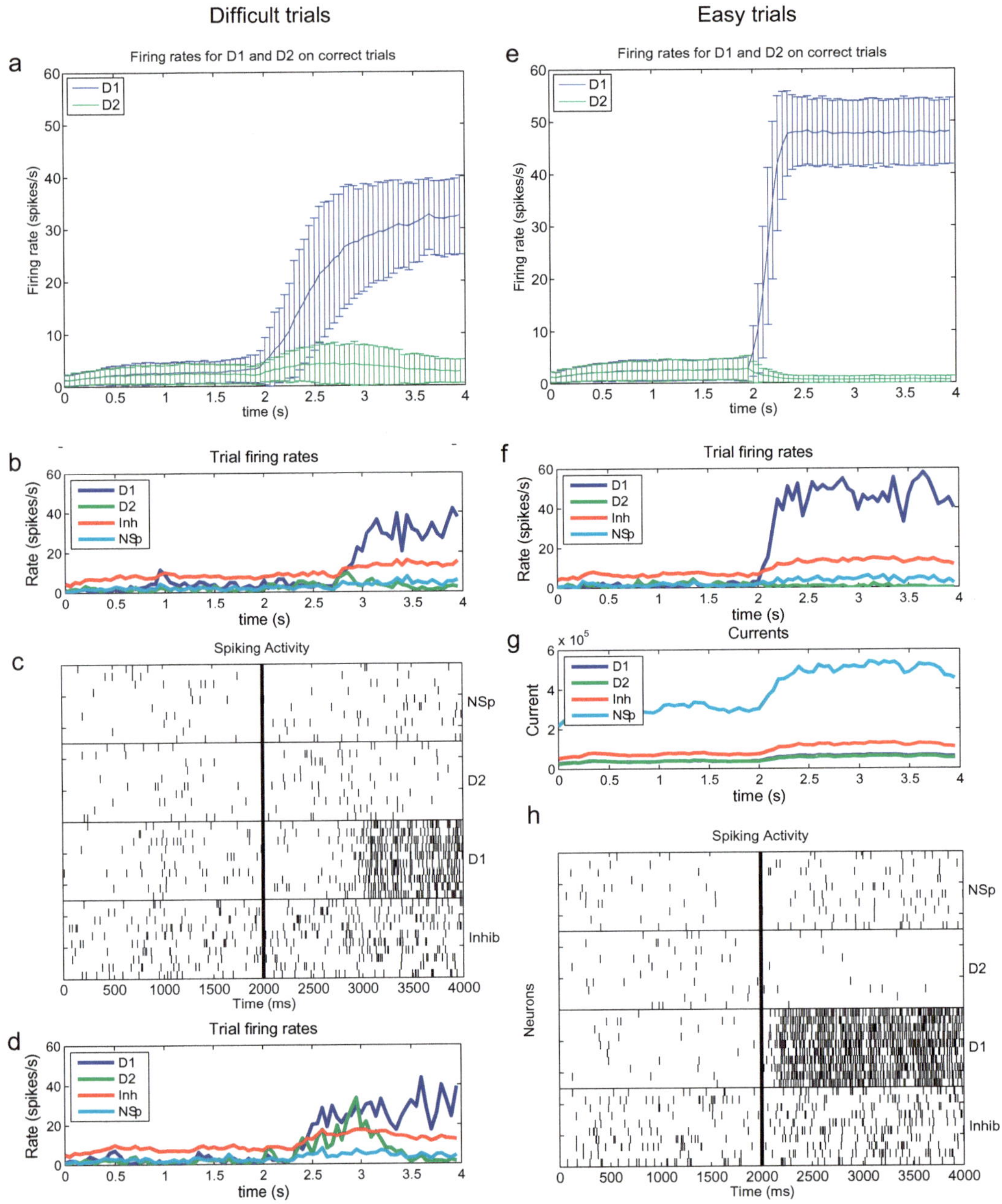

Fig. 3.21 The operation of attractor decision-making networks. (a) and (e) Firing rates (mean $\pm$ sd) for difficult (ΔI=0) and easy (ΔI=160) trials. The period 0–2 s is the spontaneous firing, and the decision cues were turned on at time = 2 s. D1: firing rate of the D1 population of neurons on correct trials on which the D1 population won. D2: firing rate of the D2 population of neurons on the correct trials on which the D1 population won. A correct trial was one in which in which the mean rate of the D1 attractor averaged > 10 spikes/s for the last 1000 ms of the simulation runs. (b) The mean firing rates of the four populations of neurons on a difficult trial. Inh is the inhibitory population that uses GABA as a transmitter. NSp is the non-specific population of neurons (see Fig. 3.19). (c) Rastergrams for the trial shown in b. 10 neurons from each of the four pools of neurons are shown. (d) The firing rates on another difficult trial (ΔI=0) showing prolonged competition between the D1 and D2 attractors until the D1 attractor finally wins after approximately 1100 ms. (f) Firing rate plots for the 4 neuronal populations on a single easy trial (ΔI=160). (g) The synaptic currents in the four neuronal populations on the trial shown in f. (h) Rastergrams for the easy trial shown in f and g. 10 neurons from each of the four pools of neurons are shown. (Data from Rolls E T, Grabenhorst F, and Deco G 2010 Choice, difficulty, and confidence in the brain. NeuroImage 33: 694–706.)

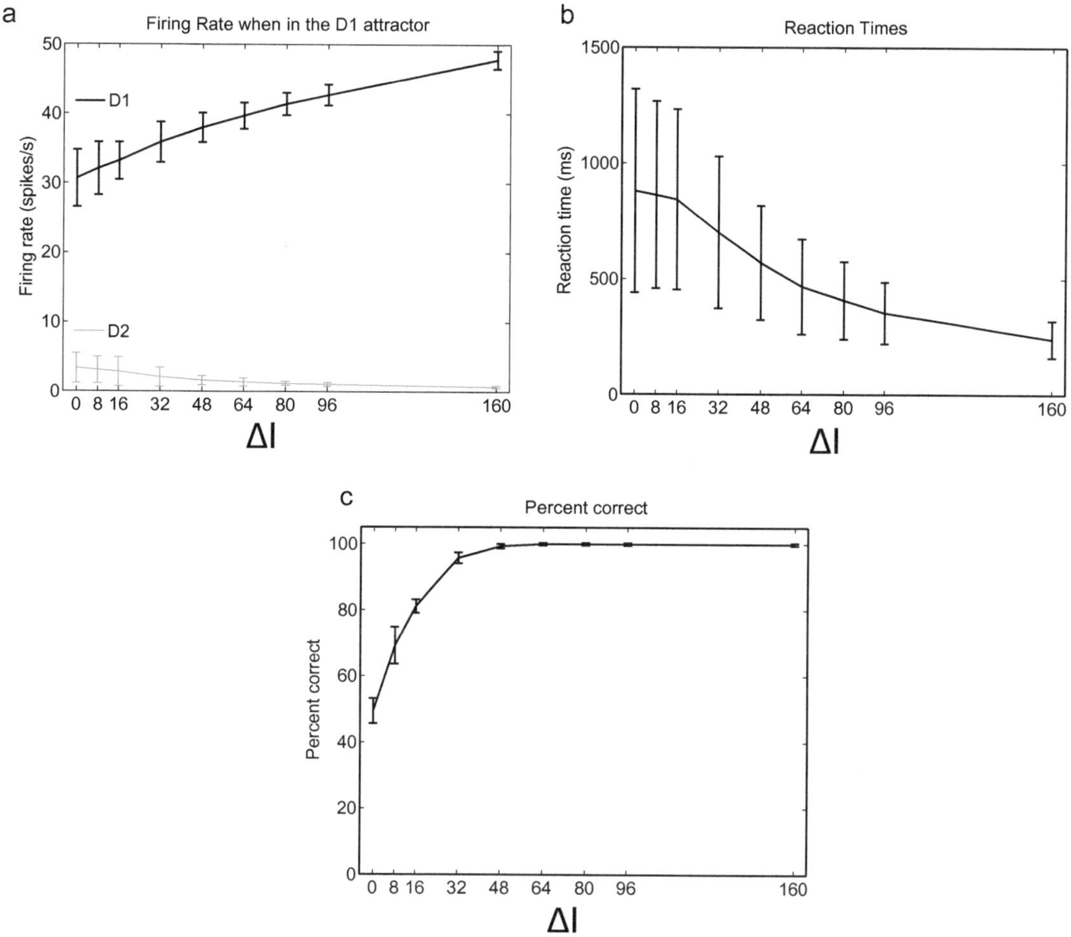

Fig. 3.22 The operation of attractor decision-making networks. (a) Firing rates (mean $\pm$ sd) on correct trials when in the D1 attractor as a function of ΔI. ΔI=0 corresponds to difficult, and ΔI=160 spikes/s corresponds to easy. The firing rates for both the winning population D1 and for the losing population D2 are shown for correct trials by thick lines. All the results are for 1000 simulation trials for each parameter value, and all the results shown are statistically highly significant. (b) Reaction times (mean $\pm$ sd) for the D1 population to win on correct trials as a function of the difference in inputs ΔI to D1 and D2. (c) Per cent correct performance, i.e. the percentage of trials on which the D1 population won, as a function of the difference in inputs ΔI to D1 and D2. The mean was calculated over 1000 trials, and the standard deviation was estimated by the variation in 10 groups each of 100 trials. (Data from Rolls E T, Grabenhorst F, and Deco G (2010) Choice, difficulty, and confidence in the brain. NeuroImage 33: 694–706.)

in Fig. 3.22c the network is starting to perform at 100% correct (see Rolls et al. (2010a,b); Rolls (2023c, 2026d)).

At ΔI=0, there is no influence on the network to fall more into attractor D1 representing decision 1 than attractor D2 representing decision 2, and its decisions are at chance, with approximately 50% of decisions being for D1. As ΔI increases, the proportion of trials on which D1 is reached increases. The relation between ΔI and percentage correct is shown in Fig. 3.22c.

These analyses of decision-making in the integrate-and-fire model thus provide a clear account of why decision-making is probabilistic, and show how the model can account for many differences in the decision-making for easy trials (when confidence will be high) vs difficult trials (when confidence will be low). As this is a biologically plausible model, it goes far beyond an artificial drift-diffusion

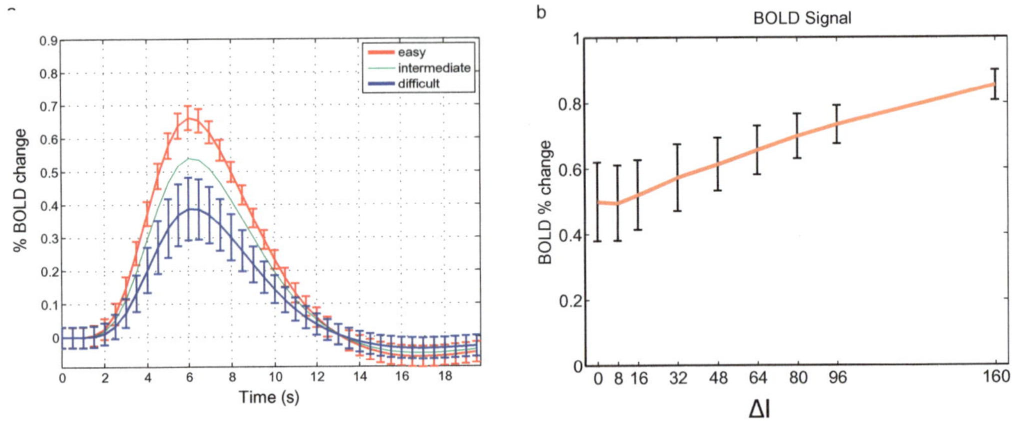

Fig. 3.23 The operation of attractor decision-making networks. (a) The percentage change in the simulated BOLD signal on easy trials (ΔI=160 spikes/s), on intermediate trials (ΔI=80), and on difficult trials (ΔI=0). The mean $\pm$ sd are shown for the easy and difficult trials. The percentage change in the BOLD signal was calculated from the firing rates of the D1 and D2 populations, and analogous effects were found with calculation from the synaptic currents averaged for example across all 4 populations of neurons. (b) The percentage change in the BOLD signal (peak mean $\pm$ sd) averaged across correct and incorrect trials as a function of ΔI. ΔI=0 corresponds to difficult, and ΔI=160 corresponds to easy. The percent change was measured as the change from the level of activity in a period of 1 s immediately before the decision cues were applied at t=0 s, and was calculated from the firing rates of the neurons in the D1 and D2 populations. The BOLD per cent change scaling is arbitrary, and is set so that the lowest value for the peak of a BOLD response is 0.5%. (Data from Rolls E T, Fabian Grabenhorst F, and Deco G 2010 Choice, difficulty, and confidence in the brain. NeuroImage 33: 694–706.)

model (Deco, Rolls, Albantakis and Romo, 2013; Steinemann, Stine, Trautmann, Zylberberg, Wolpert and Shadlen, 2024). Moreover, as will be shown, we were able with this model to make predictions about the fMRI BOLD signal during decision-making on easy vs difficult trials, which could then be compared to empirical measurements of the BOLD signal during decision-making that we made.

3.12.3 Prediction of the fMRI BOLD signals on difficult vs. easy decision-making trials from the decision-making model

It is now shown how this model makes predictions for the fMRI BOLD signals that would occur in brain areas in which decision-making processing of the type described is taking place. The BOLD signals were predicted from the firing rates of the neurons in the network (or in practice from the synaptic currents flowing in the neurons (Rolls et al., 2010a)) by convolving the neuronal or synaptic activity with the haemodynamic response function.

As shown in Fig. 3.23a, the predicted fMRI response is larger for easy (ΔI = 160 spikes/s) than for difficult trials (ΔI=0), with intermediate trials (ΔI=80) producing an intermediate fMRI response (see Rolls et al. (2010a,b); Rolls (2023c, 2026d)).

Figure 3.23b shows that the percentage change in the BOLD signal (peak mean $\pm$ sd) averaged across correct and incorrect trials increases monotonically as a function of ΔI.

3.12.4 Comparison of the fMRI BOLD signals on difficult vs. easy decision-making trials with those predicted from the decision-making model

We then performed an fMRI experiment to compare the empirically measured fMRI BOLD signal with that predicted by the model, as a test of the model. Fig. 3.24 shows experimental fMRI

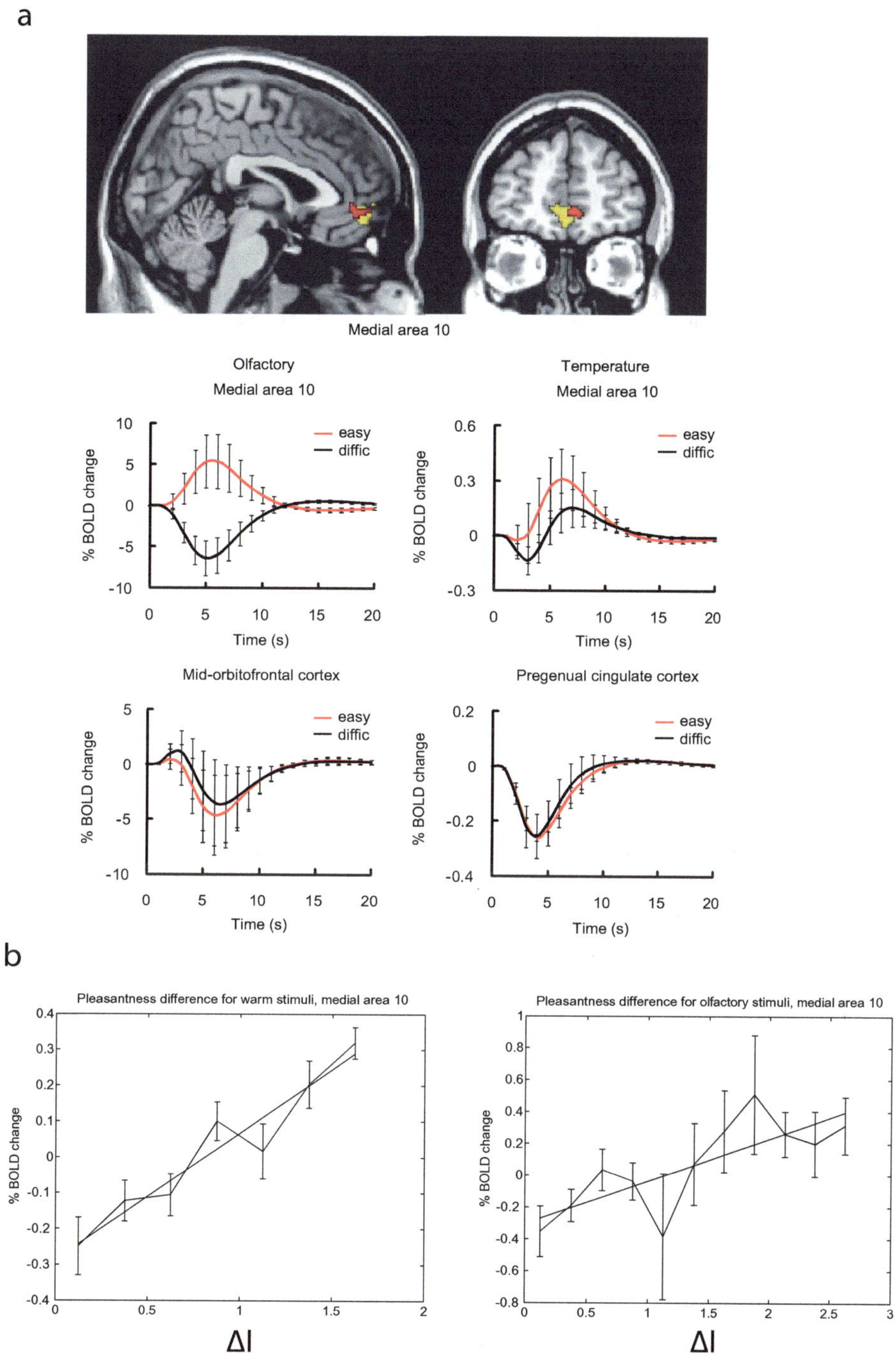

Fig. 3.24 a. Top: Experimental fMRI data showing the medial prefrontal cortex area activated on easy vs difficult trials in the olfactory pleasantness decision task (yellow) and the thermal pleasantness decision task (red). Middle: Experimental fMRI data showing the BOLD signal in medial area 10 on easy and difficult trials of the olfactory affective decision task (left) and the thermal affective decision task (right). This medial area 10 was a region identified by other criteria (see text) as being involved in choice decision making. Bottom: experimental data showing the BOLD signal for the same easy and difficult trials, but in parts of the pregenual cingulate and mid-orbitofrontal cortex implicated by other criteria (see text) in representing the subjective reward value of the stimuli on a continuous scale, but not in making choice decisions between the stimuli, or about whether to choose the stimulus in future. b. Experimental fMRI data showing the change in the BOLD signal (mean$\pm$sem, with the fitted linear regression line shown) as a function of ΔI, the difference in pleasantness of warm stimuli or olfactory stimuli about which a decision was being made, for medial prefrontal cortex area 10. (Data from Rolls E T, Fabian Grabenhorst F, and Deco G 2010 Choice, difficulty, and confidence in the brain. NeuroImage 33: 694–706.)

data indicating that fMRI signals are linearly related to the easiness of decisions, and to decision confidence, as described next.

Fig. 3.24a shows the fMRI BOLD signal measured on easy and difficult trials of an olfactory affective decision task in which participants had to choose which odor was more pleasant (left); and in a thermal affective decision task in which participants had to choose which of two thermal stimuli applied to the hand was more pleasant (right) (Rolls, Grabenhorst and Deco, 2010a). The upper records are for prefrontal cortex medial area 10 in a region identified as being involved in choice decision making in that it was a brain region in which identical stimuli show more activity when a choice decision was being made than when a rating on a continuous scale of affective value was being made. Fig. 3.24a shows for medial prefrontal cortex area 10 that there is a larger BOLD signal on easy than on difficult trials. The top diagram shows the medial prefrontal area activated in this contrast for decisions about which olfactory stimulus was more pleasant (yellow) and for decisions about whether the thermal stimulus would be chosen in future based on whether it was pleasant or unpleasant (red). The lower diagram shows the BOLD signals for these decisions on easy vs difficult trials.

Fig. 3.24b shows in the model that for both types of decision, the BOLD signal was linearly related to the difference in the pleasantness of the stimuli being compared, ΔI.

This research shows that a larger BOLD signal on easy than on difficult trials (Fig. 3.24) is an fMRI signature of a region such as medial area 10 involved in decision-making, in that the cortical regions that show this effect are areas implicated by other criteria in decision-making, including their larger activation on trials on which choice decisions must be made, rather than continuous ratings of affective value (Grabenhorst et al., 2008b; Rolls et al., 2010c).

Fig. 3.24 also shows that the orbitofrontal cortex and pregenual anterior cingulate cortex, where continuous affective value but not choice is represented (Grabenhorst et al., 2008b; Rolls et al., 2010c; Rolls and Grabenhorst, 2008), do not show this easy vs difficult effect in the BOLD signal. This is what is predicted by the model for brain areas not involved in binary choice decision-making and its emergent property confidence, but in representing pleasantness on a continuous scale, which is likely to be a precursor and prerequisite for choice decisions about pleasantness (Grabenhorst et al., 2008b; Rolls et al., 2010c; Rolls and Grabenhorst, 2008). Consistently, the orbitofrontal and pregenual cingulate cortex project to the ventromedial prefrontal cortex area 10 (Price, 2006; Rolls et al., 2022b), where the BOLD signal does show what is predicted by the model, an increase in the BOLD signal as a function of the easiness of the choice measured by ΔI, the difference in the pleasantness of the two stimuli.

This research thus makes the discovery that what is predicted by a model of decision-making is found experimentally in fMRI investigations that include findings on error trials (Rolls et al., 2010a,b), and that provides very useful support for this approach to a biologically plausible model of reward-related and other types of decision-making (Rolls and Deco, 2010; Rolls, 2023c, 2026d).

The description of decision-making implemented by this integrate-and-fire model is supplemented by a formal description of the equations in the model (see Rolls et al. (2010a,b); Rolls (2023c, 2026d)). Other approaches to decision-making, including the drift-diffusion and race models, are described by Deco, Rolls, Albantakis and Romo (2013). The application of this approach to decision-making for other brain regions is described in Rolls and Deco (2010) and Deco, Rolls, Albantakis and Romo (2013).

Overall, the research described in this section has shown how biologically realistic models of decision-making in the brain can be developed further and analyzed, and tested with empirical evidence from human brain neuroimaging. This is a major area of brain research in which our understanding of important aspects of human behavior has been developed at both the computational neuroscience and neuroimaging levels. The whole issue of stochastic neurodynamics is a key area in understanding many types of human behavior, including creativity (Sun, Liu, Rolls, Chen, Yao, Yang, Wei, Zhang, Zhang, Feng and Qiu, 2019), as set out in **The Noisy Brain: Stochastic Dynamics as a Principle of Brain Function** (Rolls and Deco, 2010) and elsewhere (Rolls, 2023c, 2026d).

3.13 A theory of depression: the orbitofrontal cortex, and attractor dynamics

3.13.1 Introduction

As I had made many discoveries about the functions of the orbitofrontal cortex in reward, emotion, motivation, and decision-making (Rolls, 2014a, 2023b,c, 2026d) (as described in the previous parts of this Chapter), and as I wish my research to have application to disease as well as health, I turned my mind to thinking about a major mental disorder, depression. That is how I developed a theory of depression that is founded on the understanding I have developed of reward and emotion systems in the brain.

My thinking is as described in this section, and some of the theory that has developed is that the non-reward-related lateral orbitofrontal cortex is too responsive and has increased connectivity leading to the sad components of depression; and the medial orbitofrontal cortex is less sensitive to reward and is underconnected, leading to the anhedonia of depression (Rolls et al., 2020b; Zhang et al., 2024a; Rolls, 2019a; Rolls et al., 2018a; Rolls, 2018b, 2016d; Cheng et al., 2016; Xie et al., 2021a; Rolls, 2023c, 2026d). What I describe here is an attractor-based theory of some of the brain mechanisms that are related to depression (Rolls, 2016d), and tests of the theory (Rolls, Cheng and Feng, 2020b; Zhang, Rolls, Wang, Xie, Cheng and Feng, 2024a).

Major depressive disorder is ranked by the World Health Organization as the leading cause of years-of-life lived with disability, with more than 300 million people having major depressive disorder (Drevets, 2007; Gotlib and Hammen, 2009; Hamilton et al., 2013; WHO, 2017). Major depressive episodes, found in both major depressive disorder and bipolar disorder, are pathological mood states characterized by persistently sad or depressed mood. Major depressive disorders are generally accompanied by: (1) altered incentive and reward processing, evidenced by amotivation, apathy, and anhedonia; (2) impaired modulation of anxiety and worry, manifested by generalized, social and panic anxiety, and oversensitivity to negative feedback; (3) inflexibility of thought and behavior in association with changing reinforcement contingencies, apparent as ruminative thoughts of self-reproach, pessimism, and guilt, and inertia toward initiating goal-directed behavior; (4) altered integration of sensory and social information, as evidenced by mood-congruent processing biases; (5) impaired attention and memory, shown as performance deficits on tests of attention set-shifting and maintenance, and autobiographical and short-term memory; and 6) visceral disturbances, including altered weight, appetite, sleep, and endocrine and autonomic function (Drevets, 2007; Gotlib and Hammen, 2009; Rolls, 2018b; Thapar et al., 2022).

I developed my thinking into an attractor theory of depression. The theory starts with the discovery that the orbitofrontal cortex contains a population of error neurons that respond to non-reward and maintain their firing for many seconds after the non-reward, providing evidence that they have entered an attractor state that maintains a memory of the non-reward (Thorpe, Rolls and Maddison, 1983; Rosenkilde, Bauer and Fuster, 1981; Rolls, 2014a, 2018b, 2019c) (Section 3.7). An example of such a neuron is shown in Fig. 3.11C. The human lateral orbitofrontal cortex is activated by non-reward during reward reversal (Kringelbach and Rolls, 2003) (Fig. 3.11), by losing money (O'Doherty, Kringelbach, Rolls, Hornak and Andrews, 2001a) or not winning (Xie et al., 2021b), and by many other aversive stimuli (Grabenhorst and Rolls, 2011). Further evidence that the orbitofrontal cortex is involved in changing rewarded behavior when non-reward is detected is that damage to the human orbitofrontal cortex impairs reward reversal learning, in that the previously rewarded stimulus is still chosen during reversal even when no reward is being obtained (Rolls, Hornak, Wade and McGrath, 1994a; Hornak, O'Doherty, Bramham, Rolls, Morris, Bullock and Polkey, 2004; Fellows and Farah, 2003; Fellows, 2011). Further, the human right lateral orbitofrontal cortex is strongly activated by non-reward in the one-trial rule-based reward reversal task (Rolls, Vatansever, Li, Cheng and Feng, 2020e) (the same task that we used in macaques when we discovered non-reward neurons (Thorpe, Rolls and Maddison, 1983)), which is the same brain region with increased functional connectivity in depression as described below.

Now it is well established that not receiving expected reward, or receiving unpleasant stimuli or events, can produce feelings of depression (Beck, 2008; Drevets, 2007; Harmer and Cowen, 2013; Price and Drevets, 2012; Pryce et al., 2011; Eshel and Roiser, 2010; Rolls, 2018b). A clear example is that if a member of the family dies, then this is the removal of reward (in that we would work to try to avoid this), and the result of the removal of the reward can be depression. More formally, in terms of learning theory, the omission or termination of a reward can give rise to sadness or depression, depending on the magnitude of the reward that is lost, if there is no action that can be taken to restore the reward (Rolls, 2005, 2013a, 2014a, 2018b) (Fig. 3.17). If an action can be taken, then frustration and anger may arise to the same reinforcement contingency (Rolls, 2014a, 2018b, 2019c,a, 2023b) (Section 3.11.1). This relates the current approach to the learned helplessness approach to depression, in which depression arises because no actions are being taken to restore rewards (Forgeard et al., 2011; Pryce et al., 2011; Maier and Seligman, 2016).

3.13.2 A non-reward attractor theory of depression

I have proposed the theory that in depression, the lateral orbitofrontal cortex non-reward / punishment attractor network system is more easily triggered, and maintains its attractor-related firing for longer (Rolls, 2016d, 2017a, 2019a, 2018b, 2023c, 2026d). The greater attractor-related firing of the non-reward / punishment system triggers negative cognitive states held on-line in other cortical systems such as the language system and in the prefrontal cortex which is implicated in attentional control. These other cortical systems then in turn have top-down effects on the orbitofrontal non-reward system that bias it in a negative direction (Rolls, 2013b) (see Section 4.3.5 and Fig. 4.16 of Rolls (2023c)), and thus increase the sensitivity of the lateral orbitofrontal cortex to non-reward and maintain its overactivity (Rolls, 2016d) (Fig. 3.25). I proposed that the interaction of non-reward and language / attentional brain systems of these types accounts for the ruminating and continuing depressive thoughts, which occur as a result of a positive feedback cycle between these types of brain system (Rolls, 2016d).

Indeed, we have shown that cognitive states can have 'top-down' effects on affective representations in the orbitofrontal cortex (De Araujo, Rolls, Velazco, Margot and Cayeux, 2005; Grabenhorst, Rolls and Bilderbeck, 2008a; McCabe, Rolls, Bilderbeck and McGlone, 2008; Rolls, 2013b) (Section 3.5.3). Further, top-down selective attention can also influence affective representations in the orbitofrontal cortex (Rolls et al., 2008a; Grabenhorst and Rolls, 2008; Ge et al., 2012; Luo et al., 2013; Rolls, 2013b), and paying attention to depressive symptoms when depressed may in this way exacerbate the problems in a positive feedback way.

The prefrontal cortical regions implicated in maintaining activity in these orbitofrontal cortex and anterior cingulate cortex emotion-related cortical regions as shown by our effective connectivity analyses (Rolls et al., 2022b) (see Figs. 3.13 and 3.15) include for the lateral orbitofrontal cortex, inferior frontal gyrus regions IFJa, IFJp, IFSa, IFSp and Broca's area proper regions 44, 45 and 47l; and for the supracallosal anterior cingulate cortex, dorsolateral prefrontal cortex regions 46 and 9-46d. These prefrontal cortex regions while not key regions involved in computing emotions, may be important as short-term memory long-loop attractor regions (Rolls, 2023c, 2026d) useful in maintaining emotional states (Fig. 3.25) (cf. (Wood et al., 2025)).

More generally, the presence of the cognitive ability that is afforded by language to think ahead and see the implications of recent events may be a computational development in the brain that exacerbates the vulnerability of the human brain to depression (Rolls, 2014a, 2018b). For example, with language we can think ahead and see that perhaps the loss of an individual in one's life may be long-term, and this thought and its consequences for our future can become fully evident.

My theory is that one way in which depression could result from over-activity in this lateral orbitofrontal cortex system is if there is a major negatively reinforcing life event that produces reactive depression and activates this system, which then becomes self-re-exciting based on the cycle between the lateral orbitofrontal cortex non-reward / punishment attractor system and the cognitive / language system, which together operate as a systems-level attractor (Fig. 3.25). (The generic cortical architecture for such reciprocal feedforward and feedback excitatory effects is illustrated by Rolls (2016b, 2026d).)

The theory is that a second way in which depression might arise is if this lateral orbitofrontal cortex non-reward / punishment system is especially sensitive in some individuals. This might be related for example to genetic predisposition, or to the effects of stress (Gold, 2015). In this case, the orbitofrontal system would over-react to normal levels of non-reward or punishment, and start the local attractor circuit in the lateral orbitofrontal cortex (Section 3.7) (Rolls, 2016d; Rolls and Deco, 2016), which in turn would activate the cognitive system, which would feed back to the over-reactive lateral orbitofrontal cortex system to maintain now a systems-level attractor with ruminating thoughts (Fig. 3.25). This is described as a 'systems-level' attractor because it includes mutual excitations between different brain regions (Rolls, 2023c, 2026d).

Given that the activations of the lateral and medial orbitofrontal cortex often appear to be reciprocally oppositely related (O'Doherty et al., 2001a; Rolls et al., 2003c; Xie et al., 2021b; Zhang et al., 2024a), the other part of the theory of depression is that in depression there may be underactivity, under-sensitivity, or under-connectivity of the (reward-related) medial orbitofrontal cortex in depression (Rolls, 2016d, 2018b; Uran et al., 2022). Rolls' theory of depression is further that under-responsiveness of the medial orbitofrontal cortex could contribute to other aspects of depression, such as anhedonia (Rolls, 2016d, 2018b; Zhang et al., 2024a).

3.13.3 The orbitofrontal cortex, and the theory of depression

There is some evidence for altered structure and function of the lateral orbitofrontal cortex in depression (Drevets, 2007; Ma, 2015; Price and Drevets, 2012; Zhang, Rolls, Wang, Xie, Cheng and Feng, 2024a). For example, reductions of grey-matter volume and cortex thickness have been demonstrated in the posterolateral OFC (BA 12/47, caudal BA 11 and the adjoining BA 45), and also in the subgenual cingulate cortex (BA 24, 25) (Drevets, 2007; Nugent et al., 2006; Schmaal et al., 2017), and in the medial orbitofrontal cortex (Schmaal et al., 2017). In depression, there is increased cerebral blood flow in areas that include the ventrolateral orbitofrontal cortex (which is a prediction of the theory), and also in regions such as the subgenual cingulate cortex and amygdala, and these increases appear to be related to the mood change, in that they become more normal when the mood state remits (Drevets, 2007; Lally, Nugent, Luckenbaugh, Niciu, Roiser and Zarate, 2015; Zhang, Rolls, Wang, Xie, Cheng and Feng, 2024a).

To discover more about cortical systems that may operate differently in depression, we performed the first brain-wide voxel-level resting state functional connectivity neuroimaging analysis of depression (with 421 patients with major depressive disorder and 488 controls). We discovered that one major

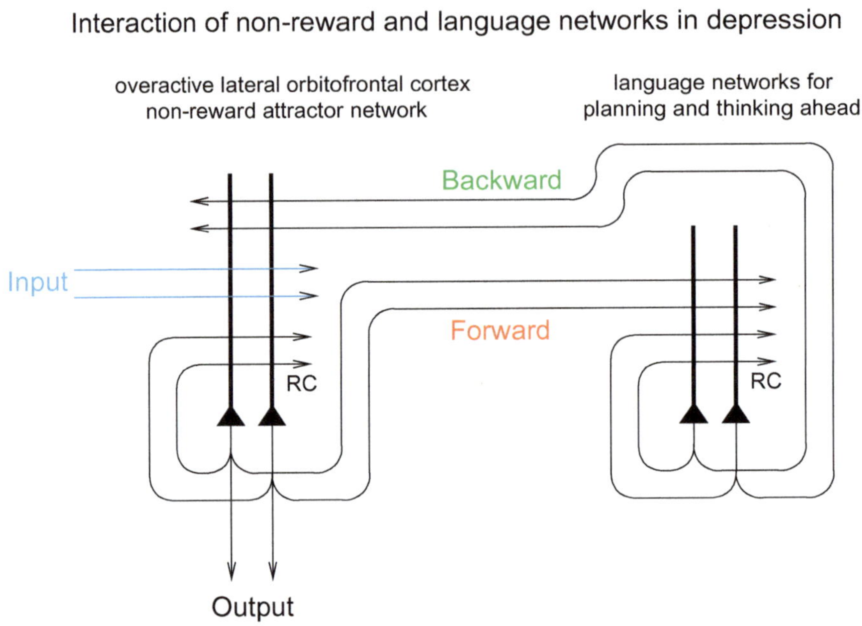

Fig. 3.25 Interaction of orbitofrontal cortex non-reward networks with language networks in depression. Illustration of how an overactive non-reward attractor network in the lateral orbitofrontal cortex could send excitatory information forward to networks for language and planning ahead (in for example the prefrontal cortex); which could in turn send excitatory 'top-down' feedback back down to the orbitofrontal non-reward network to maintain its over-activity. It is suggested that such a system with mutual 'long loop' re-excitation contributes to the persistent ruminating thoughts in depression. (Data from Rolls E T 2016 A non-reward attractor theory of depression. Neuroscience and Biobehavioral Reviews 68: 47–58.

circuit with altered functional connectivity involved the medial orbitofrontal cortex BA 13, which had reduced functional connectivity in depression with memory systems in the parahippocampal gyrus and medial temporal lobe (Cheng, Rolls, Qiu, Liu, Tang, Huang, Wang, Zhang, Lin, Zheng, Pu, Tsai, Yang, Lin, Wang, Xie and Feng, 2016) (Fig. 3.26). The lateral orbitofrontal cortex BA 12/47, involved in non-reward and punishing events, did not have this reduced functional connectivity with memory systems, so that there is an imbalance in depression towards decreased reward-related memory system functionality.

A second major circuit change we discovered was that the lateral orbitofrontal cortex area BA 12/47 had increased functional connectivity with the precuneus, the angular gyrus, and the temporal visual cortex BA 21 (Cheng et al., 2016) (Fig. 3.26). This enhanced functional connectivity of the non-reward/punishment system (BA 12/47) with the precuneus (involved in the sense of self and agency), and the angular gyrus (involved in language) is thus related to the explicit affectively negative sense of the self, and of self-esteem, in depression.

The differences in orbitofrontal connectivity with these brain regions were related to the depression by evidence that the symptoms of depression were correlated with these differences of functional connectivity; and that the lateral orbitofrontal cortex functional connectivity links described were less high if the patients were receiving antidepressant medication (Cheng et al., 2016).

The reduced functional connectivity of the medial orbitofrontal cortex, implicated in reward, with memory systems provides a new way of understanding how memory systems may be biased away from pleasant events in depression. The increased functional connectivity of the lateral orbitofrontal cortex, implicated in non-reward and punishment, with areas of the brain implicated in representing the self, language, and inputs from face and related perceptual systems provides a new way of

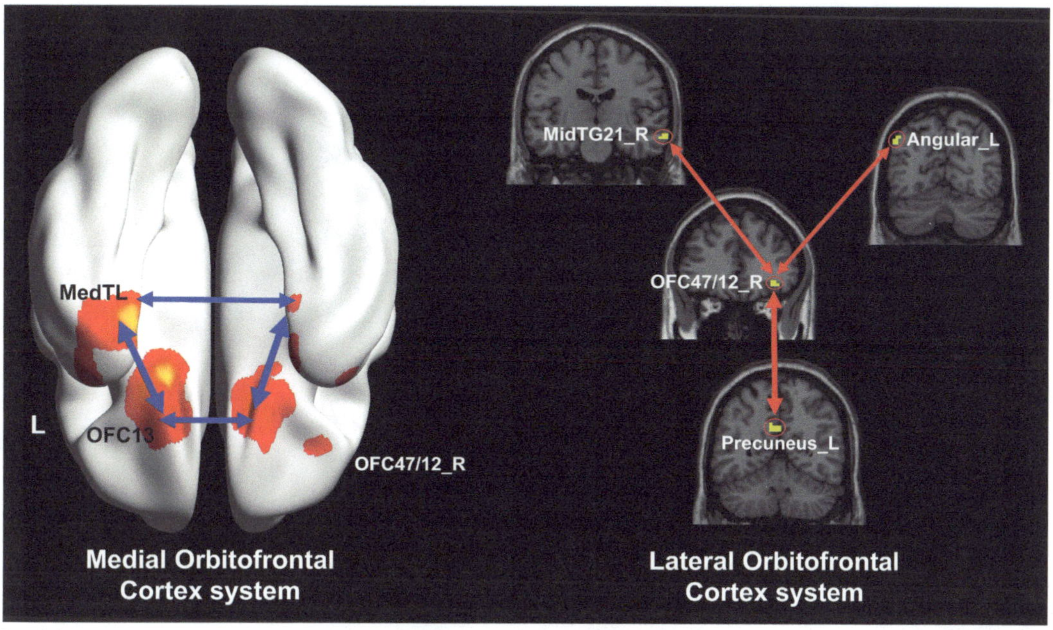

Fig. 3.26 Resting state functional connectivity in depression. The medial orbitofrontal cortex has reduced functional connectivity (blue) in depression with medial temporal lobe memory systems. The lateral orbitofrontal cortex has increased functional connectivity (red) in depression with the angular gyrus, precuneus, and middle temporal gyrus. MedTL – medial temporal lobe from the parahippocampal gyrus to the temporal pole; MidTG21R – middle temporal gyrus area 21 right; OFC13 – medial orbitofrontal cortex area 13; OFC12/47R – lateral orbitofrontal cortex area 12/47 right. The lateral orbitofrontal cortex cluster in OFC12/47 is visible on the ventral view of the brain anterior and lateral to the OFC13 clusters. (Data from Cheng W, Rolls E T et al 2016 Medial reward and lateral non-reward orbitofrontal cortex circuits change in opposite directions in depression. Brain 139: 3296–3309.

understanding how unpleasant events and thoughts, and lowered self-esteem, may be exacerbated in depression (Cheng et al., 2016; Rolls et al., 2018a).

These differences of functional connectivity are related to the orbitofrontal cortex attractor theory of depression, because increased functional connectivity of the non-reward lateral orbitofrontal cortex would increase the stability and persistence of its negative attractor mood-related states; and decreased functional connectivity of the reward-related medial orbitofrontal cortex would decrease the stability and persistence of its positive mood states (Rolls, 2018b).

These discoveries have stimulated many other large-scale voxel-level investigations of functional connectivity in depression that we have performed, described in Section 3.13.4. A large-scale fMRI activation investigation of the role of the orbitofrontal cortex in depressive symptoms is described in Section 3.13.5. Implications for treatments are considered in Section 3.13.6.

3.13.4 Altered connectivity of the orbitofrontal cortex in depression

Further investigations have provided more evidence for different functional connectivity of the orbitofrontal cortex in depression, as described next, with a summary of some of the differences shown in Fig. 3.27 (Rolls, Cheng and Feng, 2020b). **These differences of functional connectivity all relate to altered stability of the circuits involved, which help to account for some of the symptoms of depression.**

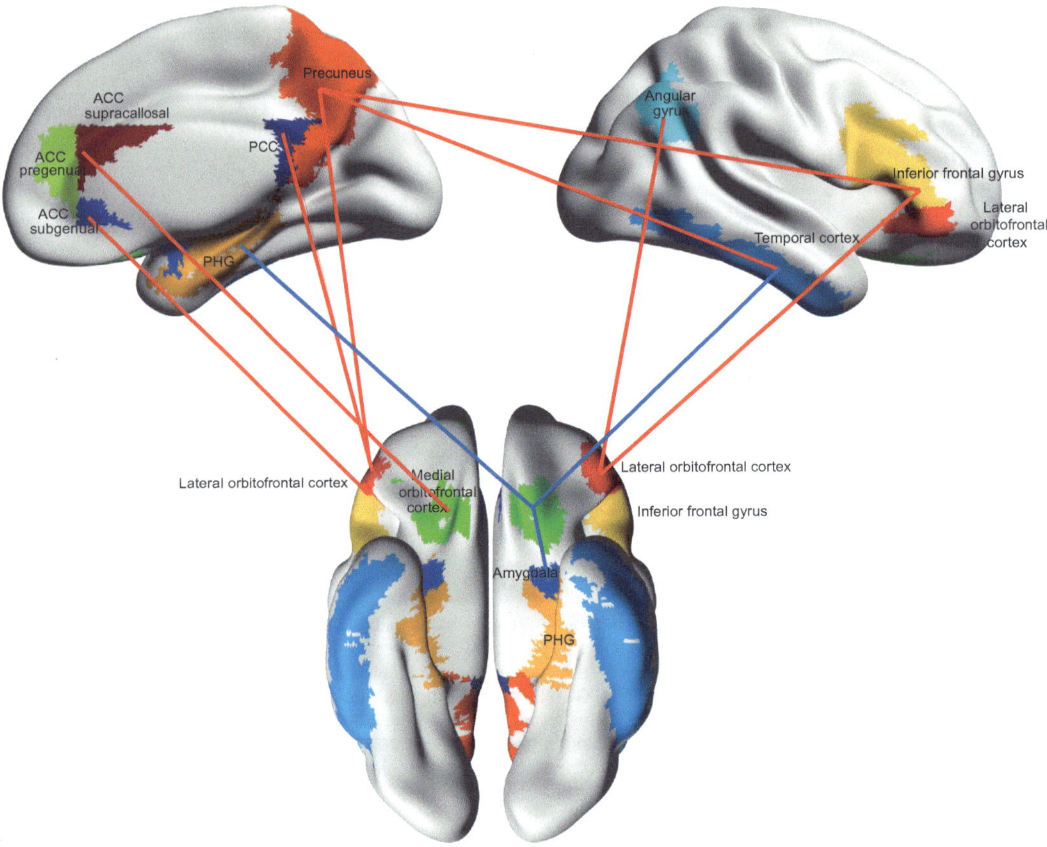

Fig. 3.27 Functional connectivity (FC) differences of the medial and lateral orbitofrontal cortex in major depressive disorder. Higher functional connectivity in depression is shown in red, and includes higher functional connectivity of the non-reward / punishment-related lateral orbitofrontal cortex with the precuneus, posterior cingulate cortex (PCC), subgenual anterior cingulate cortex (ACC subgenual), angular gyrus, and inferior frontal gyrus. Lower functional connectivity in depression is shown in blue, and includes lower functional connectivity of the medial orbitofrontal cortex with the parahippocampal gyrus memory system (PHG), amygdala, and temporal cortex. The part of the medial orbitofrontal cortex in which voxels were found with lower functional connectivity in depression is indicated in green. The areas apart from the medial orbitofrontal cortex shown are as defined in the automated anatomical labelling atlas (Rolls, Joliot and Tzourio-Mazoyer, 2015; Rolls, Huang, Lin, Feng and Joliot, 2020d), although the investigations that form the basis for the summary were at the voxel level. (Data from Rolls E T, Cheng W and Feng J. 2020 The orbitofrontal cortex: reward, emotion, and depression. Brain Communications 2: fcaa196.)

3.13.4.1 Precuneus: higher connectivity with the lateral orbitofrontal cortex

The precuneus is a medial parietal cortex region implicated in the sense of self and agency, autobiographical memory, spatial function, and navigation (Cavanna and Trimble, 2006; Freton et al., 2014). The retrosplenial cortex provides connections to and receives connections from the hippocampal system, connecting especially with the parahippocampal gyrus areas TF and TH, and with the subiculum (Bubb et al., 2017; Kobayashi and Amaral, 2007; Rolls et al., 2023g). The precuneus can be conceptualized as providing access to the hippocampus for spatial and related information from the parietal cortex (given the rich connections between the precuneus and parietal cortex) (Rolls, 2018b; Rolls et al., 2023g).

To further analyze the functioning of the precuneus in depression, we measured resting state functional connectivity in 282 patients with major depressive disorder and 254 controls (Cheng, Rolls, Qiu, Yang, Ruan, Wei, Zhao, Meng, Xie and Feng, 2018c). In 125 patients not receiving medication, voxels in the precuneus had significantly higher functional connectivity with the lateral orbitofrontal cortex (Fig. 3.27). In patients receiving medication, the functional connectivity between the lateral orbitofrontal cortex and precuneus was decreased back towards that in the controls (Cheng et al., 2018c).

These findings support the theory that the non-reward system in the lateral orbitofrontal cortex has increased effects on areas in which the self is represented including the precuneus, which could relate to the low self-esteem in depressed patients (Rolls, 2016d, 2018b; Rolls et al., 2020b).

3.13.4.2 Parahippocampal gyrus / medial temporal lobe memory system, and temporal lobe visual cortex: lower connectivity with the medial orbitofrontal cortex

We discovered that voxels in the medial orbitofrontal cortex had lower functional connectivity with the parahippocampal gyrus / medial temporal lobe memory system (Cheng et al., 2016) (Fig. 3.26), and interpreted this as resulting in fewer happy memories being recalled, as the medial orbitofrontal cortex has activations that correlate with subjective pleasantness, as described above, and the parahippocampal gyrus is a pathway in the hippocampal episodic memory system (Kesner and Rolls, 2015; Rolls, 2016b, 2018a; Rolls and Wirth, 2018; Rolls et al., 2023b, 2022a; Rolls, 2023a, 2025a, 2026c). The reduced connectivity with temporal cortex areas in which objects and faces are represented was interpreted as contributing to the reduced positive valuation of signals involved in emotion such as the sight of face expressions, and of people (Hasselmo et al., 1989a; Critchley et al., 2000).

In a further analysis, which also investigated the effects of antidepressant medication (Rolls, Cheng, Du, Wei, Qiu, Dai, Zhou, Xie and Feng, 2020a), medial orbitofrontal cortex voxels had lower functional connectivity with temporal cortex areas, the parahippocampal gyrus, fusiform gyrus, and supplementary motor area, and medication did not result in these functional connectivities being closer to controls. This is consistent with the anhedonia of depression and reduced happy memories being related to these low functional connectivities of the medial orbitofrontal cortex with temporal lobe and memory systems.

What is especially interesting is that these low functional connectivities are not normalized by treatment with typical antidepressant drugs (Rolls et al., 2020a), suggesting that one goal of future treatment for depression might be to increase the functionality of the medial orbitofrontal cortex (Rolls et al., 2020b).

3.13.4.3 Posterior cingulate cortex: higher functional connectivity with the lateral orbitofrontal cortex in depression

The posterior cingulate cortex is a region with strong connectivity in primates with the entorhinal cortex and parahippocampal gyrus (areas TF and TH), and thus with the hippocampal memory system (Vogt, 2009; Bubb et al., 2017; Rolls and Wirth, 2018; Rolls, 2019e; Rolls et al., 2023g). The posterior cingulate cortex also has connections with the orbitofrontal cortex (Vogt and Pandya, 1987; Vogt, 2009; Hsu, Rolls, Huang, Chong, Lo, Feng and Lin, 2020; Rolls, Deco, Huang and Feng, 2022b; Rolls, Wirth, Deco, Huang and Feng, 2023g), and the posterior cingulate cortex has high functional connectivity with the parahippocampal regions that are involved in memory (Cheng et al., 2018b; Rolls, 2019e; Rolls et al., 2023g). The posterior cingulate region (including the retrosplenial cortex) is consistently engaged by a range of tasks that examine episodic memory including autobiographical memory, and imagining the future; and also spatial navigation and scene processing (Leech and Sharp, 2014; Auger and Maguire, 2013). Self-reflection and self-imagery activate the ventral part of the posterior cingulate cortex (vPCC, the part with which we will be mainly concerned here) (Kircher et al., 2002, 2000; Johnson et al., 2002; Sugiura et al., 2005; Rolls, 2019e; Rolls et al., 2023g).

To analyze the functioning of the posterior cingulate cortex in depression, we performed a fully voxel-level resting-state functional connectivity neuroimaging analysis of depression of the posterior cingulate cortex, with 336 patients with major depressive disorder and 350 controls (Cheng, Rolls, Qiu, Xie, Wei, Huang, Yang, Tsai, Li, Meng, Lin, Xie and Feng, 2018b). We discovered that in depression, the posterior cingulate cortex had significantly higher functional connectivity with the lateral orbitofrontal cortex (Fig. 3.27). In patients receiving medication, the functional connectivity between the lateral orbitofrontal cortex and the posterior cingulate cortex was decreased back towards that in the controls.

These findings are consistent with the hypothesis that the non-reward system in the lateral orbitofrontal cortex has increased effects on memory systems, which contribute to the rumination about sad memories and events in depression (Cheng et al., 2018b).

3.13.4.4 Anterior cingulate cortex: higher connectivity with the orbitofrontal cortex in depression

The orbitofrontal cortex projects to the anterior cingulate cortex (Vogt, 2009; Bubb et al., 2017; Rolls, 2019e; Vogt, 2019; Rolls et al., 2022b; Zhang et al., 2024a). As shown in Chapter 12 of Rolls (2023c), the supracallosal anterior cingulate cortex is activated by many aversive stimuli, and has strong connectivity with the lateral orbitofrontal cortex (Rolls, Cheng, Gong, Qiu, Zhou, Zhang, Lv, Ruan, Wei, Cheng, Meng, Xie and Feng, 2019; Hsu, Rolls, Huang, Chong, Lo, Feng and Lin, 2020; Du, Rolls, Cheng, Li, Gong, Qiu and Feng, 2020; Rolls, Deco, Huang and Feng, 2022b). Moreover, the dorsal or supracallosal anterior cingulate cortex appears to be involved in learning actions to obtain rewards (action-outcome learning), where the outcome refers to the reward or punisher for which an action is being learned (Rolls, 2019e).

The pregenual anterior cingulate cortex is activated by many pleasant, rewarding, stimuli, and has strong functional connectivity with the medial orbitofrontal cortex (Rolls et al., 2019; Hsu et al., 2020; Du et al., 2020; Rolls et al., 2022b).

The medial orbitofrontal cortex is implicated in reward-related processing and learning, and the lateral orbitofrontal cortex in non-reward and punishment-related processing and learning. These types of learning involve stimulus-stimulus associations, where the second stimulus is a reward (or its omission), or a punisher (Rolls, 2019a,c). Now given that emotions can be considered as states elicited by rewarding and punishing stimuli, and that moods such as depression can arise from prolonged non-reward or punishment (see above), the part of the brain that processes these stimulus-reward associations, the orbitofrontal cortex, is more likely to be involved in depression than the action-related parts of the cingulate cortex. However, as an output region of the orbitofrontal cortex, the anterior cingulate cortex and other regions related to action such as the right inferior frontal gyrus might contribute to depression (Rolls et al., 2022b, 2023d). Given this background, we decided to investigate the connectivity of the anterior cingulate cortex in depression.

We found that in depression, overall the anterior cingulate cortex (ACC) had significantly higher functional connectivity with the orbitofrontal cortex, inferior frontal gyrus, superior parietal lobule, and with early visual cortical regions (Fig. 3.27) (Rolls, Cheng, Gong, Qiu, Zhou, Zhang, Lv, Ruan, Wei, Cheng, Meng, Xie and Feng, 2019). A pregenual ACC subdivision had high functional connectivity with medial orbitofrontal cortex areas, and a supracallosal ACC subdivision had high functional connectivity with the lateral orbitofrontal cortex and inferior frontal gyrus. The high FC in depression between the lateral orbitofrontal cortex and the subcallosal (subgenual) parts of the ACC may provide a mechanism for more non-reward information transmission to the ACC, contributing to depression. The high functional connectivity between the medial orbitofrontal cortex and supracallosal ACC in depression may also contribute to depressive symptoms. These higher functional connectivities in unmedicated patients were ameliorated by treatment with antidepressants (Rolls et al., 2019).

These findings are consistent with the hypothesis that the anterior cingulate cortex, with its representations of rewards and punishers received from the orbitofrontal cortex and which is involved in interfacing these to action systems (Rolls, 2019e; Rolls et al., 2022b; Wan et al., 2020; Rolls et al., 2022d), may relate to depression by a disconnection of rewards and punishers from their action-related and other outputs. The hypothesis is that this disconnection contributes to the insensitivity to the effects of rewards and punishers, amotivational states, and feelings of helplessness than can be present in depression (Rolls, Cheng, Gong, Qiu, Zhou, Zhang, Lv, Ruan, Wei, Cheng, Meng, Xie and Feng, 2019; Rolls, Cheng and Feng, 2020b; Zhang, Rolls, Wang, Xie, Cheng and Feng, 2024a).

3.13.4.5 Inferior frontal gyrus: increased connectivity with the lateral orbitofrontal cortex in depression

The lateral orbitofrontal cortex projects to the inferior frontal gyrus (Hsu et al., 2020; Rolls et al., 2022b, 2023d), and very interestingly higher functional connectivity was found in depression of voxels in the right inferior frontal gyrus with voxels in the lateral and medial orbitofrontal cortex, cingulate cortex, inferior and middle temporal gyrus and temporal pole, the angular gyrus, precuneus, hippocampus and mid- and superior frontal gyrus connectivity with the orbitofrontal cortex, temporal lobe areas, the parahippocampal gyrus and hippocampus, and motor areas (Fig. 3.27) (Rolls, Cheng, Du, Wei, Qiu, Dai, Zhou, Xie and Feng, 2020a). In medicated patients these functional connectivities of the inferior frontal gyrus were lower and towards those in controls.

The inferior frontal gyrus regions with high connectivity in depression in this investigation (Rolls et al., 2020a) include in the HCP-MMP atlas (Glasser et al., 2016a; Huang et al., 2022) parts of the lateral orbitofrontal cortex, including region 47l which has major connectivity with Broca's area regions 44 and 45 (Rolls et al., 2022c). This provides further evidence for the concept that language-related rumination of sad thoughts involves the lateral orbitofrontal cortex and language areas such as 47l in the inferior frontal gyrus (Rolls, 2016d).

We hypothesize (Rolls et al., 2020a,b) that one way in which the orbitofrontal cortex influences behavior in depression is via the right inferior frontal gyrus, which projects in turn to premotor cortical areas (Du, Rolls, Cheng, Li, Gong, Qiu and Feng, 2020; Rolls, Deco, Huang and Feng, 2023c). Consistent with the consequent hypothesis that the inferior frontal gyrus route may allow non-reward signals to have too great an effect to inhibit behavior in depression, lesions of the right inferior frontal gyrus impair stopping in the stop-signal task, and produce impulsiveness (Aron, Robbins and Poldrack, 2014; Dalley and Robbins, 2017). Also consistent with the hypothesis, successful stopping in the stop-signal task is associated with high activation of the inferior frontal gyrus and lateral orbitofrontal cortex (Deng, Rolls et al (2017)).

3.13.4.6 Amygdala: reduced connectivity with the orbitofrontal cortex in depression

The amygdala is involved in emotion, though as shown above, it may be overshadowed in humans by the orbitofrontal cortex (LeDoux and Pine, 2016; Rolls, 2019c; Rolls et al., 2023d). In a large scale study of depression, amygdala voxels had decreased functional connectivity with the medial orbitofrontal cortex (involved in reward); the lateral orbitofrontal cortex (involved in non-reward and punishment); temporal lobe areas (involved in visual and auditory perception including face expression analysis (Perrett, Rolls and Caan, 1982; Rolls, 2011b, 2012b; Leonard, Rolls, Wilson and Baylis, 1985); and the parahippocampal gyrus (involved in memory) (Fig. 3.27) (Cheng, Rolls, Qiu, Xie, Lyu, Li, Huang, Yang, Tsai, Lyu, Zhuang, Lin, Xie and Feng, 2018a). This disconnectivity of the amygdala may contribute to particularly some of the changed behavioral and autonomic responses in depression, and perhaps not to the depressed subjective feelings (Rolls et al., 2023d).

3.13.4.7 Sleep, depression, and increased lateral orbitofrontal cortex connectivity

Sleep is frequently impaired in depression (Becker et al., 2017). To advance understanding of the brain regions involved in sleep and depression, the relation between functional connectivity, depressive symptoms (the Adult Self-Report Depressive Problems scores), and poor sleep quality was measured in 1017 participants from the general population in the Human Connectome Project (Cheng, Rolls, Ruan and Feng, 2018d). The brain areas with increased functional connectivity of these common links related to both sleep and depressive scores included the lateral orbitofrontal cortex; the dorsolateral prefrontal cortex; the anterior and posterior cingulate cortex; the insula; the parahippocampal gyrus and hippocampus; the amygdala; the temporal cortex; and the precuneus. A mediation analysis showed that these functional connectivities in the brain contribute to the relation between depression and poor sleep quality.

Evidence was also found in this general population that the Depressive Problems scores were correlated with functional connectivities between areas that included the lateral orbitofrontal cortex, cingulate cortex, precuneus, angular gyrus, and temporal cortex (Cheng, Rolls, Ruan and Feng, 2018d). Part of the importance of this is that it provides strong support for a role of the lateral orbitofrontal cortex in depression in a general population in the U.S.A. in which a tendency to have depressive problems could be assessed. This cross-validation in a completely different population and in people not selected to have depression (Cheng et al., 2018d) provides support for the theory that the lateral orbitofrontal cortex is a key brain area that might be targeted in the search for treatments for depression (Rolls, 2016d). (We provided further validation in a European cohort of 1,877 adolescents (Xie, Jia, Rolls and al., 2021a).) Low sleep duration and high depressive scores are also related to lower cortical area of the orbitofrontal cortex in a sample of 11,067 participants (Cheng, Rolls, Gong, Du, Zhang, Zhang, Li and Feng, 2021).

3.13.4.8 Effective connectivity in depression

Effective connectivity measures the effect of one brain region on another in a particular direction, and can in principle therefore provide information related to the causal processes that operate in brain function, that is, how one brain region influences another.

We found in a resting state fMRI investigation that effective connectivity directed to the medial orbitofrontal cortex from areas including the parahippocampal gyrus, temporal pole, inferior temporal gyrus, and amygdala was decreased in depression (Rolls, Cheng, Gilson, Qiu, Hu, Ruan, Li, Huang, Yang, Tsai, Zhang, Zhuang, Lin, Deco, Xie and Feng, 2018a). This implies less strong positive driving influences of these input regions on the medial and middle orbitofrontal cortex, regions implicated in reward, and thus helps to elucidate part of the decreased feelings of happy states in depression (Rolls, 2016d).

The links from temporal cortical areas to the precuneus were increased in depression, and this may relate to representations of the sense of self (Cavanna and Trimble, 2006), which become more negative in depression (Rolls, 2016d; Cheng et al., 2016).

The lateral orbitofrontal cortex, implicated in non-reward and punishment, had an increased level of activity as reflected in the analysis in the depressed group.

In addition, activity in the analysis was also higher in the right and left hippocampus of patients with depression, implying heightened memory-related processing (Rolls et al., 2018a).

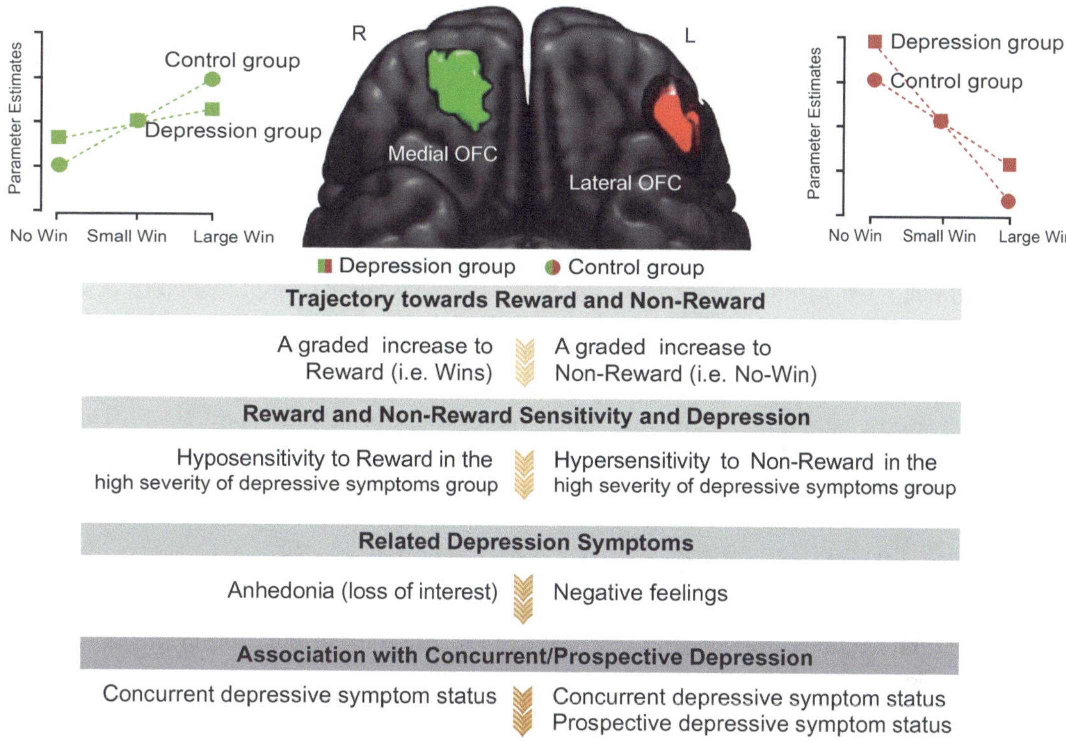

Fig. 3.28 Reduced sensitivity to reward of the medial orbitofrontal cortex, and increased sensitivity of the lateral orbitofrontal cortex to non-reward (No Win) in individuals with high scores on the Adolescent Depression Rating Scale (squares) compared to a control group (circles) in the Monetary Incentive Delay task. The summary is based on findings in 1140 adolescents at age 19 and 1877 at age 14. L, left; R, right. (Data from Xie C, Jia T, Rolls E T et al 2021 Reward versus Nonreward sensitivity of the Medial versus Lateral Orbitofrontal Cortex relates to the severity of depressive symptoms. Biological Psychiatry: Cognitive Neuroscience and Neuroimaging 6: 259-269.)

3.13.5 Increased activations to non-reward of the lateral orbitofrontal cortex, and decreased sensitivity to reward of the medial orbitofrontal cortex, are related to depression scores

I decided next to take the analysis forwards by investigating whether *activations* of the orbitofrontal cortex by rewarding and non-rewarding stimuli are different in people with depressive symptoms, as my theory of depression postulates (Rolls, 2016d).

In 1140 adolescents at age 19 and 1877 at age 14 in a monetary incentive delay task, we found that the medial orbitofrontal cortex had graded increases in activation as the reward (Win) value increased (Xie, Jia, Rolls et al (2021b)). The lateral orbitofrontal cortex had graded increases of activation as the reward value dropped to zero (the No-Win condition) (Fig. 3.28).

In a subgroup with a high score on the Adolescent Depression Rating Scale at age 19 and 14, the medial orbitofrontal cortex activations had reduced sensitivity to the different reward conditions; and the lateral orbitofrontal cortex activation showed high activation to the No-Win (i.e. Non-reward) condition (Xie et al., 2021b) (Fig. 3.28). These findings provide support for the hypothesis that those with symptoms of depression have increased sensitivity to non-reward in the lateral orbitofrontal cortex, and decreased sensitivity for differences in reward of the medial orbitofrontal cortex. Moreover, these differences are evident at an age as early as 14 years old (Xie et al., 2021b). This increase in Non-reward sensitivity of the lateral orbitofrontal cortex in depression, and decreased

Reward sensitivity of the medial orbitofrontal cortex, may act together with the altered functional connectivity of these regions just described, to make some individuals susceptible to depression (Rolls et al., 2020b; Zhang et al., 2024a).

I hypothesize that as part of the process of evolution, variation of the sensitivity of individuals to specific types of Reward and Non-Reward may be present (Rolls, 2014a). Individuals with high sensitivity to Non-Reward may be susceptible to depression, and individuals with low sensitivity to Non-Reward may be impulsive because they are little affected by non-reward (Rolls, 2014a, 2018b). Individuals with high sensitivity to Reward may be sensation-seekers (Wan, Rolls, Cheng and Feng, 2020) and risk-takers (Rolls, Wan, Cheng and Feng, 2022d) (with increased functional connectivity of the medial orbitofrontal cortex with for example the anterior cingulate cortex, and for that reason also impulsive (Wan, Rolls, Cheng and Feng, 2020; Rolls, Wan, Cheng and Feng, 2022d)); and individuals with low sensitivity to Reward may have reduced goal-seeking behavior and reduced motivation (Rolls, 2014a, 2018b; Zhang, Rolls, Wang, Xie, Cheng and Feng, 2024a). These types of natural variation may be important foundations for different types of personality (Rolls, 2014a, 2012d, 2023b).

3.13.6 Implications, and possible treatments for depression

The differences of functional connectivity described in this section (3.13) all relate to altered stability of the circuits involved, which help to account for some of the symptoms of depression. I now consider some of the implications of the discoveries described above (Rolls, Cheng and Feng, 2020b; Zhang, Rolls, Wang, Xie, Cheng and Feng, 2024a).

One implication is that the increased functional connectivity of the lateral orbitofrontal cortex with several cortical regions in depression described above is likely for example to make the activity in these non-reward related circuits more stable and hence prolonged, contributing for example to increased sad rumination in depression.

Another implication is that because the lateral orbitofrontal cortex responds to many punishing and non-rewarding stimuli (Grabenhorst and Rolls, 2011; Rolls, 2014a,b, 2019c; Zhang et al., 2024a; Rolls, 2023b) that are likely to elicit autonomic/visceral responses, as does the supracallosal anterior cingulate cortex, and in view of connections from these areas to the anterior insula which is implicated in autonomic/visceral function (Critchley and Harrison, 2013; Rolls, 2016c), the anterior insula would also be expected to be overactive in depression, which it is (Drevets, 2007; Hamilton et al., 2013; Ma, 2015).

A notable discovery is that ketamine, a N-methyl-D-aspartate (NMDA) receptor antagonist, in subanaesthetic doses, produces rapid (within hours) antidepressant responses in patients who are resistant to typical antidepressants, and that the effects may last for two weeks or longer (Zanos and Gould, 2018; Iadarola et al., 2015; Maltbie et al., 2017; Zanos et al., 2016; Wilkinson et al., 2018; Rhee et al., 2022; Johnston et al., 2024). Clinically, ketamine may be useful with a single dose, or doses may be repeated (Rhee et al., 2022). The short-term effects of ketamine include blocking excitatory NMDA receptors on cortical pyramidal cells which reduces the excitatory effect produced by the excitatory transmitter glutamate; and blocking excitatory receptors on GABA inhibitory neurons, which will tend to decrease GABAergic neuron firing, resulting in a potential increase in pyramidal cell firing (Johnston et al., 2024). However, ketamine produces further effects, such as inducing synaptogenesis on excitatory neurons, increased glutamate transmission, reversing the synaptic deficits caused by chronic stress, and effects mediated by a ketamine metabolite hydroxynorketamine (Duman and Aghajanian, 2012; Ghasemi et al., 2017; Zorumski et al., 2016; Abdallah et al., 2016; Aleksandrova et al., 2017; Zanos and Gould, 2018; Zhang et al., 2024a). Another way in which ketamine may be effective in depression is by reducing inflammatory processes, which are sometimes related to depression (Ghasemi et al., 2017).

Ketamine may act in part by quashing the attractor state in the lateral orbitofrontal cortex at least for a number of days. Evidence consistent with this is that the activity (metabolism) of the inferior frontal gyrus and orbitofrontal cortex was decreased by a single dose of ketamine, which reduced the anhedonia of depression (see further below) (Lally et al., 2015). This NMDA receptor blocker may act at least in part by decreasing the high firing rate state of attractor networks by reducing transmission in the recurrent collateral excitatory connections between the neurons (Rolls, 2016b; Rolls, Loh, Deco and Winterer, 2008d; Rolls and Deco, 2010; Rolls, 2012a; Deco, Rolls, Albantakis and Romo, 2013; Rolls and Deco, 2015a). Treatment with conventional antidepressant drugs (e.g. SSRIs) decreases the activity and connectivity of this lateral orbitofrontal cortex system (Ma, 2015; Rolls et al., 2020b; Rolls, 2018b; Rolls et al., 2020b). Electroconvulsive therapy may have antidepressant effects, and might knock the non-reward system out of its attractor state.

Given that a ketamine metabolite, hydroxynorketamine, may be related to the antidepressant effects of ketamine and may act via facilitating effects mediated by AMPA receptors (Zanos et al., 2016), the effects of ketamine might be mediated by increasing the activity of the medial orbitofrontal cortex reward-related system (which tends to be reciprocally inversely related to the lateral orbitofrontal cortex non-reward system), or the functional connectivity of the medial orbitofrontal cortex reward system with the hippocampal system which is reduced in depression (Fig. 3.26). There is now some evidence for an action of ketamine to increase activity or connectivity in the medial orbitofrontal cortex, as follows (Zhang, Rolls, Wang, Xie, Cheng and Feng, 2024a).

A pharmacological investigation tested 10 treatment resistant depression (TRD) patients with a monetary incentive delay task before and after ketamine infusions. The study showed that ketamine administration induced a sustained increase in medial orbitofrontal cortex and nucleus accumbens activation during both the anticipation of positive reward cues and during the receipt of positive feedback (the outcome phase). The enhanced reactivity of the medial orbitofrontal cortex and nucleus accumbens was accompanied by decreased depression symptoms and better behavioral performance to positive items (Sterpenich et al., 2019). Consistent with this, in one of the first clinical studies to investigate the anti-anhedonic mechanisms of ketamine in depression, 52 TRD patients received a single ketamine infusion. The primary outcome indicated that ketamine rapidly reduced levels of anhedonia, with a substantial effect within 40 minutes which lasted up to 3 days post-infusion (Lally et al., 2015). Importantly, in a subgroup of patients who received PET scans, the single ketamine infusion decreased glucose metabolism within medial orbitofrontal cortex regions, and the reduced magnitude of the orbitofrontal cortex metabolism was positively correlated with the changes in the anhedonia symptoms (i.e., changes in SHAPS scores) (Lally et al., 2015). In addition, a double-blind, placebo-controlled, crossover trial in TRD and healthy controls also showed that ketamine infusion increased the functional connectivity between the right medial orbitofrontal cortex and ventral rostral putamen for the TRD patients toward the levels observed in healthy controls, and the increased magnitude of orbitofrontal cortex-putamen connectivity was correlated with improved levels of anhedonia symptoms (i.e., reductions in SHAPS scores) (Mkrtchian, Evans, Kraus, Yuan, Kadriu, Nugent, Roiser and Zarate, 2021). Taken together, these findings provide evidence that ketamine has acute and chronic anti-anhedonic effects in major depressive disorder by acting in part to restore activity and sensitivity to reward, and increasing functional connectivity, of the medial orbitofrontal cortex (Zhang, Rolls, Wang, Xie, Cheng and Feng, 2024a).

A very interesting possibility arises. Ketamine may act at least in part to increase sensitivity to reward and connectivity of the medial orbitofrontal cortex, and thereby ameliorate the anhedonia and motivational symptoms of depression (Zhang, Rolls, Wang, Xie, Cheng and Feng, 2024a). Conventional antidepressants such as the selective serotonin reuptake inhibitors may act in part on the non-reward-related lateral orbitofrontal cortex to decrease its elevated sensitivity to non-reward and connectivity in depression, and thereby treat the symptoms of sadness related to non-reward over-efficacy (Rolls et al., 2020b; Zhang et al., 2024a). It is suggested that low doses of both types of treatment would be useful to explore.

A key issue here is that we find in the studies described above that the elevated functional connectivity of the lateral orbitofrontal cortex in depression is treated by selective serotonin reuptake inhibitors, but the reduced connectivity of the medial orbitofrontal cortex is not. New treatments are needed that increase the functional connectivity and reward sensitivity of the medial orbitofrontal cortex. There should be a search for such treatments. Ketamine, might work in this way. We do not know if psilocybin, a serotonergic psychedelic which acts as a 5-HT(2) receptor agonist, has this action, though it may be useful in the treatment of depression (Nutt et al., 2024; Wingert et al., 2025).

Deep brain stimulation of the orbitofrontal cortex may be useful in the treatment of mood disorders and depression. The macaque orbitofrontal cortex is a key brain site at which deep brain electrical stimulation is rewarding (Rolls et al., 1980c; Rolls, 2005, 2019c) (see Chapter 2). Electrical stimulation of the human orbitofrontal cortex can also produce reward and raise mood (Rao et al., 2018), and many of the sites were in the middle part of the orbitofrontal cortex, areas 13 and 11, which are categorized as medial orbitofrontal cortex, the area activated by rewards (Rolls, 2019c). It is likely that these medial orbitofrontal cortex sites will produce better reward in humans than stimulation in the lateral orbitofrontal cortex BA12/47, for these lateral sites are activated by unpleasant stimuli and by not obtaining expected rewards. The medial (/ middle) orbitofrontal cortex may for the reasons described here and elsewhere (Rolls, 2019c) be a key area of interest for deep brain stimulation to help relieve depression (Rolls et al., 2020b).

The anterior cingulate cortex, including the subcallosal cingulate cortex, is a key brain region to which the orbitofrontal cortex projects (Hsu et al., 2020; Du et al., 2020; Rolls et al., 2019; Rolls, 2019c; Rolls et al., 2022b). It is possible that brain stimulation of the subcallosal cingulate cortex might be useful in the treatment of at least some patients with depression (Riva-Posse et al., 2018; Holtzheimer et al., 2017; Dunlop et al., 2017; Lujan et al., 2013; Johansen-Berg et al., 2008), and it is possible that the subcallosal cingulate stimulation affects pathways that connect with the orbitofrontal cortex (Riva-Posse et al., 2018; Dunlop et al., 2017; Lujan et al., 2013; Johansen-Berg et al., 2008; Elias et al., 2022b,a). Given that the anterior cingulate cortex is an output region of the orbitofrontal cortex, it may be that treatments of the orbitofrontal cortex, where emotion-related processing is implemented, may be a better target for potential treatments for depression (Rolls et al., 2020b).

This non-reward / punishment attractor network sensitivity theory of depression has implications for treatments. These implications can be understood and further explored in the context of investigations of the factors that influence the stability of attractor neuronal networks with integrate-and-fire neurons with noise introduced by the close to Poisson spiking times of the neurons (Wang, 2002; Rolls, 2016b; Deco et al., 2009; Rolls and Deco, 2010; Deco et al., 2013; Loh et al., 2007a; Rolls and Deco, 2015a; Rolls, 2016b, 2023c, 2026d).

One implication is that antianxiety drugs, by increasing inhibition, might reduce the stability of the high firing rate state of the non-reward attractor, thus acting to quash the depression-related attractor state.

A second implication is that it might be possible to produce agents that decrease the efficacy of NMDA receptors in the lateral orbitofrontal cortex, thereby reducing the stability of the depression-related attractor state. The evidence that there are genes that are selective for NMDA receptors for the neurons in different populations is that there are separate knock-outs for NMDA receptors in the CA3 and CA1 regions of the hippocampus (Nakazawa et al., 2002; Tonegawa et al., 2003; Nakazawa et al., 2003, 2004). The present theory suggests that searching for ways to influence the attractor networks in the lateral orbitofrontal cortex by decreasing excitatory or increasing inhibitory transmission in this region may be of considerable interest. It should be noted that the present theory is a theory specifically of non-reward and punishment-related attractor networks in the lateral orbitofrontal cortex and related areas in relation to depression, and of reduced efficacy of the medial orbitofrontal cortex, and that alterations of attractor networks in other cortical areas may be related to other psychiatric disorders (Rolls, 2012a, 2021f, 2024b).

In terms of the implications of the attractor-based aspect of Rolls' theory of depression, an important point is that the attractor dynamics must be kept stable in the face of the randomness or noise introduced into the system by the almost Poisson firing times of neurons for a given mean firing rate. Moreover, the spontaneous firing rate state of the non-reward attractor must be maintained stable when no non-reward inputs are present (or otherwise the non-reward attractor would jump into a high firing rate non-reward state for no external reason, contributing to depression). The inhibitory transmitter GABA may be important in maintaining this type of stability (Rolls and Deco, 2010). Moreover, the high firing rate state produced by non-reward must not reach too high a firing rate, as this would cause overstability of the non-reward / depression state. In a complementary way, if the high firing rate attractor state is insufficiently high, then that attractor state might be unstable, and the individual might be relatively insensitive to non-reward, not depressed, and impulsive because of not responding sufficiently to non-reward or punishment. The excitatory transmitter glutamate acting at NMDA or AMPA receptors may be important in setting the stability of the high firing rate attractor state. In this respect and in this sense, the tendency to become depressed or to be impulsive may be reciprocally related to each other. Predictions for treatments follow from understanding these noisy attractor-based dynamics (Rolls and Deco, 2010; Rolls, 2012a, 2021f, 2024b).

Strengths of Rolls' theory of depression are that it has its foundations in an understanding of the brain systems involved in normal emotions and mood states (Rolls, 2014a, 2018b, 2019a; Rolls et al., 2020b; Rolls, 2023b; Zhang et al., 2024a); and that it incorporates concepts of the stability of attractor networks in the cortex, which are a key feature of cortical design (Rolls and Deco, 2010; Rolls, 2016b, 2023c, 2026d).

3.14 Evaluation

At the time that my research described in this Chapter started, almost nothing was known about reward neurons in the brain, about the functions of the orbitofrontal cortex, and about the brain mechanisms that relate to depression. Some of the key areas in which the discoveries described in this Chapter were made are the following, relate especially to the primate including human brain, and are therefore very relevant to understanding human brain function in health and disease:

1. We discovered reward value neurons in the lateral hypothalamus and orbitofrontal cortex, using criteria such as devaluation and rapid learning and reversal of stimulus-reward associations. We showed that the representations were of stimulus reward value, and were not related to behavioral responses or actions such as reaching with the arm to obtain a reward.
2. We discovered reward-specific (sensory-specific) satiety, and showed that it is represented in the orbitofrontal cortex. We produced a theory of how it is computed, pre-synaptic adaptation.
3. We discovered non-reward neurons in the orbitofrontal cortex.

4. We mapped the visual and taste pathways through their hierarchies of processing in primates including humans, and discovered that reward value is represented in the orbitofrontal cortex but not at earlier stages such as the inferior temporal visual cortex and insular primary taste cortex.
5. We discovered that oral texture, including fat texture, and viscosity, are represented in the insular primary taste cortex and a region to which we proved it projects, the orbitofrontal cortex.
6. We discovered face expression and face identity neurons in the orbitofrontal cortex, and related these to the functions of the orbitofrontal cortex in social behavior.
7. We extended these discoveries of reward and non-reward systems to humans using fMRI, and, importantly in relation to emotion, showed that the activations of the medial orbitofrontal cortex are linearly related to subjectively experienced pleasure.
8. We discovered that there is topology within the human orbitofrontal cortex, with many rewards, including monetary reward, represented in the medial orbitofrontal cortex, and many punishers, and non-reward, represented in the lateral orbitofrontal cortex.
9. We discovered that cognition and attention have top-down effects that modulate reward value representations in the human orbitofrontal cortex.
10. We showed how the connectivity of the human orbitofrontal cortex relates to these functions, with especially interestingly, major connectivity in part via the anterior and posterior cingulate cortex, to the hippocampal episodic memory system. This not only provides a route for reward value and affective components to be incorporated into episodic memory, but also indicates how the orbitofrontal cortex may play a role in memory consolidation. We also showed how the human orbitofrontal cortex connects to different parts of the anterior cingulate cortex, providing a route for the orbitofrontal cortex to provide reward outcome information for action-outcome learning in the anterior cingulate cortex.
11. We discovered how the emotional changes associated with damage to the human orbitofrontal cortex can be related to the rapid reward-related reversal leaning that it implements, and to its functions in representing rewards and punishers, including socially relevant reinforcers such as face and voice expression. Part of the importance of these investigations is that they provide evidence that the human orbitofrontal cortex is causally involved in the functions described.
12. We showed that the amygdala in primates including humans may be less important in emotion than the orbitofrontal cortex, with the amygdala having less clear reward representations, less rapid reward reversal learning, and less connectivity in humans back to neocortical regions. In contrast, the lateral orbitofrontal has effective connectivity with cortical language regions in Broca's area, which provide a route for emotional states to become reported.
13. These discoveries provide key foundations for Rolls' theories of emotion and motivation (Rolls, 2014a, 2023b, 2025e).
14. Part of these theories is that in humans and perhaps some other related animals there are two main routes to action, an emotion-related route with the goals set by our genes, and a rational, reasoning route to action that can act in the interests of individuals.
15. We developed a theory of depression based on these discoveries about the functions of the lateral orbitofrontal cortex in non-reward, and of the medial orbitofrontal cortex in reward. The theory illustrates how it is possible to build an understanding of brain function in health and disease by combining research from different disciplines, including neurophysiology, connectivity, neuroimaging including pioneering studies of functional connectivity at the voxel level in depressed people vs controls, task-related functional neuroimaging, neuropsychology, and computational neuroscience.

16. We developed computational neuronal network theories and models for how rule-based, one-trial, reward reversal learning (Deco and Rolls, 2005b), and the detection of non-reward (Rolls and Deco, 2016), could be computed in the orbitofrontal cortex.
17. We also utilized integrate-and-fire neuronal network models of decision-making to make predictions about the fMRI BOLD signals in decision-making brain regions, and performed fMRI investigations that found such signals in a region implicated in reward-related decision-making, the ventromedial prefrontal cortex (Rolls et al., 2010a,b). We also developed integrate-and-fire neuronal networks models in which the decision could be predicted (from the stochastic fluctuations) before the decision stimuli were applied (Rolls and Deco, 2011a), and how the confidence in a decision is reflected in the firing rate in the winning attractor of a decision-making network (Insabato et al., 2010). We developed these concepts about the importance of stochastic dynamics as a principle of brain function in the book **The Noisy Brain: Stochastic Dynamics as a Principle of Brain Function** (Rolls and Deco, 2010), with these concepts developed more and applied to different mental disorders (Rolls, 2026d).

These pioneering and multidisciplinary discoveries about reward systems in the primate including human brain are supported by the work of others:

For example, Wolfram Schultz and colleagues have recorded from neurons in the primate orbitofrontal cortex, and their results (Tremblay and Schultz, 1999; Schultz et al., 2000; Tremblay and Schultz, 2000; Cromwell et al., 2018; Pastor-Bernier et al., 2021) are consistent with our discoveries about the representation of value and expected value.

My discovery of sensory-specific satiety made when I saw the responses of hypothalamic neurons (Rolls, 1981b; Rolls, Murzi, Yaxley, Thorpe and Simpson, 1986) which led me to the first experiments confirming the importance of sensory-specific satiety in humans (Rolls, Rolls, Rowe and Sweeney, 1981a; Rolls, Rowe, Rolls, Kingston, Megson and Gunary, 1981b; Rolls, Rowe and Rolls, 1982a; Rolls, Rolls and Rowe, 1983; Rolls and Rolls, 1997; Rolls and de Waal, 1985; Rolls, Zhang and Feng, 2025b) has been extensively followed up by others (Rolls, 2016a; Hetherington, 1996; Cunningham, Roe, Keller and Rolls, 2023).

Our discovery that sensory-specific satiety (also termed reward-specific satiety) is implemented in the orbitofrontal cortex (Thorpe, Rolls and Maddison, 1983; Rolls, Sienkiewicz and Yaxley, 1989c; Critchley and Rolls, 1996a; Rolls, Critchley, Browning, Hernadi and Lenard, 1999a), has been confirmed by others (Pastor-Bernier, Stasiak and Schultz, 2021).

Our neuroeconomic discoveries that reward value including monetary reward value are represented in the primate including human orbitofrontal cortex have been interestingly extended to the economic tradeoff between the quality vs the amount of reward that is on offer (Padoa-Schioppa and Assad, 2006; Padoa-Schioppa, 2007; Padoa-Schioppa and Assad, 2008; Padoa-Schioppa, 2011; Padoa-Schioppa and Conen, 2017; Ballesta et al., 2020; Zhang et al., 2024b).

Our discoveries about the effects of damage to the human orbitofrontal cortex on reward-related learning and especially reversal, and on behavioral responses to emotional stimuli (Section 3.8), have been confirmed by others (Fellows and Farah, 2003, 2005; Heberlein, Padon, Gillihan, Farah and Fellows, 2008; Camille, Tsuchida and Fellows, 2011).

Much evidence has been described in this Chapter that the orbitofrontal cortex is involved in emotion, including in the subjective human experience of emotion. There is an interesting contrast here with the distinguished research of Joseph LeDoux, who despite a lifetime investigating the roles of the amygdala in emotion with much research on rodents (LeDoux et al., 1988; LeDoux, 1995, 2000), has more recently concluded that the amygdala has little to do with emotional feelings, but is involved in autonomic and some bodily responses to stimuli that can produce emotion (LeDoux,

2012; LeDoux et al., 2018; Taschereau-Dumouchel et al., 2022). He has not yet as far as I know proposed any alternative brain system to the amygdala that is involved in emotion, and in particular has not come to the view that the orbitofrontal cortex is the key brain processing system in emotion in primates including humans, despite the wealth of evidence now available (Rolls, 2023b; Rolls et al., 2023d; Rolls, 2023c, 2026d, 2022b; Rolls et al., 2020b; Rolls, 2019a, 2014a,b, 2026d).

An important implication of the research described here and elsewhere (Rolls, 2023c, 2026d, 2023b, 2014a, 2026d) is that it is the representation of the value of stimuli, or in neuroeconomic terms 'goods', that is represented in the primate including human orbitofrontal cortex. Behavioral actions are not strongly represented in the orbitofrontal cortex. That implication is supported by others (Padoa-Schioppa and Assad, 2006; Wallis and Miller, 2003; Grattan and Glimcher, 2014; Cai and Padoa-Schioppa, 2019), and there is even evidence that the cost of an action does not influence the representation of the reward value of goods in the orbitofrontal cortex (Cai and Padoa-Schioppa, 2019). Further, at the Orbitofrontal Cortex meeting in Paris in 2024, there was general agreement among those performing research in macaques that the reward value of stimuli is represented in the primate orbitofrontal cortex, and that actions or behavioral responses are not a key part of what is represented in the orbitofrontal cortex (Padoa-Schioppa and Assad, 2006, 2008; Padoa-Schioppa, 2011; Padoa-Schioppa and Conen, 2017; Wallis and Kennerley, 2011; Knudsen and Wallis, 2022; Rudebeck and Murray, 2014; Rudebeck et al., 2017; Murray and Rudebeck, 2018). Instead, we and others have argued that the anterior cingulate cortex is involved in action-outcome learning, with the reward outcomes reaching the anterior cingulate cortex from the orbitofrontal cortex (Rushworth et al., 2011, 2012; Rolls, 2023b; Rolls et al., 2022b; Rolls, 2019e).

This is important, for the rodent orbitofrontal cortex seems different, with representations of actions (Sharpe, Wikenheiser, Niv and Schoenbaum, 2015; Izquierdo, 2017). Much other evidence indicates that reward systems in rodents operate very differently to those in primates including humans, making rodents a poor model of reward value processing and emotion in primates including humans (see Section 29.4 of Rolls (2026d)).

The approach that I have taken to depression, to aim to understand it in terms of the functions of the brain, and especially therefore of the orbitofrontal cortex, in emotion (Section 3.13), is being followed up by us and by others (Zhang et al., 2024a; Rolls, 2018b; Rolls et al., 2020b; Rolls, 2016d, 2019a, 2026d). One important implication of the research is that we are suggesting a focus on the medial and lateral orbitofrontal cortex in understanding and potentially treating depression (Zhang et al., 2024a; Rolls, 2018b; Rolls et al., 2020b; Rolls, 2016d, 2019a), as this may provide an important complement to the research on the subcallosal anterior cingulate cortex in depression (Mayberg et al., 2005; Lozano et al., 2012; Lujan et al., 2013; Mayberg et al., 2016; Holtzheimer et al., 2017; Riva-Posse et al., 2018, 2020; Siddiqi et al., 2021; Zhu et al., 2021).

4 The discoveries of face and object cells with invariant representations in the inferior temporal visual cortex, and of how information is represented in the brain, together with a theory of face and object perception

4.1 Introduction

The neurons that we discovered in the orbitofrontal cortex and lateral hypothalamus that respond to visual stimuli associated with reward (Rolls et al., 1976, 1980c; Thorpe et al., 1979, 1983; Rolls, 2014a, 2023b,c, 2026d) (Chapter 3) led me to ask where this visual input comes from, and whether reward value is represented there or not. The inputs to the orbitofrontal cortex might come from for example the inferior temporal visual cortex at the end of the visual pathway for objects and faces (Fig. 3.2). I investigated this by performing recordings from single neurons in the inferior temporal visual cortex, as described in this Chapter, and made a whole series of discoveries that provide a foundation for understanding processing in this ventrolateral cortical visual pathway that represents 'What' object or face is being looked at.

Those empirical discoveries led me to ask how the face and object neurons that we discovered in the inferior temporal cortex could be built: what mechanisms could account for these remarkable neurons, which we discovered can respond selectively to faces and objects relatively independently of transforms, including location on the retina, size, and even for some neurons view. Given these remarkable properties, I developed a biologically plausible model for how neurons with transform invariant representations to faces and objects could be built in the ventrolateral visual cortical pathway to the inferior temporal visual cortex. This VisNet model remains one of the few biologically plausible theories and models of how transform-invariant representations can be built in the brain.

Before I started this research, I knew that anatomically there is a projection from the primary visual cortex V1 > V2 > V4 to the inferior temporal visual cortex (IT) (Seltzer and Pandya, 1978) in macaques. We knew that in V1, neurons could respond to bars or edges (Hubel and Wiesel, 1977). I knew that some neurons with visual responses had been recorded in the inferior temporal visual cortex that responded to visual stimuli such as hands, but with rather variable and poor responses under anaesthesia (Gross, Bender and Rocha-Miranda, 1969; Gross, Rocha-Miranda and Bender, 1972). I decided to record from inferior temporal cortex neurons in the behaving macaque, and we discovered very robust neuronal responses to visual stimuli (Rolls, Judge and Sanghera, 1977). I believe that we were one of the first groups to record from the inferior temporal visual cortex in the behaving animal, and I was pleased at how robust the responses were, as described below. The recordings were made in the anterior inferior temporal cortex of the macaque, area TE, for if we moved further posterior to area TEO, the neurons were more difficult to analyze during behavior, probably because the receptive fields were small, and therefore consistent and accurate visual fixation was needed to obtain repeatable responses. The area we recorded in, TE in the macaque, is probably equivalent to regions TE2p and TE1p in the human brain shown in Fig. 4.1, which are the last mainly unimodal visual cortical temporal lobe regions in humans (Rolls et al., 2023b,e).

First, to provide a foundation for understanding the cortical processing of visual inputs, I describe more recent discoveries about how the visual cortical pathways are organised in humans.

4.2 The visual cortical pathways in humans: Two 'What', two 'Where'

The discoveries about how the visual cortex pathways are organised in humans are based on the combination I introduced of using the Human Connectome Project Multimodal Parcellation of the human brain (Glasser, Coalson, Robinson, Hacker, Harwell, Yacoub, Ugurbil, Andersson, Beckmann, Jenkinson, Smith and Van Essen, 2016a) with three methods for analyzing connectivity: effective connectivity, complemented by diffusion tractography and functional connectivity (see Section 6.1) (Huang, Rolls, Hsu, Feng and Lin, 2021; Ma, Rolls, Huang, Cheng and Feng, 2022; Rolls, Deco, Huang and Feng, 2022a, 2023b; Rolls, Deco, Zhang and Feng, 2023e; Rolls, Yan, Deco, Zhang, Jousmaki and Feng, 2024c). The HCP-MMP atlas was used because it provides 360 cortical regions in humans each described by anatomical, functional connectivity, and task-related fMRI, and each of these cortical regions may be performing different functions (see Appendix A).

The cortical visual pathways in humans that we have analyzed are shown in Fig. 4.1. They comprise two 'What' pathways involved in representing what objects and/or faces are being viewed; and two 'Where' pathways involved in representing where visual stimuli are in space, and in performing actions in visual space (Rolls, 2024c).

The Ventrolateral Cortical Visual Stream (blue), a 'What' stream, connects V1 > V2 > V3 V4 > V8 > FFC > TE1p and TE2p, with the latter the two last mainly visual cortical regions for invariant face and object representation (Fig. 4.1). Beyond that there is connectivity via lateral parahippocampal gyrus TF, perirhinal cortex PeEc and entorhinal cortex EC to provide 'What' inputs to the hippocampus (Hipp). Beyond TE1p and TE2p there is connectivity to semantic regions in the anterior lateral temporal lobe regions TE1m, TE2a, TE1a and then to the temporal pole TGd and TGv.

The Superior Temporal Sulcus Visual Cortical Stream (black), a 'What' Stream, in STSvp and STSva receives face and object inputs from ventrolateral visual stream regions in the anterior temporal lobe (TE1a, TE1a, TE2a), and visual motion inputs from inferior parietal PGp which in turn receives inputs from intraparietal regions (e.g. LIPd) and parietal area 7 regions; and also has connectivity with visual inferior parietal PGi. Neurons in this system respond to socially relevant stimuli such as turning the head and eyes to make or break social contact, and face expression (Fig. 4.1).

The Dorsal Cortical Visual Stream (red), a 'Where' stream for motion and visuomotor action in space, connects V1> V2 > V3 > LO3 + MT > MST > FST which then connects to intraparietal regions (e.g. LIPd, LIPv, VIP, MIP) and parietal area 7 regions (e.g. 7PC) (Fig. 4.1). There is also connectivity V3 > V3A, V3B, V7 > V6A > intraparietal regions. The dorsal visual stream regions have connectivity via inferior parietal PGp to the medial parahippocampal PHA1-3 regions, and may be involved in self-motion update of spatial view representations.

The Ventromedial Cortical Visual Stream (green), a 'Where' stream, connects V1 > V2 > the ProStriate region ProS and the Dorsal Visual Transitional region DVT where the Retrosplenial Scene

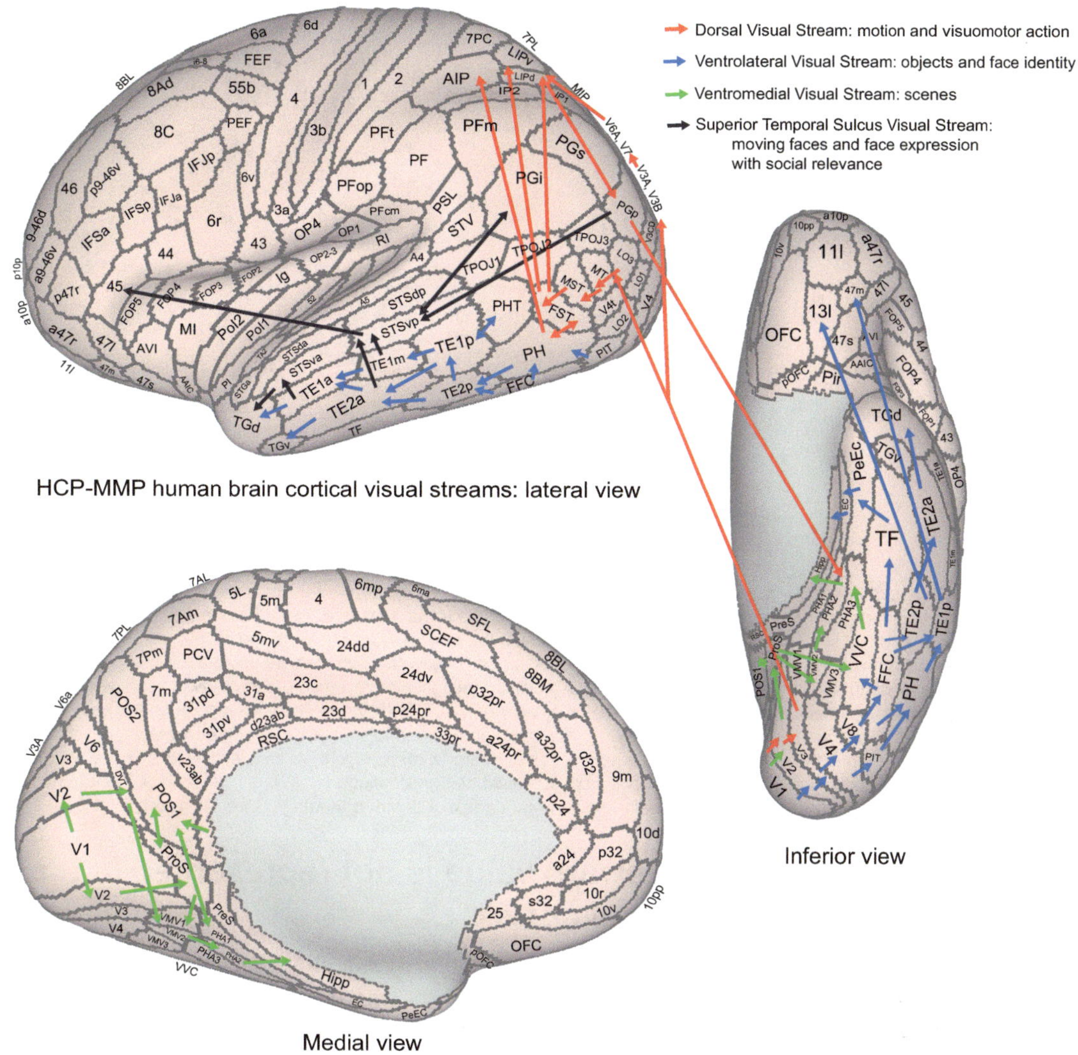

Fig. 4.1 Schematic diagram showing two 'What' and two 'Where' cortical visual streams. The cortical parcellation is the Human Connectome project MultiModal Parcellation (Glasser et al., 2016a), with the HCP figure with the most opened sulci. These pathways are described in the text. The full connectivity maps are provided elsewhere (Rolls et al., 2023e,b,a). (Data from Rolls E T. 2024 Two What, Two Where, Visual Cortical Streams in Humans. Neuroscience and Biobehavioral Reviews 160: 105650.)

Area is located (Sulpizio et al., 2020) > VMV1-3 and VVC > medial parahippocampal cortex PHA1-3 where the ParaHippocampal Scene or Place Area is located (Sulpizio et al., 2020; Rolls et al., 2024b,e) > the hippocampus Hipp (Fig. 4.1). This stream provides evidence in primates including humans about where is being viewed in scenes, using spatial view cells (Rolls, 2023a, 2025b) (see Chapter 5). The full connectivity maps are provided elsewhere (Rolls et al., 2023e,b,a, 2024c).

In addition, we have developed a new graph-theoretical approach that helps to show that the ventromedial visual cortical 'Where' pathway for spatial scenes is separate from the ventromedial visual cortical 'What' pathway for objects and faces (Rolls and Turova, 2025).

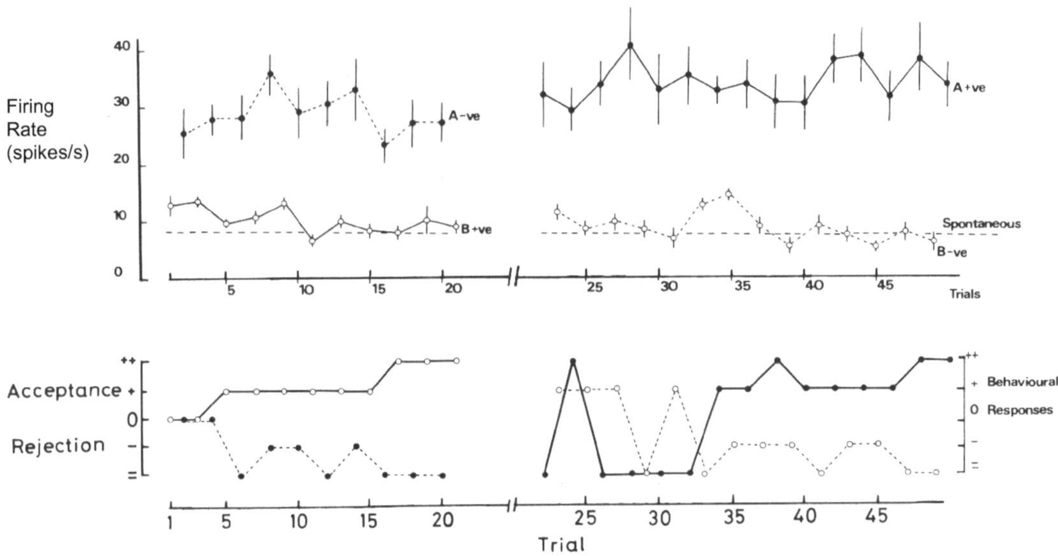

Fig. 4.2 Examples of the responses of a neuron in the inferior temporal visual cortex, showing that its responses (firing rate in spikes/s, upper panel) do not reverse when the reward association of the visual stimuli reverses. For the first 21 trials of the visual discrimination task, visual stimulus A was aversive (–ve, because if the monkey licked he obtained saline), and visual stimulus B was associated with reward (+ve, because if the monkey licked when he saw this stimulus, he obtained fruit juice). The neuron responded more to stimulus A than to stimulus B. After trial 21, the contingencies reversed (so that A was now +ve, and B –ve). The monkey learned the reversal correctly by about trial 35 (lower panel). However, the inferior temporal cortex neuron did not reverse when the reinforcement contingency reversed – it continued to respond to stimulus A after the reversal, even though the stimulus was now +ve. Thus this, and other inferior temporal cortex neurons, respond to the physical aspects of visual stimuli, and not to the stimuli based on their reinforcement association or the reinforcement contingency. (Data from Rolls E T, Judge S J, and Sanghera M K. 1977 Activity of neurones in the inferotemporal cortex of the alert monkey. Brain Research 130: 229–238.)

4.3 The discovery that reward value is not represented in the inferior temporal visual cortex

We soon made the discovery (possible in the behaving animal) that inferior temporal cortex neurons represent objects but not their reward value, in that they kept responding to visual stimuli even after devaluation by feeding to satiety, and in that they did not reverse their responses to follow reward value in the reversal of a visual discrimination task (Rolls, Judge and Sanghera, 1977). This was confirmed in a later investigation in which the reduced receptive field sizes in natural visual scenes of inferior temporal cortex neurons were investigated (Rolls, Aggelopoulos and Zheng, 2003a).

The discovery was made in visual discrimination reversal experiments in which inferior temporal cortex neurons respond to an object independently of its association with reward vs punishment value, as illustrated in Fig. 4.2. It was also shown that inferior temporal visual cortex neurons do not alter their responses after the reward value of a food-related visual stimulus is reduced to zero, by feeding to satiety (Rolls et al., 1977).

Importantly, this is in contrast to what happens one synapse further on, in the orbitofrontal cortex, in the same test situations (Thorpe, Rolls and Maddison, 1983; Rolls, Critchley, Mason and Wakeman, 1996a; Critchley and Rolls, 1996a). This clear separation of perceptual representations in the inferior temporal visual cortex from reward value (see Chapter 3) is a key feature of the organization of primate perceptual systems (it is found in the taste and olfactory systems too, see Chapter 3), and ensures that we do not go blind to objects when their reward value is decreased to zero; and that we can learn about for example the location of objects (such as fruit on a tree) even when their reward value is currently zero.

That was an important discovery in enabling us to understand cortical functional architecture, for it showed that regions at the top of visual processing hierarchies represent objects and faces, and not their reward value, and that the orbitofrontal cortex is a key stage of cortical processing in representing visual (and many other) stimuli in terms of their reward and hence emotional value (Thorpe et al., 1983; Rolls, 2023c, 2026d, 2014a).

A key issue here is that the valence, positive or rewarding, vs negative or punishing, is not represented in the inferior temporal visual cortex. But just because rewarded or punished visual stimuli will be looked at more and with more attention than neutral stimuli, this may contribute to stronger representations of both rewarded and punished stimuli than of neutral stimuli in sensory/perceptual cortical regions such as the inferior temporal visual cortex, though without encoding the valence. The mechanisms for this could include backprojections from the orbitofrontal cortex to the inferior temporal visual cortex, which would strengthen representations there without transferring the representation itself to earlier cortical regions. (It is a property of cortico-cortical backprojection connections that they do not just transfer what is represented at a higher level to a lower level (Rolls, 2016b, 2023c, 2026d).) Another potential mechanism is activation of cholinergic basal forebrain neurons from the orbitofrontal cortex (Rolls et al., 2022b), with the cholinergic projection to neocortex enhancing memory consolidation (Rolls, 2022a).

The systems-level functional architecture in primates including humans with a clear separation of reward value from perceptual representations is probably very different from rodents, see Section 29.4 of Rolls (2026d), so the study of this in primates including humans is important.

4.4 The discovery of face cells in the inferior temporal visual cortex

While recording from the temporal cortical visual regions in macaques to perform the investigation just described in Section 4.3, we were keen to discover what types of visual stimuli are encoded by inferior temporal cortex neurons. We therefore tested these and amygdala neurons with hundreds of 'junk' (everyday) objects including toys, and model animals, and faces. While doing this testing with natural real-word visual stimuli, we discovered face neurons, that is, neurons with selective responses to the sight of faces (Sanghera, Rolls and Roper-Hall, 1979; Perrett, Rolls and Caan, 1979, 1982; Rolls, 1984) (see Rolls (2011b)). The following shows how we made these discoveries (Rolls, 2011b, 2025a).

During the recordings described in Section 4.3, we found neurons that responded to complex stimuli such as faces, and reported in the paper that "neurones were found which could apparently only be excited by more complicated stimuli" (than simple geometrical stimuli) (Rolls, Judge and Sanghera, 1977), and cited earlier observations of Charles Gross and colleagues (Gross, Rocha-Miranda and Bender, 1972) on the effectiveness of complex stimuli for some of these neurons in anesthetized macaques. (They reported that "For at least three TE units, complex colored patterns (e.g., photographs of faces, trees) were more effective than the standard stimuli, but the crucial features of these stimuli were never determined" (Gross et al., 1972).) In 1977, we (Rolls, Judge and Sanghera, 1977) needed more evidence before we were ready to describe face-selective neurons, with visual neurophysiology so dominated by the inspiring discoveries of Hubel and Wiesel on neurons responding to bars and edges in the primary visual cortex (Hubel and Wiesel, 1968).

By 1979 we (Sanghera, Rolls and Roper-Hall, 1979), with an abstract in 1977 (Sanghera, Rolls and Roper-Hall, 1977), were sufficiently convinced to report in the journal *Experimental Neurology* the presence of face-selective neurons in the amygdala and inferior temporal visual cortex, as follows: "It should also perhaps be noted that nine of the amygdaloid neurons with visual responses were found

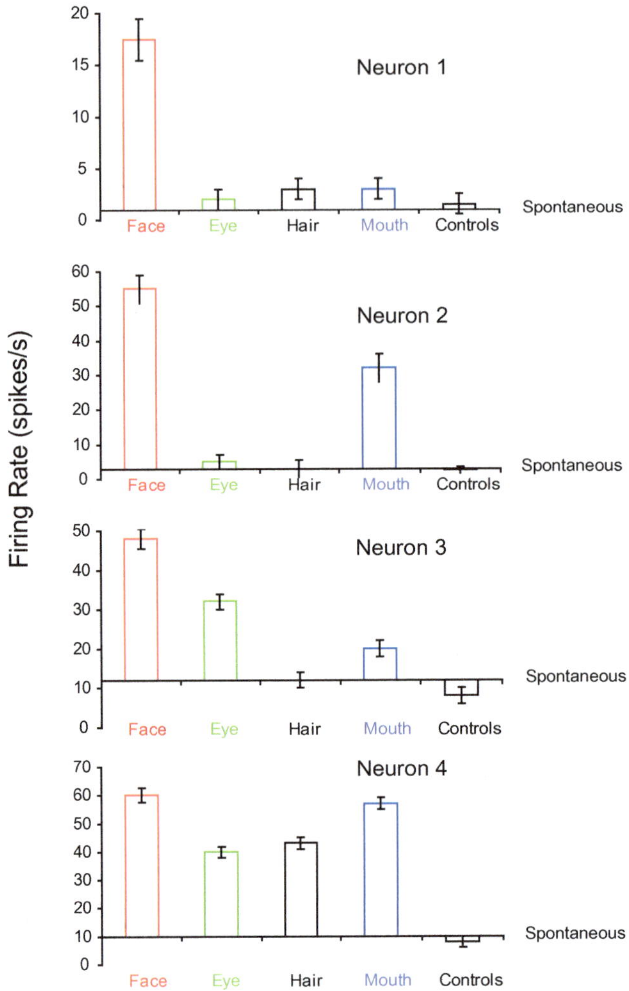

Fig. 4.3 Responses of four inferior temporal visual cortex face neurons discovered in macaques. Some neurons responded to any of several features in a face (bottom neuron). Some neurons responded to one or more features in a face (the middle two neurons). Some neurons required all these features (eye, hair and mouth) to be present (and we showed in the correct spatial configuration) in order to respond to a face (top neuron). The mean and standard error of the firing rate to each type of stimulus, and the spontaneous baseline firing rate (the horizontal line) are shown. (Data from Perrett D I, Rolls E T and Caan W. 1982 Visual neurons responsive to faces in the monkey temporal cortex. Experimental Brain Research 47: 329-342.)

to respond primarily to faces or photographs of faces, and two more to respond to all stimuli except these. It is possible that innate or learned factors contributed to the responses of these neurons, as well as to the responses of a comparable group of neurons from which we have recorded in the anterior inferotemporal cortex" (Sanghera, Rolls and Roper-Hall, 1979). A neuron with good responses to faces is illustrated in Fig. 1 of that paper (Sanghera, Rolls and Roper-Hall, 1979), even if that neuron was chosen to illustrate different tuning than that of a face-selective neuron.

Further reports on these amygdala face-selective neurons, and on comparable face-selective neurons in the temporal lobe cortex within the superior temporal sulcus, soon appeared (Rolls, 1981a,c, 1984). For example, Rolls (1981a,c) reported that in the cortex in the superior temporal sulcus eighteen neurons "responded selectively to faces. ... Masking out, or presenting isolated, parts of faces showed that some cells responded on the basis of different features. Some required eyes, some hair, some the mouth, and others showed parallel responses to each of a number of features. Some cells responded more strongly when such component features of faces were combined. Presenting a face in profile failed to elicit a response from some cells. Transformations of the face such as isomorphic rotation or alterations of color, size, or distance did not greatly affect the magnitude of the neuronal responses" (Rolls, 1981a,c).

At about the time that we were making these recordings in the amygdala and inferior temporal visual cortex, David Perrett and Simon Thorpe joined my lab at the University of Oxford in the Department of Experimental Psychology, and both were very interested in the face-selective neurons in the inferior temporal visual cortex (and also reported as present in the orbitofrontal cortex (Thorpe, Rolls and Maddison, 1983)). By 1979 we (Perrett, Rolls and Caan, 1979) (with a fuller report in 1982, Perrett, Rolls and Caan (1982)) had published a report describing some of the properties of these neurons, including their face selectivity, their invariance (for size, color, distance and isomorphic rotation), their frequent tuning to a particular view of a face such as profile or front view, the responsiveness of some neurons to face parts such as eyes, mouth or hair and of other neurons only to combinations of all these features in the correct relative spatial positions (as illustrated in Fig. 4.3), and the fact that they do not respond on the basis of arousal.

Our 1979 report (Perrett, Rolls and Caan, 1979) was as follows:

"While recording from visual cells in the temporal lobe, we have observed a small number of cells which responded strongly to faces. Analysis of the responses of 50 such neurones in the vicinity of the fundus of the superior temporal sulcus in 3 hemispheres of 2 alert rhesus monkeys showed the following.

(1) All fifty neurones responded to faces (which were human or rhesus monkey, 3-D or projected, and shown to the monkey through a large-aperture shutter with visual fixation monitored), and were almost unresponsive to gratings, simple geometrical, and other complex 3-D stimuli.

(2) The neuronal responses to the sight of a face were sustained and had latencies of 110±20 ms.

(3) 32 neurones responded also to some arousing or aversive visual stimuli.

(4) 18 neurones responded selectively to faces, and had the following properties.

(5) These neurons were in general unresponsive to auditory or tactile stimulation which was aversive, or arousing as shown by the GSR, or to stimuli such as a hand which signified a human or monkey.

(6) Masking out, or presenting isolated parts of faces showed that some cells responded on the basis of different features. Some required eyes, some hair, some the mouth, and others showed parallel responses to each of a number of features.

(7) Some cells responded more when such component features were combined.

(8) Presenting the face in profile failed to elicit a response for some cells.

(9) Transformations of the face such as isomorphic rotation, or alterations of color, size or distance, did not greatly affect the magnitude of the neuronal responses. This evidence suggests that in the temporal lobe of the primate there are neurones specialized to respond to the component visual features present in faces."

In 1981, Bruce, Desimone and Gross (1981), citing our 1979 report of face-selective neurons in the cortex in the superior temporal sulcus (Perrett, Rolls and Caan, 1979), described seven neurons in the same region that confirmed our findings in that their 7 neurons "appeared to be selective for face"; "responded to a variety of human and monkey faces (real faces, slides, and photographs) differing in size, color, and movement"; "covering the eyes on the photograph reduced but did not eliminate the response"; and "scrambling the photographs eliminated the response".

Soon after our initial reports of the discoveries of face-selective neurons in the amygdala, inferior temporal visual cortex (Sanghera, Rolls and Roper-Hall, 1977, 1979; Perrett, Rolls and Caan, 1979, 1982; Rolls, 1984), and orbitofrontal cortex (Thorpe, Rolls and Maddison, 1983), others (Desimone,

Albright, Gross and Bruce, 1984; Gross, Desimone, Albright and Schwartz, 1985; Perrett, Smith, Potter, Mistlin, Head, Milner and Jeeves, 1985b; Desimone, 1991) as well as ourselves (Rolls, 1981c; Perrett and Rolls, 1983; Rolls, 1984; Baylis, Rolls and Leonard, 1985; Leonard, Rolls, Wilson and Baylis, 1985; Rolls, 1991, 1992a, 2000b, 2007b, 2008c,a) provided many further analyses of face-selective neurons, and we went on to make many further discoveries on these and other visual neurons in the temporal cortical visual areas, and related cortical regions, as described in the remainder of this Chapter and elsewhere (Rolls, 2023c, 2026d, 2025a).

4.5 The discoveries of many of the key computational properties of face and object neurons in the visual cortical regions in the temporal lobes

We discovered many of the key neurocomputational properties of these face and object encoding neurons in the inferior temporal cortical visual areas:

4.5.1 Invariant representations for size, contrast, spatial frequency, translation, and even for some neurons view invariance

We have discovered that size, contrast, spatial frequency and even in some cases view invariance, as well as reduced translation invariance in natural scenes, are properties of the neuronal representations in the inferior temporal visual cortex (Rolls, Baylis and Leonard, 1985; Rolls and Baylis, 1986; Rolls, Baylis and Hasselmo, 1987; Booth and Rolls, 1998; Hasselmo, Rolls, Baylis and Nalwa, 1989b; Tovee, Rolls and Azzopardi, 1994).

These discoveries were driven in my mind by the computational importance of transform-invariant representations of objects and faces being the output of visual cortical perceptual regions for objects and faces, for when the information reaches further cortical regions such as the orbitofrontal cortex, amygdala, and hippocampus, generalization is provided across a wide range of transforms.

In more detail, the utility of *what* is computed by the inferior temporal visual cortex for different types of memory, emotion, and any system that receives from the inferior temporal visual cortex is that the output of the visual system will represent an object invariantly with respect to position on the retina, size, etc, and this simplifies the functionality required of the receiving systems, which need then to simply associate the object representation with reward (orbitofrontal cortex and amygdala, Chapter 3), associate it with location in the viewed environment (hippocampus, Chapter 5), recognize it as familiar (perirhinal cortex (Holscher et al., 2003; Rolls et al., 2005b)), associate it with a motor response in a habit memory (basal ganglia), etc. (Rolls, 2007e, 2016b, 2012b, 2014a, 2023c, 2026d). The associations can be relatively simple, involving for example Hebbian associativity (Rolls, 2023c, 2026d). This design is highly evolutionarily adaptive. If the representations provided by the inferior temporal visual cortex were not transform invariant, these receiving systems might learn about one view of an object in one position on the retina with a given size etc, and there would be no generalization to other views, positions, sizes etc of the same objects or faces. This is a key conceptual thrust of this series of discoveries.

Computing transform-invariant representations of objects is a major computation performed by the primate including human visual system. If an object or face is seen in a different position on the retina, in a different isomorphic (in-plane) rotation, in a different size, etc, then the retinal receptors (corresponding to the pixels in a photograph) for each transform that are activated will have little correlation with each other, and it will be difficult to identify whether the set of receptors active is for the same object, or a different object. This becomes an even more acute problem for view invariance, for some objects look completely different from different views. An extreme example is a picture postcard, in which viewing one side provides little evidence about what is present on the other side. Given that neurons read information by computing whether an input set of firings received on say 10,000 synaptic inputs matches (has a correlation with) the already learned 10,000 synaptic weights, then even shifting the input by one synapse will destroy the correlation (Rolls, 2023c, 2026d).

This shows that learning transform-invariant representations is a key problem in the visual system, and we were keen to show whether such transform invariances are a property of neurons in the inferior temporal visual cortex; and if so, to understand how the computation might be performed by the brain. That fundamental scientific question led to the research described next.

4.5.1.1 Translation invariant representations

I have elucidated that there is convergence from each small part of a region to the succeeding region (or layer in the hierarchy) in such a way that the receptive field sizes of neurons (for example 1 degree near the fovea in V1) become larger by a factor of approximately 2.5 with each succeeding stage (see Fig. 4.18) (Rolls, 1992a). Such zones of convergence would overlap continuously with each other (Fig. 4.18). This convergent connectivity provides part of the basis for the fact that many neurons in the temporal cortical visual areas respond to a stimulus relatively independently of where it is in their receptive field, and moreover maintain their stimulus selectivity when the stimulus appears in different parts of their large visual receptive field (Gross, Desimone, Albright and Schwartz, 1985; Tovee, Rolls and Azzopardi, 1994; Rolls, Aggelopoulos and Zheng, 2003a). This is called translation or shift invariance.

In addition to having topologically appropriate connections, it is necessary for the connections to have the appropriate synaptic weights to perform the mapping of each set of features, or object, to the same set of neurons in the inferior temporal visual cortex. How this could be achieved is addressed in the computational neuroscience models described later in this Chapter and elsewhere (Wallis and Rolls, 1997; Rolls and Deco, 2002; Rolls and Stringer, 2006; Stringer, Perry, Rolls and Proske, 2006; Rolls, 2012b, 2016b, 2021c, 2023c, 2026d; Zhang, Rolls and Feng, 2026a).

4.5.1.2 Reduced translation invariance in natural scenes, and the selection of a rewarded object

Research on translation invariance (and most visual neurophysiology) typically considered the case in which there is only one stimulus or object in the visual field shown against a blank screen (Hubel and Wiesel, 1968; Gross, Desimone, Albright and Schwartz, 1985; Tovee, Rolls and Azzopardi, 1994).

My thinking moved on from this, and considered what happens in a cluttered, natural, environment? Do all objects that can activate an inferior temporal neuron do so whenever they are anywhere within the large receptive fields of inferior temporal neurons? If so, the output of the visual system might be confusing for brain regions that receive inputs from the temporal cortical visual areas. If one of the objects in the visual field was associated with reward, and another with punishment, would the output of the inferior temporal visual cortex to reward / emotion-related brain systems be an amalgam of both stimuli? If so, how would we be able to choose between the stimuli, and have an emotional response to one but not perhaps the other, and select one for action and not the other (see Fig. 4.4)?

I decided to investigate this, and we found that even if an effective face was present in the parafovea, an inferior temporal visual cortex (IT) neuron represented correctly (by a low firing rate) a non-effective face at the fovea (Rolls and Tovee, 1995b). This provides evidence that in situations with more than one stimulus present, IT neurons still represent what is at the fovea.

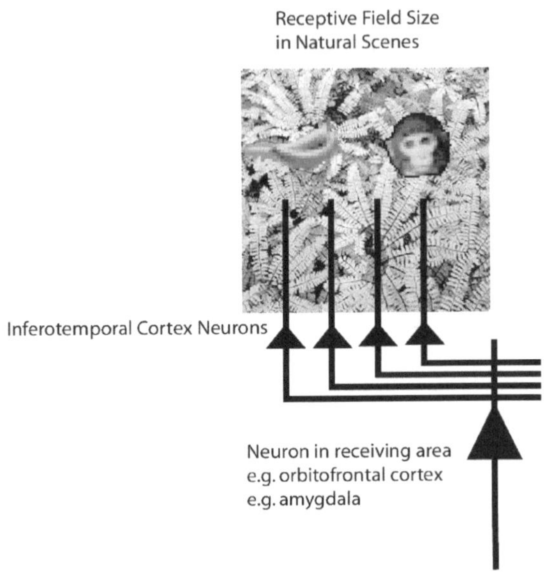

Fig. 4.4 Objects shown in a natural scene, in which the task was to search for and touch one of the stimuli. The objects in the task as run were smaller. The diagram shows that if the receptive fields of inferior temporal cortex neurons are large in natural scenes with multiple objects (in this scene, bananas and a face), then any receiving neuron in structures such as the orbitofrontal cortex and amygdala would receive information from many stimuli in the field of view, and would not be able to provide evidence about each of the stimuli separately.

This makes the interface to action simpler, in that what is at the fovea can be interpreted (e.g. by an associative memory) partly independently of the surroundings, and choices and actions can be directed if appropriate to what is at the fovea (cf. Ballard (1993)). These findings are a step towards understanding how the visual system functions in a natural environment (see also Gallant, Connor and Van-Essen (1998), Stringer and Rolls (2000), Sheinberg and Logothetis (2001), and Rolls and Deco (2006)).

To investigate further how information is passed from the inferior temporal cortex to other brain regions to enable stimuli to be selected from natural scenes for action, we analyzed the responses of single and simultaneously recorded inferior temporal visual cortex neurons to stimuli presented in complex natural backgrounds (Rolls, Aggelopoulos and Zheng, 2003a). In one situation, a visual fixation task was performed in which the monkey fixated at different distances from the effective stimulus. In another situation the monkey had to search for two objects on a screen, and a touch of one object was rewarded with fruit juice, and of another object was punished with saline (see Fig. 4.4 for a schematic overview). In both situations neuronal responses to the effective stimuli for the neurons were compared when the objects were presented in the natural scene or on a plain background. It was found that the overall response of the neuron to objects was sometimes somewhat reduced when they were presented in natural scenes, though the selectivity of the neurons remained. However, the main finding was that the magnitudes of the responses of the neurons typically became much less in the real scene the further the monkey fixated in the scene away from the object (see Fig. 4.5 and Rolls (2023c, 2026d)).

It is proposed that this reduced translation invariance in natural scenes helps an unambiguous representation of an object that may be the target for action to be passed to the brain regions that receive from the primate inferior temporal visual cortex. It helps with the binding problem, by reducing in natural scenes the effective receptive field of inferior temporal cortex neurons to approximately the size of an object in the scene.

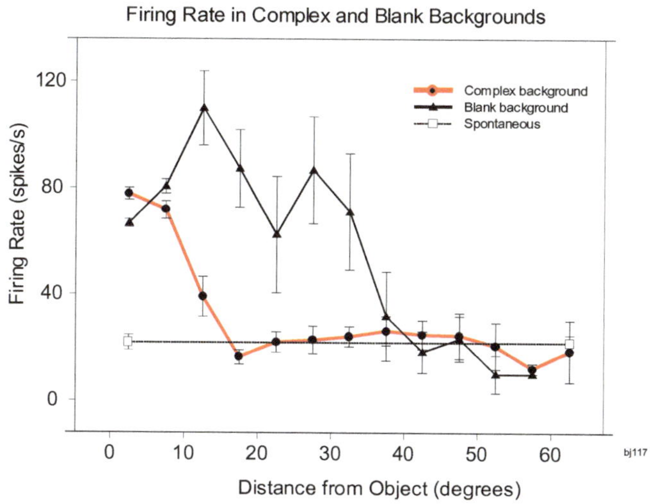

Fig. 4.5 Firing of a temporal cortex cell to an effective visual stimulus presented either in a blank background or in a natural scene, as a function of the angle in degrees at which the monkey was fixating away from the effective stimulus. The task was to search for and touch the stimulus. (Data from Rolls E T, Aggelopoulos N C and Zheng F. 2003 The receptive fields of inferior temporal cortex neurons in natural scenes. Journal of Neuroscience 23: 339–348.)

We went on to develop a computational theory and model for this which produces an advantage for what is at the fovea because of the large cortical magnification of the fovea, and shunting interactions between representations weighted by how far they are from the fovea (Rolls and Deco, 2002; Trappenberg, Rolls and Stringer, 2002; Deco and Rolls, 2004; Aggelopoulos and Rolls, 2005; Rolls and Deco, 2006; Rolls, 2023c, 2026d).

These discoveries indicate that a principle of *what* is computed in the ventral visual system is to provide strong weight to whatever is close to the fovea in natural environments. This principle of operation is very important in interfacing the visual system to action, reward, and memory systems, because the effective stimulus in making inferior temporal cortex neurons fire is in natural scenes usually at or close to the fovea. The fundamental principle revealed by these discoveries is that the spatial coordinates of where the object is in the scene do not have to be represented in the inferior temporal visual cortex, nor passed from it to the action selection system, as the latter can assume that the object making inferior temporal neurons fire is close to the fovea in natural scenes. *Thus the position in visual space being fixated provides part of the interface between sensory representations of objects and their coordinates as targets for actions in the world.* The small receptive fields that include the fovea of IT neurons in natural scenes make this possible. After this, local, egocentric, processing implemented in the dorsal visual processing stream using e.g. stereodisparity may be used to guide action towards objects being fixated (Rolls and Deco, 2002; Rolls, 2016b) (see Rolls (2023c, 2026d)).

The reduced receptive field size in complex natural scenes also enables emotions to be selective to just what is being fixated, because this is the information that is transmitted by the firing of IT neurons to structures such as the orbitofrontal cortex and amygdala that are involved in emotion as described in Chapter 3.

Another principle of what is computed is that the whole scene is not processed simultaneously by the ventral visual system, as that greatly simplifies the process of object recognition. Successive fixations to different parts of the scene then enable a representation of a whole scene to be built up, over time, and to be held effectively in memory (see Section 5.2.12).

Interestingly, although the size of the receptive fields of inferior temporal cortex neurons becomes

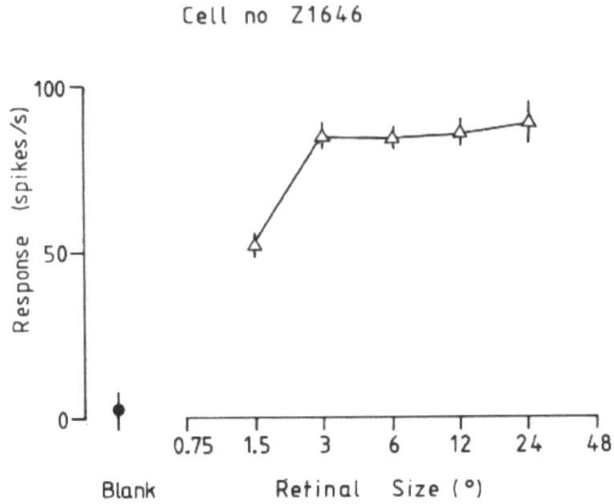

Fig. 4.6 Typical response of an inferior temporal cortex face-selective neuron to faces of different sizes. The size subtended at the retina in degrees is shown. (Data from Rolls E T and Baylis G C. 1986 Size and contrast have only small effects on the responses to faces of neurons in the cortex of the superior temporal sulcus of the monkey. Experimental Brain Research 65: 38–48.)

reduced in natural scenes so that neurons in IT respond primarily to the object being fixated, there is nevertheless frequently some asymmetry in the receptive fields (Aggelopoulos and Rolls, 2005). This provides a partial solution to how multiple objects and their position close to the fovea in a scene can be captured with a single glance (Aggelopoulos and Rolls, 2005), and is another principle of *what* is computed in the ventral visual system.

4.5.1.3 Size and spatial frequency invariance

Continuing with the aim of testing whether IT neurons have transform-invariant representations, we discovered that many neurons in the inferior temporal visual cortex and cortex in the anterior part of the superior temporal sulcus (IT/STS) respond relatively independently of the size of an effective face stimulus, with a mean size invariance (to a half maximal response) of 12 times (3.5 octaves) (Rolls and Baylis, 1986). An example of the responses of an inferior temporal cortex face-selective neuron to faces of different sizes is shown in Fig. 4.6. This is not a property of a simple single-layer network, nor of neurons in V1, which respond best to small stimuli, with a typical size-invariance of 1.5 octaves. Also, the neurons typically continued to respond to a face when the information in it had been reduced from 3D to a 2D representation in grey on a monitor, with a response that was on average 0.5 of that to a real face.

Another transform over which recognition is relatively invariant is spatial frequency. For example, a face can be identified when it is blurred (when it contains only low spatial frequencies), and when it is high-pass spatial frequency filtered (when it looks like a line drawing). We discovered that IT neurons can respond relatively invariantly with respect to spatial frequency (Rolls, Baylis and Leonard, 1985). Further analysis of these neurons with narrow (octave) bandpass spatial frequency filtered face stimuli shows that the responses of these neurons to an unfiltered face can not be predicted from a linear combination of their responses to the narrow band stimuli (Rolls, Baylis and Hasselmo, 1987). This lack of linearity of these neurons, and their responsiveness to a wide range of spatial frequencies (see also their broad critical band masking (Rolls, 2008c)), indicate that in at least this part of the primate visual system recognition does not occur using Fourier analysis of the spatial frequency components of images (cf. Campbell (1980)).

4.5.1.4 A view-invariant representation

For recognizing and representing the identity of objects and faces, it is important that an output of the visual system should be not only translation-invariant and size-invariant, but also relatively view-invariant. I decided to investigate whether there are neurons with view-invariant responses in the inferior temporal visual cortex. We discovered that some temporal cortical neurons reliably responded differently to the faces of two different individuals independently of viewing angle (Hasselmo, Rolls, Baylis and Nalwa, 1989b) (see example in Fig. 4.7 upper), although in most cases (16/18 neurons) the response was not perfectly view-independent. Mixed together in the same cortical regions there are neurons with view-dependent responses (Hasselmo, Rolls, Baylis and Nalwa, 1989b; Rolls and Tovee, 1995a; Chang and Tsao, 2017) (see example in Fig. 4.7 lower)). Such neurons might respond, for example, to a view of a profile of a monkey but not to a full-face view of the same monkey (Perrett, Smith, Potter, Mistlin, Head, Milner and Jeeves, 1985b; Hasselmo, Rolls, Baylis and Nalwa, 1989b).

These discoveries of view-dependent, partially view-independent, and view-independent representations in the same cortical regions are consistent with the hypothesis discussed below that view-independent representations are being built in these regions by associating together the outputs of neurons that have different view-dependent responses to the same individual. These findings also provide evidence that one output of the visual system includes representations of what is being seen, in a view-independent way that would be useful for object recognition and for learning associations about objects; and that another output is a view-based representation that would be useful in social interactions to determine whether another individual is looking at one, and for selecting details of motor responses, for which the orientation of the object with respect to the viewer is required (Rolls and Deco, 2002; Rolls, 2016b, 2023c, 2026d).

We then discovered further evidence that some neurons in the temporal cortical visual areas have object-based rather than view-based responses in a study of a population of neurons that responds to moving faces (Hasselmo, Rolls, Baylis and Nalwa, 1989b). For example, four neurons responded vigorously to a head undergoing ventral flexion, irrespective of whether the view of the head was full face, of either profile, or even of the back of the head. These different views could only be specified as equivalent in object-based coordinates. Further, the movement specificity was maintained across inversion, with neurons responding for example to ventral flexion of the head irrespective of whether the head was upright or inverted (see example in Fig. 4.8). In this procedure, retinally encoded or viewer-centered movement vectors are reversed, but the object-based description remains the same. This is a discovery of a remarkable computation, which as far as I know has been little followed up empirically, though we have developed a biologically plausible neuronal network model that accounts for this computation (Rolls and Stringer, 2007).

We also made another discovery consistent with object-based encoding by inferior temporal cortex neurons: we discovered a small number of neurons that respond to images of faces of a given absolute size, irrespective of the retinal image size or distance (Rolls and Baylis, 1986). This again is a discovery of a remarkable computation performed by the brain, which probably relates to the psychophysics of phenomenal regression to the real object (Thouless, 1931).

I extended these analyses to objects seen naturally in their home cages by macaques, as this potentially provided a natural opportunity to form view-invariant representations (Booth and Rolls, 1998). The stimuli were presented for 0.5 s on a color video monitor while the monkey performed a visual fixation task. The stimuli were images of 10 real plastic objects that had been in the monkey's cage for several weeks, to enable him to build view invariant representations of the objects. Control stimuli were views of objects that had never been seen as real objects. For some neurons, the responses occurred only to a subset of the objects (using ensemble encoding), irrespective of the

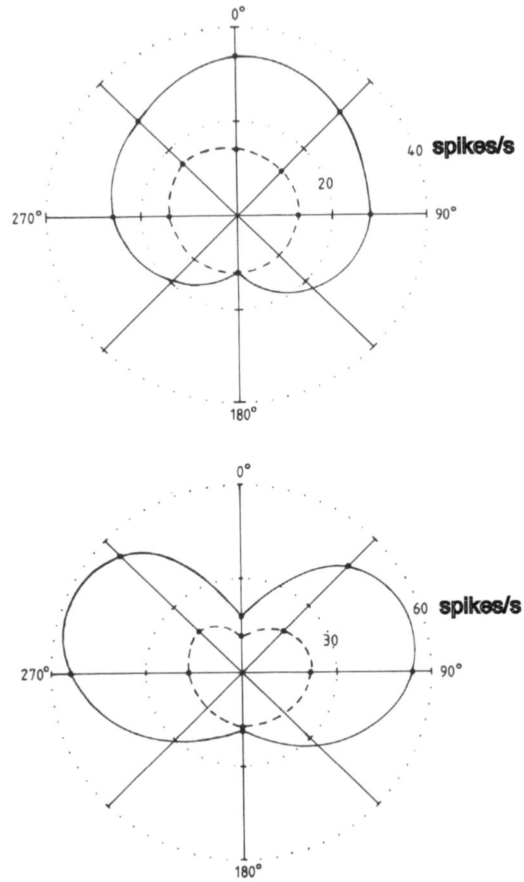

Fig. 4.7 Upper: View invariant response of an inferior temporal cortex neuron to two faces. In the polar response plot, 0^{0} is the front of the face, and 90^{0} the right profile. The firing rate in spikes/s is shown by the distance from the center of the plot. The responses to two different faces are shown by the solid and dashed lines. The neuron discriminated between the faces, apart from the view of the back of the head. Lower: View dependent response of an inferior temporal cortex neuron to two faces. The neuron responded best to the left and right profiles. Some neurons respond to only one of the profiles, and this neuron had a small preference for the left profile. (Data from Hasselmo M E, Rolls E T, Baylis G C and Nalwa V. 1989 Object-centered encoding by face-selective neurons in the cortex in the superior temporal sulcus of the monkey. Experimental Brain Research 75: 417–429.)

viewing angle. Moreover, the firing of a neuron on any one trial, taken at random and irrespective of the particular view of any one object, provided information about which object had been seen, and this information increased approximately linearly with the number of neurons in the sample (see Fig. 4.9). This is strong quantitative evidence that some neurons in the inferior temporal cortex provide a view-invariant representation of objects. Moreover, the discovery by Booth and Rolls (1998) shows that the information is available in the firing rates, and has all the desirable properties of distributed representations described below, including exponentially high coding capacity, and rapid speed of readout of the information (Rolls, 2023c, 2026d).

My theory and hypothesis was (Rolls, 1992a) and still is (Rolls, 2021c, 2023c, 2026d; Zhang et al., 2026a) that view invariant representations can be learned by associating together the different views of objects as they are moved and inspected naturally in a period that may be in the order of a few seconds. Evidence for this is described in Section 4.5.3, and the theory and model are described in Section 4.7.

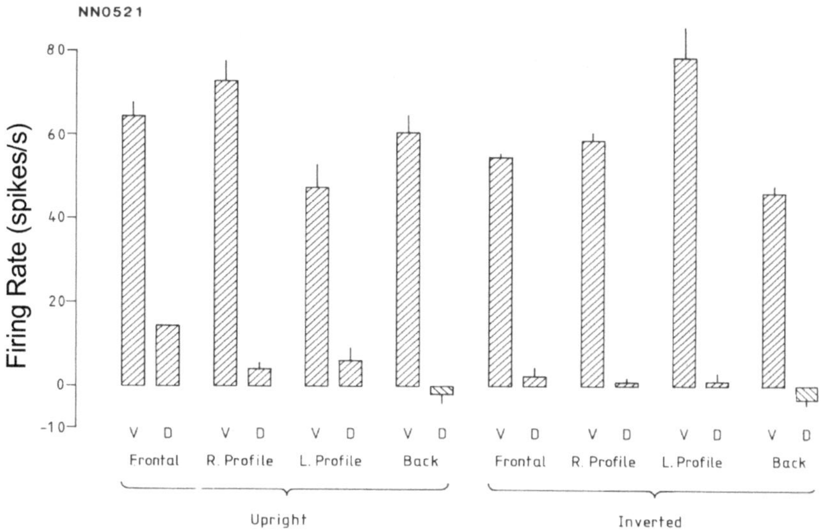

Fig. 4.8 Object-centerd encoding: Neuron responding to ventral flexion (V) of the head independently of view even when the head was inverted. Ventral flexion is tilting the head from the full frontal view forwards to face down at 45 degrees. Dorsal flexion is tilting the head until it is looking 45 degrees up. The response is shown as a change in firing rate in spikes/s ± sem from the spontaneous firing rate. (Data from Hasselmo M E, Rolls E T, Baylis G C and Nalwa V. 1989 Object-centered encoding by face-selective neurons in the cortex in the superior temporal sulcus of the monkey. Experimental Brain Research 75: 417–429.)

4.5.2 Combinations of features in the correct spatial configuration

Objects, and faces, are defined by a set of features in the correct spatial configuration with respect to each other, and it became an issue important to explore about whether these inferior temporal cortex neurons have that property. We discovered that many neurons in the temporal lobe visual cortical regions do respond to combinations of features (including objects), but not to single features presented alone, and the features must have the correct spatial arrangement. This has been shown, for example, with faces, for which it was discovered by masking out or presenting parts of the face (for example eyes, mouth, or hair) in isolation, or by jumbling the features in faces, that some cells in the cortex in IT/STS respond only if two or more features are present, and are in the correct spatial arrangement (Perrett, Rolls and Caan, 1982; Rolls, Tovee, Purcell, Stewart and Azzopardi, 1994b). Fig. 4.3 shows examples of four neurons, the top one of which responds only if all the features are present, and the others of which respond not only to the full face, but also to one or more features.

Corresponding evidence has been found for non-face cells. For example, Tanaka et al. (1990) showed that some posterior inferior temporal cortex neurons might only respond to the combination of an edge and a small circle if they were in the correct spatial relationship to each other.

The discovery that some temporal cortex neurons respond to objects or faces consisting of a set of features only if the whole combination of features is present and they are in the correct spatial arrangement with respect to each other (not jumbled) (Perrett, Rolls and Caan, 1982; Rolls, Tovee, Purcell, Stewart and Azzopardi, 1994b) is part of the evidence that some inferior temporal cortex neurons are tuned to respond to objects or faces, and are part of a feature hierarchy system as described in Rolls (2023c, 2026d).

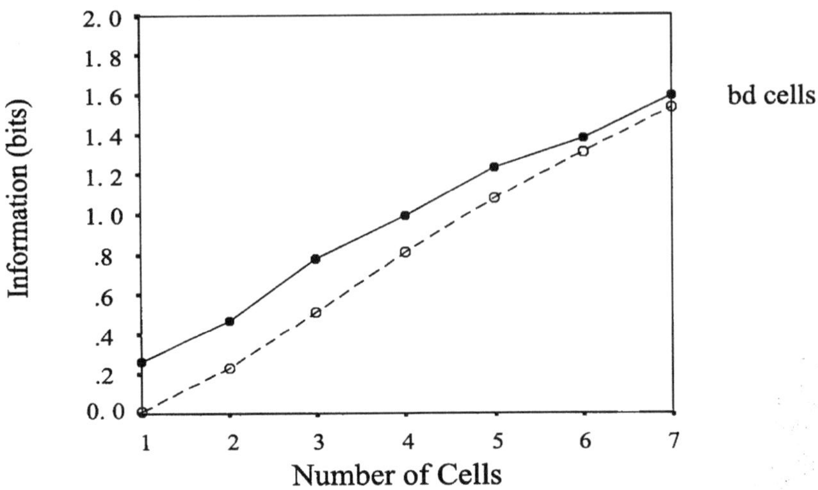

Fig. 4.9 View-independent object encoding: information in a population of different numbers of inferior temporal cortex cells available from a single trial, in which one view was shown, about which of 10 objects each with 4 different views in the stimulus set had been seen. The solid symbols show the information decoded with a Bayesian probability estimator algorithm, and the open symbols with dot product decoding. (Data from Booth M C A and Rolls E T. 1998 View-invariant representations of familiar objects by neurons in the inferior temporal visual cortex. Cerebral Cortex 8: 510–523.)

4.5.3 Learning of new representations in the temporal cortical visual areas

I next developed the hypothesis that if inferior temporal cortex neurons are important in object and face perceptual processing, they should be able to learn representations of new people or objects rapidly, just as we do. To investigate the hypothesis that visual experience might guide the rapid formation of the responsiveness of neurons so that they provide an economical and ensemble-encoded representation of items such as faces or objects actually present in the environment, we analyzed the responses of inferior temporal cortex face-selective neurons while a set of new faces was shown. We discovered that some of the neurons studied in this way altered the relative degree to which they responded to the different members of the set of novel faces over the first few (1–2) presentations of the set (Rolls, Baylis, Hasselmo and Nalwa, 1989a) (see examples in Fig. 4.10).

If in a different experiment a single novel face was introduced when the responses of a neuron to a set of familiar faces were being recorded, the responses to the set of familiar faces were not disrupted, while the responses to the novel face became stable within a few presentations. Alteration of the tuning of individual neurons in this way may result in a good discrimination over the population as a whole of the faces known to the individual. This evidence is consistent with the categorization being performed by self-organizing competitive neuronal networks, as described by Rolls (2023c, 2026d). Consistent findings have been described more recently (Freedman, 2015).

We made a further discovery that these neurons can learn new representations very rapidly in an experiment in which binarized black and white (two-tone, Mooney) images of faces that blended with the background were used (see Fig. 4.11, left). These did not activate face-selective neurons. Full grey-scale images of the same photographs were then shown for ten 0.5 s presentations (Fig. 4.11, middle). In a number of cases, if the neuron happened to be responsive to that face, when the binarized version of the same face was shown next (Fig. 4.11, right), the neurons responded to it (Tovee, Rolls and Ramachandran, 1996). This is a direct parallel to the same phenomenon that is observed psychophysically in humans, and provides dramatic evidence that these neurons are

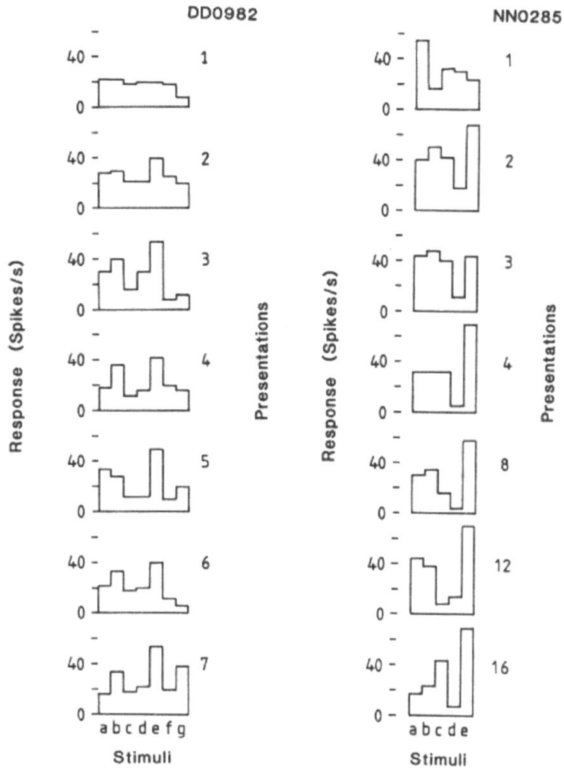

Fig. 4.10 Learning in the responses of inferior temporal cortex neurons. Results for two face-selective neurons are shown. On presentation 1 a set of 7 completely novel faces was shown when recording from neuron DD0982. The firing rates elicited on 1-sec presentations of each stimulus (a-g) are shown. After the first presentation, the set of stimuli was repeated in random sequence for a total of 7 iterations. After the first two presentations, the relative responses of the neuron to the different stimuli settled down to a fairly reliable response profile to the set of stimuli (as shown by statistical analysis). NN0285 – data from a similar experiment with another neuron, and another completely new set of face stimuli. (Data from Rolls E T, Baylis G C, Hasselmo M E and Nalwa V. 1989 The effect of learning on the face-selective responses of neurons in the cortex in the superior temporal sulcus of the monkey. Experimental Brain Research 76: 153–164.)

influenced by only a very few seconds (in this case 5 s) of experience with a visual stimulus (Tovee et al., 1996).

We went on to discover a neural correlate of this rapid visual learning in humans using similar stimuli and a similar paradigm in a PET (positron emission tomography) neuroimaging study, with a region showing an effect of the learning found for faces in the right temporal lobe, and for objects in the left temporal lobe (Dolan, Fink, Rolls, Booth, Holmes, Frackowiak and Friston, 1997). Analyzing these results and discussing them with Karl Friston triggered the development of psychophysiological interaction methods for neuroimaging (Friston, Buechel, Fink, Morris, Rolls and Dolan, 1997).

The discovery that inferior temporal visual cortex neurons can learn rapidly in an unsupervised way to categorize new objects or faces (Tovee, Rolls and Ramachandran, 1996) has been confirmed, with size invariance requiring temporal contiguity (Li and DiCarlo, 2010) as predicted by the VisNet model of memory trace rule dependent learning and related psychophysics (Perry, Rolls and Stringer, 2006), and it has been confirmed that reward is not necessary for this type of self-organising learning of invariant representations (Li and DiCarlo, 2012). The psychophysical discovery of the importance in humans of associations over time for helping to build invariant object representations (Perry, Rolls

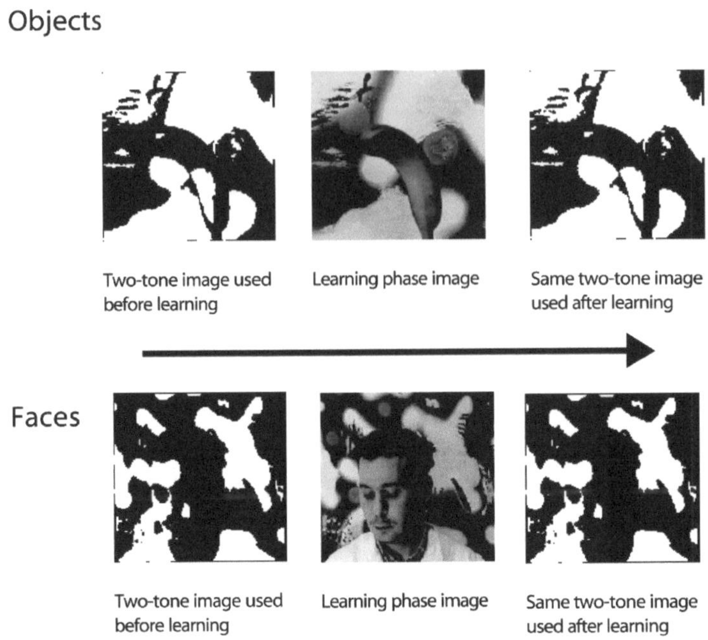

Fig. 4.11 Images of the type used to investigate rapid learning in the neurophysiological experiments of Tovee, Rolls and Ramachandran (1996) and the PET imaging study of Dolan, Fink, Rolls et al. (1997). When the black and white (two-tone) images at the left are shown, the objects or faces are not generally recognized. After the full grey scale images (middle) have been shown for a few seconds, humans and inferior temporal cortex neurons respond to the faces or objects in the black and white images (right).

and Stringer, 2006) has been confirmed in further psychophysical investigations in humans (Jia, Hong and DiCarlo, 2021).

4.5.4 Face identity is represented in the inferior temporal visual cortex, and face expression and gesture are represented in the cortex in the superior temporal sulcus

The population of neurons described so far code for face identity using sparse distributed representations as described in Section 4.6, and are in the inferior temporal cortex on the gyrus, and in areas TEa and TEm in the ventral bank of the macaque superior temporal sulcus (Hasselmo, Rolls and Baylis, 1989a).

I decided to investigate whether the face neurons were especially located in certain cytoarchitectonic areas of the macaque temporal lobe, and whether different types of neurons were found in different areas or 'patches'. We discovered that neurons with different types of visual response were found in different cytoarchitectonic areas (Baylis, Rolls and Leonard, 1987). For example, neurons in the inferior temporal cortex on the gyrus could respond well to stationary stimuli, and many neurons in parts of the cortex in the superior temporal sulcus required moving visual stimuli (Baylis, Rolls and Leonard, 1987).

I decided that the neurons in the cortex in the superior temporal sulcus deserved further investigation, and we discovered that there is a separate population in the cortex in the anterior part of the superior temporal sulcus (e.g. area TPO) that conveys information about facial expression (Hasselmo, Rolls and Baylis, 1989a) (see e.g. Fig. 4.12). Some of the neurons in this region tend to have view-based representations (so that information is conveyed for example about whether the

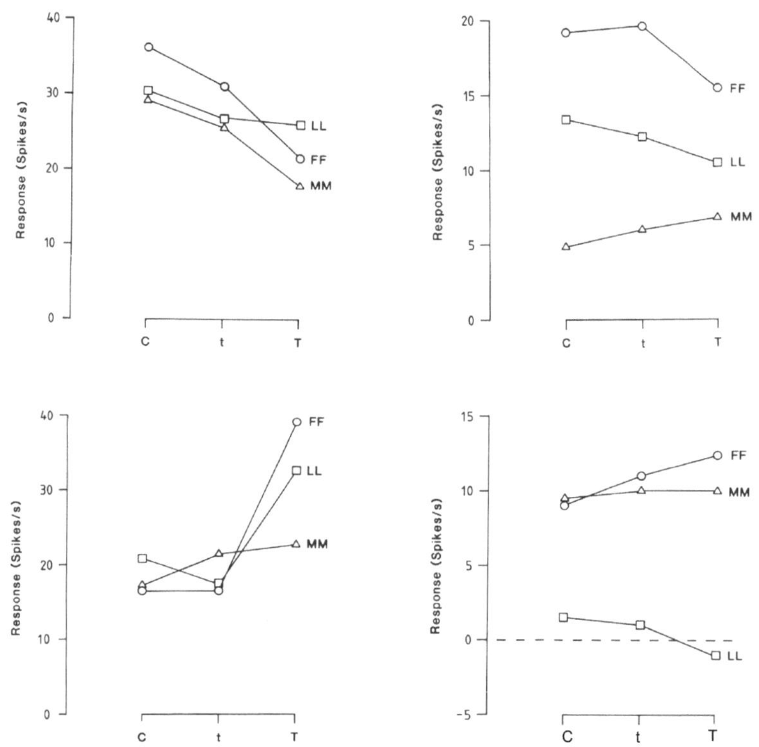

Fig. 4.12 There is a population of neurons in the cortex in the superior temporal sulcus (STS) with responses tuned to respond differently to different face expressions. The neurons in the two left panels did not discriminate between individuals (faces MM, FF, and LL), but did discriminate between different expressions on the faces of those individuals (C, calm expression; t, mild threat; T, strong threat). In contrast, the neurons in the right two panels responded differently to different individuals, and did not discriminate between different expressions. The neurons that discriminated between expressions were found mainly in the cortex in the fundus of the superior temporal sulcus; the neurons that discriminated between identity were in contrast found mainly in the cortex in the lateral part of the ventral lip of the superior temporal sulcus (areas TEa and TEm). (Data from Hasselmo M E, Rolls E T and Baylis G C 1989 The role of expression and identity in the face-selective responses of neurons in the temporal visual cortex of the monkey. Behavioural Brain Research 32: 203-218.)

face is looking at one, or is looking away), and might respond to moving faces, and to facial gesture (Hasselmo, Rolls, Baylis and Nalwa, 1989b).

The discovery of neurons in the cortex in the superior temporal sulcus (STS) that respond to face expression, face and head movement, and social signals such as breaking eye contact by closing the eyes or by turning the head to break eye contact (Baylis, Rolls and Leonard, 1987; Hasselmo, Rolls and Baylis, 1989a; Hasselmo, Rolls, Baylis and Nalwa, 1989b) led to the proposal that these cortical regions are involved in decoding face stimuli important in social behavior in a third visual cortical stream (Baylis, Rolls and Leonard, 1987; Hasselmo, Rolls and Baylis, 1989a; Hasselmo, Rolls, Baylis and Nalwa, 1989b; Rolls, 2016b). That proposal has now been accepted (Pitcher and Ungerleider, 2021). I have gone beyond that, and shown that there are (at least) two cortical visual 'What' streams, and two visual cortical 'Where' streams in humans and other primates (Rolls, 2024c) (Section 4.2).

These discoveries also led to the proposal that this third visual cortical stream involves a combination of information from the ventral and dorsal visual streams (Hasselmo, Rolls and Baylis, 1989a; Hasselmo, Rolls, Baylis and Nalwa, 1989b), and that also has been accepted (Pitcher and

Ungerleider, 2021). The connectivity that underlies this convergence in humans (Rolls et al., 2023b; Rolls, 2024c) is described in Section 4.2.

This superior temporal sulcus (STS) system has been proposed to be the source of the inputs to neurons in the primate orbitofrontal cortex that have similar responses to socially relevant face expression and gesture (Rolls, Critchley, Browning and Inoue, 2006a). These neurons in the orbitofrontal cortex are likely to be important in social and emotional responses to faces, in that we discovered that face and voice expression processing is impaired by damage to the human orbitofrontal cortex (Hornak, Rolls and Wade, 1996; Hornak, Bramham, Rolls, Morris, O'Doherty, Bullock and Polkey, 2003). We also implicated impairments of processing in this STS visual stream in autism by the lower functional connectivity of these STS cortical regions in people with autism spectrum disorder (Cheng, Rolls, Gu, Zhang and Feng, 2015; Cheng, Rolls, Zhang, Sheng, Ma, Wan, Luo and Feng, 2017; Rolls, Zhou, Cheng, Gilson, Deco and Feng, 2020f).

Thus information in inferior temporal visual cortical regions that project to the amygdala and orbitofrontal cortex includes information about face identity, and about face expression and gesture (Rolls, Critchley, Browning and Inoue, 2006a; Leonard, Rolls, Wilson and Baylis, 1985; Rolls, 2011b). Both types of information are important in social and emotional responses to other primates (including humans), which must be based on who the individual is as well as on the face expression or gesture being made (Rolls, 2014a). One output for this information is probably via the ventral striatum, for we discovered a small population of neurons in the ventral striatum with responses selective for faces (Rolls and Williams, 1987; Williams, Rolls, Leonard and Stern, 1993).

4.5.5 The semantic properties of viewed visual stimuli activate other cortical regions that are semantically relevant

The pathways in the ventrolateral cortical visual 'What' stream to the inferior temporal cortex have been followed using effective connectivity (Rolls, 2024c; Rolls et al., 2023b,e; Rolls and Turova, 2025), and it has been shown exactly which regions respond best to faces, vs scenes, vs body parts, vs tools (Rolls, Feng and Zhang, 2024b). An interesting discovery here is that even stationary visual stimuli can activate cortical regions that relate to the semantics of each stimulus type (Rolls et al., 2024b, 2026). For example, the sight of stationary tools and body parts can activate visual motion areas such as FST and parts of the inferior parietal cortex implicated in performing actions in space (Rolls et al., 2024b) (Fig. 5.5). Another example is that looking at stationary faces and body parts activates some somatosensory cortical regions including inferior parietal PF and PFt, and even somatosensory regions 1 and 2 (Rolls et al., 2024b) (Fig. 5.5), which may relate to the fact that seeing a face or body part is often associated with touch to the face or body part.

4.6 Information in the cortex is represented by sparse distributed neuronal representations, not by 'grandmother' cells

4.6.1 Sparse distributed neuronal representations are found in the cortex

A key question for understanding brain function is whether a particular object (or face) is represented in the brain by the firing of one or a few gnostic (or 'grandmother' or 'cardinal') cells (as proposed by Barlow (1972)), or whether the representation is fully distributed with half the neurons 'On' (firing) and half 'Off' (not firing) for any one stimulus which was the distribution that Hopfield analyzed for attractor networks (Hopfield, 1982). Another possibility is that the firing of a small group or

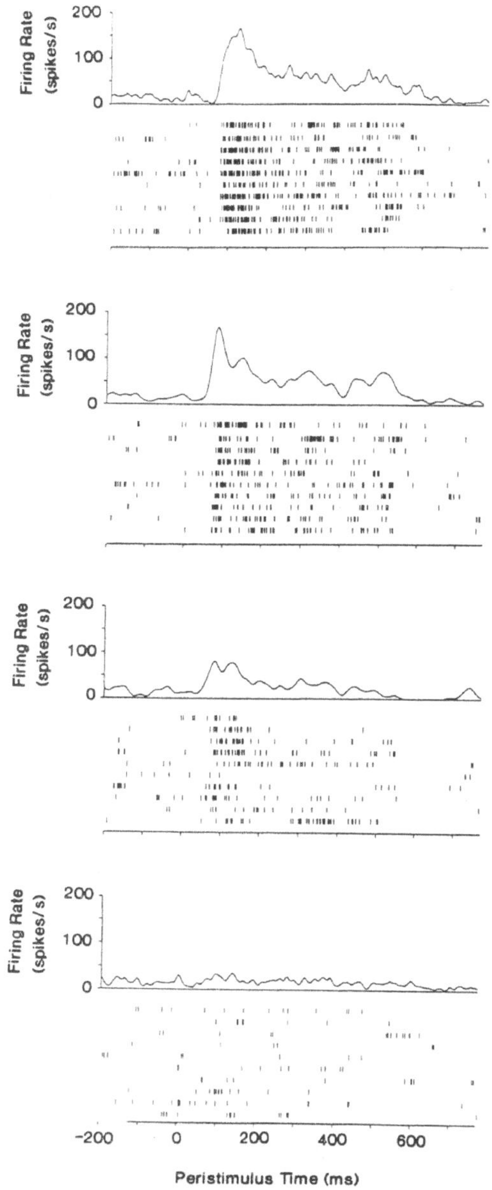

Fig. 4.13 Peristimulus time histograms and rastergrams showing the responses on different trials (originally in random order) of a face-selective neuron in the inferior temporal visual cortex to four different faces. (In the rastergrams each vertical line represents one spike from the neuron, and each row is a separate trial. Each block of the Figure is for a different face.) (Data from Tovee M J, Rolls E T, Treves A and Bellis R P. 1993 Information encoding and the responses of single neurons in the primate temporal visual cortex. Journal of Neurophysiology 70: 640–654.)

ensemble of cells each with somewhat different tuning to a subset of visual stimuli provides the representation, in what I term a 'sparse distributed' representation (Rolls, 2023c, 2026d). This is a key question, because it turns out to be a key factor that influences how many memories can be stored in a memory system in the brain, and is key to understanding many properties of brain function such as generalization between stimuli, completion of a whole memory from a part, and graceful degradation of function when the network is damaged (Rolls, 2023c, 2026d).

Very little (almost nothing) was known about how information is encoded by single neurons or populations of neurons in the higher visual cortical areas and the brain regions that receive from them such as the orbitofrontal cortex and hippocampus when we published our first paper on the inferior temporal visual cortex (Rolls et al., 1977), and I decided to take on this key issue in understanding brain function. I was interested to test whether the speculations of Horace Barlow, one of my mentors in The Physiological Laboratory at Cambridge when I was an undergraduate, about 'grandmother' or 'cardinal' cell encoding (Barlow, 1972), were correct.

We soon discovered that in the inferior temporal visual cortex and cortex in the anterior part of the superior temporal sulcus (IT/STS), responses of a group of neurons, but not of a single neuron, provide evidence about which face was shown. We showed, for example, that these neurons typically respond with a graded set of firing rates to different faces, with firing rates from 120 spikes/s to the most effective face, to no response at all to a number of the least effective faces (Baylis, Rolls and Leonard, 1985; Rolls and Tovee, 1995a; Rolls, 2016b). To provide examples, Fig. 4.13 shows typical firing rate changes of a single neuron on different trials to each of several different faces. This makes it clear that from the firing rate on any one trial, information is available about which stimulus was shown, and that the firing rate is graded, with a different firing rate response of the neuron to each stimulus. In fact, we discovered that the firing rate probability distribution of a single neuron to a set of stimuli is approximately exponential with only a few stimuli producing high firing rates, as described below and as illustrated in Fig. 4.15 (Rolls and Tovee, 1995a; Treves, Panzeri, Rolls, Booth and Wakeman, 1999; Baddeley, Abbott, Booth, Sengpiel, Freeman, Wakeman and Rolls, 1997; Franco, Rolls, Aggelopoulos and Jerez, 2007; Rolls, 2016b, 2026d).

The distributed nature of the encoding typical for neurons in the inferior temporal visual cortex is illustrated in Fig. 4.14, which shows that temporal cortical neurons typically responded to several members of a set of five faces, with each neuron having a different profile of responses to each face (Baylis, Rolls and Leonard, 1985). It would be difficult for most of these single cells to tell which of even five faces, let alone which of hundreds of faces, had been seen, so these neurons are not 'grandmother' or 'cardinal' cells. Yet across a population of such neurons, much information about the particular face that has been seen is provided, as shown below.

To provide a clear understanding of the typical firing rate distribution of neurons in the inferior temporal cortex to a set of stimuli, we decided to measure the probability distribution of firing rates for typical neurons in the inferior temporal cortex to a set of 20 face and non-face stimuli. The results are shown in Fig. 4.15, and show that the neurons typically have low firing rates to most of the stimuli, and high firing rates to a few visual stimuli from the set of 20 stimuli. In fact, we discovered that there is for many single neurons a good fit to an exponential distribution of firing rates (Franco, Rolls, Aggelopoulos and Jerez, 2007).

We next thought it useful to test whether we would obtain the same exponential firing rate distribution when a set of natural stimuli from the world was being viewed, so we played movies to the macaques while recording from inferior temporal cortex neurons. We discovered an exponential firing rate distribution in this situation too, with relatively few time periods when the firing rates were high (Baddeley, Abbott, Booth, Sengpiel, Freeman, Wakeman and Rolls, 1997).

To quantify the sparseness of the distribution for neurons with graded firing rates such as these, I decided to apply to the real neuronal data that we were recording the same sparseness measure that we developed for calculating memory storage capacities for neuronal networks (Rolls and Treves, 1990; Treves and Rolls, 1991; Rolls and Treves, 1998), to enable these two research areas to synergise. By this I mean that for the theoretical work in computational neuroscience, we needed empirical measures of the sparseness of the neuronal representations in different brain regions; and for the real neuronal data, it was useful to have a measure that enabled us to appreciate the implications of the sparseness found for memory storage in the brain. The single neuron sparseness measure a^s is shown

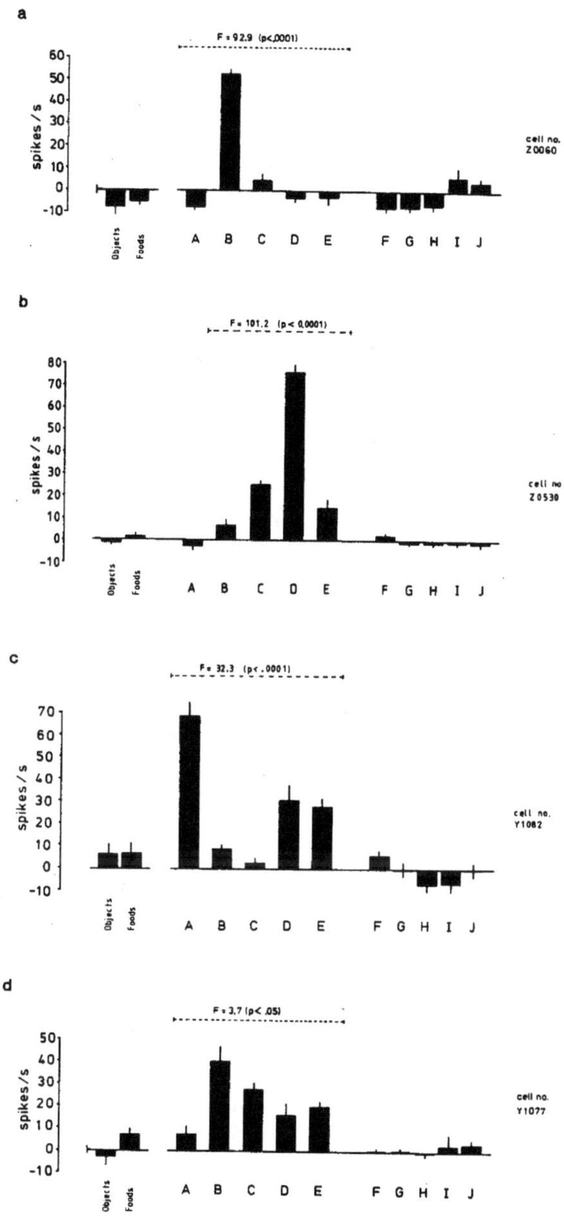

Fig. 4.14 Responses of four different temporal cortex visual neurons to a set of five faces (A–E), and, for comparison, to a wide range of non-face objects and foods. F–J are non-face stimuli. The means and standard errors of the responses computed over 8–10 trials are shown. (Data from Baylis G C, Rolls E T and Leonard C M. 1985 Selectivity between faces in the responses of a population of neurons in the cortex in the superior temporal sulcus of the monkey. Brain Research 342: 91–102.)

in Equation 4.1.

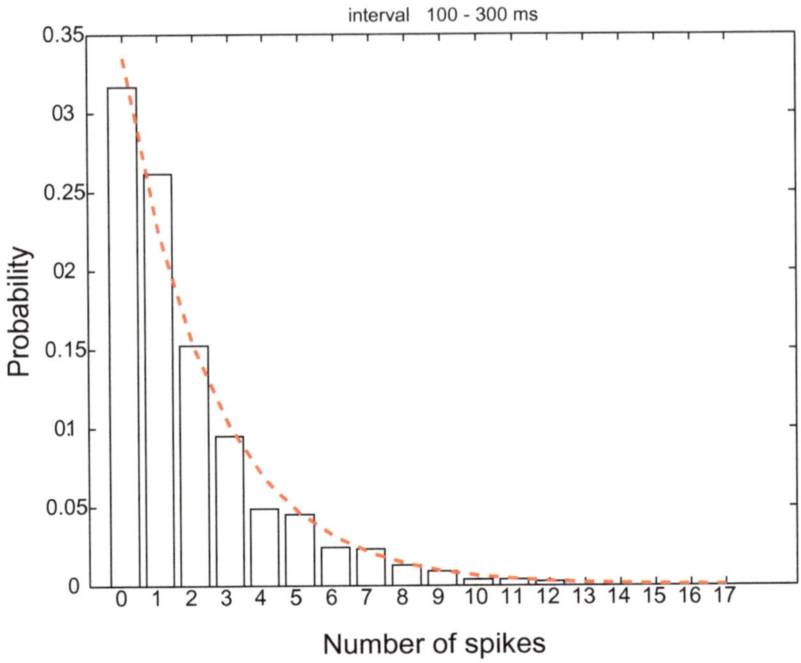

Fig. 4.15 The firing rate probability distribution of single neurons is approximately exponential. An exponential firing rate probability distribution obtained by pooling the firing rates of a population of 41 inferior temporal cortex neurons tested to a set of 20 face and non-face stimuli. The firing rate probability distribution for the 100–300 ms interval following stimulus onset was formed by adding the spike counts from all 41 neurons, and across all stimuli. The fit to the exponential distribution (dashed line) was high. (Data from Franco L, Rolls E T, Aggelopoulos N C and Jerez J M. 2007 Neuronal selectivity, population sparseness, and ergodicity in the inferior temporal visual cortex. Biological Cybernetics 96: 547–560.

$$a^{\mathrm{s}} = \frac{(\sum_{s=1}^{S} y_s/S)^2}{(\sum_{s=1}^{S} y_s^2)/S} \tag{4.1}$$

where y_s is the mean firing rate of the neuron to stimulus s in the set of S stimuli (Rolls and Treves, 1998; Rolls, 2023c, 2026d). For neurons with a binary representation, a^{s} is 0.5 for a fully distributed representation, and $1/S$ if a neuron responds to one of a set of S stimuli.

It turns out that the sparseness of the representation for a neuron with an exponential firing rate distribution such as that illustrated in Fig. 4.15 is 0.5, and that is our estimate of the sparseness of the representation for the inferior temporal visual cortex for neurons with graded firing rates (Franco, Rolls, Aggelopoulos and Jerez, 2007; Rolls, 2023c, 2026d).

These definitions, and what we discovered for cortical neuronal representations, are described further by Franco, Rolls, Aggelopoulos and Jerez (2007); Rolls and Treves (2011); Rolls (2016b, 2023c, 2026d).

The exact value of the sparseness found in any study is influenced by the set of stimuli used, but nevertheless it is interesting to see that somewhat similar values of sparseness are obtained in different sensory systems. The value of a for the inferior temporal visual cortex for sets of objects and faces is in the region 0.5 - 0.65 in the studies described here and by Rolls (2023c, 2026d). The value of a for 24 oral taste, temperature, and texture stimuli was 0.75 for the macaque insular primary taste cortex, 0.63 for the orbitofrontal cortex, and 0.81 for the amygdala (Rolls, 2023c, 2026d). For the hippocampus, for spatial view cells tested with 64 visual locations in the visually rich environment of a large laboratory, the sparseness a was lower, 0.34 (Rolls, 2023c, 2026d), consistent with the utility in a memory system of more sparse representations being useful because this increases the number of memories that can be stored and recalled, as described in Chapter 5 and by Rolls (2023c, 2026d); Rolls and Treves (2024).

The conclusion is that we discovered that sparse distributed representations, typically with exponential firing rate distributions of the type illustrated in Fig. 4.15, are used in the primate cerebral cortex. The representations may be less sparse in early cortical sensory regions where there is a need to maximize the amount of information that can be represented and transmitted (Rolls, 2023c, 2026d). The representation may be more sparse in memory systems such as the hippocampus where there is a need to maximize the number of memories that can be stored (Rolls, 2023c, 2026d). Further elaboration of these discoveries, including that the representation across a population of neurons has a similar sparseness to that as a single neuron, which indicates weak ergodicity and uncorrelated representations, is provided by Rolls and Treves (2011), Rolls (2023c, 2026d), and Franco et al. (2007).

Moreover, we have not only discovered the nature of representations as sparse and distributed within a local population of neurons in the cerebral cortex, but have also set out the very great advantages of the representations found, which in biologically plausible neuronal networks include generalization to similar stimuli, completion of a whole memory from a part, and graceful degradation of performance when the operation of the network is impaired (rather than the sudden and catastrophic loss of information in a digital computer when it is damaged) (Rolls, 2023c, 2026d, 2016b).

4.6.2 Information theory quantifies neuronal encoding

4.6.2.1 Information theory is important for understanding neuronal encoding in the brain

I have introduced the importance of information theory for understanding neuronal representations in the brain and how the brain works as follows in **Brain Computations and Principles; and AI** (Rolls, 2026d):

"Because single neurons are the computing elements of the brain and send the results of their processing by spiking activity to other neurons, we can understand brain processing by understanding what is encoded by the neuronal firing at each stage of the brain (e.g. each cortical area), and determining how what is encoded changes from stage to stage. Each neuron responds differently to a set of stimuli (with each neuron tuned differently to the members of the set of stimuli), and it is this that allows different stimuli to be represented. We can only address the richness of the representation therefore by understanding the differences in the responses of different neurons, and the impact that this has on the amount of information that is encoded. These issues can only be adequately and directly addressed at the level of the activity of single neurons and of populations of single neurons, and understanding at this neuronal level (rather than at the level of thousands or millions of neurons as revealed by functional neuroimaging) is essential for understanding brain computation.

Information theory provides the means for quantifying how much neurons communicate to other neurons, and thus provides a quantitative approach to fundamental questions about information processing in the brain. To investigate what in neuronal activity carries information, one must compare the amounts of information carried by different codes, that is different descriptions of the same activity, to provide the answer. To investigate the speed of information transmission, one must define and measure information rates from neuronal responses. To investigate to what extent the information provided by different cells is redundant or instead independent, again one must measure amounts of information in order to provide quantitative evidence. To compare the information carried by the number of spikes, by the timing of the spikes within the response of a single neuron, and by the relative time of firing of different neurons reflecting for example stimulus-dependent neuronal synchronization, information theory again provides a quantitative and well-founded basis for the necessary comparisons. To compare the information carried by a single neuron or a group of neurons with that reflected in the behavior of the human or animal, one must again use information theory, as it provides a single measure which can be applied to the measurement of the performance of all these different cases. In all these situations, there is no quantitative and well-founded alternative to information theory."

A key to understanding Shannon (1948) information theory is that information from different sources must be additive, and that this can be achieved by using a log measure, which conventionally is taken to the base 2. One bit of information enables us to select one between 2 possibilities. Two bits of information enable us to select one between 4 possibilities. Three bits of information enable us to choose between 8 possibilities[3]. More generally we define the amount I of information gained when we are informed which of k possibilities is correct as

$$I(k) = \log_2 k. \tag{4.2}$$

Then when we combine independent information, for example producing $k \times l$ possibilities from independent events with k and l possibilities respectively, we obtain

$$I(k \times l) = \log_2(k \times l) = \log_2 k + \log_2 l = I(k) + I(l). \tag{4.3}$$

An introduction to the fundamental elements of information theory is provided in Appendix 3 of Rolls (2023c, 2026d), where I also provide code for computing the information available from single neurons, and from populations of single neurons.

4.6.2.2 Some of the discoveries about neuronal encoding made using information theory

Here I summarize some of the main discoveries that we have made about neuronal encoding in the brain by using information theory. This type of analysis is essential for understanding how the brain works. These discoveries are described in Appendix 3 of Rolls (2023c, 2026d), where the original research papers are cited, with another source Rolls and Treves (2011) which makes it clear that the theoretical physicist Alessandro Treves was a wonderful colleague in helping these discoveries to be made. The discoveries we made are about neuronal encoding in the visual, hippocampal, taste, and olfactory cortical systems of behaving macaques, the closest model we have to neuronal activity in most parts of the human brain:

1. Much information is available about the stimulus presented in the number of spikes emitted by single neurons in a fixed time period, the firing rate. This is termed a rate code (Rolls, 2026d).

2. Much of this firing rate information is available in short periods, with a considerable proportion available in as little as 20 ms (Tovee and Rolls, 1995; Rolls and Tovee, 1994; Rolls et al., 1999b; Panzeri et al., 1999b). This rapid availability of information enables the next stage of processing to read the information quickly, and thus for multistage processing to operate rapidly in for example a sensory hierarchy. This time is the order of time over which a receiving neuron might be able

[3] Some understanding can be gained by noting that to choose one of 8 possibilities, three choices or guesses are needed. The first choice divides the set into four possibilities. The second choice into two possibilities. And the third choice finds which item is correct.

to utilize the information, given its synaptic and membrane time constants. In this time, a sending neuron is most likely to emit 0, 1, or 2 spikes.

3. This rapid availability of information is confirmed by population analyses, which indicate that across a population of neurons, much information is available in short time periods.

4. More information is available using this rate code in a short period (of e.g. 20 ms) than from just the first spike.

5. Little information is available by time variations within the spike train of individual neurons for static visual stimuli (in periods of several hundred milliseconds), apart from a small amount of information from the onset latency of the neuronal response (Tovee et al., 1993; Rolls et al., 2006b). A static stimulus encompasses what might be seen in a single visual fixation, what might be tasted with a stimulus in the mouth, what might be smelled in a single sniff, etc. For a time-varying stimulus, clearly the firing rate will vary as a function of time.

6. Across a population of neurons, the firing rate information provided by each neuron tends to be independent; that is, the information increases approximately linearly with the number of neurons (Rolls, Treves and Tovee, 1997d; Abbott, Rolls and Tovee, 1996). This applies of course only when there is a large amount of information to be encoded, that is with a large number of stimuli. The outcome is that the number of stimuli that can be encoded rises exponentially in the number of neurons in the ensemble. (For a small stimulus set, the information saturates gradually as the amount of information available from the neuronal population approaches that required to code for the stimulus set.) This applies up to the number of neurons tested and the stimulus set sizes used, but as the number of neurons becomes very large, this is likely to hold less well. An implication of the independence is that the response profiles to a set of stimuli of different neurons are uncorrelated.

In geometrical terms, what our population coding analysis shows is that in the multidimensional space of hundreds of stimuli, each neuron with its exponential firing rate distribution can be thought of as pointing in a direction in the high dimensional space of stimuli (Rolls and Treves, 2011). The average of the subset of the whole population of neurons that is active to a particular stimulus points towards a particular stimulus, e.g. a particular person's face, in the multidimensional space of stimuli. The subset of neurons that is active for a particular stimulus can be thought of as a population vector pointing in a particular direction in the high-dimensional stimulus space, and might be described as population vector coding, though this vector terminology is frequently used in low-dimensional spatial or motor spaces (Georgopoulos et al., 1986, 1989; Sarel et al., 2017; Tsytsarev et al., 2025), rather than in the high-dimensional stimulus spaces that we investigated (Rolls, Treves and Tovee, 1997d; Rolls and Treves, 2011). The firing rates of C neurons thus constitute a firing rate vector of dimensionality C, and this can be termed vectorial encoding with the vector of firing rates pointing in a particular direction in the C dimensional space. This type of population or vectorial coding is described by Rolls et al. (1997d), Rolls and Treves (2011), and Rolls (2026d).

Note that this is a different type of vector encoding to that implemented by single neurons that are tuned to for example a direction and a distance of a boundary (Bicanski and Burgess, 2020).

7. The information in the firing rate across a population of neurons can be read moderately efficiently by a decoding procedure as simple as a dot (i.e. inner) product (Rolls et al., 1997d). This is the simplest type of processing that might be performed by a neuron, as it involves taking a dot product of the incoming firing rates with the receiving synaptic weights to obtain the activation (e.g. depolarization) of the neuron. This type of information encoding ensures that the simple emergent properties of

associative neuronal networks such as generalization, completion, and graceful degradation (see Appendix 2 of Rolls (2023c, 2026d)) can be realized very naturally and simply.

8. There is little additional information to the great deal available in the firing rates from any stimulus-dependent cross-correlations or synchronization that may be present (Panzeri et al., 1999a; Franco et al., 2004; Rolls et al., 2004, 2003b; Aggelopoulos et al., 2005; Panzeri et al., 2022; Rolls, 2023c, 2026d). Stimulus-dependent synchronization might in any case only be useful for grouping different neuronal populations, and would not easily provide a solution to the binding problem in vision. Instead, the binding problem in vision may be solved by the presence of neurons that respond to combinations of features in a given spatial position with respect to each other (Salinas and Abbott, 1997).

9. There is little information available in the order of the spike arrival times of different neurons for different stimuli that is separate or additional to that provided by a rate code (Rolls et al., 2006b). The presence of spontaneous activity in cortical neurons facilitates rapid neuronal responses, because some neurons are close to threshold at any given time, but this also would make a spike order code difficult to implement.

10. Analysis of the responses of single neurons to measure the sparseness of the representation indicates that the representation is distributed, and not grandmother cell like (or local). Moreover, the nature of the distributed representation, that it can be read by dot product decoding, allows simple emergent properties of associative neuronal networks such as generalization, completion, and graceful degradation (see Appendix 2 of Rolls (2023c, 2026d)) to be realized very naturally and simply. The evidence becoming available from humans is consistent with this summary (Fried, Rutishauser, Cerf and Kreiman, 2014; Rolls, 2015b, 2017b).

11. The representation is not very sparse in the perceptual systems studied (as shown for example by the values of the single cell sparseness a^{s}), and this may allow much information to be represented. At the same time, the responses of different neurons to a set of stimuli are decorrelated, in the sense that the correlations between the response profiles of different neurons to a set of stimuli are low. Consistent with this, the neurons convey independent information, at least up to reasonable numbers of neurons. The representation may be more sparse in memory systems such as the hippocampus, and this may help to maximize the number of memories that can be stored in associative networks.

12. The nature of the distributed representation can be understood further by the firing rate probability distribution, which has a long tail with low probabilities of high firing rates. The firing rate probability distributions for some neurons fit an exponential distribution, and for others there are too few very low rates for a good fit to the exponential distribution. An implication of an exponential distribution is that this maximizes the entropy of the neuronal responses for a given mean firing rate under some conditions. It is of interest that in the inferior temporal visual cortex, the firing rate probability distribution is very close to exponential if a large number of neurons are included without scaling of the firing rates of each neuron. An implication is that a receiving neuron would see an exponential firing rate probability distribution.

13. The population sparseness a^{p}, that is the sparseness of the firing of a population of neurons to a given stimulus (or at one time), is the important measure for setting the capacity of associative neuronal networks. In populations of neurons studied in the inferior temporal cortex, hippocampus, and orbitofrontal cortex, it takes the same value as the single cell sparseness a^{s}, and this is a situation

of weak ergodicity that occurs if the response profiles of the different neurons to a set of stimuli are uncorrelated.

Understanding the neuronal code is fundamental for understanding how memory and related perceptual systems in the brain operate (Rolls, 2023c, 2026d), as follows:

Understanding the neuronal code helps to clarify what neuronal operations would be useful in memory and in fact in most mammalian brain systems (e.g. dot product decoding, that is taking a sum in a short time of the incoming firing rates weighted by the synaptic weights).

It clarifies how rapidly memory and perceptual systems in the brain could operate, in terms of how long it takes a receiving neuron to read the code.

It helps to confirm how the properties of those memory systems in terms of generalization, completion, and graceful degradation occur, in that the representation is in the correct form for these properties to be realized.

Understanding the neuronal code also provides evidence essential for understanding the storage capacity of memory systems, and the representational capacity of perceptual systems (Rolls, 2023c, 2026d).

Understanding the neuronal code is also important for interpreting functional neuroimaging, for it shows that functional imaging that reflects incoming firing rates and thus currents injected into neurons, and probably not stimulus-dependent synchronization, is likely to lead to useful interpretations of the underlying neuronal activity and processing. Of course, functional neuroimaging cannot address the details of the representation of information in the brain in the way that is essential for understanding how neuronal networks in the brain could operate, for this level of understanding (in terms of all the properties and working hypotheses described above) comes only from an understanding of how single neurons and populations of neurons encode information.

4.6.2.3 Examples of some of the discoveries about neuronal encoding using information theory

It has been possible to apply information theory to show that each neuron conveys on average approximately 0.4 bits of information about which face in a set of 20 faces has been seen (Tovee and Rolls, 1995; Tovee, Rolls, Treves and Bellis, 1993; Rolls, Treves, Tovee and Panzeri, 1997e). If a neuron responded to only one of the faces in the set of 20, then it could convey (if noiseless) 4.6 bits of (stimulus-specific) information about one of the faces (when that face was shown). If, at the other extreme, it responded to half the faces in the set, it would convey 1 bit of information about which face had been seen on any one trial. In fact, the average maximum (stimulus-specific) information about the best stimulus was 1.8 bits of information. This provides good evidence not only that the neuronal representation is distributed, but also that it is a sufficiently reliable representation that useful information can be obtained from it.

An important discovery is that when the information available from a population of neurons about which of 20 faces has been seen is considered, the information increases approximately linearly as the number of neurons in the population increases from 1 to 14 (Rolls, Treves and Tovee, 1997d; Abbott, Rolls and Tovee, 1996) (see Fig. 4.16 and Rolls (2023c, 2026d)). Remembering that the information in bits is a logarithmic measure, this shows that the representational capacity of this population of cells increases exponentially (see Fig. 4.17). This is the case both when an optimal, probability estimation, form of decoding of the activity of the neuronal population is used, and also when the neurally plausible dot product type of decoding is used (Fig. 4.16). (The dot product decoding assumes that what reads out the information from the population activity vector is a neuron or a set of neurons that operates just by forming the dot product of the input population vector of firing rates from a set of neurons and its synaptic weight vector – see Rolls, Treves and Tovee (1997d),

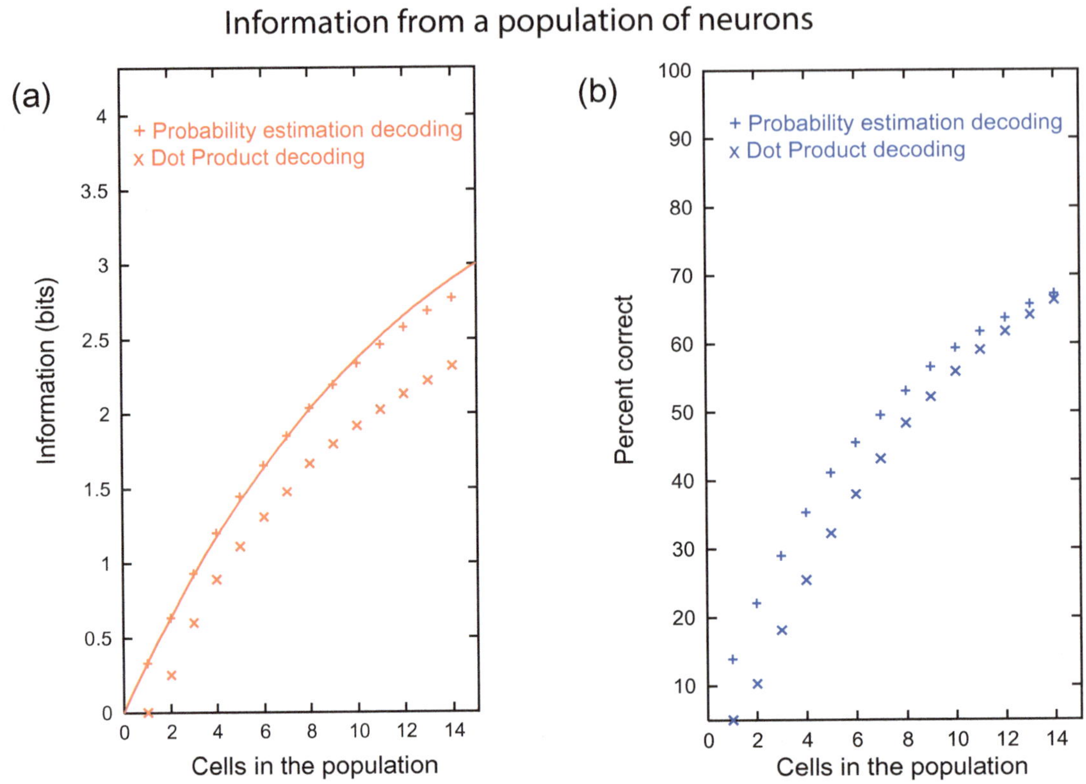

Fig. 4.16 (a) The values for the average information available in the responses of different numbers of inferior temporal visual cortex neurons on each trial, about which one of a set of 20 face stimuli has been shown. The decoding method was Dot Product (DP, $\times$) or Probability Estimation (PE, $+$). The full line indicates the amount of information expected from populations of increasing size, when assuming random correlations within the constraint given by the ceiling (the information in the stimulus set, I = 4.32 bits). (b) The percent correct for the corresponding data to those shown in (a). The measurement period was 500 ms. (Data from Rolls E T, Treves A and Tovee M J. 1997 The representational capacity of the distributed encoding of information provided by populations of neurons in the primate temporal visual cortex. Experimental Brain Research 114: 149–162.)

and Appendix 2 of Rolls (2023c, 2026d).) By simulation of further neurons and further stimuli, we have shown that the capacity grows very impressively, approximately as shown in Fig. 4.17 (Abbott, Rolls and Tovee, 1996). The result has been replicated with simultaneously recorded neurons (Rolls, Franco, Aggelopoulos and Reece, 2003b; Rolls, Aggelopoulos, Franco and Treves, 2004; Rolls, 2016b).

This result is exactly what would be hoped for from a distributed representation. This result is not what would be expected for local encoding, for which the number of stimuli that could be encoded would increase linearly with the number of cells. (Even if the grandmother cells were noisy, adding more replicates to increase reliability would not lead to more than a linear increase in the number of stimuli that can be encoded as a function of the number of cells.) Moreover, the encoding in the inferior temporal visual cortex about objects remains based on the spike count from each neuron, and not on the relative time of firing of each neuron or stimulus-dependent synchronization, when analyzed with simultaneous single neuron recording (Rolls, Franco, Aggelopoulos and Reece, 2003b; Rolls, Aggelopoulos, Franco and Treves, 2004; Franco, Rolls, Aggelopoulos and Treves, 2004) even in natural scenes when an attentional task is being performed (Aggelopoulos, Franco and Rolls, 2005). Further, we discovered that much of the information is available in short times of e.g.

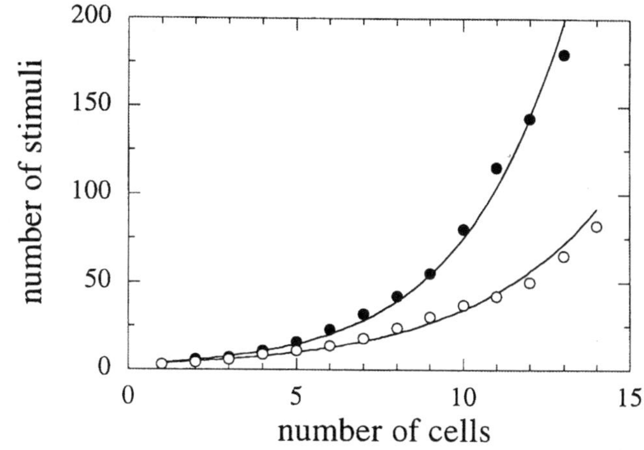

Fig. 4.17 The number of stimuli (in this case from a set of 20 faces) that are encoded in the responses of different numbers of neurons in the temporal lobe visual cortex, based on the results shown in Fig. 4.16. The solid circles are the result of the raw calculation, and the open circles correspond to the cross-validated case. (Data from Rolls E T, Treves A and Tovee M J. 1997 The representational capacity of the distributed encoding of information provided by populations of neurons in the primate temporal visual cortex. Experimental Brain Research 114: 149–162; and Abbott L F, Rolls E T and Tovee M J. 1996 Representational capacity of face coding in monkeys. Cerebral Cortex 6: 498–505.)

20 or 50 ms (Tovee and Rolls, 1995; Rolls, Franco, Aggelopoulos and Perez, 2006b), so that the receiving neuron does not need to integrate over a long time period to estimate a firing rate.

These discoveries, provided more fully and with equations in Appendix 3 of Rolls (2023c, 2026d), provide firm evidence that the encoding built at the end of the primate visual system is distributed, and that part of the power of this representation is that by receiving inputs from relatively small numbers of such neurons, neurons at the next stage of processing (for example in memory structures such as the hippocampus, amygdala, and orbitofrontal cortex) would obtain information about which of a very great number of objects or faces had been shown (Rolls and Treves, 2011; Rolls, 2023c, 2026d).

In this sense, the inferior temporal visual cortex provides a representation of objects and faces, in which information about which object or face is shown is made explicit in the firing of the neurons in such a way that the information can be read off very simply by memory systems such as the orbitofrontal cortex, amygdala, and perirhinal cortex / hippocampal systems. The information can be read off using dot product decoding, that is by using a synaptically weighted sum of inputs from inferior temporal cortex neurons (see further Rolls (2016b, 2023c, 2026d).

This representational capacity of neuronal populations has fundamental implications for the connectivity of the brain, for it shows that neurons need not have hundreds of thousands or millions of inputs to have available to them information about what is represented in another population of cells, but that instead the real numbers of perhaps 8,000–10,000 synapses per neuron would be adequate for them to receive considerable information from the several different sources between which this set of synapses is allocated.

It may be noted that it is unlikely that there are further processing areas beyond those described where ensemble coding changes into grandmother cell encoding. Anatomically, there does not appear to be a whole further set of visual processing areas present in the brain; and outputs from the temporal lobe visual areas such as those described are taken directly to limbic and related regions such as the amygdala and orbitofrontal cortex, and via the perirhinal and entorhinal cortex to the hippocampus (see Rolls (2000b, 2016b, 2011b, 2023c, 2026d)). Indeed, tracing this pathway onwards, we have found a population of neurons with face-selective responses in the amygdala, and in the majority of these neurons, different responses occur to different faces, with ensemble (not local) coding still being present (Leonard, Rolls, Wilson and Baylis, 1985). The amygdala, in turn, projects to another structure that may be important in other behavioral responses to faces, the ventral striatum, and comparable neurons have also been found in the ventral striatum (Williams, Rolls, Leonard and Stern, 1993). We have also recorded from face-responding neurons in the part of the orbitofrontal cortex that receives from the IT/STS cortex, and discovered that the encoding there is also not local but is distributed (Rolls, Critchley, Browning and Inoue, 2006a).

4.7 A biologically plausible theory and model of invariant visual face and object recognition

A key aspect of neuroscience discoveries is the ability to go beyond phenomena, empirical discoveries such as neurons that respond invariantly and selectively to faces or objects, to theories of how the computations are performed in the brain to produce these very useful types of neuronal representation. Analysis of the neuronal phenomena can show *'What'* is computed, but it is necessary and we have been able to go beyond that to produce theories of *'How'* it is computed. In this spirit, one of my books has the title **Brain Computations: What and How'** (Rolls, 2021a). The combination of these two approaches is what I believe has led to the major progress that has been achieved since the 1970s in understanding so much more than we did in 1970, which was almost nothing, about how our brains work. In this section I illustrate these fundamental points by showing how it has been possible to generate a theory and a biologically plausible computational model of how the ventral visual system builds transform-invariant representations of objects and faces (Rolls, 2021c, 2023c, 2026d, 1992a; Wallis and Rolls, 1997; Rolls, 2012b, 2023c, 2026d; Rolls, Zhang and Feng, 2025a; Zhang, Rolls and Feng, 2026a).

4.7.1 A theory and VisNet model of how invariant representations of objects and faces are computed in the ventral visual system

An overview of my theory and model of *how* the computations are performed in the ventral visual system is provided next.

Cortical visual processing for object recognition is considered to be organized as a set of hierarchically connected cortical regions consisting at least of V1, V2, V4, posterior inferior temporal cortex (TEO), inferior temporal cortex (e.g. TE3, TEa and TEm), and anterior temporal cortical areas (e.g. TE2 and TE1) (in macaques). There is convergence from each small part of a region to the succeeding region (or layer in the hierarchy) in such a way that the receptive field sizes of neurons (e.g. 1 degree near the fovea in V1) become larger by a factor of approximately 2.5 with each succeeding stage (Rolls, 1992a, 2023c, 2026d) (Fig. 4.18 right). Such zones of convergence would overlap continuously with each other (Fig. 4.18 left). (The sizes of typical parafoveal receptive fields would not be inconsistent with the calculated approximations of e.g. 8 degrees in V4, 20 degrees in TEO, and 50 degrees in the inferior temporal cortex (Boussaoud, Desimone and Ungerleider, 1991))

Soon after the time that we were discovering face and object cells and their transform invariant representations in the inferior temporal visual cortex in 1979-1982 (Section 4.4), I developed approaches to understanding how neuronal networks might form these types of representation, and reported my thinking to the 1985 Dahlem Conference on the **Neural and Molecular Bases of Learning** (Rolls, 1987). I soon developed a set of hypotheses about how invariant object and face recognition could be performed in the ventral visual stream cortex illustrated in Fig. 4.18, and

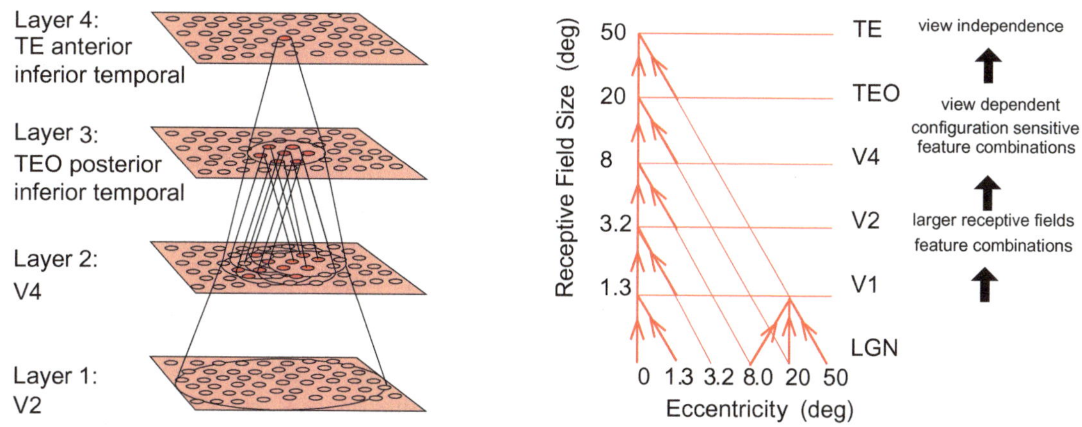

Fig. 4.18 Convergence in the visual system. Right – as it occurs in the brain. V1, visual cortex area V1; TEO, posterior inferior temporal cortex; TE, anterior inferior temporal cortex (IT). Left – as implemented in VisNet, a model of invariant visual object recognition. Convergence through the network is designed to provide fourth layer neurons with information from across the entire input retina, by providing an increase of receptive field size of 2.5 times at each stage. Layer 1 of the model corresponds to V2 in the brain. (Right: Data from Rolls E T. 1992 Neurophysiological mechanisms underlying face processing within and beyond the temporal cortical visual areas. Philosophical Transactions of the Royal Society of London B 335: 11-21. Left: Data from Wallis G and Rolls E T. 1997 Invariant face and object recognition in the visual system. Progress in Neurobiology 51: 167–194.

these in turn led to a proposed computational architecture for invariant visual object recognition, VisNet, illustrated in Fig. 4.18 (left) (Rolls, 1992a), which can be summarized as follows, with full descriptions available (Rolls, 2023c, 2026d, 2021c, 2012b, 2016b; Zhang et al., 2026a; Wallis and Rolls, 1997).

First, as illustrated in Fig. 4.18 (left), there is a series of competitive networks, organized in hierarchical layers corresponding to regions in the ventral visual pathway, that utilize mutual inhibition over a short range within each layer. These competitive networks allow combinations of features or inputs occurring in a given spatial arrangement to be learned by neurons, ensuring that higher order spatial properties of the input stimuli are represented in the network. This can be described as a feature hierarchy network. (Competitive networks are introduced in Section 1.4, and described more fully in Appendix B.4 of Rolls (2023c, 2026d).) The diluted connectivity of the neocortex via which each neuron receives from only about 10% of the neurons in the corresponding part of the preceding stage of processing helps these neurons to separate out the input representations well, and to not drift during learning (Rolls, 2023c, 2026d).

Second, the architecture of VisNet (Fig. 4.18 left) is like that of the visual pathways as illustrated in Fig. 4.18 (right), with a convergent series of connections from a localized population of cells in preceding layers to each cell of the following layer, thus allowing the receptive field size of cells to increase through the visual processing regions or layers.

Third, a modified associative (Hebb-like, see Equation 4.4) learning rule incorporating a temporal trace of each neuron's previous activity is used to enable the neurons to learn transform invariances. The hypothesis is that because objects have continuous properties in space and time in the world, an object at one place on the retina might activate feature analyzers at the next stage of cortical processing, and when the object was translated to a nearby position, because this would occur in a short period (e.g. 0.5 s), the membrane of the postsynaptic neuron would still be in its 'Hebb-modifiable' state (caused for example by calcium entry as a result of the voltage-dependent activation of NMDA receptors), and the presynaptic afferents activated with the object in its new position would thus become strengthened on the still-activated postsynaptic neuron.

I proposed that the short temporal window (e.g. 0.5 s) of Hebb-modifiability helps active neurons to learn the statistics of objects moving and more generally transforming across different retinal positions, sizes, views etc in the physical world, because the same postsynaptic firing is maintained by the temporal trace, allowing the different inputs produced during several seconds to be learned onto the same postsynaptic neuron (Rolls, 1992a, 2021c). Földiák (1991) had proposed computing an average activation of the postsynaptic neuron to assist with translation invariance, but I produced a whole theory of how a cortical hierarchy could gradually across successive cortical areas of the ventral visual system learn all transforms, to produce by the inferior temporal cortex neurons with invariant representations across any transform, with the architecture illustrated in Fig. 4.18 (left) (Rolls, 1992a). One idea here is that the temporal properties of the biologically implemented synaptic learning mechanism which uses NMDA receptors with a relatively slow time constant of 100 ms are such that they are well suited to detecting the relevant continuities in the world of real objects (Rolls, 1992a, 2021c). Another suggestion (Rolls, 1992a, 2021c) is that a memory trace for what has been seen in the last 300 ms appears to be implemented by a mechanism as simple as continued firing of inferior temporal cortex neurons after the stimulus has disappeared, as we discovered in visual masking experiments (Rolls and Tovee, 1994; Rolls, Tovee, Purcell, Stewart and Azzopardi, 1994b; Rolls, Tovee and Panzeri, 1999b; Rolls, 2003, 2016b). This continuing firing is the result of local attractor cortical networks using the recurrent connections between nearby cortical neurons, for the continuing firing is not a property of visual neurons in the lateral geniculate nucleus before the cortex is reached (Rolls, 2023c, 2026d).

A simple form of the trace update rule used to train the synaptic weights is as follows:

$$\delta w_j = \alpha \overline{y}^{\tau} x_j \tag{4.4}$$

where

$$\overline{y}^{\tau} = (1-\eta) y^{\tau} + \eta \overline{y}^{\tau-1} \tag{4.5}$$

and

x_j:	j^{th} input to the neuron.	y:	Output from the neuron.
$\overline{y}^{\tau}$:	Trace value of the output of the neuron at time step τ.	α:	Learning rate. Annealed between unity and zero.
w_j:	Synaptic weight between j^{th} input and the neuron.	η:	Trace value. The optimal value varies with presentation sequence length.

We investigated and discovered many improved versions of this trace learning rule (Rolls and Milward, 2000; Rolls and Stringer, 2001a).

A fundamental property is that VisNet is a self-organising unsupervised network: there is no teacher anywhere instructing any neuron what its response should be. Instead, VisNet uses the temporal statistics present in the natural environment to learn the transforms that correspond to a particular object or face. In contrast, AI (artificial intelligence) approaches such as deep learning using backpropagation of errors typically require prespecification of the object categories that should be learned (that is, the use of a teacher for each output neuron, which is biologically implausible). In practice these deep learning systems are trained with very large numbers of exemplars of a class, typically without systematic use of close transforms of the same object. This requires very large numbers of training trials. Deep learning networks may have up to 100 layers, again biologically implausible (see Section 4.7.6). Moreover, little is learned from these deep learning models (Cadieu, Hong, Yamins, Pinto, Ardila, Solomon, Majaj and DiCarlo, 2014; Yamins and DiCarlo, 2016) about

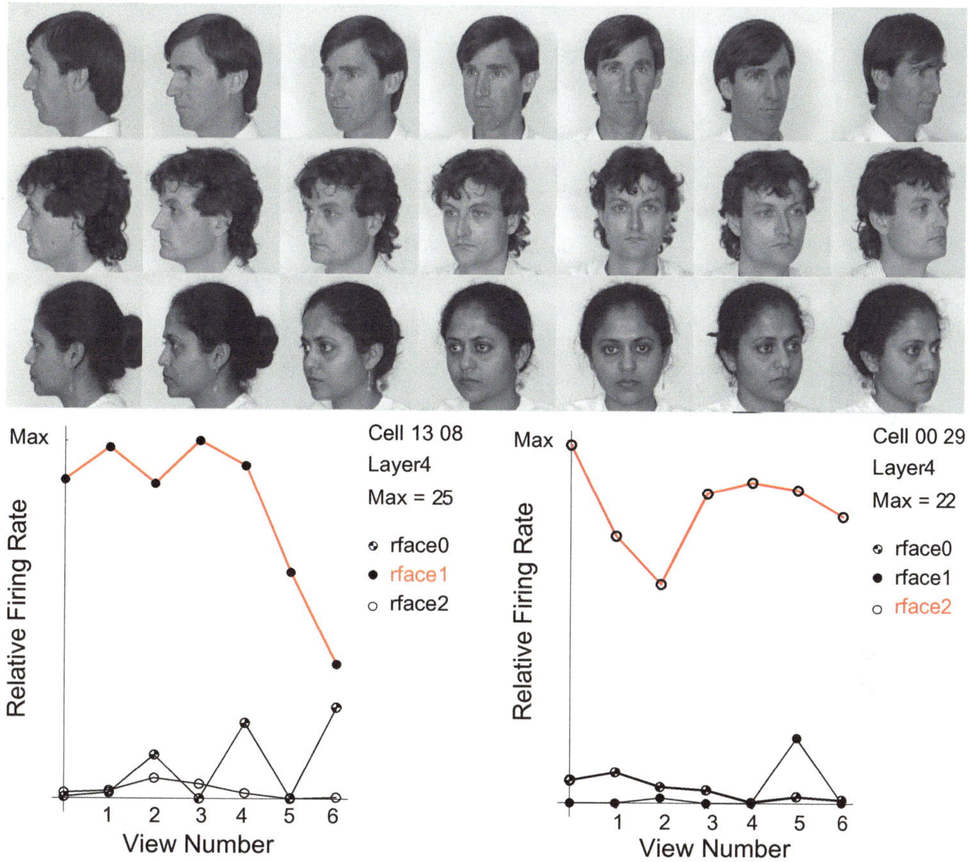

Fig. 4.19 Upper. Three faces in seven different views used as stimuli in an experiment in view-invariant recognition by VisNet. Lower: Response profiles for cells in Layer 4 of the VisNet network with seven different views of each of three faces. (Data from Wallis G and Rolls E T. 1997 Invariant face and object recognition in the visual system. Progress in Neurobiology 51: 167–194.

how the brain solves complex problems such as invariant visual object and face recognition, because deep convolutional learning does not use biological constraints such as local learning rules, and when mappings from images to categories are found, clear computational principles of how the mapping is achieved are largely not understood (Plebe and Grasso, 2019; Sejnowski, 2020). Moreover, when deep convolutional neural networks trained by error backpropagation predict the category into which an image exemplar should be placed, this is performed by methods such as using local features, whereas human perception typically uses the shape of the whole object (Bowers et al., 2022; Rolls, 2026d). The implication is that although artificial deep neural network models may learn to predict how to categorize a set of visual stimuli, they may do so in low-level ways not used by humans and may not reveal how human perception works (Bowers et al., 2022; Rolls, 2026d). VisNet is as far as I know the most biologically plausible approach to how computations in the ventral visual system for invariant object and face recognition are performed using unsupervised 'slow', memory trace-based, learning (Rolls, 2021c; Zhang et al., 2026a; Rolls, 2026d).

To illustrate how neurons in VisNet can learn to show invariant representations, Fig. 4.19 shows how different neurons in Layer 4 of VisNet can learn to respond to all views of one face, and to no views of any other face, when trained with the different views of each face being shown close together in time. The face stimuli are illustrated in Fig. 4.19 (upper).

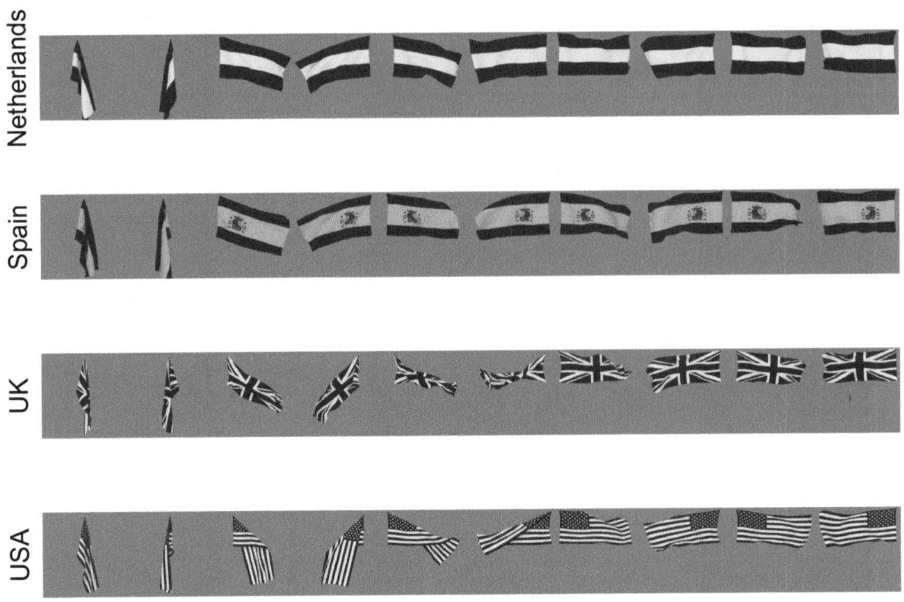

Fig. 4.20 Deformation-invariant object recognition. The flag stimuli used to train VisNet. Each flag is shown with different wind forces and rotations. Starting on the left the first pair, both the 0° and 180° views are shown for windspeed 0, and each successive pair is shown for wind force increased by 50 Blender units. (Data from Webb T J and Rolls E T. 2014 Deformation-specific and deformation-invariant visual object recognition: pose vs identity recognition of people and deforming objects. Frontiers in Computational Neuroscience 8: 37.)

4.7.2 Discoveries of how the VisNet model can solve many problems in invariant face and object recognition

Many discoveries have been made with the basic VisNet architecture and developments of it including the following, with more details elsewhere (Rolls, 2023c, 2026d, 2021c, 2012b; Zhang et al., 2026a; Rolls, 2026d) and in the original publications.

First, invariant representations can be formed for translation (Wallis and Rolls, 1997), view (Wallis and Rolls, 1997), lighting (Rolls and Stringer, 2006), and morphometric transforms such as a flag waving in a breeze or a person with different postures such as standing up vs sitting down (Webb and Rolls, 2014) (see e.g. Fig. 4.20 for some stimuli for which deformation-invariant representations could be learned by VisNet).

Second, the feature binding that is implemented enables separate representations to be learned for whole images (such as the four sides of a square) and a subset of the features (such as two sides of the square) (Elliffe, Rolls and Stringer, 2002; Rolls and Mills, 2018).

Third, an important aspect of the architecture of VisNet is that it need not be trained with every stimulus in every possible transform. For example, part of the hypothesis (Rolls, 1992a, 2023c, 2026d) is that training early layers (e.g. 1–3) with a wide range of visual stimuli will set up feature analyzers in these early layers that are appropriate later on with no further training of early layers for new objects. This generalization to new untrained transforms has been shown to be possible (Elliffe, Rolls and Stringer, 2002).

Fifth, once trained on a set of objects, VisNet can identify those objects correctly in cluttered natural environments (Stringer and Rolls, 2000). This occurs because the background clutter has not been trained as part of the object representation. We discovered that this is a property of inferior temporal cortex neurons (Rolls, Aggelopoulos and Zheng, 2003a), so it was important to discover whether VisNet could implement this.

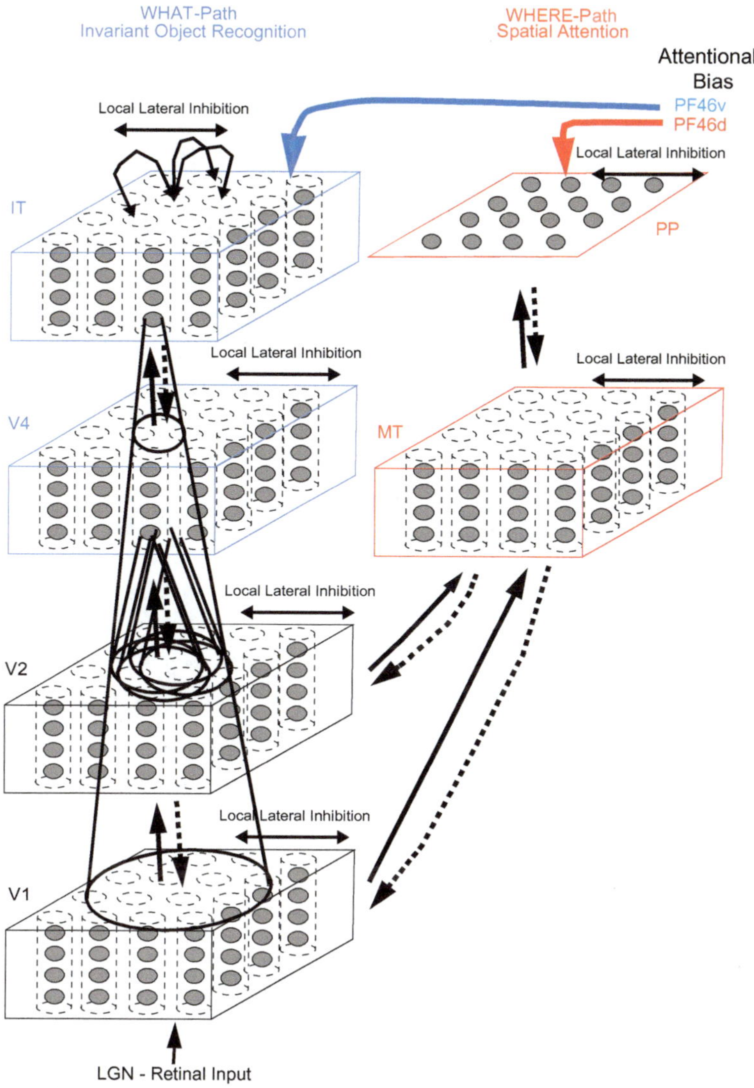

Fig. 4.21 Cortical architecture for hierarchical and attention-based visual perception. The system has five modules structured such that they resemble the two main visual paths of the primate visual cortex. Information from the retino-geniculo-striate pathway enters the visual cortex through area V1 in the occipital lobe and proceeds into two processing streams. The occipital-temporal stream leads ventrally through V2–V4 and IT (inferior temporal visual cortex), and is mainly concerned with object recognition. The occipito-parietal stream leads dorsally into PP (posterior parietal complex), and is responsible for maintaining a spatial map of an object's location. The solid lines with arrows between levels show the forward connections, and the dashed lines the top-down backprojections. Short-term memory systems in the prefrontal cortex (PF46) apply top-down attentional bias to the object or spatial processing streams. (After Deco G and Rolls E T. 2004 A neurodynamical cortical model of visual attention and invariant object recognition. Vision Research 44: 621–644.)

Sixth, we have shown that the receptive fields of Layer 4 neurons in a VisNet-like simulation can reduce in size and include mainly the fovea when objects are shown against a cluttered background (Trappenberg, Rolls and Stringer, 2002). We showed that this property relies on a greater cortical magnification factor for the fovea (Trappenberg et al., 2002), and models what we discovered for real neurons in the inferior temporal visual cortex (Rolls, Aggelopoulos and Zheng, 2003a).

Seventh, we have extended the VisNet architecture by adding top-down bias from the prefrontal

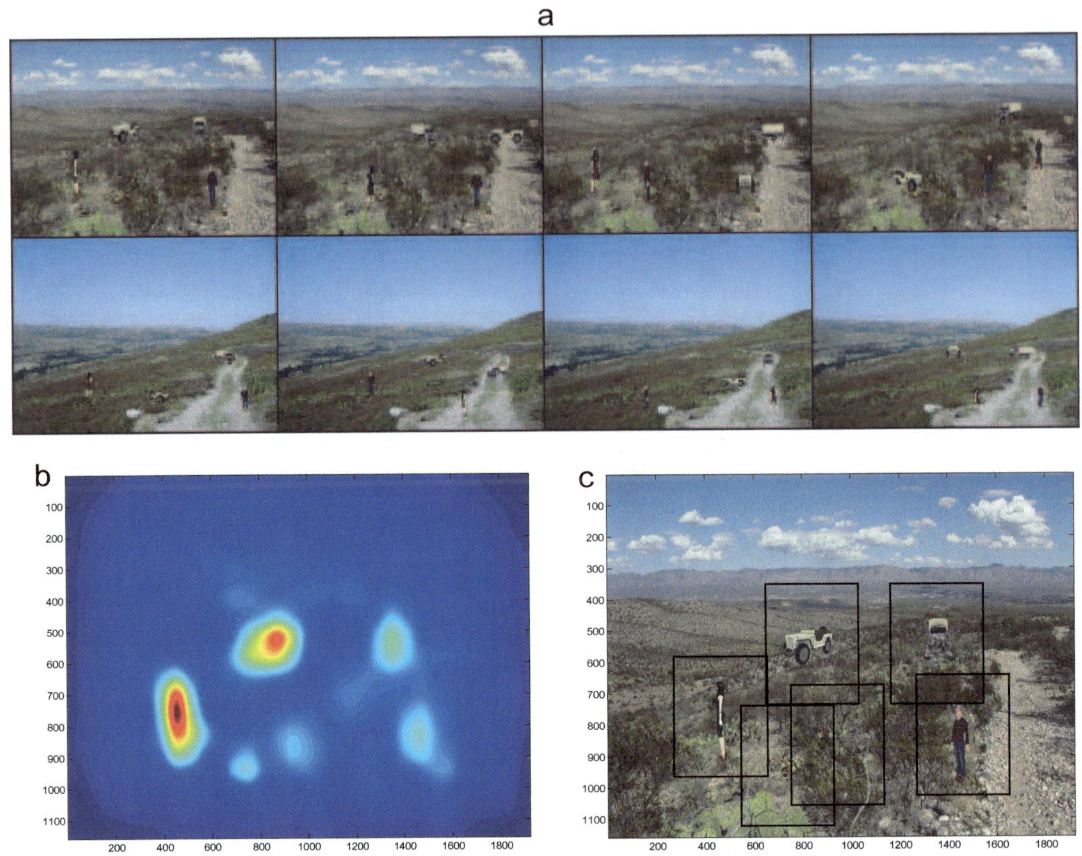

Fig. 4.22 Finding and recognizing objects in natural scenes. a. Eight of the 12 test scenes. Each scene has 4 objects, each in one of its four views.
b. The bottom up saliency map generated by the GBVS code for one of the scenes. The highest levels in the saliency map are red, and the lowest blue.
c. Rectangles (384x384 pixels) placed around each peak in the scene for which the bottom-up saliency map is illustrated in (b). (Data from Rolls E T and Webb T J. 2014 Finding and recognizing objects in natural scenes: complementary computations in the dorsal and ventral visual systems. Frontiers in Computational Neuroscience 8: 85.)

cortex to investigate effects of top-down attention as illustrated in Fig. 4.21 (Deco and Rolls, 2004). With this architecture, we were able to show that top-down spatial attention can reduce the reduction in receptive field size in a cluttered background for the object that is the target of attention (Deco and Rolls, 2004), which we discovered occurs for real neurons in the inferior temporal visual cortex (Rolls, Aggelopoulos and Zheng, 2003a).

Eighth, we have shown that the spatial continuity that can occur when the fixation point slides across an object can provide some transform invariance learning, even when there is no temporal continuity that can be used by a memory trace learning rule (Perry, Rolls and Stringer, 2006; Stringer, Perry, Rolls and Proske, 2006; Perry, Rolls and Stringer, 2010).

Ninth, we have shown that an extended version of VisNet can operate usefully to find and then recognize objects found at different locations in natural scenes such as that illustrated in Fig. 4.22 (Rolls and Webb, 2014). To achieve this, the operation of the dorsal visual system in providing a saliency map that would guide the locations to which visual fixations would occur was simulated with a bottom up saliency algorithm that is one of the standard ones that has been developed, which adopts the Itti and Koch (2000) approach to visual saliency. The approach is implemented by graph-based visual saliency (GBVS) algorithms (Harel, Koch and Perona, 2006a,b). The basis for the saliency map consists of features such as high contrast edges, and the saliency system knows nothing about objects, people, vehicles etc. This saliency system performs well, that is similarly to humans, in many bottom-up saliency tasks.

With the scene illustrated in Fig. 4.22a, the saliency map that was produced is illustrated in Fig. 4.22b. The peaks in this saliency map were used as the site of successive 'visual fixations', at each of which a rectangle (of 384x384 pixels) was placed, and was used as the input image to VisNet as illustrated in Fig. 4.22c. VisNet then had sufficient translation invariance to be able to correctly recognize each of the objects at each saliency or visual fixation location (Rolls and Webb, 2014).

Tenth, we were able to show with VisNet how the non-accidental properties of stimuli, such as concave curvature, or convex curvature, can arise and be implemented in transform invariant object recognition, as follows.

Non-accidental properties (NAPs) of images of objects are those that are relatively invariant over rotations in depth, and can be distinguished from metric properties (MPs), such as the aspect ratio of a part or the degree of curvature of a contour, which do vary continuously with rotation in depth (Biederman, 1987), as described next.

It has been proposed that non-accidental properties of objects as they transform into different views are useful for view-invariant object recognition processes, and that metric properties are much less useful (Biederman, 1987; Biederman and Gerhardstein, 1993; Kayaert, Biederman and Vogels, 2005b; Amir, Biederman and Hayworth, 2011; Kayaert, Biederman, Op de Beeck and Vogels, 2005a). Non-accidental properties (NAPs) are relatively invariant over rotations in depth (Biederman, 1987). A NAP is an image property, such as the linearity of a contour or the cotermination of a pair of contours, that is unaffected by rotation in depth, as long the surfaces manifesting that property are still present in the image (Lowe, 1985). NAPs can be distinguished from metric properties (MPs), such as the aspect ratio of a part or the degree of curvature of a contour, which do vary continuously with rotation in depth. There is psychophysical and functional neuroimaging evidence that supports this distinction (Kayaert et al., 2005b; Amir et al., 2011; Kayaert et al., 2005a). In addition, there is evidence that neurons in the lateral occipital cortex and inferior temporal visual cortex of the macaque are tuned more to non-accidental than to metric properties (Kim and Biederman, 2012; Vogels, Biederman, Bar and Lorincz, 2001).

These concepts are illustrated in Fig. 4.23. If we take object 1 as an example (Fig. 4.23), then we can see that it is approximately cylindrical but with concave sides. Its appearance when vertical is at 0 deg (close to the view illustrated of -6 and 6 deg). As the object is tilted towards or away from the viewer, the degree of curvature appears larger. The degree of curvature is thus a metric property (MP) of the object which is not very useful in transform-invariant object recognition. On the other hand, the fact that object 1 has concave sides is apparent across all these transforms, and therefore is a non-accidental property (NAP) used in identifying an object. This is made evident by a comparison with object 7, which has convex edges, which are always convex through these transforms, while the degree of curvature is a metric property and again varies as a function of the tilt transform. Further, object 4, a cylinder, always has straight sides whatever its tilt, so having straight sides is in this case also a NAP.

We tested VisNet with these sets of stimuli to investigate whether it would learn to represent the three cylinder types based on the non-accidental properties, and would treat the metric properties as transforms, and we found that VisNet did indeed operate in this way (Rolls and Mills, 2018).

Further, consideration of what is illustrated in Fig. 4.23 leads in fact to a new theory about how Non-Accidental Properties are learned, and how they are relatively invariant with respect to metric changes (Rolls and Mills, 2018; Rolls, 2021c). If we look at object 1, we see that a range of curvatures

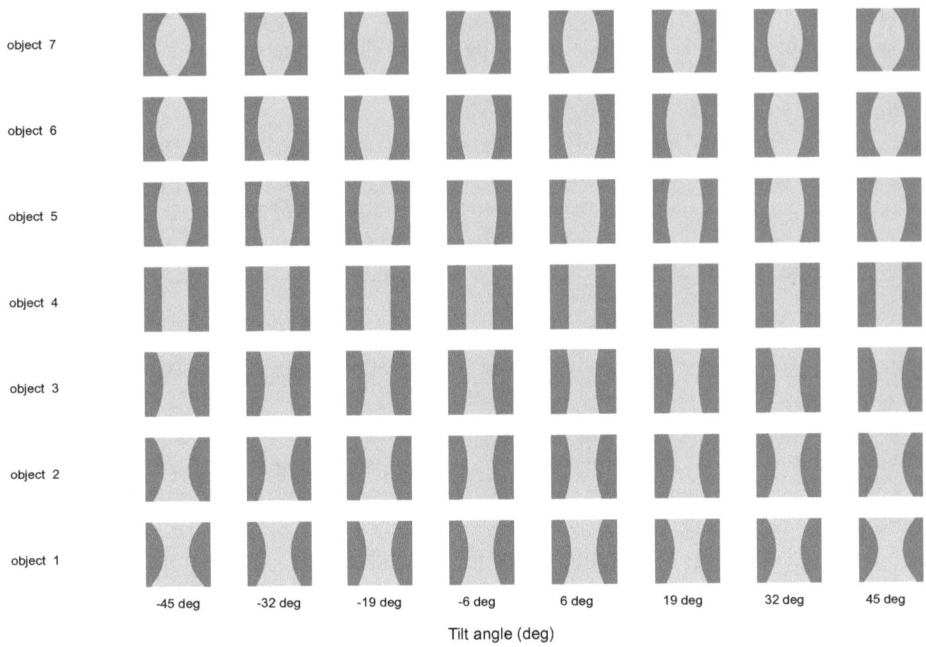

Fig. 4.23 Encoding of non-accidental properties. Stimuli used to investigate non-accidental properties (NAP) vs metric properties (MP). Each object is shown as white on a grey background. Objects 1–3 have the non-accidental property of concave edges. Object 4 has the non-accidental property of parallel edges. Objects 5–7 have the non-accidental property of convex edges. The vertical view of each object was at 0 deg, with the images at -6 and 6 deg of tilt illustrated. Different amounts of tilt of the top towards or away from the viewer are shown at the tilt angles indicated. Each object was thin, and was cut off at the top and bottom to ensure that any view of the top or bottom of the object did not appear, so that the type of curvature of the edges (concave, straight, or convex) was the main cue available. (Data from Rolls E T and Mills W P C. 2018 Non-accidental properties, metric invariance, and encoding by neurons in a model of ventral stream visual object recognition, VisNet. Neurobiology of Learning and Memory 152: 20–31.)

are shown as the object transforms. These different curvatures are all learned as properties of this class of object by the temporal trace learning rule, because those different views were presented close together in time so that the trace learning rule could learn that these were transforms of the same object. But if we then look at object 2, we see that some of the degrees of curvature in some of its views are similar to what has already been learned for object 1. So without further training of any association between object 1 and object 2, there will be some generalization across these two objects with different degrees of curvature, because some of the degrees of curvature are similar for some of the views of these two objects. *It is in this way, we proposed (Rolls and Mills, 2018), that non-accidental properties are learned by the brain, and generalize across a range of objects with the same non-accidental property (e.g. concave sides), even though the degree of curvature, the metric property, may be quite different.*

4.7.3 Increased biological plausibility in VisNet3: a weight decay synaptic learning rule, and limiting the synaptic weights

4.7.3.1 Beneficial effects of a synaptic learning rule that reduces the synaptic strength from inputs with low activity

VisNet utilizes competitive networks in each of its 4 layers (Rolls, 2023c, 2026d, 2021c, 2012b) (Fig. 4.18). In a competitive network, it is important that all neurons compete on an equal basis, so that different neurons learn to respond to different inputs, and similar inputs are allocated to the same neuron, so that categorisation is performed usefully (Rolls, 2023c, 2026d).

The usual way in which this is implemented in a competitive net is that after learning with a Hebbian associative rule or one with a short term memory trace in the post-synaptic term y such as that in Equation 4.4, the length of the vector of synaptic weights on a neuron is set to one (Rumelhart and Zipser, 1985; Hertz et al., 1991; Rolls, 2023c, 2026d). Given that a Hebbian rule will always increase synaptic weights if the presynaptic and postsynaptic firing rates are greater than 0, the synaptic modification will increase some synaptic weights. The weight normalization (setting the sum of the squares of the weights = 1) will then decrease the weights usefully (Hertz et al., 1991; Rolls, 2023c, 2026d). That is what is implemented in VisNet, by dividing the synaptic weight vector on a neuron by the length of its synaptic weight vector after its synapses have received an update (Rolls, 2012b, 2021c).

However, setting the length of the synaptic weight vector on each neuron is not very biologically plausible. So for VisNet3 we introduced an alternative method of allowing each neuron to compete on an equal basis by using a learning rule that allows synaptic weights to decrease in value if they are on a strongly activated neuron, and the current weight is larger than the presynaptic term (Zhang, Rolls and Feng, 2026a). This provides for heterosynaptic long-term depression, which as many experimentalists will know, is easier to obtain if the synaptic weights are already high. The rule we introduce for VisNet3 is

$$\delta w_j = \alpha y (x_j - w_j) \tag{4.6}$$

where δw_j is the change of the synaptic weight, α is the learning rate, y is the post-synaptic firing rate, x_j is the presynaptic firing rate, and w_j is the present value of the synaptic weight. This rule was used in different applications previously (Willshaw and von der Malsburg, 1976; Grossberg, 1976; Hertz et al., 1991), and has been termed the 'standard competitive net learning rule' (Hertz et al., 1991). The rule implements heterosynaptic long-term synaptic depression, that is, a decrease of the synaptic weight for an inactive input to an active post-synaptic neuron which can have many advantages (Rolls, 2023c, 2026d), but makes the amount of this depression greater for strong synaptic weights.

We found considerably better performance of VisNet3 when trained with the learning rule shown in Equation 4.6 than when trained on a simple Hebbian associative learning rule followed by weight normalization (Zhang, Rolls and Feng, 2026a). A reason for the better performance of the competitive net rule in Equation 4.6 compared to Hebbian learning followed by synaptic weight normalization is that the former (Equation 4.6) specifically decreases strong synaptic weights if the postsynaptic term y is high, and the presynaptic term x_j is low, thereby potentially reducing effects of other stimuli on that neuron. In contrast, with weight normalization all synaptic weights are scaled down equally after the synaptic update, so that process builds less selective neurons.

Overall, the 'standard competitive net rule' shown in Equation 4.6 is an important advance towards biological plausibility for it does not need any explicit synaptic weight normalization on each neuron, and also leads to better performance for the reason just described (Zhang, Rolls and Feng, 2026a). An advantage is that this can reduce the synaptic weight of inputs to a neuron that are not involved in activating the neuron, with the result that later that neuron will not be activated for inputs that may be from other objects. That is part of the mechanism for increasing the selectivity of each neuron in the network to only one of a few objects, and thereby reducing interference from other objects during later recognition. In contrast, synaptic weight normalization does not increase the selectivity of a neuron to some inputs, but merely scales down all the synaptic weights in the same proportion.

With both of these changes, the weight limiting, and the learning rule shown in Equation 4.6, the performance including the capacity of VisNet3 was considerably improved, and when larger versions of VisNet3 (with 256×256 neurons in each of its four layers, and 1,000 synapses onto each neuron) were tested, it was found that the capacity scaled up well to provide for transform-invariant identification of hundreds of objects, indicating that this is a biologically plausible and effective approach to invariant visual object recognition in the brain (Zhang, Rolls and Feng, 2026a).

Together, the competitive net learning rule described in this section, and the limiting in synaptic strength described in Section 4.7.3.2, may have widespread use in many different parts of the cerebral cortex, and provide what are potentially important advances in understanding how the cerebral cortex operates with biologically plausible computations (Zhang, Rolls and Feng, 2026a).

4.7.3.2 Beneficial effects of limiting the maximum value of synaptic weights on neurons

With normal Hebbian associative learning using a rule like that shown in Equation 4.4 (but without a short-term memory trace on the post-synaptic term y) some synaptic weights might continue to increase to high values, especially if some features are present in different objects. It seems biologically implausible that synaptic weights could grow without bound, so we have investigated limiting the maximum value that a synaptic weight on a neuron can reach in the updated version of VisNet, VisNet3 (Zhang, Rolls and Feng, 2026a). We in fact proposed that this could be beneficial, by encouraging neurons not to rely on a few strong synaptic weights from high-firing inputs to the neuron, but to grow synaptic weights from a number of inputs, in order to increase the sampling of information from the preceding layer of the network by producing a more distributed representation for what is learned by each neuron.

We discovered in VisNet3 that this biologically plausible limiting of synaptic strength did increase the capacity of VisNet3 to learn transform-invariant representations of large numbers (up to hundreds) of objects (Zhang, Rolls and Feng, 2026a). It is proposed that this limiting of the maximum strength of a synaptic weight may be a useful principle of the operation of some real neuronal networks in the brain.

4.7.3.3 Comparison with AI approaches to invariant object and face recognition

Given that VisNet3 is even more biologically plausible than VisNet, we may compare it with approaches that depend on deep learning with non-local synaptic modification rules that are being used as models of how visual object recognition may be implemented in the primate including human brain (Kar and DiCarlo, 2024; Rajalingham et al., 2018; Yamins and DiCarlo, 2016). These models include hierarchical convolutional deep neural networks (HCNN) (Rajalingham et al., 2018; Yamins and DiCarlo, 2016), which involve non-biologically plausible operations such as error backpropagation learning, and copying what has been set up in one part of a Layer to all other parts of the same Layer, which is also a non-local operation (LeCun et al., 2015, 2010; Bengio et al., 2017; Rolls, 2023c, 2026d).

The deep learning and convolution neural network approaches are superficially attractive (Kar and DiCarlo, 2024), for the responses of neurons in these artificial networks have similarities with the responses of neurons recorded in the primate visual cortical pathways to the inferior temporal visual cortex. But while these similarities are computationally interesting, it is not surprising if artificial neural networks as powerful as deep learning and convolution can be set up to produce these similarities, which do not show that the algorithms used in these artificial network approaches are how the brain implements invariant visual object representations.

Key issues for these artificial neural network approaches (Kar and DiCarlo, 2024) include the following (Rolls, 2026d) (see also Section 4.7.6).

One is that the learning rule is not local to the synapses, involving just the presynaptic and the post-synaptic firing rates, but depends on errors artificially computed by for example backpropagation of errors to train the synaptic weights in hierarchical convolutional deep neural networks (HCNN) (Rajalingham et al., 2018; Yamins and DiCarlo, 2016; Kar and DiCarlo, 2024). Hierarchical convolutional deep neural networks involve non-biologically plausible operations such as error backpropagation learning, and copying what has been set up in one part of a Layer to all other parts of the same Layer, which is also a non-local operation (LeCun et al., 2015, 2010; Bengio et al., 2017; Rolls, 2023c, 2026d; Kar and DiCarlo, 2024).

A second issue it that typically (though not always (Zhuang et al., 2021)) a teacher for each output neuron is needed, and nothing like this is evident in the visual cortical regions.

A third issue is lateral propagation of synaptic weights in deep convolution networks (Kar and DiCarlo, 2024; Rajalingham et al., 2018; Yamins and DiCarlo, 2016) which is not biologically plausible.

A fourth issue is that with the powerful training of artificial deep networks, the system can become fragile, with alteration of even a few pixels in an image altering the performance of the artificial network, whereas this typically (Kar and DiCarlo, 2024) has almost no effect on object recognition in humans.

A fifth issue is that the artificial network approaches do not train in a biologically plausible way in that they use just large training sets of images for which categories are typically specified, whereas VisNet capitalizes on the statistics of how images transform in the natural world, and uses these statistics that tend to be produced by the same object over a few seconds as a key part of the training of transform invariance, using the trace learning rule (Rolls, 2021c).

A sixth issue is that VisNet provides utility for the local recurrent collateral synaptic connections within a cortical region that have the potential to support attractor networks (Rolls, 2023c, 2026d) which are likely to be important in maintaining neuronal activity for periods of a few seconds while an object is being viewed with its different transforms to help implement trace rule learning, whereas local attractor dynamics do not have a key role in most artificial network approaches to visual object recognition (Kar and DiCarlo, 2024).

A seventh issue is that the cerebral cortex is able to perform its computations for invariances in networks with just 4 or 5 Layers (see Fig. 4.18). Part of the reason for this is to maximize processing speed, and minimize computation and reaction time (Rolls, 2016b, 2023c, 2026d), but it does show that networks with one hundred or more Layers as in some artificial neural networks are not needed to solve the computations involved in transform-invariant object recognition.

An eighth issue is that after an artificial neural network has been trained with error backpropagation and obtains good performance on the training set, if a few more stimuli need to be learned, this can degrade the performance of the network on the already trained images, and there is then a delay until good performance can be achieved again after much more training (Song, Millidge, Salvatori, Lukasiewicz, Xu and Bogacz, 2024). This undesirable property does not occur when real human brain networks learn new objects or faces.

A ninth issue is that when deep convolutional neural networks trained by error backpropagation predict the category into which an image exemplar should be placed, this is performed by methods such as using local features, whereas human perception typically uses the shape of the whole object (Bowers et al., 2022; Rolls, 2026d). The implication is that although artificial deep neural network models may learn to predict how to categorize a set of visual stimuli, they may do so in low-level ways not used by humans and may not reveal how human perception works (Bowers et al., 2022; Rolls, 2026d).

Nor do these artificial network approaches (Kar and DiCarlo, 2024) address important issues about how the brain works to solve these computationally demanding tasks, issues such as how synaptic weights are kept bounded in the real brain to be approximately equal on each neuron, nor the computational utility demonstrated here of having the synaptic weights on neurons limited to a biologically plausible maximum value. Other key properties of the VisNet3 model of biologically plausible invariant object recognition described here that may not be incorporated in AI models include the use of sparse connections from layer to layer (which helps each neuron to build different and stable representations); sparse representations (which help with storage capacity); and large numbers of synapses on each neuron which are important in the capacity of the system, as is the large number of neurons in potentially specialized different cortical modules for different types of stimulus such as faces, objects, etc. (Rolls, 2026d)

Another key property of the network described here is that it utilizes only 4 layers or stages of processing after V1 to model the real cortical organization where a few layers or stages of hierarchical processing (Fig. 4.18) ensures that the whole network can respond rapidly. Thus by investigating a biologically plausible network of invariant visual object recognition such as VisNet3, key issues are opened up about how invariant visual object recognition is actually implemented in the primate including human ventral visual system.

The emphasis on biological plausibility of computational models of brain function in my research is an extremely timely issue to think about in neuroscience now. This is in the context that it is of course now possible to train deep networks with backpropagation of error (rather than a biologically plausible local synaptic learning rule) to emulate some aspects of brain function, but perhaps using completely different algorithms to those used by the brain. This is a key issue for the neuroscience community raised by the research described here, that now needs to be considered by all those interested in how our brains work. My view is that to take neuroscience forward, local learning rules are biologically plausible, whereas at least for now deep learning does not seem very biologically plausible (Rolls, 2023c, 2026d), so it is sensible to see how far we can take understanding of brain function forward using biologically plausible algorithms, which may turn out to be very valuable and different from those currently used in artificial neural networks.

4.7.4 Invariant object-based motion in the dorsal visual system

A key issue in understanding the cortical mechanisms that underlie motion perception is how we perceive the motion of objects such as a rotating wheel invariantly with respect to position on the retina, and size. For example, we perceive the wheel shown in Fig. 4.24a rotating clockwise independently of its position on the retina. This occurs even though the local motion for the wheel in the different positions may be opposite. How could this invariance of the visual motion perception of objects arise in the visual system? In a sense, this is a computation of an invariant representation of object-based motion, and I decided to explore whether ideas that we had developed about the computation of invariant representations in the ventral visual system might be useful for understanding some of the computations in the dorsal visual system, such as the invariant representation of the global motion of objects.

Invariant motion representations are known to be developed in the cortical dorsal visual system. Motion-sensitive neurons in V1 have small receptive fields (in the range 1–2 deg at the fovea), and can therefore not detect global motion, and this is part of the aperture problem. Neurons in cortical region MT, which receives inputs from V1 and V2, have larger receptive fields (e.g. 5 degrees at the fovea), and are able to respond to planar global motion (Movshon et al., 1985; Newsome et al., 1989). One example is a field of small dots in which the majority (in practice as few as 55%) move in one direction. Another example is the overall direction of a moving plaid, the orthogonal grating components of which have motion at 45 degrees to the overall motion. Further on in the dorsal visual

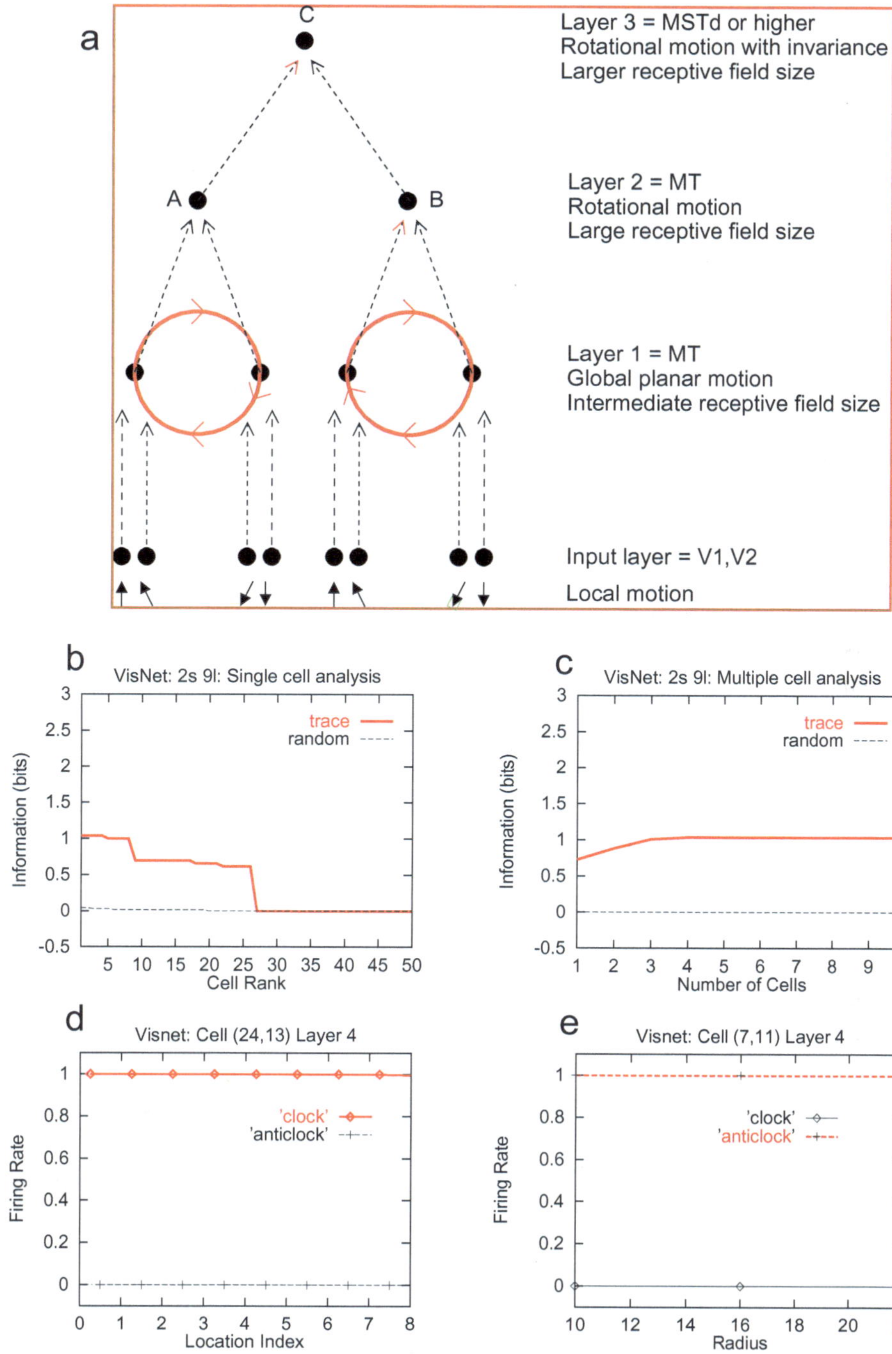

Fig. 4.24 Invariant global motion in the dorsal visual system. (a) Two rotating wheels at different locations rotating in opposite directions. The local flow field is ambiguous. Clockwise or counterclockwise rotation can only be diagnosed by a global flow computation, and it is shown how the network is expected to solve the problem to produce position invariant global motion-sensitive neurons. One rotating wheel is presented at any one time, but the need is to develop a representation of the fact that in the case shown the rotating flow field is always clockwise, independently of the location of the flow field. (b–d) Translation invariance, with training on 9 locations. (b) Single cell information measures showing that some layer 4 neurons have perfect performance of 1 bit (clockwise vs anticlockwise) after training with the trace rule, but not with random initial synaptic weights in the untrained control condition. (c) The multiple cell information measure shows that small groups of neurons have perfect performance. (d) Position invariance illustrated for a single cell from layer 4, which responded only to the clockwise rotation, and for every one of the 9 positions. (e) Size invariance illustrated for a single cell from layer 4, which after training with three different radii of rotating wheel, responded only to anticlockwise rotation, independently of the size of the rotating wheels. (Data from Rolls E T and Stringer S M. 2007 Invariant global motion recognition in the dorsal visual system: a unifying theory. Neural Computation 19: 139–169.)

system, some neurons in macaque visual area MST (but not MT) respond to rotating flow fields or looming with considerable translation invariance (Graziano et al., 1994; Geesaman and Andersen, 1996; Mineault et al., 2012), with MST especially involved in the representation of object motion in the visual field (Galletti and Fattori, 2018).

In the cortex in the anterior part of the superior temporal sulcus, which is a convergence zone for inputs from the ventral and dorsal visual systems (Rolls, 2023c, 2026d), we discovered that some neurons respond to object-based motion, for example to a head rotating clockwise but not anticlockwise, independently of whether the head is upright or inverted which reverses the optic flow across the retina (Hasselmo, Rolls, Baylis and Nalwa, 1989b) (see Section 4.5.1.4 and Fig. 4.8).

In a unifying hypothesis with the design of the ventral cortical visual system about *how* this might be computed, we (Rolls and Stringer, 2007) proposed that the dorsal visual system uses a hierarchical feedforward network architecture (V1, V2, MT, MSTd, parietal cortex) with training of the connections with a short-term memory trace associative synaptic modification rule to capture what is invariant at each stage. The principle is illustrated in Fig. 4.24a.

Simulations showed that the proposal is computationally feasible, in that invariant representations of the motion flow fields produced by objects self-organize in the later layers of the architecture (see examples in Fig. 4.24b–e). The model produces invariant representations of the motion flow fields produced by global in-plane motion of an object, in-plane rotational motion, and looming vs receding of the object. The model also produces invariant representations of object-based rotation about a principal axis, of the type illustrated in Fig. 4.8.

Thus we proposed that the dorsal and ventral visual systems may share some unifying computational principles (Rolls and Stringer, 2007). Indeed, the simulations of Rolls and Stringer (2007) used a standard version of VisNet, with the exception that instead of using oriented bar receptive fields as the input to the first layer, local motion flow fields provided the inputs.

The interesting and quite original principle proposed is that some of the same mechanisms including trace rule learning and hierarchical organization that are used in the ventral visual system to compute invariant representations of stationary objects may also be used in the dorsal visual system to compute representations of the global motion of a moving object (Rolls and Stringer, 2007; Rolls, 2021c).

My concept is that this may well be an example of a *principle of cortical operation: the re-use of the same principles of cortical operation for different computations in different cortical areas* (Rolls, 2016b, 2026d).

4.7.5 Complementary investigations of slow learning for object and place representations

Slow learning, which is used in VisNet with a short-term memory trace learning rule (Rolls, 2021c), has been used successfully in related and other applications, including the following.

In a more artificial network trained by gradient ascent with a goal function that included forming relatively time-invariant representations and decorrelating the responses of neurons within each layer of the 5-layer network, place-like cells were formed at the end of the network when the system was trained with a real or simulated robot moving through spatial environments (Wyss, Konig and Verschure, 2006). Slowness as an asset in learning spatial representations has also been investigated by others (Wiskott and Sejnowski, 2002; Wiskott, 2003; Franzius, Sprekeler and Wiskott, 2007; Schonfeld and Wiskott, 2015; Weghenkel and Wiskott, 2018), supporting the approach developed as described here.

Another application of slow learning is to the learning of semantic representations in the anterior temporal lobe. It is proposed that slow learning with a delay that covers a typical period in which different events or episodic memories are processed that are relevant to the same semantic representation may help the semantic learning. The architecture studied has the four-layer VisNet followed by a fifth layer that operates as an attractor network with a trace learning rule that operates over the period in which a semantic representation is learned (Rolls, Zhang and Feng, 2025a) (Section 6.10.5).

4.7.6 Are hierarchical convolutional deep neural networks an alternative biologically plausible mechanism for learning invariant representations of objects and faces?

A different approach has been to compare neuronal activity in visual cortical areas with the neurons that are learned in artificial models of object recognition such as hierarchical convolutional deep neural networks (HCNN) (Yamins and DiCarlo, 2016; Rajalingham, Issa, Bashivan, Kar, Schmidt and DiCarlo, 2018) (see also Section 4.7.3.3). Convolutional networks are described in Section B.14 of Rolls (2023c). They involve non-biologically plausible operations such as error backpropagation learning, and copying what has been set up in one part of a layer to all other parts of the same layer (see Section B.14 of Rolls (2023c)), which is also a non-local operation. They also (typically) require a teacher for each output neuron, which again is biologically implausible (see Section B.14 of Rolls (2023c)). The parameters of the hierarchical convolutional deep neural network are selected or trained until the neurons in the artificial neural network become similar to the responses of neurons found in the brain.

The next step of the argument then seems to need some care. The argument that appears to be tempting (Yamins and DiCarlo, 2016; Rajalingham, Issa, Bashivan, Kar, Schmidt and DiCarlo, 2018) is that because the responses of the neurons in the HCNN are similar to those in for example the inferior temporal visual cortex, the HCNN provides a model of how the computations are performed in the ventral visual system. But of course the model has been trained so that its neuronal responses do appear similar to those of real neurons in the brain. So the similarity of the artificial and real neurons is not surprising. What would be surprising is if it were proposed that the HCNN is a model of how the ventral visual stream computes (Yamins and DiCarlo, 2016; Rajalingham, Issa, Bashivan, Kar, Schmidt and DiCarlo, 2018), given that a HCNN with its non-local operation does not appear to be biologically plausible (see Section B.14 of Rolls (2023c)).

VisNet, in contrast, utilizes only local information such as the presynaptic and postsynaptic firing rates and a slowly decaying traces of previous activity (that could be implemented by a local attractor network using the recurrent collateral connections), so is a biologically plausible approach to invariant visual object recognition.

Although unsupervised training of deep convolutional neural networks has advanced (Zhuang, Yan, Nayebi, Schrimpf, Frank, DiCarlo and Yamins, 2021), these networks still implement convolution and deep learning using non-local learning rules including backpropagation of error to optimize representations, so are still biologically implausible.

Importantly, as noted earlier, comparing neurons recorded in the brain with the units found in deep learning algorithms tuned to produce neurons that resemble those in the brain (Cadieu, Hong, Yamins, Pinto, Ardila, Solomon, Majaj and DiCarlo, 2014; Yamins, Hong, Cadieu, Solomon, Seibert and DiCarlo, 2014; Yamins and DiCarlo, 2016; Rajalingham, Issa, Bashivan, Kar, Schmidt and DiCarlo, 2018; Jia, Hong and DiCarlo, 2021; Zhuang, Yan, Nayebi, Schrimpf, Frank, DiCarlo and Yamins, 2021) seems like a poorly founded exercise, for two reasons.

First, the solutions found by deep convolution networks and other machine learning algorithms for image to category mapping may be very different to those found in a system such as the brain that uses only local learning rules and is restricted to only 4–5 layers in the hierarchy (see Section 4.7.3.3).

Second, it is difficult to know what can be learned about how primate including human vision works computationally from deep learning and related machine learning algorithms, given that the understanding of exactly how these machine learning algorithms reach their mapping is almost un-understood, apart from the recipe that makes them work (Plebe and Grasso, 2019; Sejnowski, 2020).

In addition, the ways in which HCNNs solve visual object recognition problems is different from the whole-object perceptual strategy used by humans (Bowers et al., 2022), with many other differences between the operation of the real brain and artificial deep networks model described in Section 8.2.

Overall, VisNet is the most biologically plausible approach I know to how computations in the ventral visual system for invariant object and face recognition are performed using unsupervised 'slow', memory trace-based, learning (Rolls, 2021c, 2023c, 2026d; Zhang et al., 2026a).

4.7.7 Evaluation of Rolls' approach to the computations about visual object and face representations in the ventral visual pathway

An evaluation of Rolls' theory and model of invariant visual object recognition in the brain is as follows (Rolls, 2023c, 2026d,d):

1. Invariant visual object recognition is implemented in the ventral visual system using only about 5 stages of hierarchical processing, V1, V2, V4, posterior inferior temporal cortex (TEO), and anterior inferior temporal cortex (TE) (Fig. 4.18). This allows approximately 15 ms of processing time for each stage. That is just sufficient for processes implemented by recurrent collaterals such as attractor network computations to be involved (Rolls, 2023c, 2026d).
2. Although there are as many backprojections as forward projections between each pair of stages of the hierarchy, these are used for memory recall and top-down attention, and probably not for backpropagation of error learning, which is prima facie biologically implausible (Rolls, 2016b, 2023c, 2026d).
3. A biologically plausible form of learning to categorize new stimuli is competitive learning (Rolls, 2023c, 2026d), and this is at present the most plausible candidate mechanism for how object and face representations are learned over successive stages of the hierarchy, with convergence from stage to stage consistent with the increase of receptive field size from stage to stage (Fig. 4.18).
4. Because different views of an object can be quite different, a mechanism is needed to associate together different transforms of the same object. An associative synaptic learning rule that incorporates a short-term trace of previous neuronal activity is proposed in the VisNet model to enable these associations to be learned, utilizing the property of the natural world that temporally close visual stimuli are statistically likely to be different transforms of the same object (Rolls, 1992a, 2012b, 2016b, 2021c). Systems (such as HMAX (Riesenhuber and Poggio, 1999)) without this 'slow learning' capability fail in this key aspect of visual object recognition, transform-invariant representations of objects (Robinson and Rolls, 2015).
5. Training a system such as VisNet therefore benefits from the types of transforms of objects as they are found in the natural world, to facilitate this type of learning about the statistics of the environment that utilizes time, but also spatial continuity. The types of transform over which VisNet can produce transform-invariant representations of objects include position on the retina, size, spatial frequency, lighting, deformation, and view.

6. A key feature of the type of computation implemented in VisNet is that it is unsupervised, performed without a teacher. Instead, VisNet uses the temporal statistics present in the natural environment to learn the transforms that correspond to a particular object or face.
7. In contrast, AI approaches such as deep learning using backpropagation of errors typically require prespecification of the object categories that should be learned (that is, the use of a teacher for each output neuron, which is biologically implausible). In practice these deep learning systems are trained with very large numbers of exemplars of a class, typically without systematic use of close transforms of the same object. This requires very large numbers of training trials. Deep learning networks may have up to 100 layers, again biologically implausible. Moreover, little is learned from these deep learning models about how the brain solves complex problems such as invariant visual object and face recognition, because deep convolutional learning does not use biological constraints such as local learning rules, and when mappings from images to categories are found, clear computational principles of how the mapping is achieved are largely not understood (Plebe and Grasso, 2019; Sejnowski, 2020), but appear to be different from those used by humans (Bowers et al., 2022) (Section 4.7.6). There are many other ways in which these deep artificial neural networks do not provide a good model of the operation of the real brain (Section 8.2) (Rolls, 2026d). VisNet is the most biologically plausible approach I know of how computations in the ventral visual system for invariant object and face recognition are performed using unsupervised 'slow', memory trace-based, learning (Rolls, 2021c, 2026d).
8. Other proposals such as feature spaces are too simple to implement object recognition for the reasons given above, though features such as texture, and color may be usefully combined with shape information to help invariant visual object recognition in a VisNet-like system (Rolls, 2023c, 2026d). Object-based systems that utilize connections between parts also seem implausible, because of the requirement of powerful syntax to describe how all the parts are joined together.
9. VisNet implements a feature hierarchy approach to object recognition (Rolls, 2023c, 2026d). Feature hierarchy networks are biologically plausible in that they can be implemented with local synaptic learning rules (needing only the activity in the pre-synaptic and post-synaptic neurons), implemented in competitive networks (described in Section 1.4). They are evolutionarily economical in that the same type of competitive network can be utilized in each stage of the hierarchy, without the need for evolution to optimise many different types of network and the genetic specification for them; and in that they can be used in every sensory system.

4.8 Further findings consistent with the discoveries on the operation of the visual cortical pathways for object and face representation described here

We discovered different types of face neuron in different patches: for example, in some patches face identity is represented (in the inferior temporal cortex), and in other patches face expression is represented (in the cortex in the anterior superior temporal sulcus) (Hasselmo, Rolls and Baylis, 1989a). Face and head motion involved in social interactions are also represented in the cortex in the superior temporal sulcus (Hasselmo, Rolls, Baylis and Nalwa, 1989b; Perrett, Smith, Mistlin, Chitty, Head, Potter, Broennimann, Milner and Jeeves, 1985a).

These discoveries have been followed up by others (Hesse and Tsao, 2020; Chang and Tsao, 2017; Grimaldi, Saleem and Tsao, 2016; Tsao, 2014; Tsao, Schweers, Moeller and Freiwald, 2008; DiCarlo and Cox, 2007; Tsao, Freiwald, Tootell and Livingstone, 2006; Deen, Koldewyn, Kanwisher and Saxe, 2015; Liu, Harris and Kanwisher, 2010; Kanwisher, McDermott and Chun, 1997; Freiwald, 2020; Freiwald, Tsao and Livingstone, 2009; Afraz, Yamins and DiCarlo, 2014; Li and DiCarlo, 2012; Rust and DiCarlo, 2010; DiCarlo and Maunsell, 2000; Desimone, 1991; Desimone and Ungerleider, 1989; Desimone, Albright, Gross and Bruce, 1984), including Tsao, Freiwald and Kanwisher, who were awarded the Kavli prize in 2024 "for the discovery of a highly localized and specialized system for representation of faces in human and non-human primate neocortex", and many congratulations are due to them for this award.

In addition, the discovery of neurons in the cortex in the superior temporal sulcus that respond to socially relevant aspects of face stimuli such as expression and movement (Hasselmo et al., 1989b,a) has now been followed up and accepted as a third visual cortical pathway (Pitcher and Ungerleider, 2021). Moreover, I have proposed that a disconnection of these face neurons that respond to face expression and gesture from the orbitofrontal cortex emotion system may be important in understanding autism, and this has implications for treatment (Rolls et al., 2020f; Cheng et al., 2015).

The computation of translation and size invariant representations of objects and faces proposed in the VisNet model, yet in which the spatial relations between the parts are encoded, is echoed by the proposal that similar mechanisms may be involved in word recognition, with the relative position of letters in a word a key part of what is encoded (Dehaene, Cohen, Sigman and Vinckier, 2005; Agrawal and Dehaene, 2024).

Much more of what has been discovered in this research about the inferior temporal visual cortex and its functions in visual perception including the perception of faces and objects is provided in **Brain Computations and Connectivity** (Rolls, 2023c) and **Brain Computations and Principles; and AI** (Rolls, 2026d).

This whole approach to invariant visual object recognition has impact in other areas of neuroscience, for Stanislaus Dehaene has described an analogous system that he believes could be involved in reading, starting with letter strings, and building up from there (Vinckier, Dehaene, Jobert, Dubus, Sigman and Cohen, 2007; Agrawal and Dehaene, 2024).

5 The discovery of hippocampal spatial view cells for memory and navigation in primates including humans, and of their Ventromedial Visual Cortical 'Where' pathway, and the development of a theory for the hippocampal memory system

5.1 Introduction

The background to the discoveries described in this Chapter is that we had made fundamental discoveries of the representation of visual information in the primate temporal lobe visual cortical regions, and of differences in the visual representations in brain regions to which the temporal visual cortical regions project, including the orbitofrontal cortex, amygdala, and tail of the caudate nucleus, as described in Chapters 2–4 and in **Brain Computations and Principles; and AI** (Rolls, 2026d).

There was, in addition, evidence for visual pathways to the hippocampal system (Van Hoesen, 1982; Van Hoesen and Pandya, 1975), which was implicated in memory in humans (Zola-Morgan et al., 1986; Smith and Milner, 1981; Scoville and Milner, 1957). The hippocampal memory system was thus the next natural region to investigate, I thought, in order to build up a comprehensive understanding of the functions and computations performed by another output from the temporal lobe visual cortical regions, and that was potentially very important to understand, as the hippocampal system seemed so important in episodic memory, the memory for events involving typically what happened where on a particular occasion.

We did not know when we started this research (see Rolls (1987)) what we might discover for primate hippocampal neurons. One hypothesis at the time was that the hippocampus is involved in recognition memory, but we found that a recognition memory task, in which visual stimuli had to be remembered with intervening stimuli present, engaged very few primate hippocampal neurons (Rolls, Cahusac, Feigenbaum and Miyashita, 1993), and it was shown that the perirhinal cortex is instead a key cortical region in recognition memory (Buckley, 2005; Buckley and Gaffan, 2000; Waters et al., 2023). Indeed, we discovered neurons in the perirhinal cortex that respond to long-term familiarity memory (Rolls et al., 2005b; Holscher et al., 2003). We also knew that place cells, that respond when the rat is located at one place in an environment, were found in the rat hippocampus (O'Keefe, 1979; O'Keefe and Dostrovsky, 1971). We also knew that hippocampal / fornix lesions in macaques could impair tasks about spatially directed movements (Rupniak and Gaffan, 1987), and that primate hippocampal neurons become engaged in a delayed spatial response task (Watanabe and Niki, 1985).

5.2 Discoveries of hippocampal spatial view cells in primates, and of a ventromedial cortical scene pathway in humans

5.2.1 Object-spatial location neurons discovered in macaques

Taking those findings as a starting point, we set up fundamental investigations to investigate whether primate hippocampal neurons were coding for the spatial aspects of the response, or the spatial aspects of the stimuli that were shown before the delay period on the left or right to be remembered in this previous research. When we recorded from neurons in the monkey hippocampus in a conditional spatial response task, we found neurons that learned and had firing rates related to the conditional spatial responses (Cahusac, Miyashita and Rolls, 1989; Cahusac, Rolls, Miyashita and Niki, 1993; Miyashita, Rolls, Cahusac, Niki and Feigenbaum, 1989), which was later confirmed in a paper published in Science (Wirth et al., 2003).

In an object-place memory task that we designed to assess whether the location of the stimulus was being encoded by hippocampal neurons, we displayed a trial-unique object on the left or right of a monitor in front of the macaque (Cahusac et al., 1989). Then during the delay period the macaque had to remember the location and identity of the stimulus. At the end of the delay period, a stimulus was shown in the left or right location, and the macaque had to press the location if the stimulus object and the location were the same as before the delay to obtain a taste reward. Thus the task could only be solved by remembering both the identity of the object and the location in which it was shown. We discovered that some macaque hippocampal neurons respond differently for stimuli shown in different positions in space, and some respond differently when the monkey is remembering different positions in space. In addition some of the neurons responded to a combination of object and place information, in that they responded only to a novel object in a particular place (Cahusac, Miyashita and Rolls, 1989). The abstract of that paper stated that this was the first demonstration that some hippocampal neurons in the primate have activity related to the spatial location of viewed stimuli, and this was a key discovery in our understanding of hippocampal function in primates. Nothing like this had ever been found in rodents.

Remembering where trial-unique visual stimuli have been seen before in space provides a potential model of episodic memory, and there was evidence that hippocampal system damage in macaques could impair this type of memory (Mishkin, 1982; Gaffan, 1987; Parkinson et al., 1988). To investigate this key type of memory further, I then set up a task in which macaques had to remember whether they had seen before a particular trial-unique image in a particular location on a viewed monitor screen in front of them (Rolls, Miyashita, Cahusac, Kesner, Niki, Feigenbaum and Bach, 1989b). We very soon discovered that some primate hippocampal neurons responded differently to different viewed locations on the screen, and that some combined this spatial tuning with information about whether the object had been shown previously in that location on a screen (Rolls et al., 1989b). This was a key discovery that showed that locations in space "out there" are encoded by primate hippocampal neurons, and could be combined with information about what object / image was at each location, which is a model of episodic memory in which 'what' is seen at a particular viewed location ('where') must be remembered. The place where the macaque was located was not what was being encoded in these experiments, for the monkey was always in one place during the investigation.

These primate neurons are thus quite unlike rodent place cells, which respond to the place where the rodent is located (O'Keefe, 1979; McNaughton, Barnes and O'Keefe, 1983; McNaughton, Barnes, Gerrard, Gothard, Jung, Knierim, Kudrimoti, Qin, Skaggs, Suster and Weaver, 1996; Edvardsen, Bicanski and Burgess, 2020; Kubie, 2025; McNaughton, 2025; Morris, Amaral, Bliss, Duff and

O'Keefe, 2025). (When the term rodent is used here, it refers to rats and mice, the main rodent species in which hippocampal function has been investigated.)

5.2.2 The discovery that hippocampal spatial view cells encode spatial locations "out there" in an allocentric, world-based, spatial coordinate framework

The next scientific question that I thought it was important to investigate was the spatial coordinate framework utilized by these macaque hippocampal and parahippocampal gyrus spatial view neurons with responses to some but not other locations on a screen that was being viewed in front of the macaque. We performed this by moving the video monitor in front of the macaque to the left or to the right, and by moving the macaque and screen to different places in the large room. We discovered that for the majority of hippocampal spatial view cells (69%), the coding was in allocentric, world based, coordinates (Feigenbaum and Rolls, 1991). For 44% of the neurons the spatial field remained in the same location on the screen, even though the egocentric location (relative to the macaque) had changed. For the remainder of the allocentric neurons, the position of the screen in the world-based framework of the room was what was encoded. This was clear evidence that these spatial view neurons code in an allocentric framework, and for the majority this was the location in the space being viewed as defined by the framework of the screen (subtending 12 degrees) that was being viewed (Feigenbaum and Rolls, 1991). In addition, 10% of the neurons coded in egocentric coordinates.

Overall, the discovery of an allocentric (world-based) representation in primates was very interesting, for in rodents place cells are often described as coding in an allocentric framework (McNaughton et al., 1983; O'Keefe, 1979; Edvardsen et al., 2020; McNaughton et al., 1996), and this was useful consistency to help understand hippocampal computations. However, what was being encoded in an allocentric framework was completely different, with encoding of the place where the animal is for rodents in the hippocampus and medial entorhinal cortex for grid cells (McNaughton et al., 1983; O'Keefe, 1979; Edvardsen et al., 2020; McNaughton et al., 1996; Moser et al., 2017); and encoding of the location out there in allocentric space that is being viewed by the primate (Georges-François, Rolls and Robertson, 1999; Feigenbaum and Rolls, 1991; Rolls, Robertson and Georges-François, 1997b; Rolls, Treves, Robertson, Georges-François and Panzeri, 1998).

5.2.3 Spatial view vs. place for primate hippocampal and parahippocampal neurons

So far, we had discovered that hippocampal and parahippocampal gyrus neurons in macaques could represent allocentric locations in space "out there" when they were being looked at (Feigenbaum and Rolls, 1991; Rolls, Miyashita, Cahusac, Kesner, Niki, Feigenbaum and Bach, 1989b; Cahusac, Miyashita and Rolls, 1989), and could participate in one-trial object-spatial location memory (Rolls et al., 1989b).

But that left the issue open of the extent to which the primate hippocampus might in addition represent the place where the individual was located by having place cells. To investigate the relative roles of spatial view vs place for primate hippocampal neurons, we now set up a 2x2x2 m cue controlled environment within which the macaque could be moved to different places on wheels or by a robot, and could from those places view the cues on the walls of the enclosure (Rolls and O'Mara, 1995). The majority of the neurons with spatial responses responded to the location at which the macaque was looking (5.5% of the total number of neurons recorded), independently or relatively independently of the place where the macaque was located. 0.7% of the neurons had responses that were related to the place where the individual was located, and these place-related neurons were also influenced by where the individual was looking or by whole-body motion (Rolls and O'Mara, 1995). Thus I have never taken the position that there are no place cells in the primate hippocampus, but have instead emphasized the importance of the completely new type of representation provided for primates by spatial view neurons (Rolls, 2023a,c, 2026d, 2025a).

5.2.4 Primate spatial view hippocampal and parahippocampal neurons during free locomotion

At about this time (1992-1995) I led a Human Frontier Science Program grant to investigate hippocampal function, with the members including Bruce McNaughton, Carol Barnes, Alessandro Treves, and Jay McClelland. Bruce McNaughton suggested that as place cells in rodents can be influenced by whether the rat can locomote, then it would be useful to know what would happen when primates could locomote. Perhaps primate spatial view cells would turn into place cells was perhaps an idea.

I was skeptical that this would make a great deal of difference, for primates with their foveate vision can effectively actively explore a spatial environment by moving their eyes, without having to locomote to explore. But I set up to record from the hippocampus in freely moving and locomoting macaques (Rolls, Robertson and Georges-François, 1997b). I did this by making a special lightweight primate chair that could be moved into a horizontal position and had wheels, and that allowed the macaque's four feet to be on the ground so that he could run around freely using his four legs. He was able to run fast, like a small dog, with high linear and angular velocity, as shown in the videos in the Supplementary Material of Rolls (2025a). The head was free to move in the horizontal plane at high angular velocity, because as the animal rotated the head, the lightweight chair moved with his head. The head did not move in the pitch and roll planes. A camera above the individual was used to measure the exact location and head direction of the macaque, using two LEDs (light emitting diodes) on the chair, one flashing at twice the frequency of the other. The eye position was measured to within 2 degrees by field coils attached to the chair, and a scleral search coil. We could therefore monitor in continuous time exactly where the monkey was visually fixating in the environment. The environment was a large open lab space, and this was important, for many more spatial view cells were recorded in this rich spatial environment than in the cue-controlled environment used by Rolls and O'Mara (1995). (The importance of a rich spatial environment, and the ability to locomote freely in this with vestibular and proprioceptive information etc, in order to obtain the best responses from many primate spatial view cells, has recently been confirmed, in that macaque spatial view cells were much better activated during active locomotion in a rich spatial environment than in a Virtual Reality model of the same environment (Yan and Mao, 2024).) Our recording setup enabled the macaque to walk freely in a 2x2 m space in the middle of the lab, and to turn as much as the individual wished (Rolls et al., 1997b) (although after about 20 minutes we sometimes had to stop the monkey to unwind the wires going to the ceiling high above the head).

With this setup for free locomotion, we were able to record spatial view cells while the macaques actively and freely locomoted (Rolls, Robertson and Georges-François, 1997b; Georges-François, Rolls and Robertson, 1999; Rolls, Treves, Robertson, Georges-François and Panzeri, 1998; Robertson, Rolls and Georges-François, 1998). When analyzing whether the neurons responded to spatial view, or the place where the individual was located, great care was used to implement a factorial design in which all spatial views could be seen from all places in the analysis. Unless a factorial design of this type is used, some views may only be visible from some places, which would produce a confound between analyses of whether a neuron is coding for view, or place, or has some mixed selectivity. In my opinion, this type of factorial design is essential for reliable evidence about

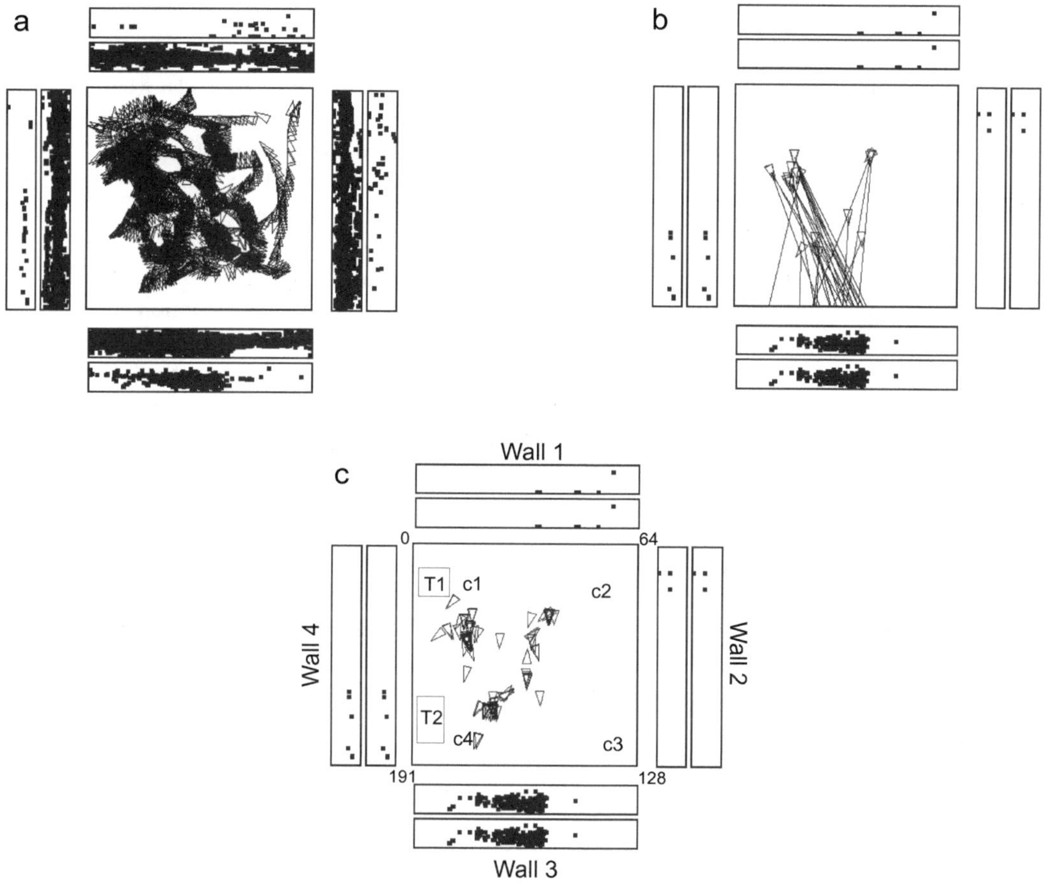

Fig. 5.1 A hippocampal spatial view cell (az033) recorded while a monkey walked around in an open field area 2.5x2.5 m shown as the square within a rich and large laboratory environment. In (a) every time that the cells fired is shown by a spot in the outer rectangles each of which represents one of the four walls of the room. The inner rectangles show where the monkey looked on the walls. The neurons has a spatial view field on wall 3. The places to which the monkey walked are shown by the triangles, with the pointed end showing the head direction. (b) shows some of the many different places at which the monkey was located when the neuron fired, and the lines show where the monkey was fixating when the spatial view cell fired. (c) provides more evidence about the places where the monkey was located when the cell fired because he was looking at the view field on wall 3. This helps to show that the neuron responds to spatial view, and not to the place where the monkey was located. C1 to C4 are cups containing food to encourage the monkey to forage. T1 was a trolley and T2 a table. Data from Georges-Francois P, Rolls E T and Robertson R G. 1999 Spatial view cells in the primate hippocampus: allocentric view not head direction or eye position or place. Cerebral Cortex 9: 197-212.

whether neurons code for spatial view, or place, or have mixed spatial view / place encoding (Rolls, 2023a).

With this design, we were able to measure with rigorous information theoretic approaches (Rolls, Treves, Robertson, Georges-François and Panzeri, 1998) how much information single hippocampal and parahippocampal cortex neurons encoded about spatial view versus place during free locomotion in a rich visual environment of a laboratory. The discovery was that clearly most of the information from most cells is about spatial view, and not the place where the individual is located, nor head direction, nor eye position in the head (left/right and up/down). The mutual information across locations was measured, and the mean across neurons about spatial view was 0.47 bits, about place was 0.033 bits, about head direction was 0.054 bits, and about eye position was 0.017 bits (Georges-

François et al., 1999). An example of a spatial view cell recorded during active locomotion in a macaque is shown in Fig. 5.1, and videos to illustrate the firing are provided in the Supplementary Material of Rolls (2025a). Full details of these discoveries are provided elsewhere (Georges-François, Rolls and Robertson, 1999; Rolls, Treves, Robertson, Georges-François and Panzeri, 1998; Robertson, Rolls and Georges-François, 1998; Rolls, Robertson and Georges-François, 1997b), and some of their implications have been described (Rolls and Wirth, 2018; Rolls, 2023a,d, 2025a, 2026c). These spatial view neurons are tuned to vertical as well as horizontal locations in the viewed spatial environment (Georges-François et al., 1999).

A feature of these discoveries is that I ensured that we could combine recordings during active locomotion in a rich open field environment in a naturalistic type of investigation, with carefully controlled complementary investigations in which the macaque was stopped. When stopped, the firing of cells was measured when the macaque was in different places, with different head directions, and with different eye positions, enabling a very clear distinction to be made that the neurons were responding to where in the environment the individual was looking, and not just the facing direction, with an example shown in Fig. 5.2 (Georges-François et al., 1999; Rolls et al., 1998; Robertson et al., 1998; Rolls et al., 1997b).

In a different investigation, visual spatial cells were confirmed in the primate hippocampus that respond to viewed locations “out there” (Mao, Avila, Caziot, Laurens, Dickman and Angelaki, 2021). A small difference is that it was proposed that hippocampal visual spatial neurons code for the direction in which the individual is facing rather than where the macaque is looking (Mao et al., 2021), but this could be related to the fact that typically when macaques navigate, they face straight ahead in the direction in which they are most commonly looking. In situations such as this, it is helpful to perform the more controlled type of experiment illustrated in Fig. 5.2, for this allows different head directions, places, and eye positions to be systematically explored, which as shown in Fig. 5.2 led to the clear conclusion that it is where in the environment the macaque is looking that activates spatial view neurons (Rolls, 2023d).

This neuronal encoding is of course highly adaptive, in that the viewed spatial location in the environment can be recognized later, even if the individual is in a different place, and with a different head direction, which would enable a ‘good’ (i.e. reward) at that location to be recalled, whereas it would be missed if facing direction was the encoding scheme, unless the individual was at the same place with the same head direction (Rolls, 2023d). That, is, spatial view cells, because they respond allocentrically to locations ‘out there’ in the environment, implement spatial representations that are invariant with respect to facing direction, head direction, eye position (in the head), gaze (i.e. compass direction of the eyes), and place where the individual is located (Rolls, 2023d; Georges-François et al., 1999; Rolls, 2025b). (I have discussed this with Dun Mao, the first author of the paper referred to above, and my understanding is that he accepts this point.)

5.2.5 The discovery that primate spatial view hippocampal and parahippocampal neurons show idiothetic (self-motion) update

I knew that rat place cells could be updated by self-motion (McNaughton, Barnes, Gerrard, Gothard, Jung, Knierim, Kudrimoti, Qin, Skaggs, Suster and Weaver, 1996; McNaughton, Battaglia, Jensen, Moser and Moser, 2006), and decided to test this for primate spatial view cells. This led to another key discovery about hippocampal spatial view neurons, that they can be updated idiothetically (i.e. by self-motion) in the dark or when the view is obscured (Robertson, Rolls and Georges-François, 1998). For example, spatial view neurons fire in the dark whenever the macaque looks towards the spatial view field, as illustrated in Fig. 5.3. The update is based on for example eye movements being made by a macaque. This is idiothetic update in that there is some drift over time, and the update

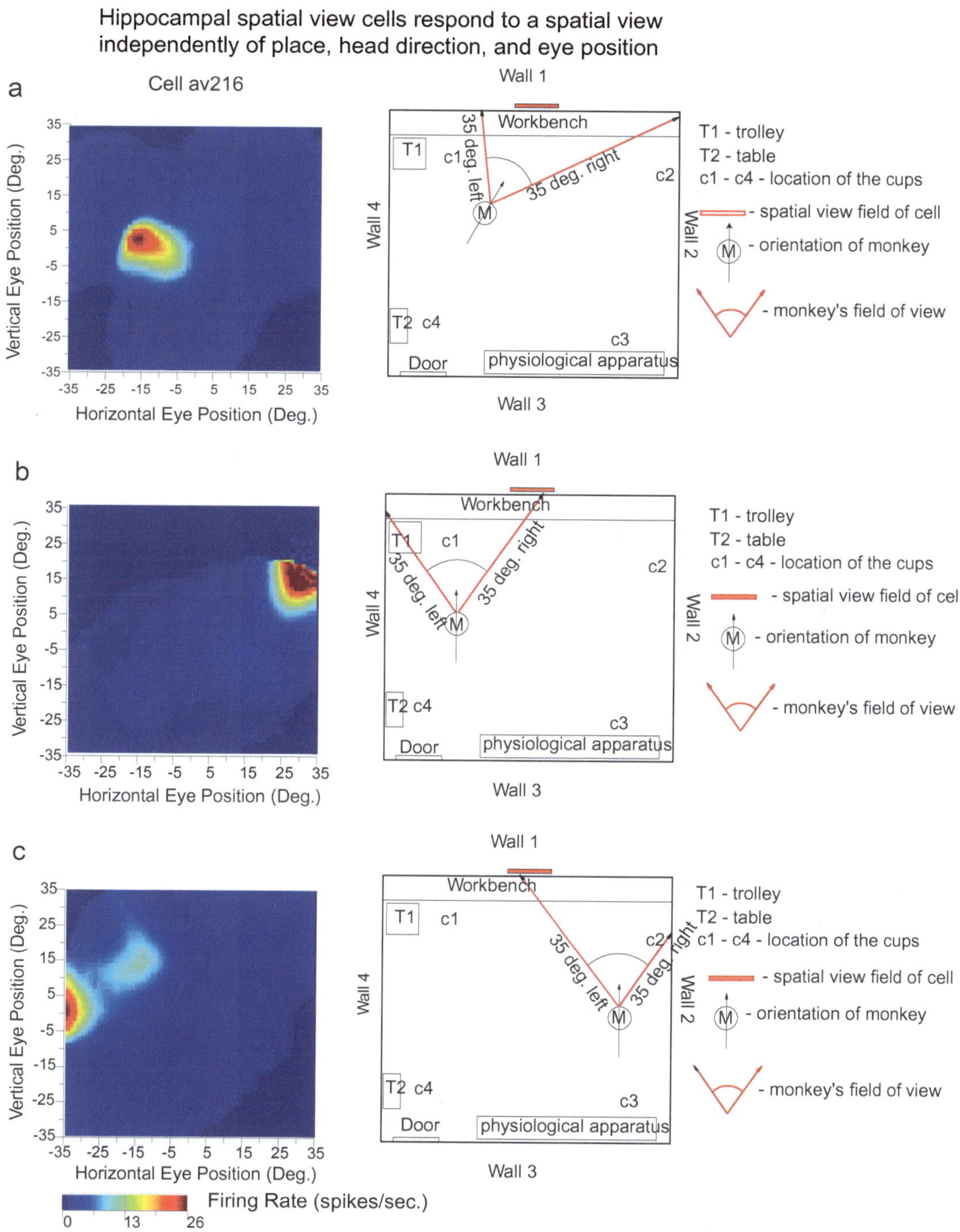

Fig. 5.2 Testing of a hippocampal spatial view neuron (av216) to show that it has allocentric encoding, and that the response does not depend on where the monkey is located. The firing rate is shown as a function of the horizontal and vertical eye position, where positive values indicate right or up. The neuron responded when the monkey looked towards its view field (indicated with a hatched bar) relatively independently of place, eye position, or head direction. ANOVAs and information theory analyses performed on the same data cast in different ways confirmed this: for spatial view, the ANOVA was $p<0.001$ with 0.217 bits in a 500 ms period for the average Shannon mutual information; for place $p=0.9$ with 0.001 bits; for head direction $p=0.5$ with 0.0 bits; and for eye position $p=0.8$ with 0.006 bits. (Data from Georges-Francois P, Rolls E T and Robertson R G. 1999 Spatial view cells in the primate hippocampus: allocentric view not head direction or eye position or place. Cerebral Cortex 9: 197-212.)

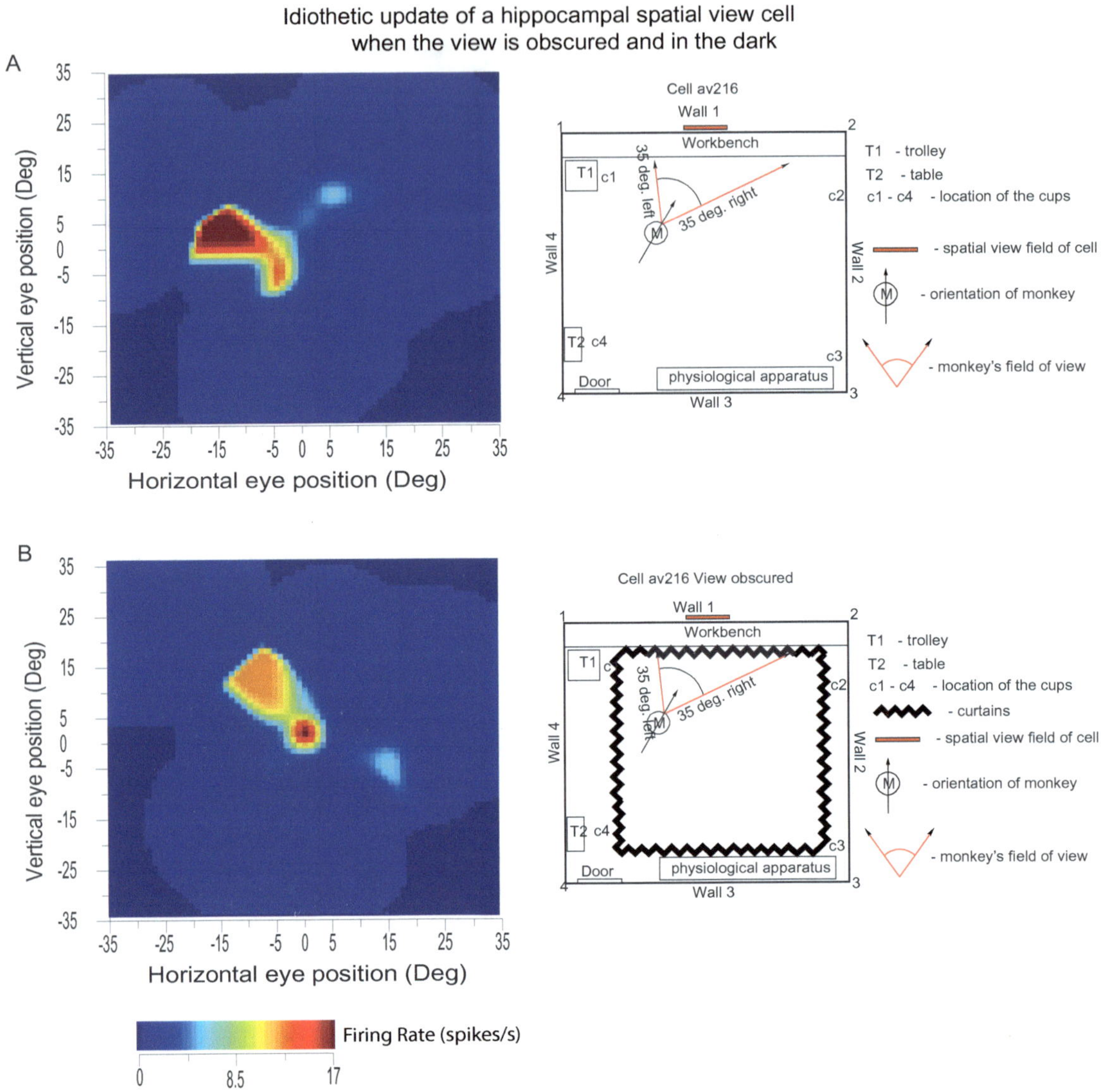

Fig. 5.3 Idiothetic update (by self-motion of the eyes) of the firing of a hippocampal spatial view cell. This cell fired when the macaque looked towards the effective spatial view even when the view was obscured with curtains. The firing is shown with the monkey stationary with his head facing in the direction indicated by the arrow when the curtains were drawn open (A) or were drawn closed (B). Left: firing rate of the cell in spikes/s is indicated by the color (a calibration bar in spikes/s is shown below) projected onto the monkey's field of view. The two-dimensional firing rate profile of the cell was smoothed for clarity using a 2-dimensional Gaussian spatial filter. The space adequately sampled by the eye movements of the monkey is indicated by light blue. The eye position in degrees is shown. Right: a plan view of the room to indicate the monkey's view of the wall is shown. M, position of the monkey. (Data from Robertson R G, Rolls E T and Georges-Francois P. 1998 Spatial view cells in the primate hippocampus: Effects of removal of view details. Journal of Neurophysiology 79: 1145-1156.)

becomes gradually lost over 3-5 minutes (by which time idiothetic update in the experimenters started to fail, and they had to feel their way towards a light switch) (Robertson et al., 1998). This idiothetic update is not only useful for navigation when the view details are obscured (Rolls, 2021b), but is also interesting as it is parallelled by idiothetic update for place in rodents (McNaughton et al., 2006, 1996), providing evidence that the rodent and primate hippocampal systems have considerable

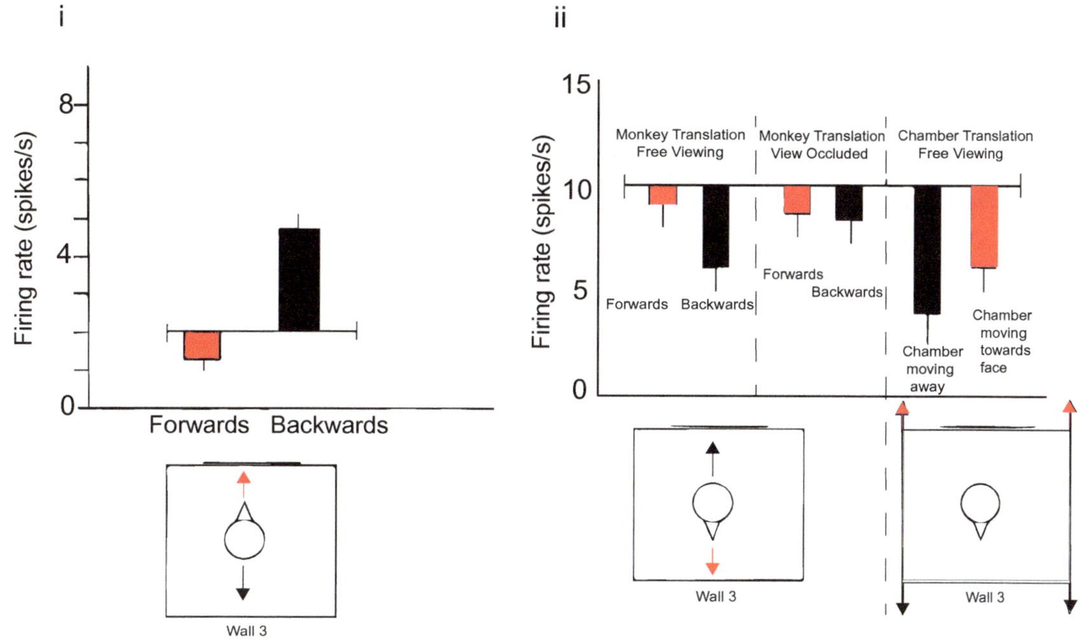

Fig. 5.4 Speed cells. i. Activity of a primate hippocampal whole body motion (speed) neuron that increased its firing rate during backward linear translation, and decreased its firing rate during forward linear translation. ii. Activity of another primate hippocampal whole body motion (speed) neuron that decreased its firing rate during linear translation of the monkey away from wall 3, and that decreased its firing rate more during linear translation of the test chamber (with wall 3 being viewed) away from the macaque. The neuron did not respond during linear translation if the view was occluded (middle panel). This neuron was thus activated by visual cues of motion, and not by vestibular cues. (Data from O'Mara S M, Rolls E T, Berthoz A and Kesner R P. 1994 Neurons responding to whole-body motion in the primate hippocampus. Journal of Neuroscience 14: 6511-6523.)

similarities computationally, but use largely different types of input, namely about place in rodents and about spatial view in primates including humans.

However, the mechanisms for path integration may be outside the hippocampus, and performed by different systems, with grid cells in the rodent medial entorhinal cortex implicated in path integration for place (Moser, Moser and McNaughton, 2017; Moser, Rowland and Moser, 2015; Moser, Moser and Roudi, 2014; Giocomo, Moser and Moser, 2011; Brandon, Bogaard, Libby, Connerney, Gupta and Hasselmo, 2011), whereas the dorsal visual system leading to the parietal cortex of primates is implicated in transforms from retinal coordinates to head based coordinates and even to some type of allocentric encoding and in idiothetic update (Rolls, 2020b; Snyder, Grieve, Brotchie and Andersen, 1998; Rolls, Deco, Huang and Feng, 2023b; Rolls, 2024c). The implication of idiothetic update in primates is that there is a model of space and that the location in space is updated by eye movements made by the individual. Further support for idiothetic update of hippocampal neurons (Robertson, Rolls and Georges-François, 1998) is that primate hippocampal neurons can respond to a spatial view field towards which a macaque saccades even before the virtual reality system has updated to show the view to which the macaque is making a saccade (Wirth et al., 2017).

5.2.6 The discovery of speed cells in the hippocampal system for navigation

We have just seen the self-motion of the eyes can idiothetically update the firing of spatial view cells (Section 5.2.5, Robertson, Rolls and Georges-François (1998)). We also knew at this time that rodent place cells can be updated by self-motion (McNaughton et al., 1991, 1996). In discussions with Alain Berthoz who is interested in path integration (Israel, Chapuis, Glasauer, Charade and Berthoz, 1993; Wiener, Berthoz and Wolbers, 2011), we decided to measure whether neurons in the primate hippocampus respond in relation to speed, and in particular to linear and angular velocity of the individual moving through space, which are signals needed for path integration during navigation.

Alain Berthoz came to Oxford from the College de France to work with us on this, and brought with him a French robot (Robulab made by Robosoft S.A.), which I was able to program to produce quantitative velocities for angular (0-100°/s) and linear motion forwards or backwards (with a standard value of 0.2 m/s), with velocities of 0°/s and 0 m/s as the baseline. The macaque sat on top of the robot, which moved within a 2x2x2 m cue-controlled test chamber. The whole chamber could also be rotated round the macaque (O'Mara, Rolls, Berthoz and Kesner, 1994).

We discovered that some of these hippocampal neurons respond to linear motion (with for example some responding to forward motion of the macaque, and others to backward motion), and others to axial whole body rotation of the macaque (with some responding to clockwise and others to anticlockwise rotation). Some of these speed neurons respond to vestibular information, in that they respond to speed when the view of the world is obscured. Other hippocampal speed cells neurons respond to the optic flow that induces a sense of motion. These hippocampal neurons were termed 'whole-body motion' neurons , but can also be termed 'speed' cells though that term describes less well that the neurons encode whole body motion that includes the head (O'Mara, Rolls, Berthoz and Kesner, 1994; Rolls, 2025c).

Figure 5.4i illustrates the activity of a primate hippocampal whole body motion (speed) neuron that increased its firing rate during backward linear translation, and decreased its firing rate during forward linear translation (O'Mara et al., 1994).

Fig. 5.4ii illustrates the activity of a different primate hippocampal whole body motion (speed) neuron that decreased its firing rate during linear translation of the monkey away from wall 3, and that also decreased its firing rate more greatly during linear translation of the test chamber (wall 3) away from the macaque's view of the chamber. The neuron did not respond during linear translation if the view was occluded (middle panel). This shows that optic flow can influence the activity of some primate hippocampal whole body motion (speed) cells (O'Mara et al., 1994; Rolls, 2025c).

Since these discoveries (O'Mara, Rolls, Berthoz and Kesner, 1994), speed cells have now also been confirmed in rodents, in for example the medial entorhinal cortex (Kropff, Carmichael, Moser and Moser, 2015), though angular velocity, and the contribution of vestibular vs optic flow cues to self-motion were not described. These rodent 'speed cells' could be used for idiothetic update of place (Moser, Rowland and Moser, 2015; Moser, Moser and Roudi, 2014; Giocomo, Moser and Moser, 2011). Hippocampal speed (/whole body motion) cells, for angular and translational motion, have also now been confirmed in primates (marmosets) (Piza, Corrigan, Gulli, Do Carmo, Cuello, Muller and Martinez-Trujillo, 2024). At least some of these primate speed cells may be interneurons, for some have higher firing rates than is typical of hippocampal pyramidal cells (O'Mara et al., 1994; Piza et al., 2024).

The primate speed or whole body motion neurons that we discovered (O'Mara, Rolls, Berthoz and Kesner, 1994) are probably involved in the idiothetic (self-motion) update of primate hippocampal neurons that respond to spatial view (Rolls, 2023a; Robertson, Rolls and Georges-François, 1998). It is proposed that these primate hippocampal system whole body motion (speed) neurons are involved in navigation especially when the view details are obscured. It is also proposed that these neurons are involved in episodic memory, enabling world-based recall of locations of objects and people in scenes independently of eye position, head direction, or the place where the individual is located, especially when the view details are obscured (Rolls, 2023a, 2026d).

5.2.7 Supporting evidence for neurons responding to spatial view in the primate including human hippocampal system

Relatively recently, supporting evidence related to our discovery of spatial view neurons in primates has accumulated, with the indication that spatial view encoding is a key type of representation found in primates including humans, and that differs from what has been the primary focus in rodents, place cells. Some of this evidence is in papers in a Special Issue of the journal **Hippocampus** entitled **Hippocampal system neurons encoding views in different species** (Rolls and Wirth, 2023).

One example is that in a virtual reality investigation of navigation in a star maze, hippocampal neurons were found that responded to where the macaque was gazing in the virtual reality space (Wirth et al., 2017). Some neurons also responded to a combination of where the macaque was and where the macaque was looking in space, but it is not clear that a balanced factorial design was used in which only spatial locations that could be seen from all places equally well (and sufficiently often) were involved in that type of analysis. Unless all spatial views can be seen from all places in the data analyzed, the data may appear to show mixed encoding to combinations of view and place (Rolls, 2023a).

Further supporting evidence about spatial view cells in primates comes from a study of them in which the room cues were changed (Baraduc, Duhamel and Wirth, 2019). In rodents, if the room cues are changed, but the spatial structure of the environment is not changed, place cells rapidly learn to adjust to the change of the room cues, and the spatial representations provided by the place cells remains unchanged. This well-known phenomenon is referred to as rate remapping (Leutgeb, Leutgeb, Treves, Meyer, Barnes, McNaughton, Moser and Moser, 2005a; Leutgeb, Leutgeb, Barnes, Moser, McNaughton and Moser, 2005b; Battaglia and Treves, 1998b; McNaughton, Barnes, Gerrard, Gothard, Jung, Knierim, Kudrimoti, Qin, Skaggs, Suster and Weaver, 1996; Samsonovich and McNaughton, 1997). In contrast, when the spatial structure of the environment is altered, from for example a square to a circular environment, place cells show global remapping, in which a new chart or map must be formed, and the locations of place cells in the new chart or map are completely different, to provide a different map of the different space (Leutgeb et al., 2005a,b; Battaglia and Treves, 1998b). Rate remapping was shown to also apply to spatial view cells in primates (Baraduc, Duhamel and Wirth, 2019), although the term ‘schema’ was used (Baraduc, Duhamel and Wirth, 2019), rather than the conventional terms ‘chart’ or ‘map’ (Leutgeb et al., 2005a,b; Battaglia and Treves, 1998b; McNaughton et al., 1996; Samsonovich and McNaughton, 1997). The term ‘schema’ is more usually reserved for describing neocortical semantic representations that are more than a hippocampal spatial chart or map, and might refer to for example the whole semantic representation / knowledge of what a doctor’s surgery is like (Gilboa and Marlatte, 2017; Wagner, van Buuren, Kroes, Gutteling, van der Linden, Morris and Fernandez, 2015; Ghosh, Moscovitch, Melo Colella and Gilboa, 2014; Ghosh and Gilboa, 2014).

Another example of supporting evidence for the discovery of spatial view cells is that some primate hippocampal neurons code for the location viewed “out there” on a screen in object-location memory tasks (Chen and Naya, 2020; Yang and Naya, 2020; Yang, Chen and Naya, 2023; Yang and Naya, 2023).

Another example is that in freely moving macaques, hippocampal neurons were found to respond to the spatial view towards which the macaque was looking (Mao, Avila, Caziot, Laurens, Dickman and Angelaki, 2021). Very interestingly, such spatial view neurons are better activated when freely moving in a real environment than in Virtual Reality of the same environment, and are well activated when the macaque is performing an object-viewed location task (Yan and Mao, 2024). Those findings are of course consistent with the good responses of spatial view neurons that we found when macaques were freely locomoting in an open environment (Rolls et al., 1997b; Robertson et al., 1998; Rolls

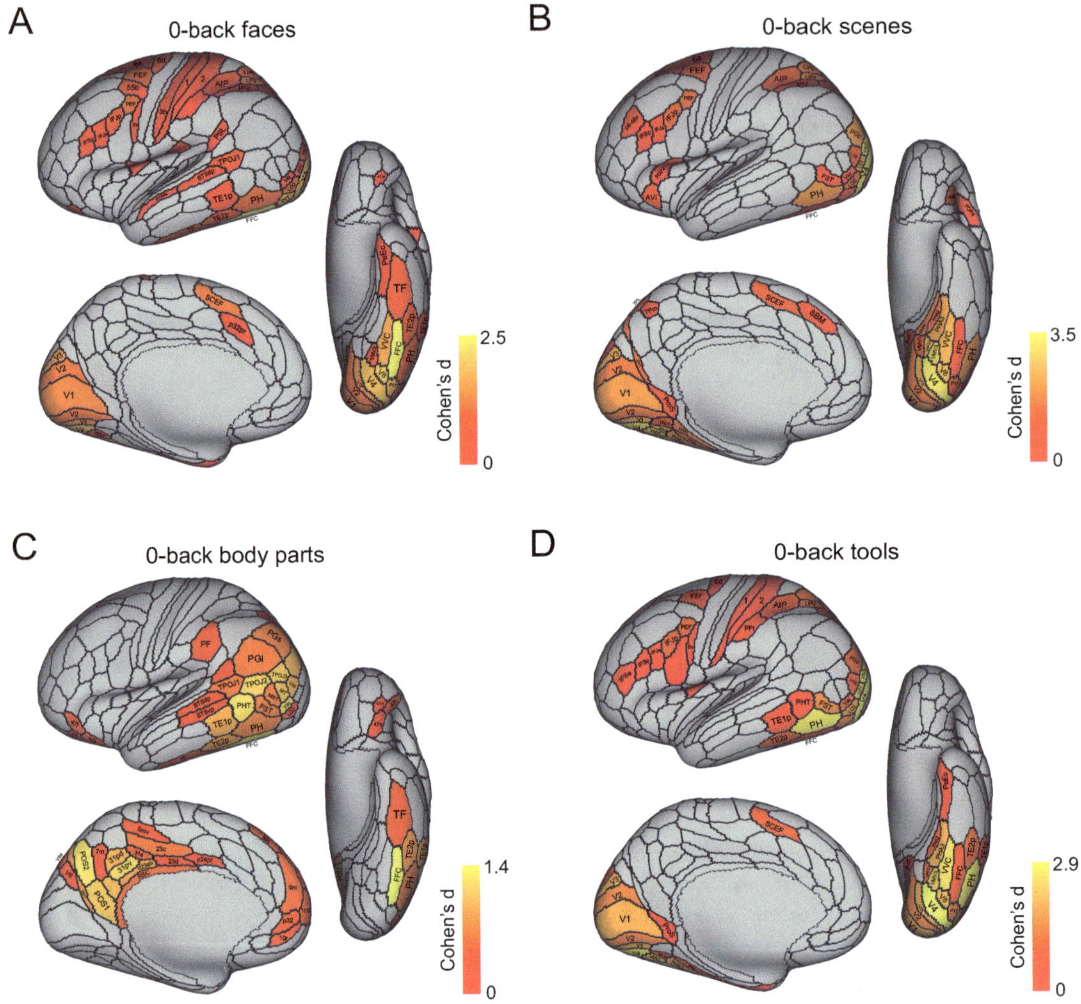

Fig. 5.5 The cortical regions exhibiting significant activations of the fMRI BOLD signal for faces, scenes, body parts and tools in 956 Human Connectome Project participants. The activations are shown after Bonferroni correction (p=0.05) for multiple comparisons, and using the Human Connectome Project Multimodal Parcellation atlas (HCP-MMP) (Glasser et al., 2016a). The effect size as indicated by Cohen's d is indicated. The activations are shown in red to yellow. This illustrates that when humans are in one place but viewing different scenes, ventromedial visual cortical stream regions VMV1-3 and PHA1-3 leading to the hippocampus are activated. (Data from Rolls E T, Feng J and Zhang R. 2024. Selective activations and functional connectivities to the sight of faces, scenes, body parts and tools in visual and non-visual cortical regions leading to the human hippocampus. Brain Structure and Function 229: 1471-1493.)

et al., 1998; Georges-François et al., 1999); and with the object - scene location neurons that we discovered in a prototypical hippocampal episodic memory object - viewed spatial location task (Rolls, Xiang and Franco, 2005c).

Another example of supporting evidence for the discovery of spatial view cells is that in marmosets freely moving in a 3D environment, approximately 1/3 of hippocampal neurons responded to the view towards which the individual was looking (Piza, Corrigan, Gulli, Do Carmo, Cuello, Muller and Martinez-Trujillo, 2024).

Another example is that in a virtual reality experiment involving navigation towards goals, some macaque hippocampal neurons responded to where the animal was looking, rather than the place where the animal was located (Buffalo, 2025).

Also, for humans there is now some evidence for medial temporal lobe neurons with properties like those of spatial view cells (Ekstrom et al., 2003; Miller et al., 2013). For example, in the study by Ekstrom and colleagues, some medial temporal lobe neurons were found to represent views of landmarks (Ekstrom et al., 2003). In another study of human medial temporal lobe neurons, it was found that in a Treasure Hunt game, some neurons respond to the sight of remote locations rather than the subject's own place (Tsitsiklis et al., 2020). Just like macaque spatial view cells, these neurons in humans respond when the spatial location is seen with different bearings. Also in humans some medial temporal lobe neurons reflect the learning of paired associations between views of places, and people or objects (Ison et al., 2015), and this implies that views of scenes are important for human hippocampal function.

Consistent with these neuron-level investigations, human functional neuroimaging studies do show hippocampal or parahippocampal activation when different scenes or parts of scenes are viewed even when the human is fixed in one place for neuroimaging (Burgess, 2008; Zeidman and Maguire, 2016; Maguire, 2014; Chadwick et al., 2013, 2010; Hassabis et al., 2009; Brown et al., 2016; Epstein and Kanwisher, 1998; O'Keefe et al., 1998; Epstein and Baker, 2019; Rolls et al., 2024b,e, 2026). An example of how viewing different scenes activates medial temporal lobe regions including VMV1-3 and PHA1-3 leading to the hippocampus is shown in Fig. 5.5 (Rolls et al., 2024b).

The discovery of spatial view cells in primates (Georges-François et al., 1999; Rolls et al., 1998; Robertson et al., 1998; Rolls et al., 1997b, 1989b; Cahusac et al., 1989; Rolls and O'Mara, 1995; Feigenbaum and Rolls, 1991) that is leading to a revolution in understanding hippocampal function in primates including humans (Rolls, 2023d,a, 2025a, 2026c) is now therefore supported by a great deal of more recent evidence. The spatial view representations are so different from the representations provided by place cells in the rodent hippocampus of the place where the rodent is located that there are major implications for understanding hippocampal functions in primates including humans (Rolls, 2026c). These implications are considered in Section 5.2.13, but first I describe more discoveries on spatial view neurons that draw out further some of the ways in which they may be important, including how spatial view cells may be important in episodic memory.

5.2.8 Hippocampal spatial view cells are involved in object – spatial location episodic memory

The next issue that I addressed is how the primate hippocampus is involved in an example of episodic memory, remembering where objects have been seen in scenes with rapid learning and later recall.

We discovered that in an object-location memory task in which macaques had to rapidly learn and remember where objects were shown in a room, some hippocampal neurons (10%) responded differently to different objects independently of location; other neurons (13%) responded to the spatial view independently of which object was present at the location; and some neurons (12%) responded to a combination (conjunction) of a particular object and the place where it was shown in the room (Rolls, Xiang and Franco, 2005c). An example of a hippocampal neurons with conjunctive responses to Object 1 in Place 1 is shown in Fig. 5.6 (Rolls et al., 2005c). The results show that there are separate as well as combined representations of objects and their locations in space in the primate hippocampus in this type of object–viewed location memory task. This is a key property of an episodic memory system, for which associations between objects and the locations where they are seen are prototypical. Similarly, it has been shown more recently in humans that some hippocampal neurons code for the conjunction of the different elements that make up an episode (Kolibius et al., 2023).

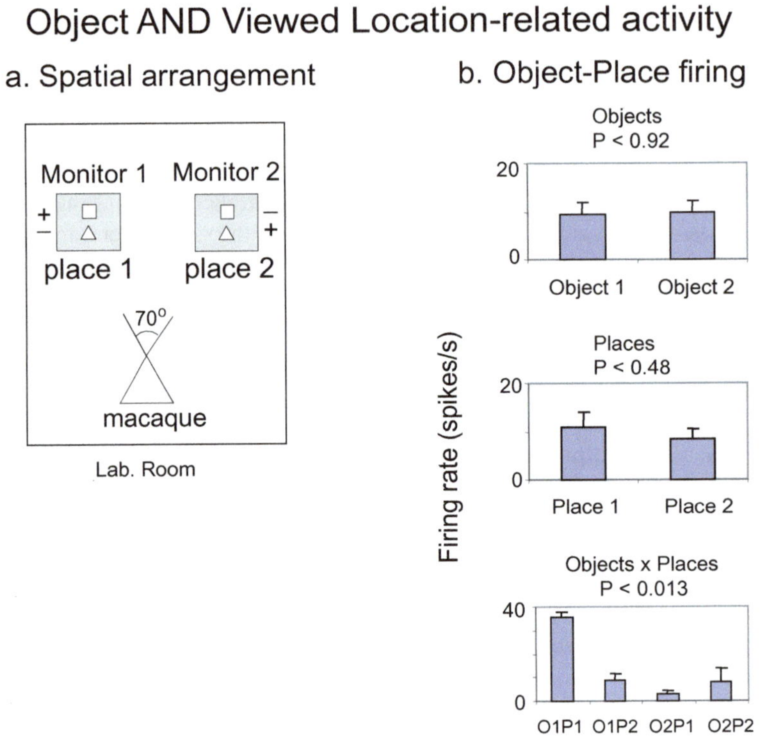

Fig. 5.6 Object–spatial view associations in the primate hippocampus. A hippocampal neuron (BL003c8a) with firing occurring to a particular combination of an object and place being viewed (Object 1–Place 1, O1P1). Note that 'Place' refers here to the viewed location in space. (a) Schematic diagram showing the spatial arrangement used in the Go–NoGo object–place combination task in this experiment. + indicates that a Go (lick) response to that object–place combination will be rewarded, and a minus that it will be punished. Only one object was shown on the screen at a time. The room was approximately 4 m x 4 m, and the distance of the macaque from the monitors was typically 1.5–2 m in the different experiments. The triangle labelled macaque shows the position of the macaque, and the angle shows the approximate angle subtended by the video display monitors. (b) The firing rate response of the neuron on different trials sorted according to the object that was shown on the trial, the place where the object was shown, and the combination of which object was shown with the place where it was shown. In the object*position bar histograms, O1P1 = Object 1 in Place 1, etc. The mean responses ± sem are shown. The neuron showed a conjunctive response to Object 1 in Place 1, O1P1. (Data from Rolls E T, Xiang J-Z and Franco L. 2005 Object, space and object-space representations in the primate hippocampus. Journal of Neurophysiology 94: 833–844.)

We also discovered that in a one-trial object – spatial location task, some hippocampal neurons responded during recall when an object cue signaled that the spatial view location should be recalled, and that other neurons responded during recall when the object was recalled by a spatial location cue (Rolls and Xiang, 2006). That makes spatial view neurons highly relevant to episodic memory, and in particular to how episodic memories can be recalled by an incomplete recall cue. This discovery of how recall can be implemented in the hippocampal memory system is complemented by more recent investigations (Yang and Naya, 2020, 2023).

5.2.9 Hippocampal spatial view cells are involved in reward/goal – spatial location memory

The next question I addressed was whether spatial view neurons can learn where rewards, goals, are in viewed space, for that is important in both episodic memory, and in navigation towards goals. To investigate how this affective input may be incorporated into primate hippocampal function, we recorded hippocampal neuronal activity while macaques performed a reward-place association task

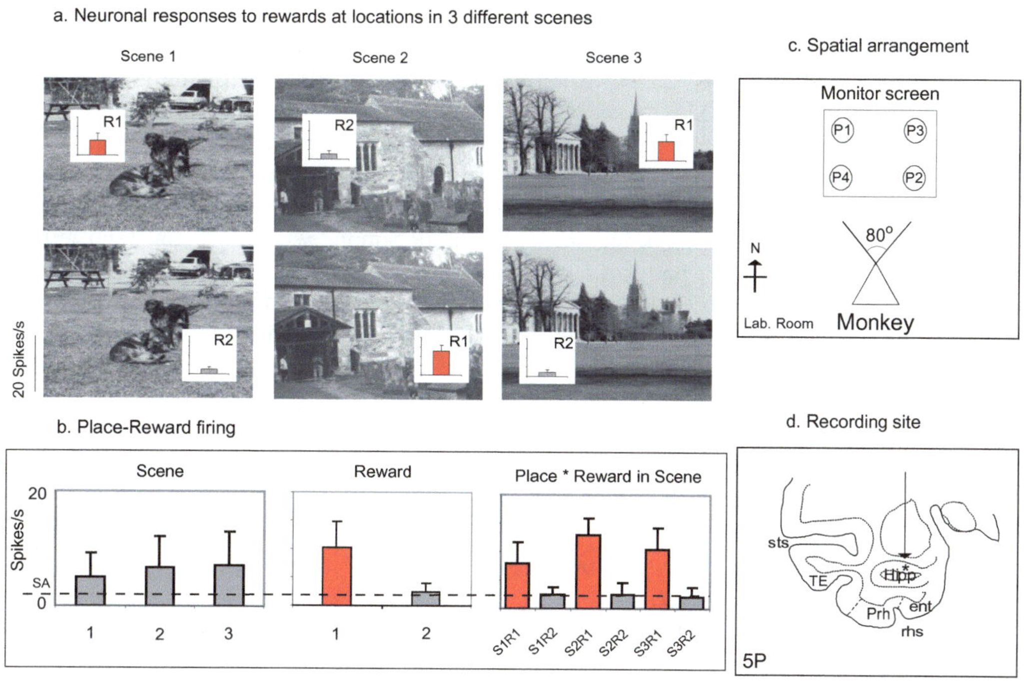

Fig. 5.7 Reward–spatial view associations in the primate hippocampus. A hippocampal neuron that encoded the particular rewards available at different locations in different scenes. On each trial the monkey could touch a circled location in the scene, and, depending on the location, received either a preferred juice reward or a less preferred juice reward. (a) Firing rate inserts to show the firing in 3 different scenes (S1–S3) of the locations associated with reward 1 (R1, preferred, shown in red) and reward 2 (R2). The mean responses ± sem are shown. SA, spontaneous firing rate. (b) The firing rates sorted by scene, by reward (1 vs 2), and by scene-reward combinations (e.g. scene 1 reward 1 = S1R1). (c) The spatial arrangement on the screen of the 4 spatial locations (P1–P4). (d) The recording site of the neuron. ent–entorhinal cortex; Hipp–hippocampal pyramidal cell field CA3/CA1 and dentate gyrus; Prh–perirhinal cortex; rhs–rhinal sulcus; sts–superior temporal sulcus; TE–inferior temporal visual cortex. (Data from Rolls E T and Xiang J-Z. 2005 Reward-spatial view representations and learning in the primate hippocampus. Journal of Neuroscience 25: 6167–6174.

in which each spatial scene shown on a video monitor had one location which if touched yielded a preferred fruit juice reward, and a second location which yielded a less preferred juice reward (Rolls and Xiang, 2005). Each scene had different locations for the different rewards. Of 312 hippocampal neurons analyzed, 18% responded more to the location of the preferred reward in different scenes, and 5% to the location of the less preferred reward (Rolls and Xiang, 2005). An example of one of these neurons, responding to the spatial location in each scene of a preferred reward, is shown in Fig. 5.7.

Further, we discovered that when the reward values are reversed, these spatial view reward neurons reverse the locations in the scene to which they respond (Rolls and Xiang, 2005).

What appear to be similar neurons have since been described that respond in humans to the viewed location of the goal in a Treasure Hunt game (Tsitsiklis et al., 2020). Further, what may be analogous hippocampal neurons have more recently been described that respond when a chickadee looks at a particular food cache location (Chettih, Mackevicius, Hale and Aronov, 2024). Interestingly, these neurons in the avian hippocampus do not respond when the cache location is viewed when the food has been removed from the cache so that it is no longer a reward location, which is very analogous to the primate spatial view and reward neurons that stop responding to the high value location in a scene when it is devalued in reward reversal (Rolls and Xiang, 2005). The avian evidence supports the point that good foveal vision is useful for setting up spatial view cells (De Araujo, Rolls and

Stringer, 2001) (Fig. 5.9), and also for the evidence that a random new set of CA3 neurons is chosen for each episodic memory (Rolls and Treves, 2024; Rolls, 2016e, 2013c), termed in the chickadee a "bar code" (Chettih et al., 2024).

The discovery of this reward-related functionality of primate spatial view neurons (Rolls and Xiang, 2005) is important, for it is very valuable to primates including humans to remember where in the environment they have seen rewards (or for that matter aversive stimuli). Interestingly, the spatial view – reward neurons are a feature of the hippocampus with its spatial (and temporal) processing, for when macaques perform an object-reward association task, hippocampal neurons are rather unresponsive, and we discovered that this type of object-reward association learning and reversal is implemented in the primate orbitofrontal cortex (Thorpe, Rolls and Maddison, 1983; Rolls, Critchley, Mason and Wakeman, 1996a; Rolls, Vatansever, Li, Cheng and Feng, 2020e; Rolls, 2023c, 2026d) (Chapter 3).

These spatial view – reward neurons are also very likely to be important in navigation to goal locations, for which the representation of where goals are in viewed space is important (see Section 5.2.13).

5.2.10 A ventromedial cortical Where visual stream to the human hippocampus for spatial view

Another key issue that I thought was important to investigate is how the primate including human hippocampus receives information about spatial view. One concept (Bicanski and Burgess, 2018) has been that the dorsal 'Where' visual cortical stream to the parietal lobe, which has been called a 'Where' pathway (Ungerleider and Haxby, 1994; Mishkin, Ungerleider and Macko, 1983), is the route for spatial information to reach the hippocampus. Another concept is that medial entorhinal cortex grid cells in rodents provide a spatial input to the hippocampus for 'Where' place representations (Moser, Moser and McNaughton, 2017).

We addressed this issue of how spatial view information reaches the human hippocampus by measuring effective connectivity in the human brain to analyze visual cortical processing. Effective connectivity measures the effects of each pair of brain regions on each other, and is computed in the Hopf algorithm that we use based on functional connectivity and delayed functional connectivity, with fMRI data from 171 Human Connectome Project participants in the resting state at 7T, validated by 956 participants at 3T in the resting state, and with fMRI activation data from 956 Human Connectome Project participants when viewing scenes, faces, tools and body parts (Rolls et al., 2022c,b, 2023g,d,b,c,a, 2024b,a; Rolls, 2024c; Rolls et al., 2026). These effective connectivity analyses are all complemented by functional connectivity and by diffusion tractography, and also included magnetoencephalography which because of its fast timecourse can provide clear evidence about the directionality of the effective connectivity (Rolls et al., 2024c, 2023e) (see Section 6.1 for further background about connectivity measurements).

With these approaches, we have been able to identify a ventromedial cortical visual pathway in humans illustrated in Fig. 5.8 with connectivity from V1 the primary visual cortex and V2 to the ProStriate cortex and nearby regions where the retrosplenial scene area that responds to scenes is located (Rolls et al., 2024b; Sulpizio et al., 2020). The ProStriate region then projects forwards to ventromedial visual cortex VMV1-3 and VVC, which in turn project to the medial parahippocampal cortex PHA1-3 where the parahippocampal scene area is located (Rolls et al., 2024b; Sulpizio et al., 2020), which in turn connect to the hippocampus (Huang et al., 2021; Rolls et al., 2023e,b, 2022a; Ma et al., 2022; Rolls et al., 2024c; Rolls, 2024c). The discovery of this pathway and associating it with spatial view cells (Huang et al., 2021; Rolls et al., 2023e,b, 2022a; Ma et al., 2022; Rolls et al., 2024c; Rolls, 2024c) is new, but is supported by earlier research showing that what were termed

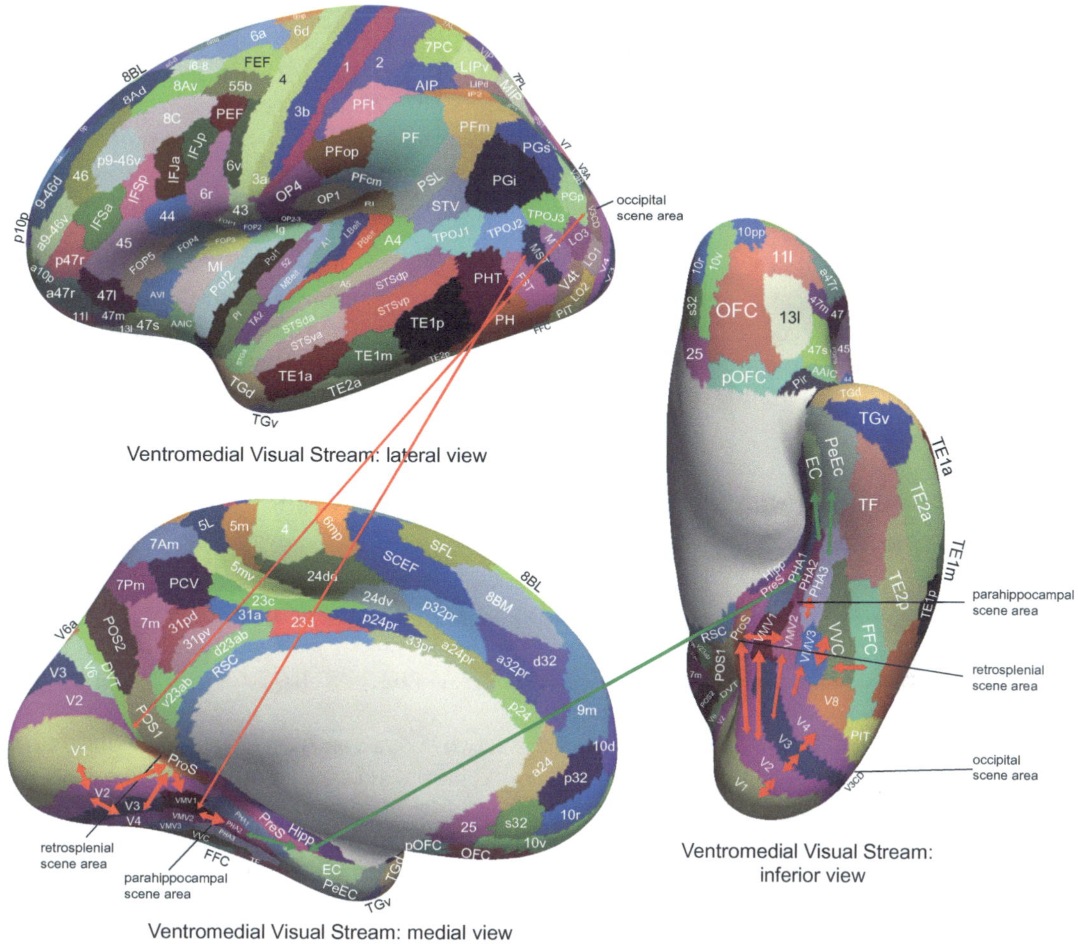

Fig. 5.8 Effective connectivity of the human Ventromedial Visual Cortical Stream for spatial scenes which reaches the parahippocampal gyrus PHA1 - PHA3 regions via ventromedial (VMV) and Ventral Visual Complex (VVC) and ProStriate regions: schematic overview. Visual scenes are represented in the anterior parts of VMV and the posterior parts of PHA1 - PHA3 (Sulpizio et al, 2020) in what is termed the Parahippocampal Place Area, PPA (Epstein and Baker 2019). (The PPA might better be termed the Parahippocampal Scene Area PSA as it is the scene being viewed, not the place where the individual is located, that is represented (Rolls, 2023c, 2026d).) The retrosplenial scene area is in a band of cortex in the Prostriate cortex PRoS and Dorsal Visual Transitional cortex DVT that is posterior to region RSC (Sulpizio et al, 2020). The occipital scene area is in V3CD and borders V4 (Sulpizio et al, 2020). The green arrows show how the Ventromedial Visual Stream provides 'where' input about locations in scenes to the hippocampal memory system from the medial parahippocampal gyrus PHA1-PHA3 region (which corresponds to TH in macaques). The connectivity from PGp to PHA regions is suggested in the text to be involved in self-motion (idiothetic) update of locations in scenes. The widths of the lines and the size of the arrowheads indicate the magnitude and direction of the effective connectivity. (Data from Rolls ET, Deco G, Huang CC, and Feng J. 2023b Multiple cortical visual streams in humans. Cerebral Cortex 33: 3319-3349; and Rolls ET, Yan X, Deco G, Zhang Y, Jousmaki V, and Feng J. 2024c. A ventromedial visual cortical 'where' stream to the human hippocampus for spatial scenes revealed with magnetoencephalography. Communications Biology 7: 1047.)

'places' activate a parahippocampal 'place' area PPA (Epstein and Kanwisher, 1998; Epstein and Julian, 2013; Epstein and Baker, 2019). To distance this human cortical area from rodent place cells, and because scenes are being looked at, with the place where the individual is located held constant, I have introduced use of the term Parahippocampal Scene Area PSA for this brain region (Huang et al., 2021; Rolls et al., 2023e,b, 2022a; Ma et al., 2022; Rolls et al., 2024c; Rolls, 2024c, 2026d,c).

What is quite remarkable is this discovery that a 'Where' representation for locations in viewed spatial scenes is encoded by a ventromedial cortical visual pathway, not by the dorsal visual stream (Rolls, Deco, Huang and Feng, 2023b; Rolls, 2023a; Rolls, Yan, Deco, Zhang, Jousmaki and Feng, 2024c; Rolls, 2024c; Rolls, Feng and Zhang, 2024b; Rolls, Zhang, Deco, Vatansever and Feng, 2024e; Rolls, Feng and Zhang, 2026; Rolls and Turova, 2025; Rolls, 2026c,d). The proposal is that computations of the type typical of ventral visual stream cortical processing (Rolls, 2012b, 2021c, 2023c, 2026d) are used to form feature combinations of the visual features present in different parts of scenes to represent scenes based on the features found in scenes, a proposal that has been developed further as shown in Section 5.2.12 (Rolls, 2025b). This is different to anything proposed for rodents, and emphasises how fundamental discoveries about the brains of humans and other primates in health and disease are being made in this research, by pioneering a research strategy on the hippocampal system that is different to previous research that has been based on rodents, which have very different systems-level brain organization (see Section 29.4 in **Brain Computations and Principles; and AI** (Rolls, 2026d)).

5.2.11 A ventrolateral cortical What visual stream to the human hippocampus for faces and objects

These connectivity analyses have been extended to show how 'What' information about faces and objects reaches the human hippocampus by the ventrolateral visual cortical stream (Rolls, 2024c; Rolls et al., 2023e,b; Rolls, 2023a) (Figs. 6.1 and 4.1); and how reward /goal information reaches the hippocampus from the orbitofrontal cortex (Rolls et al., 2022b; Rolls, 2022a, 2024c) (Figs. 3.14, 3.15, and 3.16).

For comparison, the four visual cortical pathways, two 'What', and two 'Where', are illustrated in Fig. 4.1 to show how the Ventromedial 'Where' and Ventrolateral 'What cortical visual pathways fit into our current understanding of cortical visual pathways in humans (Rolls, 2024c, 2023c, 2026d).

These connectivity investigations have been complemented by fMRI activation studies that add function to these pathways in further discoveries. We have shown in 956 Human Connectome Project (HCP) participants that the medial parahippocampal PHA1-3 and ventromedial visual VMV1-3 are more activated when viewing scenes; and that the fusiform face cortex FFC and lateral parahippocampal region TF are more activated when viewing faces (Fig. 5.5) (Rolls, Feng and Zhang, 2024b, 2026). Further, we have shown that these spatial scene regions become selectively activated when performing episodic memory tasks involving object – viewed spatial location memory or reward – viewed spatial location memory (Rolls, Zhang, Deco, Vatansever and Feng, 2024e).

Interestingly, we have even been able to extend this to sex differences in humans, with the ventromedial cortical spatial scene 'Where' pathway more strongly activated by scenes in males, and the ventrolateral cortical face 'What' pathway more strongly activated by faces in females (Zhang, Rolls and Feng, 2026b). These cortical specializations are likely to be relevant to differences for females compared to males in the processing of socially relevant stimuli such as faces and body parts; and to differences for males compared to females in the processing of spatial scenes and tools (Zhang, Rolls and Feng, 2026b).

5.2.12 How spatial view cells are built by self-organizing learning to represent locations in scenes

I have taken on another key issue in understanding the primate including human hippocampal system for memory and navigation, by considering how spatial view cells, and scene representations, are built computationally in the cerebral cortex.

A first approach to understanding the computations involved in forming primate spatial view

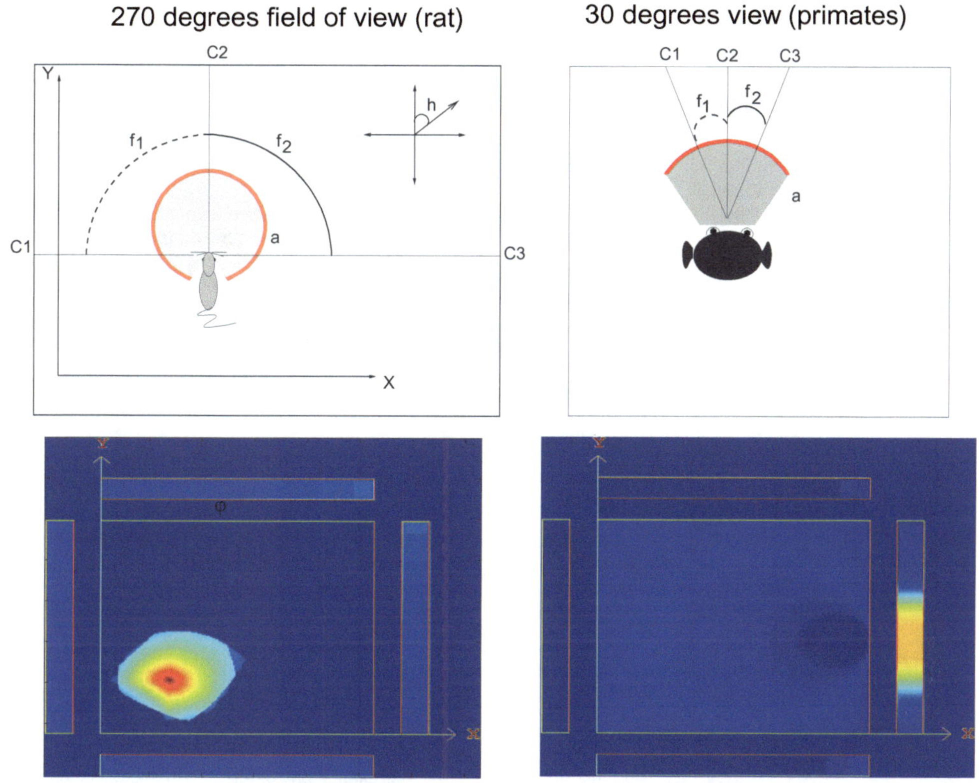

Fig. 5.9 Simulation of rodent place cells (left) vs primate spatial view cells (right). The agent moved through a grid of all 200x200 places x, y. At each place the head direction h was rotated by 5 degree increments. Hippocampal cells are activated by a set of 3 or more landmark visual cues within the field of view a of the agent. The firing rates of the hippocampal neurons depended on the angles f subtended by the landmarks. The top left shows that for a rodent with a 270° field of view a combination of such cues defines a place. The top right shows that for a primate with a 30° field of view the combination of cues defines a spatial view. The sizes of the fields of view are shown by shading. The bottom left shows that in the simulations place fields arise with a 270° field of view, and the bottom right that spatial view fields arise on one of the walls indicated by the rectangles when the field of view is 30°. High firing rates are indicated by yellow-red. (Data from De Araujo I E T, Rolls E T and Stringer S M. 2001 A view model which accounts for the spatial fields of hippocampal primate spatial view cells and rat place cells. Hippocampus 11: 699-706.)

neurons rather than rodent place cells is illustrated in Fig. 5.9 (De Araujo, Rolls and Stringer, 2001). The concept builds on the fact that rodents have no fovea and a field of view in the order of 270 degrees, whereas primates have a fovea that provides high spatial resolution for a small part of the visual field, and a highly developed cortical visual system that results in representations of objects and faces that subtend a few degrees at the retina in complex natural scenes (Rolls, Aggelopoulos and Zheng, 2003a). If rodents form feature combinations of widely spaced landmarks in their 270 degree visual field, a place would be represented (Fig. 5.9 left). If primates form feature combinations of features close to where they are fixating with their fovea in a visual scene, then the feature combinations formed by the neurons would represent a location in a scene, providing a model for spatial view cells (Fig. 5.9 right). We discovered that self-organizing learning would form these types of place or spatial view representations, depending on how wide the field of view was (Fig. 5.9) (De Araujo, Rolls and Stringer, 2001).

A second approach that develops these concepts further is illustrated in Fig. 5.10, which shows a model of the ventromedial cortical visual stream for building spatial view cells by using visual feature combinations present in the scene, VisSceneNet (Rolls, 2025b). Fig. 5.10 shows the architecture of

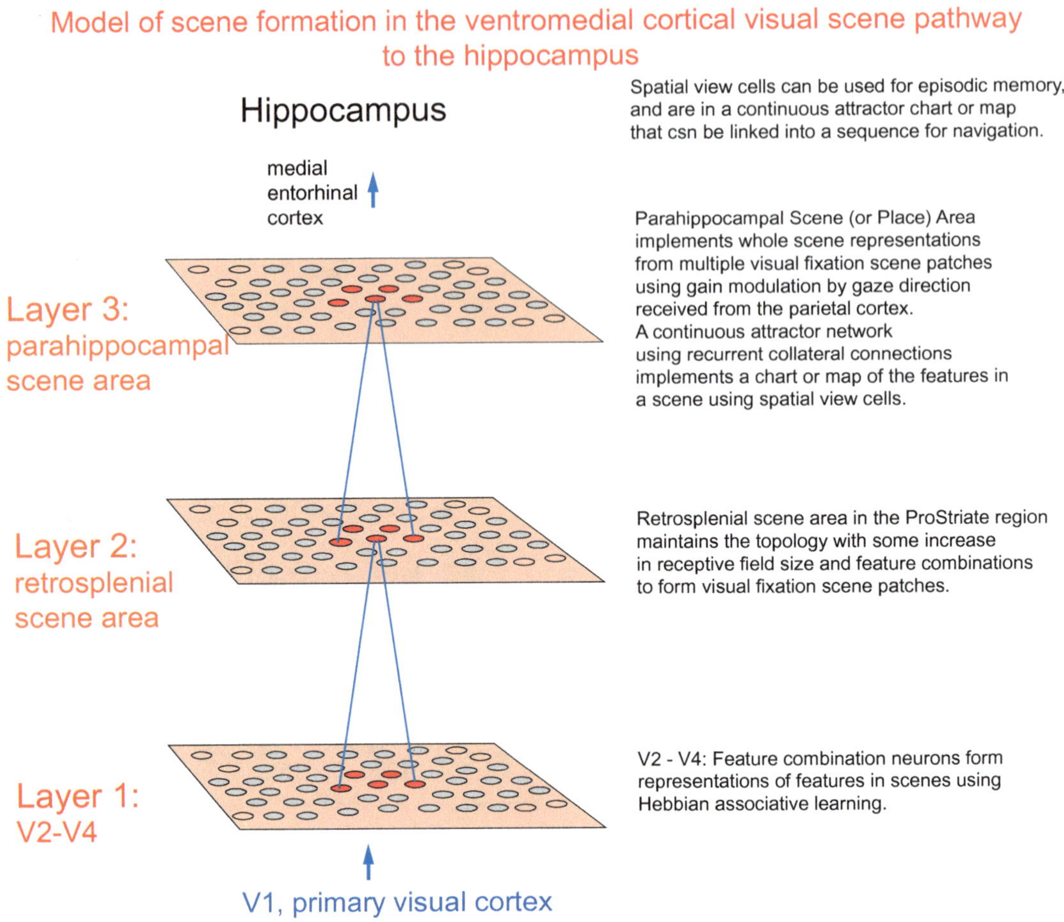

Fig. 5.10 The theory and model VisSceneNet for building hippocampal spatial view cells and scene representations along the ventromedial cortical visual scene pathway. This is a three-layer feedforward network, with competitive learning implemented using the forward synaptic connections at each layer, and short-range convergence from layer to layer. The input to Layer 1 is from the primary visual cortex V1, and is produced from a scene by Gabor filtering to produce a V1-like representation. Visual fixation scene patches for a single fixation of a scene are produced by Layer 2. Layer 3 implements gain modulation by world-based gaze direction to map visual fixation scene patches into a whole scene representation in Layer 3. The whole scene representation in Layer 3 makes use of associatively modifiable recurrent collateral connections to form a continuous attractor network for the whole scene. Nearby features in a scene are more strongly linked in the continuous attractor because they are more likely to be co-active. (Data from Rolls E T. 2025 A theory and model of scene representations with hippocampal spatial view cells. Hippocampus 35: e70013.)

the feedforward self-organizing network. There is a series of layers with competitive learning at each stage like VisNet (Rolls, 2012b, 2021c, 2023c, 2026d) (Chapter 4), but unlike VisNet with only limited convergence from stage to stage in order to maintain topography in the representations. Layer 3 uses recurrent collateral connections between the neurons to build a continuous attractor network so that the distance between the features can be represented in both the horizontal and vertical dimensions. In Layer 3, gain modulation by gaze direction to map visual fixation scene patches to the correct part of the whole scene representation is included. This process may be helped by inputs from the dorsal visual system (Fig. 4.1) to the medial parahippocampal gyrus that take into account gaze direction (the sum of eye and head direction) to reflect the distances between different locations in a scene, and consistent with this, some hippocampal system neurons in primates are influenced by eye movements (Nowicka and Ringo, 2000; Sobotka and Ringo, 1997; Sobotka et al., 1997; Ringo et al., 1994; Buffalo, 2025). Further, gaze direction may be represented in the primate hippocampus (Dun Mao, personal communication, 2024).

What is revolutionary about these concepts (Rolls, 2025b, 2026c) is that feature combinations are used in a type of processing typical of ventral visual pathways to build 'Where' representations of scenes in a ventromedial visual cortical pathway in humans and other primates (Rolls, 2026c). The theory and model (Rolls, 2025b) show how the cortical regions such as the retrosplenial scene area and the parahippocampal scene (or place) area can build scene representations in these regions (Sulpizio et al., 2020; Natu et al., 2021; Kamps et al., 2016; Epstein and Kanwisher, 1998; Epstein and Baker, 2019; Epstein and Julian, 2013; Epstein, 2008, 2005; Rolls et al., 2024b) that can represent locations in scenes using spatial view cells.

5.2.13 Implications of the discovery of spatial view cells for understanding memory and navigation in primates including humans

The discovery of spatial view cells in primates (Rolls et al., 1989b; Cahusac et al., 1989; Rolls and O'Mara, 1995; Georges-François et al., 1999; Feigenbaum and Rolls, 1991; Rolls et al., 1997b, 1998; Robertson et al., 1998; Rolls and Xiang, 2006; Rolls et al., 2005c; Rolls and Xiang, 2005) is revolutionizing our understanding of hippocampal functioning in primates including humans (Rolls, 2023d,a,c, 2026d; Rolls and Treves, 2024; Rolls, 2025a, 2026c).

First, human episodic memory typically involves remembering where people, objects, and rewards or goals are in viewed scenes on particular occasions. We have shown that hippocampal spatial view cells are involved in the rapid association learning and then recall required for this type of episodic memory for objects and their locations in viewed scenes (Rolls and Xiang, 2006; Rolls, Xiang and Franco, 2005c), and rewards and their locations in viewed scenes (Rolls and Xiang, 2005). In contrast, in rodents the focus has been on the representation of the place in which the rodent is located, and on path integration between places for navigation. The rodent findings are consistent with their much poorer visual systems, no fovea, and very wide field of view, and greater reliance on local cues such as touching and exploring local cues with the vibrissae, and olfaction (Rolls, 2025a). For primates, spatial view cells are naturally suited to the representations of locations being viewed, and the importance of vision in representing the identity of objects, people, and rewards such as fruit or food, so that the 'What', the 'Where', and the reward value can be associated together where they are combined, in the single attractor network in hippocampal CA3 (Rolls and Treves, 2024). This provides enormous adaptive value in finding rewards, people, and objects in the future, by recalling particular memories of where they have been seen. The ability of the primate hippocampus to associate viewed locations with reward value (Rolls and Xiang, 2005) also has great importance in enabling the emotional or affective value to become a key part of a typical human episodic memory, and in navigation towards goals in locations "out there".

Second, the focus in rodents of research on the hippocampal system has been on the places where the individual is located as represented by place cells, and on navigation between those places using path integration that takes into account the head direction and distance moved (O'Keefe and Dostrovsky, 1971; Burgess and O'Keefe, 1996; McNaughton, Barnes and O'Keefe, 1983; Morris, Garrud, Rawlins and O'Keefe, 1982; O'Keefe, 1979; McNaughton, Barnes, Gerrard, Gothard, Jung, Knierim, Kudrimoti, Qin, Skaggs, Suster and Weaver, 1996; Samsonovich and McNaughton, 1997; McNaughton, Battaglia, Jensen, Moser and Moser, 2006; Moser, Moser and McNaughton, 2017; Kropff, Carmichael, Moser and Moser, 2015; Moser, Moser and Roudi, 2014). However, the discovery of spatial view cells in primates opens up new vistas in understanding navigation in primates including humans, for now navigation can be based on vision, including of landmarks in scenes (Vijayabaskaran et al., 2025).

Indeed, a simple model of how navigation towards a sequence of landmarks does provide for one important type of human navigation or piloting, with the hippocampus as a system well suited to memory helping to provide the memory for the sequence (Rolls, 2021b). If the view details are obscured, path integration over eye position, head direction, and place is proposed to take place in the dorsal visual system (Rolls, 2020b), to produce a representation in world-based coordinates suitable for updating hippocampal spatial view cells by the pathways from the dorsal visual cortical regions including PGp to the medial parahippocampal cortex (Rolls, 2024c, 2023a, 2025b). Remembering a sequence of spatial locations is a type of memory in which the hippocampus is involved (Kesner and Rolls, 2015; Weeden et al., 2009), and may involve associations with time cells (Howard and Eichenbaum, 2015; Eichenbaum, 2014; MacDonald et al., 2011; Rolls and Mills, 2019; Kraus et al., 2013). Landmarks are of course also potentially useful for navigation even when they are not approached.

In contrast, navigation in rodents might be described as 'blind navigation', when it is described as involving path integration using head direction and distance travelled (see above). In fact, path integration is notoriously imperfect and likely to fail within a very few minutes and a short distance travelled, and most humans would not trust themselves to navigate from place to place with their eyes covered for more than a few metres and turns. **Indeed, it is notable that much of the current research on the brain and navigation is based on what is found in rats and mice, nocturnal animals with poor vision that navigate in dark underground tunnels. In this situation, the discovery of spatial view cells does open up new vistas in understanding navigation, and finding goals, in humans and other primates with excellent vision and the ability to utilize visual representations of space and visible landmarks (Rolls, 2021b, 2020b, 2026c,d).**

Third, the neural pathways to the hippocampus, and the information represented in the hippocampus, are very different in rodents compared to primates (Rolls and Treves, 2024; Rolls, 2024c, 2023c, 2026d), though the actual computations performed by the hippocampal circuitry may be similar in primates and rodents (Rolls and Treves, 2024; Rolls, 2023c, 2026d; Rolls and Wirth, 2018). In rodents, the representations provided by place cells may utilize properties such as local texture, and local olfactory cues (Itskov et al., 2011; Weeden et al., 2014; Kesner et al., 2011), and it appears that hippocampal place cells may be updated by path integration by medial entorhinal cortex grid cells (Moser et al., 2017). In contrast, we have discovered that in primates including humans, the pathways to the hippocampus, and accordingly the information received by the hippocampus, are very different (Rolls and Treves, 2024; Rolls et al., 2024b; Rolls, 2024c, 2025a, 2026c,d), as follows:

(1) In humans, there is the ventromedial visual cortical pathway to the hippocampus for spatial scenes that uses spatial view cells, as described in Sections 5.2.10 and 5.2.12.

(2) In humans, there is also a ventrolateral visual cortical pathway via the fusiform face cortex and the lateral parahippocampal gyrus TF for face and object, 'What', information to be brought to the hippocampal memory system (Rolls and Treves, 2024; Rolls et al., 2024b; Rolls, 2024c; Rolls et al., 2023b,e) (Section 6.2). Neither of these pathways has anything closely comparable in rats and mice with their poor vision.

(3) In humans, there are major pathways carrying reward information from the orbitofrontal cortex to the hippocampus, that involve regions such as the ventromedial prefrontal cortex, pregenual anterior cingulate cortex, and posterior cingulate cortex (Rolls, Deco, Huang and Feng, 2023d, 2022b; Rolls, Wirth, Deco, Huang and Feng, 2023g). Not only is the rodent orbitofrontal cortex very much less well developed than in primates (Passingham, 2021; Rolls, 2023c, 2026d), but it also is not directly comparable in what it computes, with actions apparently represented in the rodent but not primate orbitofrontal cortex (Rolls, 2023c, 2026d) (Section 3.14), and most of these cortical regions on these

pathways are poorly developed or absent in rodents (e.g. the posterior cingulate cortex) (Rolls, 2023c, 2026d, 2019c; Vogt, 2009).

5.3 A computational theory for the hippocampus and related systems

A second key set of discoveries is about the computations that may be performed by the hippocampus, and in the pathways that bring information to it, together with how the information is represented, and about how information can be recalled from the hippocampus back to the neocortex (Rolls, 1989a; Rolls and Treves, 2024; Rolls, 2023c, 2026d).

5.3.1 Background to computations performed by the hippocampus including an attractor network in CA3

The background is that by the 1980s, we had made many discoveries about what is represented in the hippocampus and in other parts of the brain (Rolls, 2023c, 2026d) (see also Chapters 2–4), but I felt the need for a framework to understand how each part of the brain was performing its functions, and how brain regions operate together to achieve a useful result. The answer I could see lay in understanding the computations being performed by the neuronal networks in each brain region, and how the regions are connected together to perform useful functions as a system.

Key insights came when a colleague mentioned that the book **Parallel Models of Associative Memory** edited by Hinton and Anderson (1981) had just been published. This contained Chapters on autoassociation (later termed attractor) networks (Kohonen et al., 1981) and on competitive networks (see Rumelhart and Zipser (1985)), and I read the stimulating books by Kohonen (1977; 1984). I very quickly programmed up autoassociation networks, competitive networks, and pattern association networks, and delivered a paper in 1985 to the *Dahlem Conference on The Neural and Molecular Bases of Learning*. Eric Kandel thought that this was sufficiently important to stop the separate workgroups of the conference, and brought everyone together for me to address the meeting about the potential of neural networks for understanding brain computations, including those performed by the hippocampus, including that hippocampal CA3 could be considered as an autoassociation network for episodic memory, that it could be used to store object-location memories, and that the pathways from the hippocampus back to the neocortex are an important route for the information to be retrieved for long-term storage in the neocortex. (The paper appeared in print (Rolls, 1987), and is on my website https://www.oxcns.org. In the same year that this paper appeared in print, another paper appeared that also proposed that the CA3 cells operate as an autoassociation network (McNaughton and Morris, 1987). The important earlier work of David Marr (Marr, 1971) was cited by both of these papers, though Marr had not identified CA3 as an autoassociation matrix.)

The research that we developed on how the hippocampal system computes is described in the remainder of Section 5.3.

5.3.2 Background to a theory of the recall of memories from the hippocampus back to the neocortex

The next key step in the development of the theory of the hippocampus was to produce a theory of how information could be recalled from the hippocampus back to the neocortex. David Marr (1971) had promised a theory for this, but never produced a theory. My theory is that the backprojection pathways from the hippocampus to the neocortex (shown in green in Fig. 5.11) are active during the

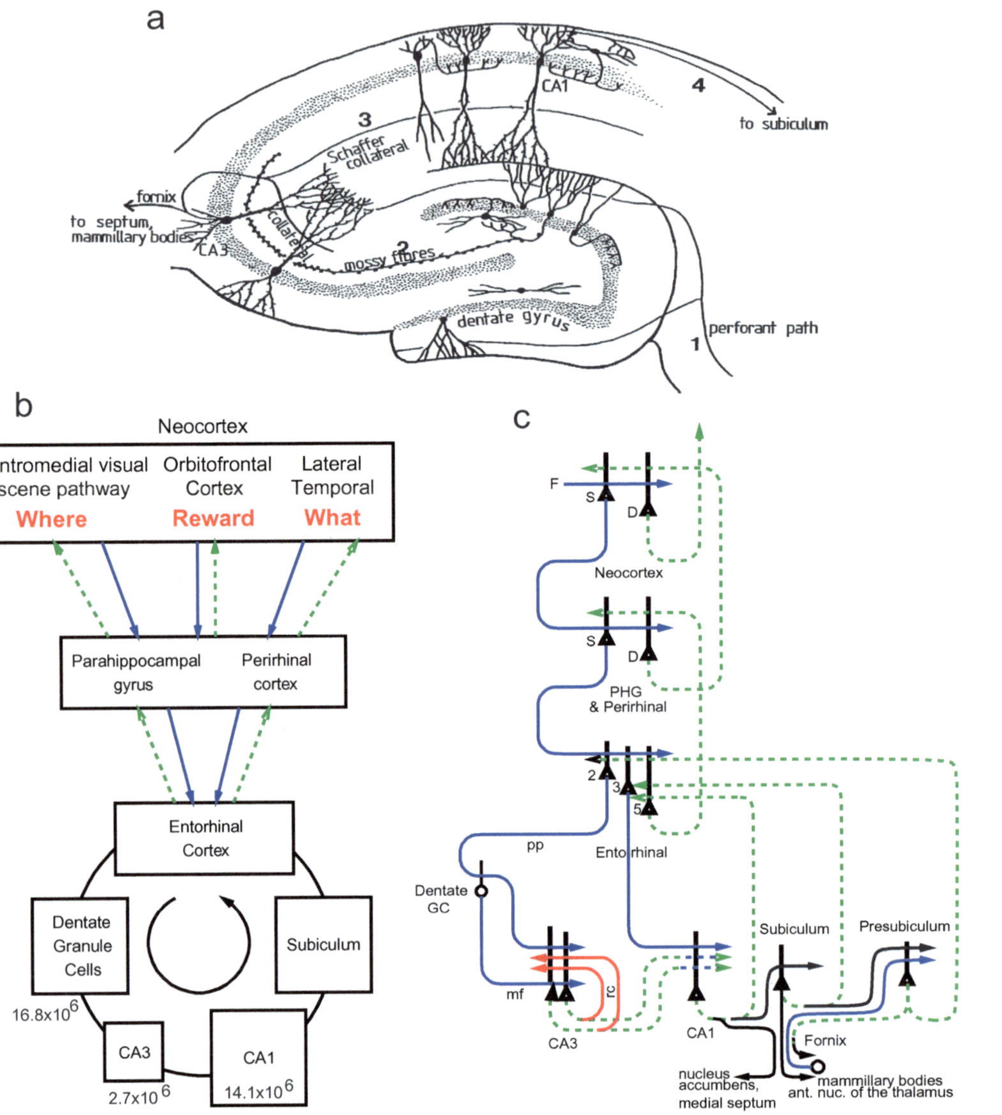

Fig. 5.11 a. Connections within the hippocampus. Inputs reach the hippocampus through the perforant path (1) which makes synapses with the dendrites of the dentate granule cells and also with the apical dendrites of the CA3 pyramidal cells. The dentate granule cells project via the mossy fibres (2) to the CA3 pyramidal cells. The well-developed recurrent collateral system of the CA3 cells is indicated. The CA3 pyramidal cells project via the Schaffer collaterals (3) to the CA1 pyramidal cells, which in turn have connections (4) to the subiculum. b-c. The human / primate hippocampus receives neocortical input connections (blue) not only from the 'what' temporal lobe and 'where' ventral visual scene areas, but also from the 'reward' prefrontal cortex areas (orbitofrontal cortex, vmPFC, and anterior cingulate cortex) for episodic memory storage; and has return backprojections (green) to the same neocortical areas for memory recall. There is great convergence via the parahippocampal gyrus, perirhinal cortex, and dentate gyrus in the forward connections down to the single network implemented in the CA3 pyramidal cells, which have a highly developed recurrent collateral system (red) to implement an attractor episodic memory by associating the what, where and reward components of an episodic memory. a: Block diagram. b: Some of the principal excitatory neurons and their connections in the pathways. Time and temporal order are also important in episodic memory, and may be computed in the entorhinal-hippocampal circuitry (Rolls and Mills, 2019). Abbreviations - D: Deep pyramidal cells. Dentate GC: Dentate Granule cells. F: Forward inputs to areas of the association cortex from preceding cortical areas in the hierarchy. mf: mossy fibres. PHG: parahippocampal gyrus and perirhinal cortex. pp: perforant path. rc: recurrent collateral of the CA3 hippocampal pyramidal cells. S: Superficial pyramidal cells. 2: pyramidal cells in layer 2 of the entorhinal cortex. 3: pyramidal cells in layer 3 of the entorhinal cortex. The thick lines above the cell bodies represent the dendrites. The numbers of neurons in different parts of the hippocampal trisynaptic circuit in humans (Rogers Flattery et al., 2020) are shown in (b), and indicate very many dentate granule cells, consistent with expansion recoding and the production of sparse uncorrelated representations prior to CA3 (Rolls, 2016e, 2021h, 2023c, 2026d). Data from Rolls E T. 2026 Hippocampal Revolutions. Neuroscience and Biobehavioral Reviews 180: 106492.

learning of the episodic memory, and form associatively modifiable synapses onto the neocortical neurons that are active during the learning (Rolls, 1989d,c,b,a). Each stage of the backprojection pathways is thus formally a pattern association network. (Descriptions of autoassociation / attractor, pattern association, and competitive networks, and Matlab and Python programs to illustrate their operation, are provided in **Brain Computations and Principles; and AI** (Rolls, 2026d).) Later, during recall, the neocortex provides a partial recall cue (such as the object, then the system needs to recall the location of the object), completion of the whole 'What' - 'Where' memory occurs in the CA3 autoassociation / attractor network, and then the correct 'Where' neurons are recalled by the multiple stage backprojection pattern association pathways back to the neocortex that are illustrated in green in Fig. 5.11 (Rolls, 1989a). Rolls' theory of recall from the hippocampus to the neocortex (Rolls, 1989d,c,b,a) was developed quantitatively (Treves and Rolls, 1994; Rolls and Treves, 1994), and adopted later by McClelland, McNaughton and O'Reilly (1995), though they pointed out that the pattern association process for the backprojections might not need to be learned simultaneously during the storage at every backprojection stage as long as it occurred at one or more stages during the storage, if the other stages had already been set up to implement the same pattern association process for the backprojection synapses (see Rolls and Treves (2024)).

The research that we developed on how the backprojections from the hippocampus to the neocortex implement the recall of episodic memories is described in Section 5.3.5.

5.3.3 The development of quantitative approaches to understanding the storage and recall of information in hippocampal and other cortical networks with sparse distributed coding, graded firing rates, and asymmetric diluted connectivity

At about this time (1990), the theoretical physicist Daniel Amit, who wrote an important book on attractor networks (Amit, 1989), visited Oxford and gave a research seminar on attractor networks. In discussion after the seminar, I pointed out to him that neurons in the brain have sparse distributed representations (with a sparseness often close to 0.1), graded firing rates (typically in the range 0-100 spikes/s), and diluted connectivity (typically 0.04 - 0.1), whereas the attractor networks analyzed by theoretical physicists generally used binary neurons with firing rates of for example either +1 or -1, full connectivity guaranteeing symmetric connections between neurons with use of an associative learning rule and hence a stable basin of attraction, and a sparseness of 0.5 (i.e. 50% of the binary neurons active for any one vector of firing rates to be stored in the network) (Hopfield, 1982; Amit, 1989; Hertz et al., 1991). I emphasized that it would be very important if the physics-based approaches to attractor networks could be extended to networks with sparse representations, diluted connectivity, and graded firing rates, in order to produce quantitative estimates of the storage capacity, stability, etc of biologically realistic attractor networks. I give great credit to Daniel Amit for somehow arranging for his wonderful student the theoretical physicist Alessandro Treves to visit my lab at Oxford for several years as a postdoctoral scientist and to develop the theory of the hippocampus further to become fully quantitative, when the theory of attractor networks and similar networks was extended to deal with the biologically realistic case of asymmetric diluted connectivity, sparse representations, and graded firing rates.

These were key advances that provide a quantitative foundation for the analysis of biologically plausible networks in the brain, as described below. **I believe that this research was and still is very important and foundational, because it goes beyond the classical theoretical physics approaches (Hopfield, 1982; Amit, 1989; Hertz et al., 1991; Dayan and Abbott, 2001; Wang, 2025), by showing that biologically plausible autoassociation networks with diluted asymmetric**

connectivity and therefore no formal energy function, as well as graded firing rates and sparse representations, can operate as attractor networks, and that the capacity is then set by the number of recurrent collateral connections C onto each neuron, not by the number of neurons N in the network, as well as by the sparseness of the representation.

We started with pattern association networks (see Fig. 1.3), used for association learning in the backprojection pathways from the hippocampus, and in other brain regions such as the orbitofrontal cortex (Rolls, 2023c, 2026d). We were able to show that the maximum number of patterns, $p_{\max}$, that can be stored and correctly retrieved with sparse representations, and graded firing rates, in a pattern association network is approximately

$$p_{\max} = C/[a_o \ \log(1/a_o)] \tag{5.1}$$

where C is the number of synaptic connections onto each neuron, and a_o is the sparseness of the output representation (Rolls and Treves, 1990). The population sparseness a_o in Equation 5.1 would be for binary neurons the proportion of the population of neurons with high firing rates, and has been generalized to the case with graded firing rate neurons, which are found in the brain (Rolls and Treves, 2011; Franco et al., 2007; Rolls and Tovee, 1995a; Rolls, 2023c, 2026d), to

$$a = \frac{(\sum_{i=1}^{N} y_i/N)^2}{\sum_{i=1}^{N} y_i^2/N} \tag{5.2}$$

where y_i is the firing rate (e.g. spikes/s, typically in the range 0-100 spikes/s) of the ith neuron in the set of N neurons (Rolls and Treves, 1990; Treves and Rolls, 1991). The sparseness ranges from $1/N$, when only one of the neurons responds to a particular stimulus (a local or grandmother cell representation (Rolls and Treves, 2011)), to a value of 1.0, attained when all the neurons are responding at the same rate to a given stimulus (Treves and Rolls, 1991; Franco, Rolls, Aggelopoulos and Jerez, 2007; Rolls and Treves, 2011; Rolls, 2023c, 2026d).

Next, the storage capacity of attractor (autoassociation) networks (see Fig. 1.3) with asymmetric diluted connectivity, sparse representations, and graded firing rates was investigated, and it was shown that the memory capacity of an attractor network in terms of p the maximum number of patterns that can be stored and correctly retrieved is approximately

$$p = Ck/[a \log(1/a)] \tag{5.3}$$

where C is the number of recurrent collateral synaptic connections onto each neuron, a is the sparseness of the representation, and k is a factor that depends weakly on the detailed structure of the firing rate distribution, on the connectivity, etc, but is roughly in the order of 0.2-0.3 (Treves and Rolls, 1991; Treves, 1991b,a; Rolls, 2023c, 2026d). For C=12,000 synapses per neuron as in the rat hippocampal CA3, Equation 5.3 evaluates to approximately 12,000 when the sparseness a is close to 0.1.

This research advances the Hopfield model (Hopfield, 1982) by incorporating sparse representations, diluted connectivity, and graded firing rates, and does not rely on the binary synaptic weights considered by Marr (1971), which is often referred to as the Willshaw model (Willshaw, Buneman and Longuet-Higgins, 1969; Rolls and Treves, 1998). An advantage of neurons with graded firing rates is that this prevents or limits the chaotic attractor states that can arise with binary neurons and with asymmetric dilution of the connectivity (Treves, 1991b; Treves and Rolls, 1992), and this is a point that is still insufficiently appreciated. In addition, the approach applies for diluted connectivity,

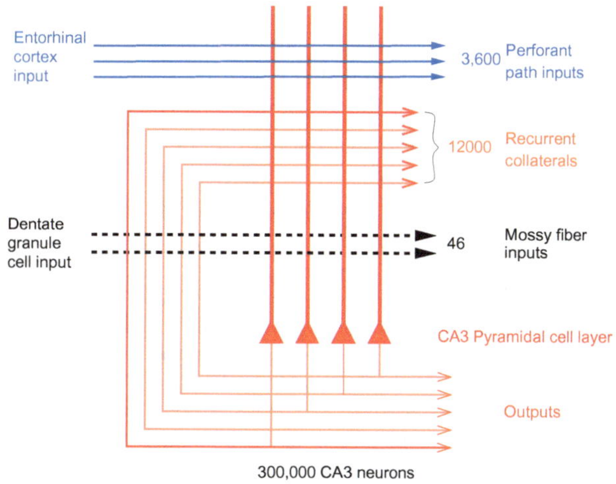

Fig. 5.12 The numbers of connections from three different sources onto each CA3 cell from three different sources in the rat. (Data from Amaral et al., 1990; Treves and Rolls, 1992; Rolls and Treves, 1998)

in which N the number of neurons in the network is greater than the number of connections C onto any one neuron, and N does not set the limit on the memory capacity p. Further details of these and related analyses are described elsewhere (Treves, 1991b,a, 1990; Rolls and Treves, 1998; Rolls et al., 1997c; Rolls, 2023c, 2026d; Rolls and Treves, 2024).

5.3.4 The development of a quantitative theory of the operation of hippocampal CA3 as an attractor network for memory storage and recall

The results shown in Equations 5.3 and 5.1 provide key foundations of the analysis included in the quantitative theory of hippocampal function that we developed next (Treves and Rolls, 1994, 1992; Rolls and Treves, 1994, 2024). To help develop the hippocampal theory quantitatively, I visited David Amaral's lab, and asked him what estimates he might be able to provide for the numbers of synaptic connections per neuron, and the numbers of neurons, in different parts of the hippocampal circuit. David Amaral was very helpful, and this resulted in the values shown in Fig. 5.12, which were later published (Amaral, Ishizuka and Claiborne, 1990).

The quantitative theory of CA3 and its inputs from the dentate gyrus and entorhinal cortex includes the following key concepts (Treves and Rolls, 1992, 1994; Rolls and Treves, 1994, 2024), which were complemented by empirical tests of the hippocampal theory (Kesner and Rolls, 2015) :

5.3.4.1 The storage capacity of CA3

The storage capacity of CA3 is set by the number of connections C per neuron for the recurrent collaterals, not by the number of neurons N, and given a sparseness of about 0.1, is about 12,000 memories in the rat, given that the number of recurrent collateral connections per CA3 neuron is approximately 12,000 in the rat (Fig. 5.12, though now updated to 14,300 in the rat with new data from Watson et al. (2025)). I knew that it would be very useful indeed to know this number C in humans, for that would help to provide an estimate of the number of memories that can be stored in the human hippocampus, and I encouraged several colleagues to investigate this. That information has now been obtained, and the estimate is that C=17,500 for the number of CA3 recurrent collateral connections onto any one CA3 neuron in humans (Watson et al., 2025). This is important, for beyond that limit, an inability sets in to retrieve memories from an attractor network, which needs therefore to forget by for example gradually over-writing previous memories with each new random set of CA3 neurons allocated to a new memory, and this is very relevant to understanding the gradient of retrograde amnesia, and why some forgetting is so important (Rolls and Treves, 1998; Rolls, 2023c, 2026d; Rolls and Treves, 2024) (see also McClelland et al. (1995); Kim and Fanselow (1992)).

The important point here is that the estimate for humans of C=17,500 CA3 recurrent collateral connections onto any one CA3 neuron (Watson et al., 2025) gives an indication that, according to our theory of the operation of the hippocampus (Rolls and Treves, 2024), the number of episodic memories that could be stored in the human hippocampus is in the order of 17,500.

5.3.4.2 The dentate granule cell / mossy fibre connectivity to CA3 selects a new random subset of CA3 neurons for each new memory to be stored in the hippocampus

Part of the computational utility of a new random subset of CA3 neurons for each new memory is that this minimizes the correlations between the sets of CA3 neurons that are active for each memory pattern to be stored, which is important in order not to diminish the memory capacity of autoassociation / attractor networks (Rolls, 2023c, 2026d). Part of what I have termed this pattern separation (Rolls, 1989a) is achieved by the low contact probability of the mossy fibres with CA3 neurons, which is 46 synapses per CA3 neuron / 10^6 dentate granule cells = 0.000046 (Treves and Rolls, 1992). (This can now be updated to 220 mossy fibre inputs per CA3 neuron in humans from 15×10^6 dentate granule cells (Watson et al., 2025), for which the same arguments apply.) Other contributions to this pattern separation include the competitive learning performed by the dentate granule cells, the very sparse representation in the dentate granule cells which can be considered to perform expansion recoding, and the sparse coding in CA3 (Rolls, 1989a, 2016e, 2013c, 2023c, 2026d; Rolls and Treves, 2024; Strehler, 1991) (see also McNaughton et al. (1991); Hasselmo and Wyble (1997)). This random new set of CA3 neurons selected for each new memory is what in the caching chickadee was termed a bar code for each new cache or memory (Chettih, Mackevicius, Hale and Aronov, 2024). This randomizing process is likely to be facilitated by neurogenesis of dentate granule cells (Rolls, 2016e; Gage, 2025).

5.3.4.3 The entorhinal cortex to CA3 synapses are used to retrieve the memory in CA3

We showed that the dentate mossy fibre to CA3 synapses are too few (46 synapses per CA3 neuron in the rat, Fig. 5.12) to recall more than a few memories (Treves and Rolls, 1992). We further showed that the entorhinal cortex to CA3 synapses if associatively modified at the time of memory storage could later be used to recall the random set of CA3 neurons that had been set up for a particular memory (Treves and Rolls, 1992). The memory capacity as shown in Equation 5.1 for a pattern association network would be moderately high, approximately 3,600 memories given the 3,600 entorhinal to CA3 synapses per CA3 neuron and a sparseness of the CA3 neuron representation in the region of 0.1. We are now exploring whether this number sets any limit on the number of memories that could be recalled, in the context that this is a pattern association computation, but with diluted connectivity.

5.3.5 A quantitative theory of the recall of memories from the hippocampus to the neocortex

The theory I developed (Rolls, 1989a,d,c,b) was that recall from the hippocampus to the neocortex was implemented after the recall of the whole set of neurons in CA3 for a memory by:
(1) competitive learning in CA1 to make an efficient recall cue in which the parts of a memory, necessarily separate in CA3 for the autoassociation, could be combined to make a more efficient

memory recall cue; and
(2) the CA1 acting on a series of cortical stages back to the neocortex involving pattern association learning for the backprojections shown in green in Fig. 5.11. The multistage pathway for the backprojections from CA1 to the neocortex was understood to provide a way, stage by stage, of increasing the numbers of neurons that could be activated during recall, in order to reach the vast number of neurons in the neocortex (Rolls, 1989a, 2023c, 2026d). (If CA1 were to connect directly to the final neocortex regions, each CA3 neuron would have to send in the order of 10,000 synapses to each of hundreds of millions of neocortex neurons, which is biologically implausible, hence the need for a multistage backprojection pathway from the hippocampus to the neocortex (Rolls, 1989a) (Fig. 5.11).

But the next question that it was important to address was: what is the memory recall capacity of a series of pattern association networks used in the backprojection pathways (Fig. 5.11) (Treves and Rolls, 1994)? How many memories could be recalled from the hippocampus to the neocortex? This held Alessandro Treves and I up a little, until one evening just before Christmas the idea emerged that a multistage series of pattern association networks could be analyzed with the quantitative methods used to analyze the capacity of an attractor network, for each backprojection stage could be considered as one iteration round an autoassociation (attractor) network (Treves and Rolls, 1994). The result was that the memory recall capacity p of the multistage backprojection system could now be expressed as:

$$p = Ck/[a \log(1/a)] \tag{5.4}$$

where C is now number of backprojection synapses onto any one neuron in the backprojection pathway, a is the sparseness of the representation in now the backprojection pathway (green in Fig. 5.11), and k is a factor that depends weakly on the detailed structure of the firing rate distribution, on the connectivity, etc, but is roughly in the order of 0.2-0.3 (Treves and Rolls, 1994)[4].

The implication is that there should be many backprojection synapses on each neuron for the pathway shown in green in Fig. 5.11. To enable the same number of memories to be recalled to neocortex as could be stored in CA3, with its 12,000 synapses per neuron for the recurrent collaterals (Fig. 5.12), the number of backprojection synapses on each neuron in the cortical backprojections should be close to 12,000. In fact, the number of backprojection synapses onto each cortical neuron is high, with estimates in the order of 10,000 per neuron (Rolls, 2023c, 2026d).

This theory (Treves and Rolls, 1994; Rolls and Treves, 1994) remains the only quantitative theory of the recall of memories from the hippocampus to the neocortex (Rolls and Treves, 2024). The hippocampus can be seen as providing a pointer, or in fact pointers (Rolls, Zhang and Feng, 2024d), back to the neocortex, and overall sets up an arbitrary set of CA3 neurons for each memory, and then uses the backprojections as pointers back to the neocortex to recall the neocortical neurons in different cortical regions that were active during storage back into activity during recall (Rolls, 2023c, 2026d). The concept that the hippocampus might operate as a pointer system reminds one of the pointer or indexing functionality discussed previously but with no theory and mechanism for how anything like that could work (Teyler and DiScenna, 1986).

This theory provides a quantitative account of why there are as many backprojections between each pair of cortical regions as there are forward connections, and is the only quantitative theory for this fundamental property of neocortex architecture (Rolls, 2016b, 2023c, 2026d). These backprojections are used not only for memory recall, but also for top-down attention, another key feature of neocortical design that is considered in my research (Rolls, 2016b, 2023c, 2026d).

[4] I remark in this context that for interdisciplinary research to flourish, I have found it very useful for scientists to work and discuss together in the same place for some considerable time, for then understanding develops that can lead to new insights when their different expertise can be combined.

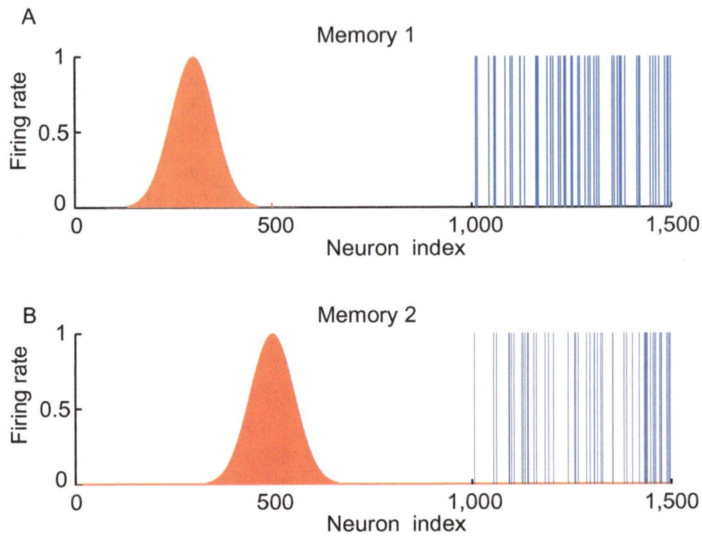

Fig. 5.13 The types of firing patterns stored in continuous attractor networks are illustrated for the patterns present on neurons 1-1000 for Memory 1 (when the firing is that produced when the spatial state represented is that for location 300), and for Memory 2 (when the firing is that produced when the spatial state represented is that for location 500). The continuous nature of the spatial representation results from the fact that each neuron has a Gaussian firing rate that peaks at its optimal location. This particular mixed network also contains discrete representations that consist of discrete subsets of active binary firing rate neurons in the range 1001-1500. The firing of these latter neurons can be thought of as representing the discrete events that occur at the location. Continuous attractor networks by definition contain only continuous representations, but this particular network can store mixed continuous and discrete representations, and is illustrated to show the difference of the firing patterns normally stored in separate continuous attractor and discrete attractor networks. For this particular mixed network, during learning, Memory 1 is stored in the synaptic weights, then Memory 2, etc, and each memory contains part that is continuously distributed to represent physical space, and part that represents a discrete event or object. The spatial and object representations are bound together by being simultaneously present when the event is stored. (Data from Rolls E T, Stringer S M and Trappenberg T P. 2002 A unified model of spatial and episodic memory. Proceedings of the Royal Society B 269: 1087-93.)

5.3.6 Is the CA3 network a memory system or a self-motion path integration system for navigation?

Some previous approaches have suggested that the hippocampal CA3 recurrent network might operate as a path integration system for self-motion update from place to place (McNaughton, Barnes, Gerrard, Gothard, Jung, Knierim, Kudrimoti, Qin, Skaggs, Suster and Weaver, 1996; Samsonovich and McNaughton, 1997). One problem with this is that if CA3 is also used for episodic memories, to associate location with a continuous spatial representation, and objects with a discrete representation with a random set of neurons firing for each object (Rolls et al., 2002) (Fig. 5.13), then the energy landscape of the network would be too 'bumpy' to support a continuous attractor which utilizes a flat energy landscape to enable movement freely and continuously from location to location (Rolls and Treves, 2024; Rolls, 2023c, 2026d).

But in any case, it appears that path integration in rodents may be performed before the hippocampus, in the medial entorhinal cortex, by grid cells (Moser et al., 2017, 2014; Giocomo et al., 2011).

And in primates including humans in which eye movements are required to be part of the path integration system, it is likely that path integration occurs in the dorsal visual system, and is then communicated to the medial parahippocampal cortex via the intraparietal visual cortical regions and PGp (Rolls, 2023c, 2026d; Rolls and Treves, 2024; Rolls, 2024c, 2023a, 2020b).

Given these points, it is proposed that the hippocampal CA3 system may be much more involved in episodic memory than in the path integration that can be useful in the dark for navigation

(Rolls, 2023c, 2026d; Rolls and Treves, 2024; Rolls, 2024c, 2023a, 2025a, 2026c). The hippocampal episodic memory system could of course be useful in this scenario for navigation using visual cues and landmarks, by providing a sequence memory for the locations in spatial scenes to be used for navigation (Rolls, 2021b, 2023c, 2026d, 2025a).

5.3.7 New developments in understanding hippocampal computation

I have continued to develop further all of these and many other concepts about hippocampal function (Kesner and Rolls, 2015; Rolls, 2018a, 2023a, 2025a), and proposed to Alessandro Treves that 30 years after some of our original publications (Treves and Rolls, 1994; Rolls and Treves, 1994), we might provide an update. We worked together on this, and further new developments in the theory of hippocampal function, and updates on storage capacity calculations for attractor networks with spatial representations, include the following (Rolls and Treves, 2024).

5.3.7.1 'What' inputs to the primate including human hippocampus for episodic memory

As summarized in Chapter 4 and in detail elsewhere (Booth and Rolls, 1998; Rolls, 1984, 2000b, 2021c, 2023c, 2026d; Hasselmo et al., 1989b,a; Rolls et al., 1987; Rolls, 2012b), we have discovered that there is a pathway to the macaque inferior temporal visual cortex for developing representations of objects and faces that are invariant with respect to translation, size, contrast, spatial frequency, and even in some cases to view. The invariance in these representations is ideal for an input to a memory system such as the hippocampus, for any view etc of the person or object seen later will act as an appropriate input to the hippocampal memory system for retrieval of the whole memory with its 'Where' and reward components.

Moreover, we have shown that the identity of faces and objects is in an excellent form with sparse distributed representations in which the information increases linearly with the number of neurons (Rolls and Treves, 2011; Rolls et al., 1997e,d; Rolls, 2025a) for an input to a memory system such as the hippocampus.

Further, we have shown that the connectivity of the 'What' system in humans reaches the hippocampus via the fusiform face cortex FFC, and lateral parahippocampal cortex region TF (Rolls, 2024c; Rolls et al., 2023b; Rolls and Turova, 2025; Rolls et al., 2024b).

Very little that is equivalent has been described in rats and mice, yet this is a key component of the episodic memory system that I have described in primates including humans (Rolls, 2023c, 2026d), contributing to the revolution in our understanding of hippocampal function in primates including humans (Rolls, 2025a, 2026c).

5.3.7.2 'Where' inputs to the primate including human hippocampus for episodic memory

As summarized in Section 5.2 and in detail elsewhere (Rolls, 2023c, 2026d, 2023a, 2024c), the 'Where' input to the hippocampus is about the viewed location in a spatial scene, which is ideal for a human episodic memory system that prototypically enables storage and later recall of where we have seen objects, people, and rewards (goals), in space "out there" that is being viewed. The allocentric location in the viewed scene is again an ideal input to the hippocampal episodic memory system, for the input is invariant with respect to the eye position, head direction, facing direction, and place when the location "out there" in the scene is being viewed.

Moreover, I have described with colleagues the effective and functional connectivity and tractography of a new ventromedial cortical visual stream to the human hippocampus for scene information (Rolls et al., 2024c; Rolls, 2024c; Rolls et al., 2023b), shown how the cortical regions along this pathway are activated by scenes compared to faces, tools and body parts with data from 956 participants (Rolls, Feng and Zhang, 2024b), and have developed a theory and model of how the scene representations may be built along this pathway (Rolls, 2025b).

Very little that is equivalent has been described in rats and mice, yet this is a key component of the episodic memory system that I have described in primates including humans (Rolls, 2023c, 2026d,d), contributing to the revolution in our understanding of hippocampal function in primates including humans (Rolls, 2025a).

5.3.7.3 Reward / affective inputs to the primate including human hippocampus for episodic memory

The reward / affective input is a key component of episodic memory, and we have shown that this is represented in the primate hippocampus and can be combined with location information (Rolls and Xiang, 2005).

We have also shown the pathway in humans by which reward information from the orbitofrontal cortex (Rolls, 2023b, 2019c, 2014a) reaches the hippocampal episodic memory system is in part via the ventromedial prefrontal cortex, pregenual anterior cingulate cortex, and posterior cingulate cortex (Rolls, Deco, Huang and Feng, 2022b; Rolls, Wirth, Deco, Huang and Feng, 2023g). The computation performed by the human hippocampus can thus be considered to include episodic memory storage and recall with 'What', 'Where', and reward / affective / goal value components (Rolls, 2023c, 2026d, 2025a) (Fig. 5.11).

These pathways to the primate hippocampal system are also proposed to be important in providing the goals for navigation (Rolls, 2023c, 2026d, 2025a).

5.3.7.4 The roles of reward in memory consolidation

I have also led research that shows that there is an input from the orbitofrontal cortex reward / emotion system to the basal forebrain / septal system where there are cholinergic neurons that project to the neocortex and hippocampus (Rolls, Deco, Huang and Feng, 2022b), and have generated the theory that this contributes to memory consolidation (Rolls, 2022a) by increasing long-term potentiation (Zaborszky et al., 2018; Hasselmo and Sarter, 2011; Hasselmo and Giocomo, 2006; Hasselmo and McGaughy, 2004; Hasselmo, 1999). This theory is supported by evidence that primate basal forebrain neurons can be activated by rewarding, punishing, or novel visual stimuli, which signal times when the storage of episodic memories may be especially useful (Wilson and Rolls, 1990c,b,a; Rolls, 2023c, 2026d).

In addition, I have proposed that when an episodic memory is recalled to the neocortex from the hippocampus with a reward (or punishment) value component, that episodic memory may be rehearsed and thought about more because it may be of value, and that extra processing may enhance memory consolidation into semantic or long-term memory (Rolls, 2022a).

5.3.7.5 Re-entrant neocortex-hippocampus-neocortex processing

In an integrate-and-fire model of the dynamics of neocortical-hippocampal-neocortical processing, we found that after a partial cue has retrieved memory back to the neocortex (Fig. 5.14), the neocortex then in this continuous time model sends input back to the hippocampus, which returns it back to the neocortex, in a continuous positive feedback loop (Rolls, Zhang and Feng, 2024d). A theory is being developed that this may enable several different neocortical representations (e.g. What, Where, and Reward) to be kept active by what are effectively backprojection pointers back to specific neocortical representations in several different cortical regions. Keeping these several neocortical representations active for a short period of perhaps seconds may help the neocortex to build long-term semantic

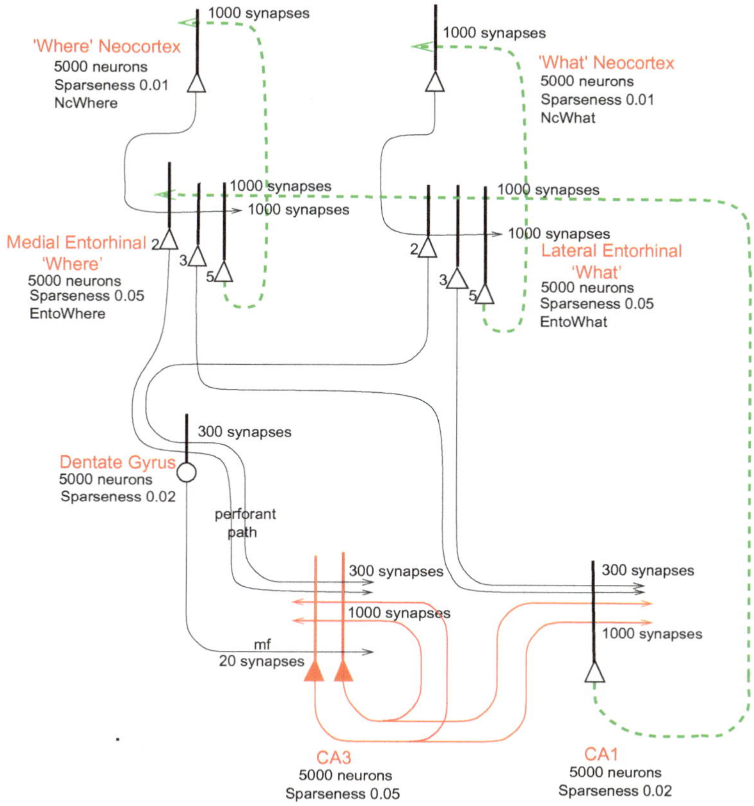

Fig. 5.14 Simulation of neocortical 'what' and 'where' inputs to the hippocampus for the storage of episodic memory, and for the recall of 'what' (object or face) and 'where' (spatial view) information back to the 'what' and 'where' neocortex. The pyramidal cells bodies are shown as triangles, the dendrites as the thick lines above the cell bodies, and the axons as lines terminated with an arrow. The numbers of synapses shown are the numbers on any one neuron. The backprojection pathways for memory recall are shown in dashed green lines, and in red the CA3 recurrent collaterals via which 'what' and 'where' representations present at the same time can be associated during episodic memory storage, and via which completion of a whole memory from a part can occur during recall. All synapses are associatively modifiable except for the Dentate Gyrus (DG) mossy fibre (mf) synapses on the CA3 pyramidal cells. The dentate granule cells, the CA1 cells, and the entorhinal cortex inputs from the neocortex operate as competitive networks. The CA3 cells operate as an autoassociation attractor network to implement completion. The backprojection connections shown in green operate as pattern association networks (Rolls, 2023c, 2026d). (Data from Rolls E T, Zhang C, and Feng J. 2024 Hippocampal storage and recall of neocortical what - where representations. Hippocampus 34: 608-624.)

representations in, for example, the anterior temporal lobe semantic regions (Rolls, 2023c, 2026d; Rolls, Deco, Huang and Feng, 2022c; Rolls, Zhang and Feng, 2024d, 2025a) (see Section 6.10.5).

5.3.7.6 How the hippocampus may contribute to learning new semantic representations

Another new development is directed towards how semantic representations, of knowledge about the world, are learned. The hippocampus has been implicated in helping in this learning, and new discoveries about relevant connectivity of the human hippocampus with semantic cortical regions, and new computational hypotheses about how the hippocampus may contribute to the learning of new semantic representations, are described in Section 6.10.5 (Rolls et al., 2025a).

A key point is that in humans it appears possible to take forward our understanding of hippocampal function and how this may relate to something as cognitive as forming new semantic categories by recalling previous related episodic memories, and this appears to be another difference beyond research in rodents focussing on place cells and maps, for semantic memories become highly developed in humans and are part of the language system, and involve the anterior temporal lobe, which is very highly developed in humans (Rolls et al., 2022c; Rolls, 2023c, 2026d). This is one of the revolutions (Rolls, 2026c) in the previous understanding of hippocampal function (Morris, Amaral, Bliss, Duff and O'Keefe, 2025) that are described in the next section.

5.4 Hippocampal revolutions

The discoveries and research described here are leading to revolutions, revolutionary turning round of concepts, in our understanding of hippocampal functions in primates including humans (Rolls, 2026c), which go beyond place cells and cognitive maps in rodents (O'Keefe and Nadel, 1978; O'Keefe, 1979; Morris et al., 1982; Andersen et al., 2007; O'Keefe and Krupic, 2021; Moser et al., 2017; Morris et al., 2025). A point being made is that although the research on the rodent hippocampus is extremely interesting, we are now in a position where we can build on and go beyond that to understand how the hippocampal system operates in primates including humans.

First, research in rodents has emphasized hippocampal place cells and medial entorhinal cortex grid cells for places (O'Keefe, 1990; Hartley et al., 2014; Moser et al., 2017; Morris et al., 2025), but in primates including humans, and in some birds, hippocampal system neurons respond to spatial view, to where the individual looks in space (Rolls, 1999a; Rolls and Wirth, 2018; Rolls, 2023a; Chettih et al., 2024; Rolls, 2023c, 2026d, 2025a, 2026c).

Second, research in rodents has emphasized navigation as moving from hippocampal encoded place to place using self-motion, which is almost 'blind' navigation, with a path integration place system that when it drifts can be reset by for example allocentric boundary vector cells (McNaughton et al., 1996, 2006; Burgess et al., 2000). In primates including humans, the highly developed foveal visual system and hippocampal spatial view cells allow the use of visual landmarks, which leads to completely different strategies for navigation (Rolls and Wirth, 2018; Rolls, 2021b, 2025a, 2026c).

Third, research in primates including humans emphasizes the role of the hippocampus in episodic memory, in which objects including faces, viewed spatial locations, and rewards are associated together to form a unique episodic memory, which can later be recalled from any of its components, using associations made in hippocampal CA3 (Rolls and Treves, 2024; Rolls, 2023c, 2026d,c, 2022a).

Fourth, in primates including humans the pathways to the hippocampus are now becoming understood, with a ventromedial cortical visual pathway for scenes using spatial view cells; with the ventrolateral cortical visual pathway for visually fixated faces and objects; and with reward inputs from the highly developed orbitofrontal cortex reaching the hippocampus via the ventromedial prefrontal cortex and anterior cingulate cortex (Rolls et al., 2023b, 2022b; Rolls, 2023a, 2024c, 2026c). Similar equivalent and highly developed cortical pathways in rodents are not described or are much less clear.

Fifth, all of this fits well in primates including humans with the quantitative theory of hippocampal CA3 circuitry as being useful for episodic memory, and with the highly developed pathways from the hippocampus to the neocortex being appropriate to recall the same number of episodic memories as can be stored in the hippocampus (Rolls, 2023c, 2026d; Rolls and Treves, 2024).

Sixth, in humans, the major connectivity from the hippocampal episodic memory system to the anterior temporal lobe semantic cortical regions is now becoming understood, and is stimulating new approaches to how the hippocampus may help semantic cortical regions build semantic memories (Rolls et al., 2025a) (see Section 6.10.5), with no clear equivalent yet understood in rodents.

In proposing revolutions in our thinking about hippocampal function, I have thus built on the evidence on hippocampal function in rodents, but have gone beyond that by showing that key revolutions in our understanding of the primate including human hippocampal system are taking place that are very important in understanding what the primate including human hippocampus does, that is, what and how it computes (Rolls, 2026c).

These discoveries are important, for they help in the understanding of the episodic memory, memory consolidation, and other problems that can follow medial temporal lobe damage and damage to related systems such as the ventromedial prefrontal cortex in humans (Zola-Morgan et al., 1986; Smith and Milner, 1981; Scoville and Milner, 1957; Squire and Wixted, 2011; Dede et al., 2016; Renoult et al., 2019; Moscovitch et al., 2016; Feigenbaum and Morris, 2004; De Luca et al., 2019; McCormick et al., 2018; Bonnici and Maguire, 2018; Rolls, 2022a).

This research also further highlights the differences at the systems neuroscience level of computations in many brain systems in primates including humans compared to rodents (Rolls, 2023c, 2026d).

5.5 Evaluation

Research in nocturnal animals such as rats and mice that live in underground tunnels and have poor vision has made advances in understanding 'blind' navigation from place to place using path integration over the direction and distance travelled, but is not useful in humans for most of the navigation that we perform. We can move beyond the concept of the hippocampus as a cognitive map, and representations of the place where the individual is located, and navigation as self-motion update from place to place.

Navigation in primates and humans can be guided by the sight of distant landmarks encoded by spatial view cells, opening up new vistas for understanding navigation in primates including humans.

Consistent with the importance of foveal vision and spatial view cells in primates including humans, primate including human episodic memory typically involves associating objects and people with locations that are being viewed.

The discoveries described here thus emphasize the roles of vision in the functions of the hippocampus in memory and in navigation in primates including humans.

Moreover, reward and emotion-related information reach the hippocampus from the orbitofrontal cortex (in part via regions such as the anterior and posterior cingulate cortex) via pathways that are highly developed in primates and humans as described here, but much less developed in rodents. This helps to place emphasis on the roles of reward and emotion as key components of primate including human episodic memory, and also of their potential importance in memory consolidation, as described here and elsewhere (Rolls, 2022a).

It is anticipated that the empirical and computational discoveries described here and elsewhere (Rolls, 2023c, 2025a, 2026d) provide the foundation for new vistas and revolutions in the understanding of brain systems involved in memory and navigation in humans and other primates (Rolls, 2026c).

6 The human functional and computational brain connectome

6.1 Introduction

To understand how the brain functions, I have developed an approach in which it is important to be able to analyze *'what'* is computed by each brain region, and *'how'* it is computed (Rolls, 2016b, 2021a, 2023c, 2026d). Part of this computational neuroscience approach is to show how the different brain regions analyzed in this way are connected to each other, for what and how a region computes depends on the inputs that it receives. This shows that in order to understand how the brain computes anything, a key part of the theory must include how different brain regions are connected to each other. The importance of analyzing how different brain regions are connected to provide a framework for understanding what is computed at each stage of the framework is described for the discoveries on reward and emotion systems in Chapter 3; on cortical visual systems in Chapter 4; and on memory systems in Chapter 5. However, detailed evidence on the connectivity of all the regions of the human brain and especially the human cerebral cortex has been lacking, because of the difficulty of performing anatomical tract-tracing studies in humans that have been very useful in non-human animals (Rolls, 2023c, 2026d).

Because of the importance of knowing the connectivity of the human brain to provide a framework for understanding how it functions, I have performed a whole series of investigations to see what discoveries can be made about the connectivity of the human brain. These studies are very important for the human brain, because the human brain has some parts that have developed beyond what is present in non-human primates such as macaques (Rolls, 2023c, 2026d), including the prefrontal cortex (Passingham, 2021; Rolls et al., 2023c), the anterior temporal semantic memory system (Rolls et al., 2025a), the language system (Rolls et al., 2022c), the orbitofrontal cortex (Rolls et al., 2022b), and the inferior parietal cortex (Rolls et al., 2023a). The discoveries on the connectivity of different systems in the human cerebral cortex provide a framework which I have then built on by considering what is known from functional neuroimaging, neurophysiology, and neuropsychology about each of the connected cortical regions in each brain system to help develop an understanding of what is computed in each cortical region, and how the computations take place in the connected stages of each cortical system (Rolls, 2026d). The advances, discoveries and new concepts about the functions of the different cortical systems in the human brain made with this approach are described in this Chapter.

The cortical regions between which we measure the connectivity in humans are those defined in the Human Connectome Project Multimodal Parcellation atlas (HCP-MMP) (introduced in Section 1.3), for it aims to define anatomically and functionally different regions of the human cerebral cortex using anatomical architecture (T1w/T2w myelin content and cortical thickness maps that can be identified with MRI), resting-state functional connectivity, and task-based fMRI activation in seven tasks (Glasser, Coalson, Robinson, Hacker, Harwell, Yacoub, Ugurbil, Andersson, Beckmann, Jenkinson, Smith and Van Essen, 2016a). Its use of multiple criteria makes it a better parcellation for our purposes than any other atlas that I know. It defines 180 cortical regions in each hemisphere. We have added 66 subcortical regions to an extended version of the HCP-MMP, the HCPex, that can be used easily with volumetric data (Huang, Rolls, Feng and Lin, 2022). This parcellation and the divisions or groups of the 360 cortical regions has been introduced in Section 1.3. The cortical regions in this parcellation are shown in Figs. 1.1 and 1.2, with a list of the cortical regions in Appendix A in Tables A.1–A.6 and of the subcortical regions in HCPex in Table A.7. Coronal slices to further help visualization of the locations in the brain of these cortical regions are provided by Huang, Rolls, Feng and Lin (2022).

We have used three types of connectivity measurement that are all non-invasive and that are complementary, and for which data can be obtained using functional magnetic resonance imaging (fMRI) or magnetoencephalography (MEG). MEG has lower spatial resolution than fMRI, but is very fast, with data collection with a resolution of 1 ms or better which can be very helpful when measuring the directionality of signal flow between cortical regions (Rolls et al., 2023e, 2024c).

Diffusion tractography uses MRI to measure the diffusion of water in the human brain, and this tends to occur along fibre tracts, so provides some evidence about direct connections between brain regions (Catani and Thiebaut de Schotten, 2008; Lerch et al., 2017; Maier-Hein et al., 2017; Jeurissen et al., 2019).

Diffusion tractography can provide evidence about fibre pathways linking different brain regions. Diffusion tractography shows only direct connections, so comparison with effective connectivity (described below) can help to suggest which effective connectivities may be mediated directly vs trans-synaptically. Diffusion tractography does not provide evidence about the direction of connections. The term 'connections' is appropriate for use for diffusion tractography, for it provides evidence about anatomical connections.

Diffusion tractography as a method for following connections and measuring their strength is supported by comparisons made in macaques between diffusion tractography and conventional anatomical methods with tracers (van den Heuvel et al., 2015; Donahue et al., 2016).

The methods used in the application of diffusion tractography in the investigations included in this book are described by Huang, Rolls, Hsu, Feng and Lin (2021). Diffusion tractography cannot follow fibres where they become unmyelinated within the cortex, and that is one of its limitations. Another is that it does not work well for long connections across the midline of the human brain (Huang et al., 2021).

Functional connectivity is measured by the Pearson correlation between the BOLD (Blood Oxygenation-Level Dependent) fMRI signal in each pair of brain regions, and can provide evidence that may relate to interactions between brain regions, while providing no evidence about causal direction-specific effects. A high functional connectivity may in this scenario thus reflect strong physiological interactions between areas.

Effective connectivity estimates the effect of one brain region on another, and utilizes differences detected at different times in the signals in each connected pair of brain regions to infer effects of one brain region on another. The method used here (Rolls et al., 2023b,f, 2022b,a, 2023g,c,a, 2022c, 2023d) was developed by Gustavo Deco from a Hopf algorithm to enable measurement of effective connectivity between many brain areas (Deco, Cruzat, Cabral, Tagliazucchi, Laufs, Logothetis and Kringelbach, 2019; Kringelbach, Perl, Tagliazucchi and Deco, 2023; Ponce-Alvarez and Deco, 2024; Rolls, Deco, Huang and Feng, 2022a; Rolls, Yan, Deco, Zhang, Jousmaki and Feng, 2024c). A principle is that the functional connectivity is measured at time t and time $t + \tau$, where τ is typically 2 s to take into account the time within which a change in the fMRI BOLD signal can occur, and then the effective connectivity model is trained by error correction until it can generate the functional connectivity matrices at time t and time $t + \tau$. Accordingly this is sometimes called 'generative effective connectivity'. Full details of the algorithm and its validation are provided elsewhere (Rolls et al., 2022a, 2024c), with a more full introduction in Rolls (2023c) and Rolls (2026d). The directionality described in our papers has been investigated and validated with analyses using magnetoencephalography (MEG), which operates on a much faster time scale of ms (Rolls et al., 2023e, 2024c). This is the only algorithm we know that can measure effective connectivity between all 360 cortical regions in the HCP-MMP atlas. Because we are the only neuroscientists measuring effective connectivity as well as the easier functional connectivity and diffusion tractography between all 360 cortical regions in the HCP-MMP atlas, we have been able to make many discoveries about the organization of the human brain connectome.

We calculate and typically display the connectivities as matrices showing the connectivity between every pair of brain regions in the analysis (Huang et al., 2021; Ma et al., 2022; Rolls et al., 2022a,b,c, 2023a,c,b,d,e,f,g, 2024a,c,b,e, 2025a). This is a major new way of analyzing brain connectivity, for traditional anatomical techniques though very valuable for providing precise information relied on investigations in animals in which the connections from a single brain region with other regions could be investigated (Carmichael and Price, 1996; Saleem, Miller and Price, 2014; Witter and Amaral, 2021), whereas now we measure the connectivities between all brain regions in humans in the analysis, which can include all 360 cortical regions in the HCP-MMP atlas. Examples of these connectivity matrices are shown in the original papers just cited with an example shown in Fig. 3.13.

Much of the data that we have used comes from the Human Connectome Project, including in our sample from 171 individuals imaged in the resting state at 7T, and 956 individuals at 3T (Glasser et al., 2016b), but we have performed our own data collection when we needed for example MEG data using visual stimuli about scenes (views of places) (Rolls, Yan, Deco, Zhang, Jousmaki and Feng, 2024c), or when we wished to analyze connectivity during episodic memory tasks (Rolls, Zhang, Deco, Vatansever and Feng, 2024e).

Most of the data we have analyzed have been in the resting state, while individuals are in the scanner not doing a task. Data collected in the resting state are sometimes described as measuring the basic connectivity of the human brain when it is not performing tasks, but when being scanned in the resting state individuals are often following their own internally generated thoughts (Rolls et al., 2026). It is also very interesting and revealing to measure how the functional and effective connectivities change when particular tasks are being performed, for this provides evidence about what sets of brain regions are recruited for particular tasks and how they are connected when performing these tasks (Rolls, Feng and Zhang, 2024b; Rolls, Zhang, Deco, Vatansever and Feng, 2024e; Rolls, Yan, Deco, Zhang, Jousmaki and Feng, 2024c; Rolls, Feng and Zhang, 2026; Zhang, Rolls and Feng, 2026b), which is important for understanding brain computations.

In the following sections of this Chapter, interesting aspects of new discoveries made about human cortical connectivity in humans are highlighted, with the details of the connectivities available in the original papers cited. We start with evidence about connectivity in cortical sensory processing pathways, and follow that with connectivity in the human memory, emotion, and language systems, which are greatly developed from those in non-human primates. I do emphasize that in addition to knowing the connectivity between brain regions it is important to know about the neuronal activity in those brain regions, and the effects of damage to those brain regions, and that is the more broad approach aimed at in Brain Computations and Connectivity (Rolls, 2023c) and **Brain Computations and Principles; and AI** (Rolls, 2026d). There is of course a great deal remaining to be discovered, but we can build on the foundations that are described in this book and in Rolls (2026d).

6.2 The ventrolateral visual cortical system for objects and faces

An overview of four visual cortical pathways, two 'What' and two 'Where', that we have analyzed in humans, is provided in Section 4.2 and Fig. 4.1 (Rolls, 2024c). Here I provide some highlights of what has been discovered for the ventrolateral visual cortical pathway involved in face and object representations using fMRI (Rolls, Deco, Huang and Feng, 2023b; Rolls, Feng and Zhang, 2024b) and MEG (Rolls, Deco, Zhang and Feng, 2023e), after a brief description of the pathways involved.

The human Ventrolateral Visual Cortical Stream is characterized by effective connectivities that lead towards the inferior temporal visual cortex (Fig. 6.1). After a progression from V1 > V2 > V3 > V4 [5], V4 (and to some extent V3) have effective connectivity to HCP-MMP ventral visual division regions such as PIT (a posterior inferior temporal cortex region identified in humans) and V8. PIT, and to some extent V8, then have connectivity to FFC, the 'fusiform face cortex' (Fig. 6.1), which is in the fusiform gyrus. FFC in turn has connectivity to posterior inferior temporal cortex regions TE2p and PH (and PH also connects to TE2p). PH and TE2p project to PHT, which in turn projects to TE1p, and to an anterior inferior temporal cortex region TE2a. The posterior inferior temporal cortex regions TE1p and TE2p project to anterior inferior temporal cortex regions TE2a, TE1m and TE1a (Fig. 6.1).

TE1a then projects anteriorly to temporal pole region TGd which is involved in semantic representations for language (Rolls et al., 2022c) (Section 6.10). Other regions related to language, temporo-parieto-occipital regions TPOJ1, TPOJ2 and TPOJ3 (Rolls et al., 2022c), also receive effective connectivity from ventrolateral visual cortical stream regions including the FFC, providing a route for face and word form inputs to reach language-related regions (Figs. 6.7 and 6.10). We therefore have a route through the ventrolateral visual stream as shown schematically in Fig. 6.1 which can now be shown to reach human language-related cortical regions. This ventrolateral cortical visual stream also provides 'What' visual outputs (Rolls et al., 2023b) to the orbitofrontal cortex / vmPFC reward/punishment emotion-related system (Rolls et al., 2022b); to the hippocampal memory system as a 'What' component of episodic memory (Rolls et al., 2022a); to the prefrontal cortex for short-term memory and planning (Rolls et al., 2023c); and to the cortex in the superior temporal sulcus (STS) involved in multimodal representations relevant to social behavior and to semantics (Rolls et al., 2023b).

One highlight is that when visual stimuli are shown, inputs from intermediate visual cortical regions in the hierarchy FFC, PIT and V8 reach (have effective connectivity to) the posterior inferior temporal visual cortex regions TE1p and TE2p, but not further anterior (as shown by the connectivity matrix in Fig. 3 of Rolls et al. (2023e)). Instead, the anterior temporal lobe regions TE1a, TE1m, TE2a and temporal pole TGd and TGv, receive visual inputs from TE1p and TE2p, but also receive inputs from other modalities including auditory, somatosensory (opercular and frontal opercular), and orbitofrontal cortex (Rolls et al., 2023b,e, 2024c). The implication is that TE1p and TE2p are the last mainly unimodal visual cortical regions in the ventrolateral visual cortical stream, and the very large expanse of the anterior temporal lobe in front of TE1p and TE2p is multimodal, and is devoted to semantic representations, as shown in Section 6.10. This anterior temporal lobe region is large in humans, and occupies a much larger proportion of the temporal lobe than in macaques.

Another highlight is that in the fusiform gyrus, scenes are represented medially, then objects and faces (Rolls, Feng and Zhang, 2024b), and more laterally, words in a visual word form area (Dehaene and Cohen, 2011; Vinckier, Dehaene, Jobert, Dubus, Sigman and Cohen, 2007; Woolnough, Donos,

[5] '>' indicates the general progression and does not exclude some connectivity that is more than between strictly adjacent levels in the hierarchy.

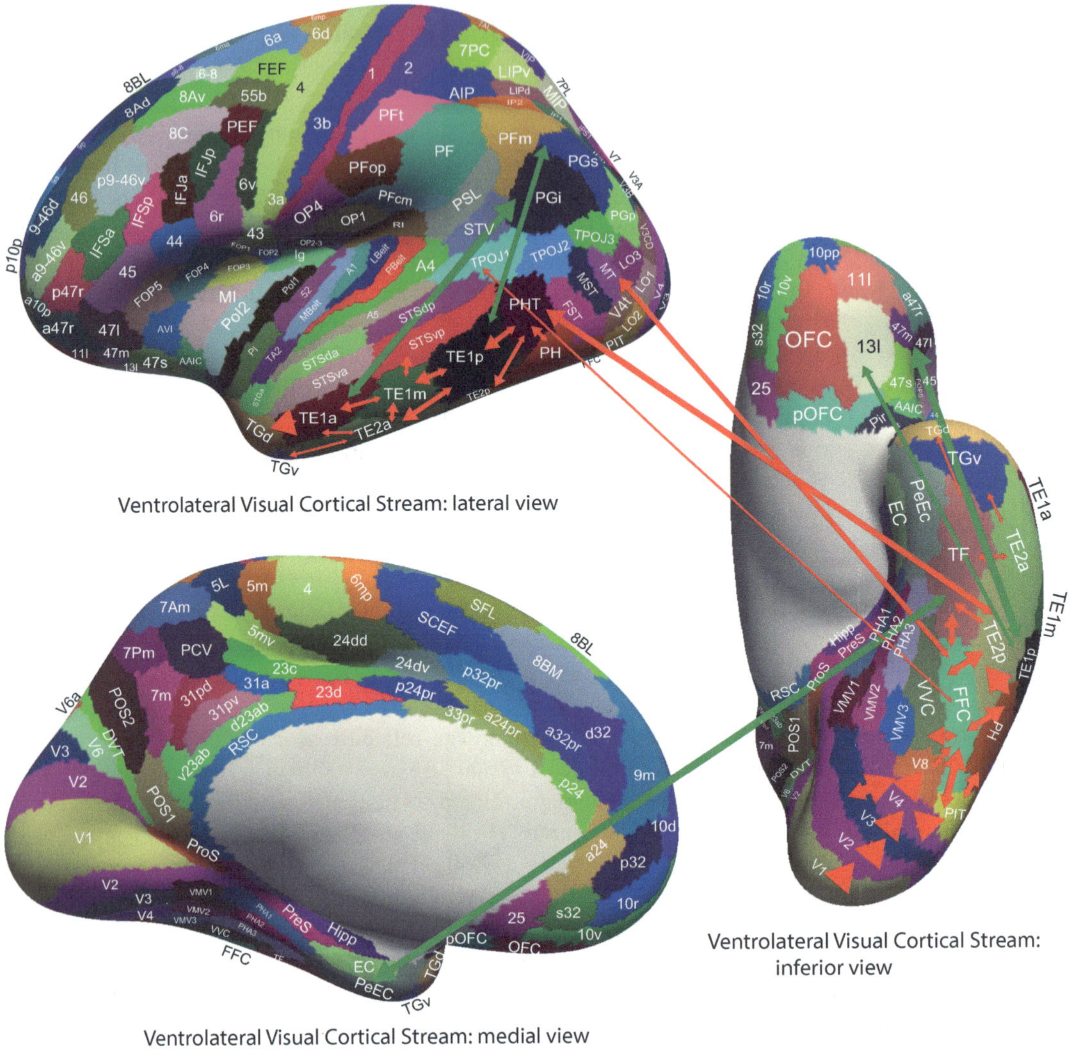

Fig. 6.1 Hierarchical organization of the Ventrolateral Visual Cortical 'What' Stream for faces and objects measured with effective connectivity, functional connectivity, and diffusion tractography with Human Connectome project data for fMRI (Rolls et al., 2023a) and MEG (Rolls et al., 2023f). First V1 > V2 > V3 connect to V4. Second, V4 has connectivity to FFC (fusiform face cortex), V8 and PIT. Third, FFC, V8 and PIT have connectivity to TE1p and TE2p, which are the last mainly unimodal visual inferior temporal cortex regions where faces and objects are represented. Fourth, TE1p and TE2p have connectivity to TE1m, TE2a, TE1a and the temporal pole TGd and TGd. This fourth level anterior temporal lobe region is multimodal, in that it also receives auditory cortex input from auditory cortex regions such as the Belt regions; and from somatosensory regions in the frontal operculum (FOP) and opercular (OP) regions. A green arrow shows how the Ventrolateral Visual Stream provides 'What' input to the hippocampal memory system via lateral parahippocampal gyrus TF to perirhinal PeEc, hippocampus etc. Two green arrows show how posterior inferior temporal visual cortex regions TE1p and TE2p and more anterior temporal cortical regions have effective connectivity to the lateral orbitofrontal cortex (47m, 47l, a47r) and medial orbitofrontal cortex (13l). The Ventrolateral Visual Stream also has connectivity with the semantic regions, such as STSvp in the ventral bank of the superior temporal sulcus, and inferior parietal visual regions PFm, PGs and PGi as indicated by 2 green arrows. The widths of the lines and the size of the arrowheads indicate the magnitude and direction of the effective connectivity. (Data from Rolls E T, Deco G, Huang C-C and Feng J. 2023 Multiple cortical visual streams in humans. Cerebral Cortex 33: 3319-3349.)

Rollo, Forseth, Lakretz, Crone, Fischer-Baum, Dehaene and Tandon, 2021; Agrawal and Dehaene, 2024). This can be considered as a series of representations each at a different spatial scale of size from medial to lateral. Further evidence on this is provided in Fig. 5.5 which shows the activations produced when faces, scenes (viewed places), body parts, and tools are being viewed (Rolls et al., 2024b). Fig. 5.5 shows that scenes activate most medially (PHA1, PHA2, PHA3, VVC, VMV1, VMV2); and faces more laterally (FFC, VVC, TE2p, TE1p, TF and perirhinal PeEC). Body parts activate FFC, TE2p, TE1p and TF, close to but perhaps a little lateral to faces in the medial temporal lobe; and tools (some of which were quite large) activate VVC most highly in a region between the face and scene regions. Although a size gradient has been noted before in the medial temporal lobe (Kravitz et al., 2013), this research anchors the representations to the well-defined HCP-MMP atlas, to the functional connectivity between these cortical regions (Rolls, Feng and Zhang, 2024b), to laterality differences (Rolls et al., 2026), and to sex differences (Zhang et al., 2026b).

Another highlight of the representations of visual stimuli identified with this approach is that the activations and connectivity produced by stationary visual stimuli extend beyond the classical ventrolateral visual cortical pathway shown in Fig. 6.1 to include other cortical regions that are I propose related to the semantic representations raised by seeing different types of visual stimuli. Indeed, Fig. 6.1 shows connectivity not only to anterior temporal lobe semantic regions, but also to visual motion regions (e.g. between TE2p and MT), but also to inferior parietal visual cortical regions PGs and PGi implicated in actions in space and spatial processing (Rolls et al., 2023b).

This is also brought out by the activations produced by faces, scenes, body parts and tools shown in Fig. 5.5, with for example the sight of tools activating visual motion areas such as FST and LO2 even though the tools were stationary, and somatosensory cortical areas such as regions 1, 2, and PFt (semantically relevant it is proposed as tools are typically held in the hands), and visuomotor regions for actions in space such as parietal AIP, LIPd and PGp (which are highly semantically relevant for tools).

Another example in Fig. 5.5 is that faces and tools both activate regions of cortex in the superior temporal sulcus (e.g. STSdp, STSvp, STSda), which are implicated in representing moving faces and body parts (Hasselmo et al., 1989b; Perrett et al., 1985a; Jellema and Perrett, 2003) (Section 6.3).

Another example in Fig. 5.5 is that stationary body parts activate inferior parietal cortex regions PGi and PGs implicated in representations of space and actions in space (Rolls, Deco, Huang and Feng, 2023a). These activations are reflected too in the functional connectivities that are present when these different types of visual stimuli are being viewed (Rolls et al., 2024b).

The recruitment of these cortical regions outside the classical ventrolateral cortical visual pathway by certain types of visual stimuli depending on their semantic associations (Rolls et al., 2024b) is a very interesting way in which semantics are represented in the brain (see further Section 6.10).

6.3 A visual cortical 'What' stream to the cortex in the superior temporal sulcus for moving faces and objects

The visual cortical 'What' stream to the cortex in the superior temporal sulcus for moving faces and objects has been introduced in Section 4.2 and is included in Fig. 4.1 (Rolls, 2024c). This Superior Temporal Sulcus Visual Cortical Stream (black in Fig. 4.1), a 'What' Stream, in STSvp and STSva receives face and object inputs from ventrolateral visual stream regions in the anterior temporal lobe (TE1a, TE1a, TE2a), and visual motion inputs from inferior parietal PGp which in turn receives inputs from intraparietal regions (e.g. LIPd) and area 7 regions; and also has connectivity with visual inferior parietal PGi (Fig. 4.1). Neurons in this system respond to socially relevant stimuli such as turning the head and eyes to make or break social contact, and face expression, as described in Section 4.5.4.

In more detail, the visual inputs to regions STSda and STSdp reach STSdp from PGi (Rolls et al., 2023b) (Fig. 6.10), and provide a route for moving visual stimuli / objects analyzed in the parietal cortex to reach STS regions (Rolls et al., 2023a). Visual inputs also reach STSdp from the Superior Temporal Visual (STV) region which receives from both MT and FFC (Fig. 6.10), and which as described below could contribute to the neuronal activity in the cortex in the STS that has been shown to respond to moving heads, faces and objects in macaques (Hasselmo, Rolls, Baylis and Nalwa, 1989b; Hasselmo, Rolls and Baylis, 1989a). STSda and STSdp also have connectivity from STSva and STSvp which have strong connectivity with the Ventrolateral Visual Stream (Figs. 6.1 and 6.10). STSda and STSdp also receive auditory effective connectivity from A5 (Fig. 6.10) (Rolls et al., 2023f). STSdp has connectivity directed towards Broca's area 44 and 45, and related areas (47l), and to the PeriSylvian Language region PSL, and to the Superior Frontal Language region SFL (Rolls et al., 2022c).

Discoveries in macaques provide an indication for what is represented in these STS regions. We discovered that single neurons in the macaque STS respond to face expression and also to face and head movement to encode the social relevance of stimuli (Hasselmo, Rolls, Baylis and Nalwa, 1989b; Hasselmo, Rolls and Baylis, 1989a). For example, a neuron might respond to closing of the eyes, or to turning of the head away from facing the viewer, both of which break social contact (Hasselmo et al., 1989b,a). Some neurons respond to the direction of gaze (Perrett et al., 1987). It was found that many of the neurons in the STS respond only or much better to moving faces or objects (Hasselmo et al., 1989a), whereas in the anterior inferior temporal cortex neurons were discovered that respond well to static visual stimuli, and are tuned for face identity (Perrett, Rolls and Caan, 1982; Rolls, 1984, 2000b; Rolls and Treves, 2011; Rolls, Treves, Tovee and Panzeri, 1997e; Rolls, Treves and Tovee, 1997d; Hasselmo, Rolls and Baylis, 1989a). It has been proposed that PGi, with its inputs from PGs which has connectivity with superior parietal and intraparietal regions that encode visual motion, is part of this processing stream for socially relevant face-related information (Rolls et al., 2023a). Consistent with this, the effective connectivity is strong from PGi to STS regions (Rolls et al., 2023b).

In humans, representations of this type could provide part of the basis for the development of systems to interpret the significance of such stimuli, including theory of mind. Consistent with this proposal, activations in the temporo-parietal junction region are related to theory of mind (DiNicola et al., 2020; Buckner and DiNicola, 2019; Schurz et al., 2017). Signals of this type are important in understanding the meaning of seen faces and objects. Indeed evidence about moving objects present in the STS cortex may reach it from PGs and PGi, which in turn receive connectivity from the intraparietal sulcus regions (Rolls et al., 2023a) in which neurons respond to visual motion and to grasping objects which are important in tool use (Maravita and Romano, 2018), which is another fundamental aspect of the meaning or semantics of stimuli.

We proposed that the cortex in the STS in which neurons respond to moving faces, eyes, etc and to changing facial expression enables ventral stream 'What' information to be combined with dorsal stream motion information to form a third visual stream, and that this could be useful for social functions (Hasselmo, Rolls, Baylis and Nalwa, 1989b; Hasselmo, Rolls and Baylis, 1989a), especially as this system projects to the orbitofrontal cortex / vmPFC where similar types of neuronal response are found (Rolls, Critchley, Browning and Inoue, 2006a). The concept that this STS cortical system is important in social behavior has more recently gained support by others (Pitcher and Ungerleider, 2021; Pitcher, Ianni and Ungerleider, 2019). Moreover, neurons in macaques can respond to auditory stimuli such as vocalization both in the STS regions (Baylis, Rolls and Leonard, 1987) and in the orbitofrontal cortex (Rolls, Critchley, Browning and Inoue, 2006a), with such auditory stimuli also important in social and emotional behavior.

Our connectivity and neuron-level discoveries described here help to provide a functional

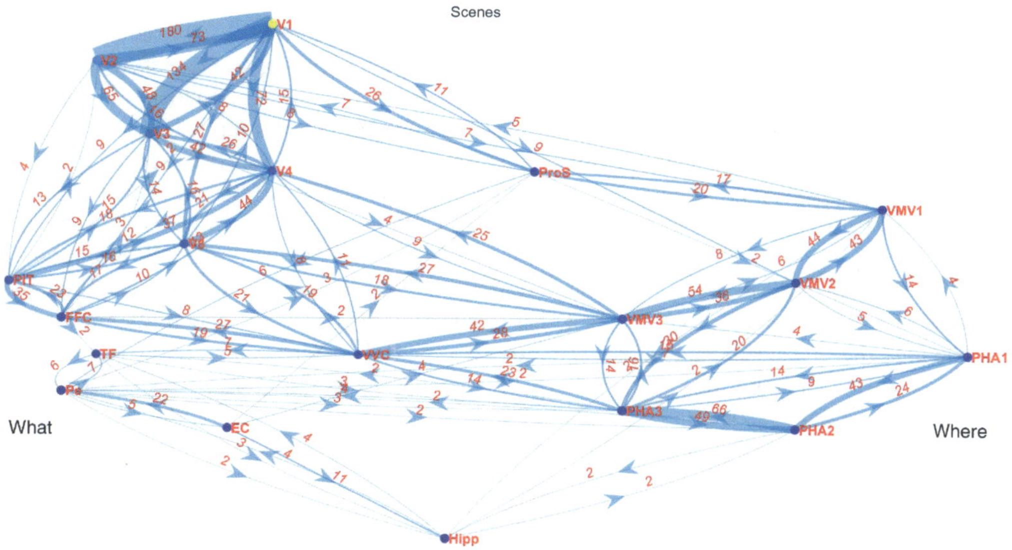

Fig. 6.2 Visual cortical networks when viewing spatial scenes in 956 Human Connectome Project participants. The flows F measured at time=40 are indicated by arrows with numbers close to the arrows, and the width of the arrows represents the strength of the flow. The flows were multiplied by 1,000 for ease of reading. The flows were thresholded at 2 to clarify the Figure by not showing small flows. The flows in each direction are indicated by the numbers close to the curved edges, and if one of the flows is <2, a single straight arrow is shown between a pair of nodes. The activations S in each of the nodes are indicated by color from yellow as the highest to dark blue as the lowest. The ventrolateral cortical visual 'What' stream is on the left, and the ventromedial cortical visual 'Where' stream is on the right. (Data from Rolls E T and Turova T. 2025 Visual cortical networks for 'What' and 'Where' to the human hippocampus revealed with dynamical graphs. Cerebral Cortex 35: bhaf106.)

framework for the processing streams involved in social and emotional function, from STS regions to the orbitofrontal cortex (Rolls, 2026d).

6.4 The ventromedial visual cortical system for spatial scenes and navigation

A ventromedial cortical 'Where' pathway for visual scenes has been described in Section 5.2.10 and illustrated in Fig. 5.8.

As described in Section 5.2.10, a highlight of this discovery is that a 'Where' representation for locations in viewed spatial scenes is encoded by a ventromedial cortical visual pathway using spatial view cells, not by the dorsal visual stream (Rolls et al., 2023b; Rolls, 2023a; Rolls et al., 2024c; Rolls, 2024c, 2025b). The proposal is that computations of the type typical of ventral visual stream cortical processing (Rolls, 2012b, 2021c, 2023c, 2026d) are used to form feature combinations of the visual features present in different parts of scenes to represent scenes based on the features found in scenes, a proposal that has been developed further as shown in Section 5.2.12 (Rolls, 2025b).

This is different to anything proposed for rodents, and emphasizes that fundamental discoveries about the brains of humans and other primates in health and disease are being made in this research, by pioneering a research strategy on the hippocampal system that is different to previous research that has been based on rodents, which have very different systems-level brain organization (see Section 29.4 in **Brain Computations and Principles; and AI** (Rolls, 2026d)).

Another highlight is that it has been possible to obtain evidence that the ventrolateral visual cortical pathway for faces and objects is separate from the spatial scene pathway described in Section 5.2.10. The evidence comes from effective connectivity measured when viewing stationary faces, scenes, tools and body parts, and then using a graph theory approach to estimate the flow of signal between cortical regions (Rolls and Turova, 2025). The flows between 19 visual cortical and hippocampal system regions when viewing scenes are shown in Fig. 6.2. There are high flows between nodes in the ventromedial cortical regions shown on the right of Fig. 6.2, and region VVC was strongly coupled to these regions. It was found that the flows were different when faces were viewed, with for example VVC coupled more to the ventrolateral cortical visual regions when viewing faces (Rolls and Turova, 2025). These analyses with effective connectivity and graph theory to calculate the flows of signal between the nodes (cortical regions) thus help to provide evidence that the whole cortical networks activated by faces vs scenes are different (Rolls and Turova, 2025).

The conceptual advance here is that the graph theory approach that we describe enables whole networks to be identified, which is beyond pairwise connectivity between cortical regions (Rolls and Turova, 2025).

6.5 The human Dorsal Visual Cortical Stream for visual motion leading to the intraparietal visual areas, and then to parietal area 7 regions for actions in space

The human Dorsal Visual Cortical Stream pathways described here leads via dorsal visual system cortical regions to regions in the intraparietal sulcus and parietal cortex area 7. This stream has visual motion outputs to the superior temporal sulcus (STS) semantic systems (Section 6.10). This stream includes a dorsal 'where' / action system in which idiothetic update of spatial representations is performed.

The effective connectivity of the human Dorsal Visual Stream is shown in Fig. 6.3 (Rolls et al., 2023b). Effective connectivity from regions such as V2, V3, V3A and V3B, V6A and V7 reach the MT+ complex regions (FST, LO1, LO2, LO3, MST, MT, PH, V3CD and V4t) (Fig. 6.3). The MT+ complex regions then have effective connectivity to the intraparietal regions (AIP, LIPd, LIPv, MIP, VIP IP0, IP1 and IP2), which in turn have effective connectivity to the area 7 regions. Interestingly, in humans there are some inputs to this Dorsal Visual Stream from Ventrolateral Visual Stream regions such as FFC and TE2p, and these are shown with dashed lines in Fig. 6.3.

6.5.1 Dorsal stream visual division regions

V3A and V6 are motion-sensitive areas that project to the regions in the MT+ Complex division (Table A.1) in what is described in macaques as a dorsolateral visual stream which eventually reaches the cortex in the superior temporal sulcus (Galletti and Fattori, 2018). V6 responds to coherent optic flow stimuli and may thus be useful in egomotion (Sulpizio et al., 2020). In macaques, V6 may also project to V6A which is involved in the fast control of prehension and in selecting appropriate postures during reach-to-grasp behaviors (Pitzalis et al., 2015, 2013; Monaco et al., 2011; Tosoni et al., 2015). V6A then projects to what is described as a dorsomedial visual stream directed towards the intraparietal areas such as MIP which are also involved in reach to grasp behavior (Galletti and Fattori, 2018). Neurons in V6A and VIP are invariant with respect to eye position, that is, they can respond in head-based coordinates (Galletti and Fattori, 2018). Coordinate transforms of this type can be implemented by gain modulation (Salinas and Sejnowski, 2001) (in which for example eye direction modulates retinotopic position) which is greatly helped by slow learning to capture the

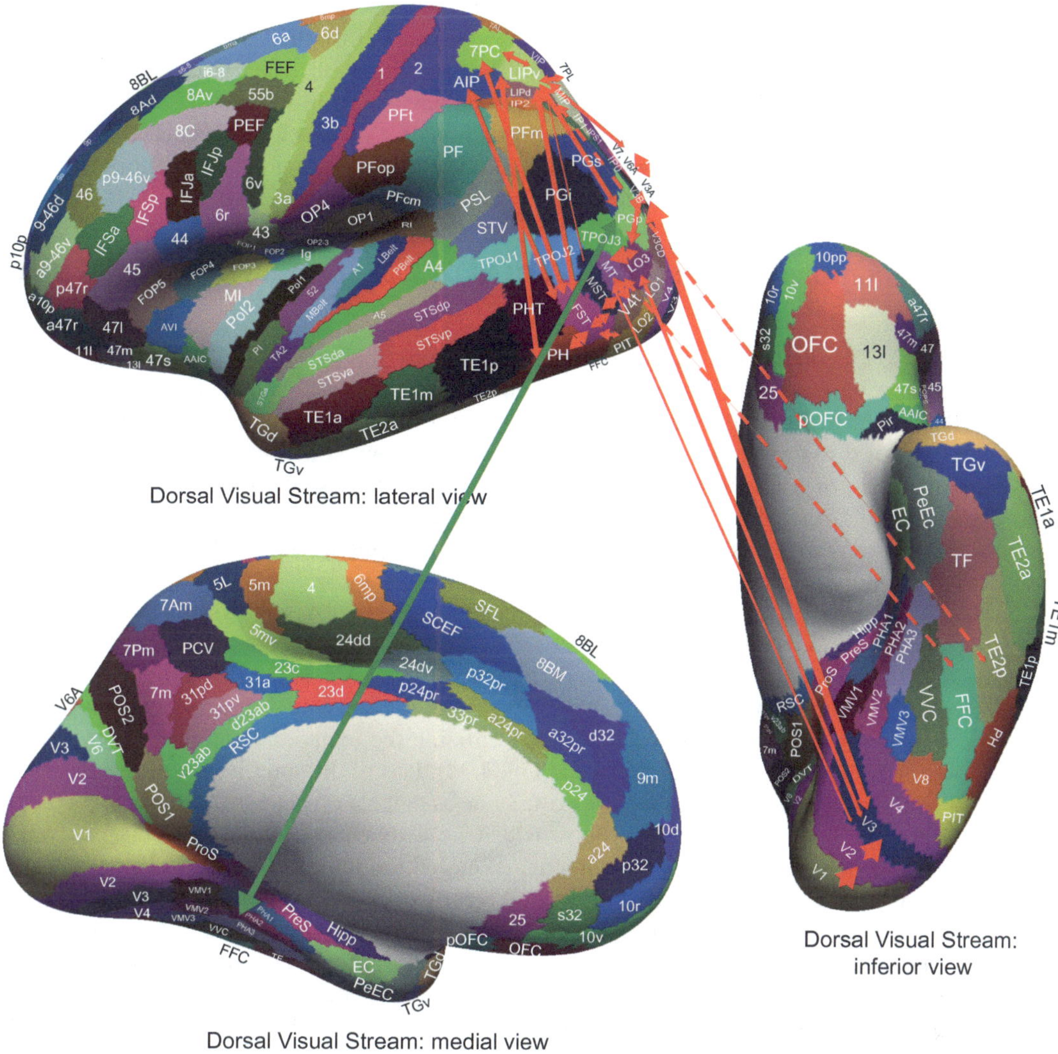

Fig. 6.3 Effective connectivity of the human Dorsal Visual Cortical Stream which reaches (partly via V3, V3A and LO3) the MT+ complex regions (FST, LO1, LO2, LO3, MST, MT, PH, V3CD and V4t), and then the intraparietal regions (AIP, LIPd, LIPv, MIP, VIP IP0, IP1 and IP2) and then the area 7 regions: schematic overview. Connectivity to the inferior parietal cortex region PGp, which in turn has effective connectivity to the parahippocampal scene area in PHA1-3 (Rolls et al., 2023a) is shown. Inputs to this stream from ventral stream regions such as FFC and TE2p are shown with dashed lines. (Data from Rolls E T, Deco G, Huang C-C and Feng J. 2023 Multiple cortical visual streams in humans. Cerebral Cortex 33: 3319-3349.)

statistical continuities of an object in a fixed position relative to the head when the eyes move (Rolls, 2020b). The same principle can be extended to account for the next transform to allocentric direction in space, and then to a further transform to allocentric location in space independently of the place where the individual is located (Rolls, 2020b), as represented in the parahippocampal gyrus and hippocampus by spatial view cells (Rolls, 2023a, 2025a).

This dorsomedial part of the dorsal visual system provides for idiothetic (self-motion) update of allocentric spatial representations of the type provided by for example primate spatial view cells (Robertson, Rolls and Georges-François, 1998), and is a key computation performed in the dorsal visual 'where' system enabling idiothetic update of the VentroMedial 'Where' Visual Cortical

Stream for building scene representations in the VMV and PHA1-PHA3 parahippocampal areas (see Figs. 6.3 and 5.10, and Section 5.2.12) (Rolls et al., 2023b,a; Rolls, 2023a, 2025b). These scene representations are likely to be involved in human navigation from viewed location to viewed location, which is frequently how navigational instructions are provided in humans (Rolls, 2021b). The idiothetic update of scene representations is likely to be important for navigation if the view details are temporarily obscured, or in the dark (Rolls, 2021b, 2023a, 2025a,b).

6.5.2 MT+ complex regions (FST, LO1, LO2, LO3, MST, MT, PH, V3CD and V4t)

In macaques, MT neurons compute global motion aided by their receptive fields being about 10 times larger than V1 neurons (Newsome, Britten and Movshon, 1989; Zaharia, Goris, Movshon and Simoncelli, 2019). In MST, the receptive fields are larger (e.g. 40°), and some neurons respond to more complex types of motion such as expanding optic flow consistent with moving forwards in an environment, or rotation of the optic flow in a clockwise vs anticlockwise direction (Wild and Treue, 2021). These neurons are therefore useful in analyzing self-motion or the motion of an object, disparity may be encoded for stereoscopic depth perception, and vestibular inputs related to self-motion can influence some neurons (Wild and Treue, 2021).

In a unifying hypothesis with the design of the ventral cortical visual system about how these types of encoded motion might be computed, Rolls and Stringer (2007) proposed that the dorsal visual system uses a hierarchical feedforward network architecture (V1, V2, MT, MST, parietal cortex) with training of the synaptic connections with a short-term memory trace associative synaptic modification rule to compute what is invariant at each stage by utilizing the temporal continuity inherent in the statistics of the visual inputs (Section 4.7.4 and Fig. 4.24). Part of the significance of this theory and model is that the computational architecture is the same as that for object perception as implemented by VisNet, except that the input to VisNet provided by Gabor spatial filters is replaced by local motion inputs for the dorsal visual stream model of global motion processing (Rolls and Stringer, 2007). This may reflect economy in the design of the cortex (Rolls, 2016b, 2026d).

The same principle (Rolls and Stringer, 2007) can account for the responses of neurons in the cortex in the superior temporal sulcus (which may receive from these dorsal visual stream regions) which it was discovered can respond to object-based rotation about a principal axis (see Fig. 4.24 and Section 4.7.4), for example to a head rotating clockwise, invariantly with respect to whether the head was upright or inverted (Hasselmo, Rolls, Baylis and Nalwa, 1989b).

6.5.3 Intraparietal sulcus posterior parietal cortex regions (AIP, LIPd, LIPv, MIP, VIP; with IP0, IP1 and IP2)

Neurons in macaques in region VIP represent the direction and speed of visual motion (Colby et al., 1993), and may be useful in for example tracking moving visual stimuli, and encoding optic flow which can be useful in assessing self-motion and thereby in navigation. These neurons do not respond in relation to saccades. Neurons in macaques in LIP are active in visual, attentional and saccadic processing (Gnadt and Andersen, 1988; Colby et al., 1996; Munuera and Duhamel, 2020). The ventral part of LIP (LIPv) has strong connections with two oculomotor centers, the frontal eye field and the deep layers of the superior colliculus, and may be especially involved in the generation of saccades (Chen et al., 2016). The dorsal part (LIPd) may be more involved in visual processing, responding for example to visual targets for saccades (Chen et al., 2016). Neurons in MIP (which may be the parietal reach region, PRR (Connolly et al., 2003)) are related to arm movement preparation and execution (Passarelli et al., 2021; Prado et al., 2005; Filimon et al., 2009; Cavina-Pratesi et al.,

2010; Gallivan et al., 2011). They are implicated in the sensory-to-motor transformations required for reaching toward visually defined targets (Gamberini et al., 2020). AIP is activated by grasping objects, for which shaping of the hand based on the visual properties of objects is needed (Culham et al., 2006, 2003; Orban et al., 2021).

As shown in Fig. 6.3 (Rolls et al., 2023b), in humans these intraparietal sulcus cortical regions have strong effective connectivity from early visual cortical areas including from several MT+ complex visual regions in which neurons respond to global optic flow (MST, FST, PH and V3CD), intraparietal sulcus area 1 (IPS1), V3B, V6A (which is a visuo-motor region involved in grasping seen objects (Gamberini et al., 2020)), V7, and superior parietal area 7 regions involved in visuo-motor actions. These intraparietal regions also receive in humans from ventral stream visual cortical areas including the fusiform face cortex (FFC), and from inferior temporal cortex regions PIT, PHT, TE1p and TE2p (Fig. 6.3) (Rolls et al., 2023b). These ventral stream regions are likely to bring shape / visual form information to the intraparietal cortex regions important in shaping the hand to grasp and manipulate objects and tools. Consistent with this, neurons in regions such as AIP in macaques can respond to 2D and 3D shape information which could be helpful in tool use (Kastner et al., 2017; Urgen and Orban, 2021).

These intraparietal and also area 7 regions may provide information to the inferior parietal cortex visual cortical regions such as PFt, PF and PGi that have developed so greatly in humans and that are involved in human tool use (Rolls et al., 2023a). These intraparietal regions also have connectivity with the inferior frontal gyrus and with the dorsolateral prefrontal cortex (especially 46, 8C, a9-46, i6-8 and p9-46v), which are likely to be important when there is a delay between the visual input and when the action can be performed (Funahashi, Bruce and Goldman-Rakic, 1989). These connectivities are stronger towards these prefrontal areas (Rolls et al., 2023b), as is appropriate for the operation of short-term memory systems so that the memory does not dominate sensory inputs (Rolls, 2016b, 2023c, 2026d). There is also connectivity directed towards the parahippocampal TH cortex (PHA3) which may be useful in providing information about visuo-motor actions to the hippocampal memory system. There is also connectivity with the frontal pole p10p, which is likely to be important when sequencing and planning is involved in actions (Shallice and Cipolotti, 2018; Shallice and Burgess, 1996; Gilbert and Burgess, 2008).

The connectivity from the intraparietal regions is strongly towards premotor cortical regions including especially 6a, and to the frontal eye field FEF and prefrontal eye field PEF (from especially AIP, LIPd, LIPv and VIP), which provide action-related outputs from these visuo-motor intraparietal regions (Rolls et al., 2023b). There is also connectivity especially for IP1 and IP2 from the orbitofrontal cortex (medial regions, 11 and OFC) which may provide reward feedback (Rolls, 2019c,a) of potential utility in learning whether actions made are correct, and with the frontal pole. There is also connectivity with inferior parietal regions including PGp and PGs, with PGp having effective connectivity to the parahippocampal scene area PHA1-3 to provide, it is proposed, for idiothetic update of parahippocampal and hippocampal spatial view representations (Rolls et al., 2023a; Rolls, 2023a). Interestingly, this intraparietal part of the parietal cortex has relatively little effective connectivity with the posterior or anterior (or mid) cingulate cortex. The connectivity from MT+ regions such as MT and MST to intraparietal regions from auditory cortical regions such as A4 and A5, and from somatosensory regions such as 5L and 5m, provides a basis for the auditory and somatosensory responses evident in some macaque intraparietal neurons that also respond to visual stimuli (Kastner et al., 2017).

The functional connectivity is largely consistent with the effective connectivity (Rolls et al., 2023b), but indicates more interactions with early visual cortical areas; with somatosensory/premotor areas; with the hippocampal system; with TE1p and TE2p, with posterior cingulate including DVT; and with supragenual anterior cingulate 33pr and p24pr (Rolls et al., 2022b). The diffusion tractography (Rolls et al., 2023b) is also consistent, but suggests in addition connections with auditory cortex that may be useful in orienting gaze towards sounds.

The intraparietal cortical regions in humans thus are likely in terms of their connectivity (Rolls et al., 2023b) to perform visuomotor functions (without substantial somatosensory processing unlike area 7 regions), and the extensive research on these regions in macaques provides a guide to their functions in humans, including the control of eye movements to acquire and track visual stimuli given the outputs to the FEF and PEF. There are also outputs to premotor cortex regions 6a and 6r that might produce head and body movements to help stabilize images for the visual system. The output to the posterior inferior temporal visual cortex PHT is of interest, and might be involved in the stabilization of images for processing in later parts of the ventral visual system (Rolls et al., 2023b).

6.5.4 Posterior parietal area 7 regions

Area 7 regions in the posterior parietal cortex receive from the intraparietal areas, have connectivity to premotor and somatosensory cortical regions, and are involved in actions in space (Gamberini et al., 2020; Passarelli et al., 2021; Orban et al., 2021). Their connectivity in humans is described by Rolls et al. (2023a) and in Chapter 10 of Rolls (2026d).

6.6 Somatosensory cortical regions and connectivity in humans

It has become possible to understand much better the organization of the human somatosensory cortical pathways by use of the Human Connectome Project - Multimodal Parcellation (HCP-MMP, Section 1.3), and measuring effective connectivity, complemented by functional connectivity and diffusion tractography (Section 6.1) with data from the Human Connectome Project, as described next (Rolls et al., 2023c). This has implications for understanding what is computed in different somatosensory cortical regions, including their connectivity via the insula to regions in the inferior parietal cortex that develop greatly in humans (Rolls et al., 2023c,a). Indeed, a key advance to be described here is that in humans a ventral somatosensory stream has been found that connects from the primary somatosensory cortical regions 3b and 1 via Frontal Opercular (FOP) regions to the insula, which in turn has connectivity to the inferior parietal cortex PF regions, as illustrated in Fig. 6.4 (Rolls et al., 2023c). The concept is advanced that the human somatosensory system can thus be considered as a vast hierarchically organised system with multiple stages ending in inferior parietal PF regions after connectivity through the opercular and insular somatosensory regions (Rolls et al., 2023c).

Physiological investigations in primates show that regions 3b and 1 represent cutaneous somatosensory information with area 1 having larger receptive fields, that area 3a represents proprioceptive information, and that neurons in area 2 can respond to cutaneous or proprioceptive information or both (Delhaye et al., 2018). As in humans, the legs are represented at the superior end of the somatosensory cortices 3a-2, and the face at the inferior end close to the operculum (Delhaye et al., 2018). In macaques, ventral to 3a-2 in the operculum are areas S2 and 'parietal ventral' which are described as a ventral somatosensory stream (Delhaye et al., 2018; Mishkin, 1979). In macaques the pathways to 5 and 7b are described as a dorsal somatosensory stream to the posterior parietal cortex (Delhaye et al., 2018; Gardner, 2008; Mishkin, 1979). It has been suggested that several somatosensory processing streams may be present in humans (de Haan and Dijkerman, 2020).

6.6.1 A ventral somatosensory stream via the insula to the inferior parietal cortex

First, the new evidence from our investigation (Rolls, Deco, Huang and Feng, 2023c) shows that in humans there is a ventral somatosensory stream that connects via Frontal Opercular (FOP) regions to the insula, which in turn has connectivity to the inferior parietal cortex PF regions, as illustrated in Fig. 6.4 (Rolls et al., 2023c). The connectivity from the primary somatosensory cortex 3b to the middle insula is shown with green arrows in Fig. 6.4. The onward connectivity of the middle insula (MI) to for example inferior parietal region PF, the PeriSylvian Language region PSL, and the supracallosal anterior cingulate cortex p32pr, a32pr and a24pr is shown with red arrows. The connectivity of the posterior insular region PoI2 is very similar to that of MI. The pathway in more detail is as follows, where > reflects an effective connectivity but does not exclude effects across levels:
(3b + 3a <> 1 + 2) <> OP4 <> 43 <> FOP1 <> FOP3 <> (MI + PoI2 + PoI1) <> PF + PFop.

One of the fascinating features of the connectivity shown in Fig. 6.4 is that it draws out the point that some of the opercular and the frontal opercular (FOP) regions can be considered as a continuation ventrally of somatosensory cortex regions 3a, 3b, 1 and 2; with then a further ventral extension of the cortical sheet into the insula. Another important feature is that this ventral stream is probably a 'what' stream, for it does not have major connectivity with visuo-spatial action regions in area 7, but it does have connectivity with PFop and PF, in which representations appear to be about the properties of objects and one's own body (Rolls et al., 2023a). Indeed part of the function of these PF areas may be related to somatosensory / body image and the sense of body ownership and of self that this provides, consistent with evidence that anosognosia and other disorders of awareness of the body can be produced by PF damage in humans (Ronchi et al., 2018).

Another interesting feature is that PF may on its connectivity evidence (Rolls et al., 2023a) not only be the top of a somatosensory hierarchy, but it adds visual and reward inputs to form semantic representations of felt objects, enabling recognition of for example a tool not only by its touch, but also by its sight. Also consistent with the insula to PFop and PF stream being a 'what' stream is that it also has access to language systems, via the PeriSylvian Language Area (Rolls et al., 2022c), thus enabling verbal declarations to be made about what is felt, with declarative systems being 'what' systems in the brain, whereas action systems are typically procedural and not declarative (Squire and Zola, 1996; Squire, 1992; Milner and Goodale, 1995; Goodale and Milner, 1992).

Another feature of the connectivity is the very directional input from the reward-related (Rolls, 2019c,a; Grabenhorst and Rolls, 2011) medial orbitofrontal cortex region 11l to the middle insula (Fig. 6.4), leading to the hypothesis that the somatosensory insula may encode some aspects of the reward or aversive value of somatosensory stimuli. Consistent with this, activation of the somatosensory insula was found by painful touch to the hand (Rolls, O'Doherty, Kringelbach, Francis, Bowtell and McGlone, 2003d), with a dorsal posterior insular region implicated in pain (Segerdahl et al., 2015; van Strien and Hollmann, 2025).

An interesting property of insular function is that the insula was activated by touch to the arm, but not by the sight of touch to the arm, whereas the somatosensory cortical areas 1–3 responded to the sight of touch to the arm, as well as to touch to the arm (McCabe, Rolls, Bilderbeck and McGlone, 2008). It was therefore suggested that the insula provides evidence that it is one's own body that is being touched, which might give it special status in body representations (McCabe et al., 2008; Craig, 2011, 2009). Consistent with this, most of the insula (MI, PoI1, PoI2 and also the FOP regions) does not receive connectivity from early cortical visual regions, whereas 3b, 3a, 1, 2, 5L, 5m and 5mv do (Rolls et al., 2023c). Another difference is that the somatomotor and paracentral areas have connectivity with parietal area 7 regions, whereas the insula does not; and the insula

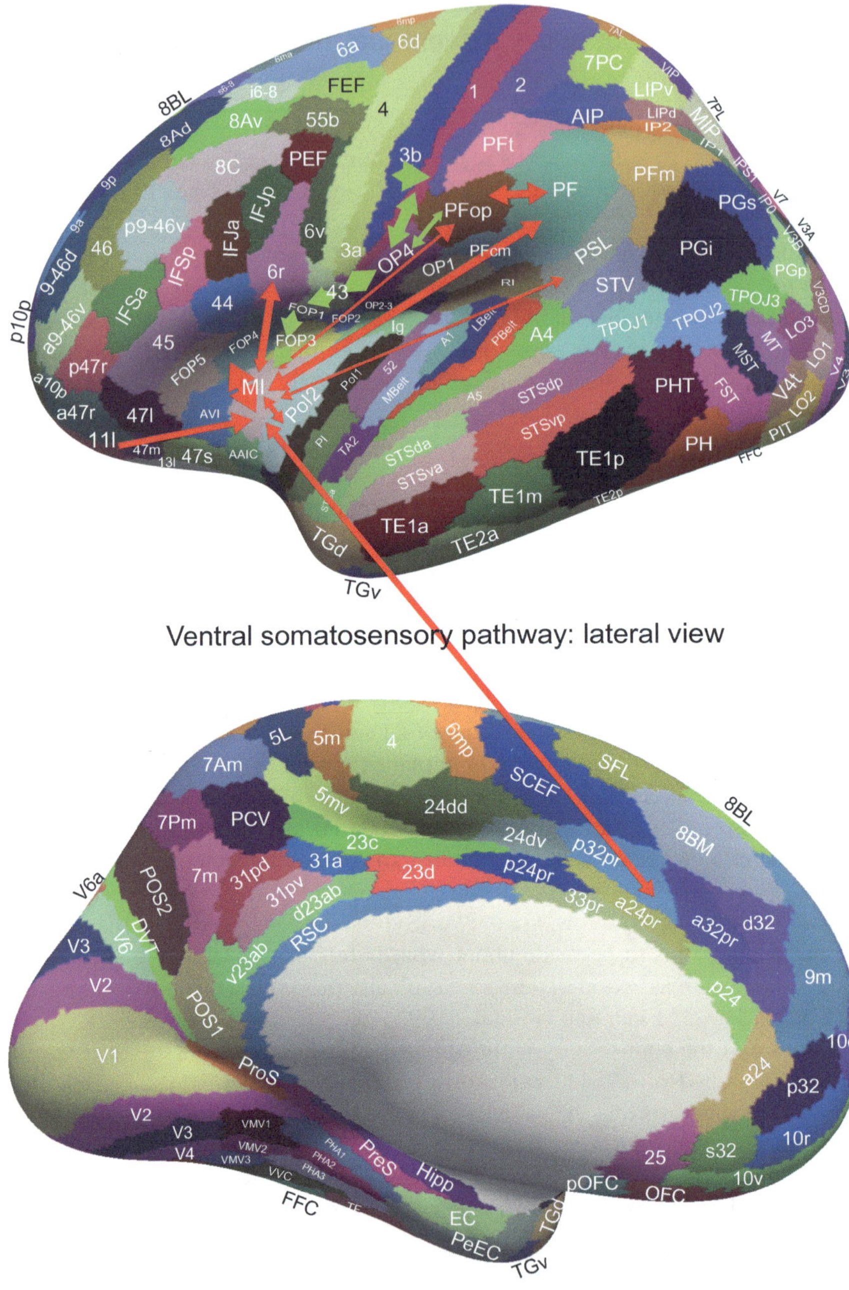

Fig. 6.4 Ventral somatosensory / insula cortical stream effective connectivity. The connectivity from the primary somatosensory cortex 3b to the middle insular area is shown with green arrows. The onward connectivity of the middle insular (MI) to for example inferior parietal region PF, the PeriSylvian Language region PSL, and the supracallosal anterior cingulate cortex p32pr, a32pr and a24pr is shown with red arrows. The connectivity of the posterior insular region PoI2 is very similar to that of MI. The width of the arrows reflects the effective connectivity with the size of the arrowheads reflecting the connectivity in each direction. The pathway in more detail is as follows, where > reflects an effective connectivity but does not exclude effects across levels: (3b + 3a <> 1 + 2) <> OP4 <> 43 <> FOP1 <> FOP3 <> (MI + PoI2 + PoI1) <> PF + PFop. The abbreviations are listed in Appendix A. (Data from Rolls E T, Deco G, Huang C-C and Feng J. 2023 Prefrontal and somatosensory-motor cortex effective connectivity in humans. Cerebral Cortex 33: 4939-4963.)

has more connectivity with inferior parietal PF regions than do the somatomotor and paracentral regions. Another difference is that as mentioned above the insula has connectivity directed towards language-related areas (the PeriSylvian Language area PSL, STV, TPOJ1, and TPOJ2 (Rolls et al., 2023c, 2022c)), whereas the somatomotor and paracentral regions do not.

Our research supports the hypothesis that the insula is part of a ventral (or inferior) 'What' stream of somatosensory processing that continues into inferior parietal cortex PF regions (Fig. 6.4); whereas as described next a dorsal (or superior) 'action' stream of processing connects from somatomotor (areas 3b, 3a, 1, 2) and paracentral (area 5) cortical regions to posterior parietal cortex area 7 and thereby intraparietal regions (Fig. 6.5) (Rolls et al., 2023c).

The supracallosal part of the anterior cingulate cortex is implicated in action-outcome learning, with information about the actions that have been performed available from the somatosensory/motor regions described here, and about the reward outcome received after an action being provided via the pregenual anterior cingulate cortex (Rolls et al., 2022b). The connectivity described by Rolls et al. (2023c) implicates the insula in this circuitry, and the effective connectivity from the reward-related medial orbitofrontal cortex 11l to the insula could provide reward outcome information to enable the insular cortex to also be involved in action-outcome learning. Consistent with this hypothesis, the insular cortex does have some connectivity to premotor cortex 6 regions, and to the mid-cingulate motor region (Rolls et al., 2023c).

6.6.2 A dorsal somatosensory stream

A diagram of a dorsal somatosensory stream connecting via somatosensory area 5 regions to posterior parietal area 7 regions is provided in Fig. 6.5. The area 5 regions in humans appear to be displaced dorsally with respect to their location in macaques such that the area 5 regions roll over to the medial wall of the hemispheres (Fig. 6.5).

In Fig. 6.5 the connectivity from the primary somatosensory cortex area 3b partly via areas 1 and 2 to area 5 somatosensory regions 5m > 5L > 5mv is shown with green arrows. The onward connectivity of 5mv to for example posterior parietal 7AL, 7Am and 7PC which have connectivity with intraparietal regions such as LIP is shown with red arrows. The pathway in more detail is as follows, where > reflects an effective connectivity but does not exclude effects across levels:
(3b + 3a <> 1 + 2) <> 5m <> 5L <> 5mv <> 7AL, 7Am and 7PC.

5mv also has connectivity to 6mp, TPOJ2 and the supracallosal anterior cingulate cortex p32pr, a32pr and a24pr. Given that area 7 is involved in actions in space (Snyder et al., 1998; Rolls, 2020b, 2021b; Dean and Platt, 2006; Vedder et al., 2017; Andersen, 1995; Avila et al., 2019; Orban et al., 2021; Passarelli et al., 2021; Gamberini et al., 2020; Rolls et al., 2023a), this dorsal somatosensory stream may be characterized as an 'action' 'where' stream.

Another feature is that the somatosensory cortical areas do have connectivity directed towards motor (area 4), premotor (area 6), or mid-cingulate premotor cortex. This is a principle established in macaques (Rizzolatti and Sinigaglia, 2016), and presumably allows appropriate movements to be made to the somatosensory/ proprioceptive properties analyzed in each somatosensory / proprioceptive region.

6.7 Auditory cortical regions and connectivity in humans

It has become possible to understand much better the organization of the human auditory cortical pathways by use of the Human Connectome Project - Multimodal Parcellation (HCP-MMP, Section 1.3), and measuring effective connectivity, complemented by functional connectivity and diffusion

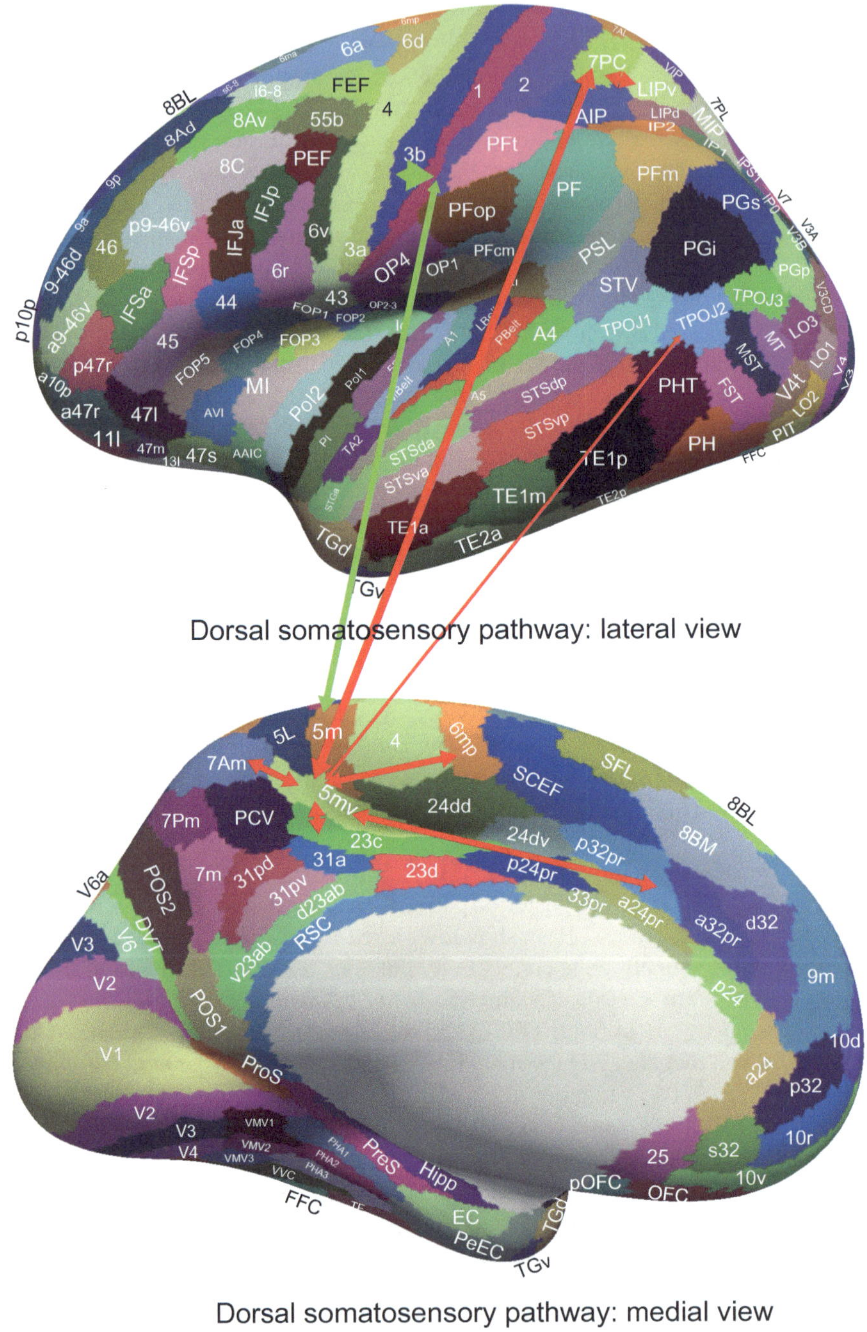

Fig. 6.5 Dorsal somatosensory stream effective connectivity. The connectivity from the primary somatosensory cortex 3b partly via 1 and 2 to area 5 somatosensory regions 5m > 5L > 5mv is shown with green arrows. The onward connectivity of the 5mv to for example posterior parietal 7AL, 7Am and 7PC which have connectivity with intraparietal regions such as LIP is shown with red arrows. The width of the arrows reflects the effective connectivity with the size of the arrowheads reflecting the connectivity in each direction. The pathway in more detail is as follows, where > reflects an effective connectivity but does not exclude effects across levels:

(3b + 3a <> 1 + 2) <> 5m <> 5L <> 5mv <> 7AL, 7Am and 7PC.

5mv also has connectivity to 6mp, TPOJ2 and the supracallosal anterior cingulate cortex p32pr, a32pr and a24pr. The abbreviations are listed in Appendix A. (Data from Rolls E T, Deco G, Huang C-C and Feng J. 2023 Prefrontal and somatosensory-motor cortex effective connectivity in humans. Cerebral Cortex 33: 4939-4963.)

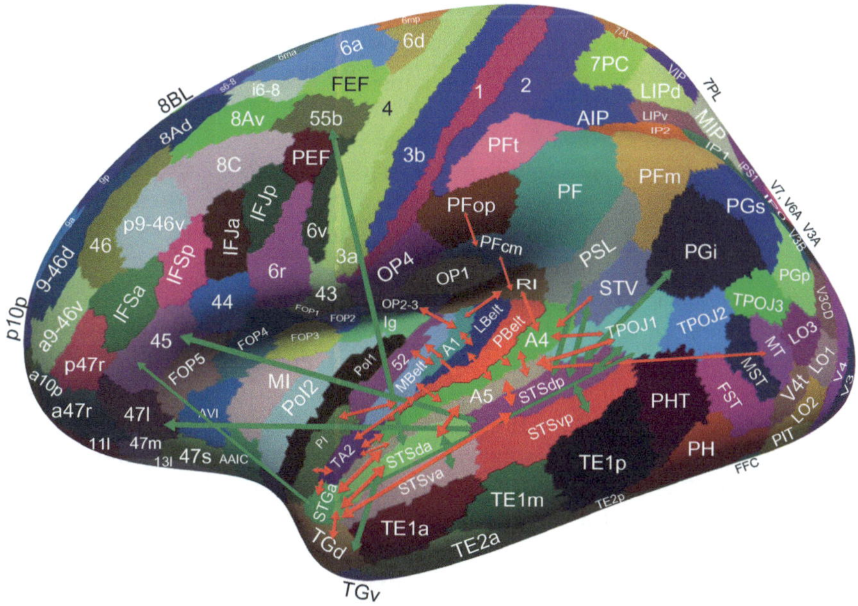

Auditory system: lateral view

Fig. 6.6 Effective connectivity of human auditory cortical regions shown schematically. The widths of the lines and the size of the arrowheads indicate the magnitude and direction of the effective connectivity. Red arrows show the main auditory division connectivity, and green arrows further connectivity. A simplified schematic organization is that region A1 > LBelt, MBelt and 52 > PBelt > A4 > A5 > STGa, STSda, STSdp > TPOJ1, STV, PGi. (Here > indicates has effective connectivity to, though some connectivity may cross levels of the hierarchy.) Somatosensory cortical regions have connectivity with A1, RI, TA2 etc. MT has connectivity to A5. TPOJ1, STV, PSL, STSdp and STSda are involved in language as analyzed elsewhere (Rolls et al., 2022c). Connections between early cortical auditory regions and area 44 in what may be a dorsal language related auditory stream are found. Lbelt, Pbelt, A4 and A5 have effective connectivity with MT/MST regions as indicated, and this may be part of a dorsal 'where' stream leading to intraparietal and area 7 regions. The abbreviations are listed in Appendix A. (Data from Rolls E T, Rauschecker J, Deco G, Huang C-C and Feng J. 2023 Auditory cortical connectivity in humans. Cerebral Cortex 33: 6207-6227.)

tractography with data from the Human Connectome Project, as described next (Rolls, Rauschecker, Deco, Huang and Feng, 2023f). This has implications for understanding what is computed in different auditory cortical regions, and their connectivity to semantic and language systems involved at least in speech understanding.

A schematic overview of the effective connectivity of the human auditory cortical systems is shown in Fig. 6.6 (Rolls et al., 2023f). A simplified schematic organization is that regions A1 and 52 > LBelt and MBelt > PBelt > A4. That part of the connectivity appears to be hierarchical.

In both humans and nonhuman primates, the auditory cortex is described as having core (corresponding to primary and primary-like auditory cortex), then belt (which surrounds the core), and then parabelt fields, all of which contain several subfields (Rauschecker, 2015). Neurons in the core show responses with narrow tuning to tone frequency; belt neurons respond best to band-passed noise of a specific frequency and bandwidth; and parabelt neurons respond to increasingly complex sounds (Rauschecker, 2015; Leaver and Rauschecker, 2016; Tian et al., 2001).

After A4 there may be a split in the ventral auditory cortical pathways in humans (Rolls et al., 2023f).

In one ventral stream, A4 has effective connectivity to anterior temporal lobe region TA2, and TA2 given its location is likely to be a ventral stream region (Fig. 6.6). TA2 has interesting connectivity to the parainsular region PI, which also receives visual and somatosensory inputs (Rolls et al., 2023c).

TA2 has some effective connectivity to anterior temporal gyrus region STGa. However, STGa cannot be regarded as a region that mainly receives from TA2, for the connectivity to STGa from A5 is ten times stronger than that from TA2. TA2 also has effective connectivity to visual motion regions MT, MST, V6, V3A etc; and with region 1 (somatosensory cortex).

In a second ventral stream, A4 has effective connectivity to A5. A5 is higher in the hierarchy than A4, in that A4 receives from A1 and 52; and A5 receives from PBelt, LBelt, and A4 (Rolls et al., 2023f). A5 then connects to dorsal bank superior temporal sulcus (STS) regions STGa, STSda, and STSdp, which then have onward effective connectivity to TPOJ1, STV, PSL, TGv, TGd and PGi (Fig. 6.6), which are language-related regions (Rolls et al., 2022c).

The dorsal bank superior temporal sulcus regions (STSda and STSdp) also receive visual inputs about, for example, the movements made by the face and mouth during speaking, in that some neurons in these cortical regions of macaques respond perfectly to the small lip and mouth movements made by humans when they speak (Hasselmo, Rolls, Baylis and Nalwa, 1989b; Hasselmo, Rolls and Baylis, 1989a). We also discovered neurons in the same cortical regions that respond to auditory stimuli including vocalization (Baylis, Rolls and Leonard, 1987), and indeed some neurons in this cortical region respond to both visual and auditory stimuli (Khandhadia et al., 2021; Belin, 2019). This makes these dorsal bank STS regions important for the decoding of articulator speech movements.

These dorsal bank STS regions are therefore important in linking motion-related changes in the face being viewed to the dynamically changing auditory input, and this is likely to be useful for identifying who in a group is making the vocalization, in maximizing the ability to decode information in noisy environments in order to understand speech, in maximizing the information in social signals by combining the auditory and visual components, etc. In the sense that this processing provides evidence about what the message is, this could be considered as a type of ventral stream 'What' semantic processing (Rolls, 2023c, 2026d).

In addition, there is evidence (Rolls et al., 2023f) for effective connectivity of some auditory cortical regions to the hippocampus, and to the scene-related ventromedial visual region VMV1 (Sulpizio et al., 2020), and the lateral parahippocampal TF, with both types of connectivity involved in episodic memory (Rolls et al., 2023b, 2022a; Rolls, 2023a, 2022a). These auditory inputs may provide 'What' inputs for episodic memory, but also 'Where' spatial inputs for episodic memory and the representation of space, including space behind the individual that is inaccessible by vision in humans. This suggests interesting new research, on what potentially could be a hippocampal auditory representation of space behind the individual.

There also appear to be dorsal auditory pathways in humans, though they are less well understood (Rolls et al., 2023f; Rolls, 2023c, 2026d).

6.8 The hippocampal connectome for memory and navigation

The connectivity of the human hippocampal system for episodic memory and navigation and its implications are described in Sections 5.2.10 and 5.2.11. The connectivity of the human hippocampus with the semantic knowledge systems are described in Section 6.10.

6.9 The orbitofrontal cortex connectome for reward and emotion

The connectivity of the human orbitofrontal cortex and related systems for reward valuation, decision-making, and emotion is described in Section 3.10.

6.10 A semantic cortical system

The connectivity between brain regions involved in language is a topic of importance for understanding what is probably the least computationally well-understood aspect of human brain function, and in order to minimize language impairments during neurosurgery (Milton et al., 2021; Ding et al., 2016; Hickok and Poeppel, 2015, 2007; Hickok, 2022; Bornkessel-Schlesewsky et al., 2015; DeWitt and Rauschecker, 2013; Rauschecker, 2012; Chang et al., 2015; Bouchard et al., 2013; Battistella et al., 2020). To help take this endeavor forward, I decided to analyze the connectivity between cortical systems implicated in language using the HCP-MMP atlas as the basis for investigating the connectivity of the language systems in the human brain (Rolls et al., 2022c).

A useful starting point for this research was that in a previous investigation utilizing the Human Connectome Project multimodal parcellation (HCP-MMP) atlas (Glasser et al., 2016a), 157 task-based fMRI studies were analyzed to identify brain regions in the HCP-MMP atlas related to semantics based on categorization of visual words and objects, and auditory words and stories (Milton, Dhanaraj, Young, Taylor, Nicholas, Briggs, Bai, Fonseka, Hormovas, Lin, Tanglay, Conner, Glenn, Teo, Doyen and Sughrue, 2021). The following regions were identified as part of the semantic network: 44, 45, 55b, IFJa, 8C, p32pr, SFL, SCEF, 8BM, STSdp, STSvp, TE1p, PHT and PBelt, and were shown with tractography to be interconnected (Milton et al., 2021).

I started by performing a community analysis on the 360x360 effective connectivity data for both hemispheres, and a community of 26 cortical regions was identified (Rolls et al., 2022c) that included most of those or very similar regions noted above (Milton et al., 2021) related to language. I then performed analyses on the connections, functional connectivity, and effective connectivity of these 26 left hemisphere cortical regions with the 360 cortical regions in the HCP-MMP atlas (Rolls et al., 2022c).

Having identified 26 cortical regions related to language in the left hemisphere, a further community analysis (Rubinov and Sporns, 2010) was performed on the effective connectivities of the 26 language-related ROIs from all 180 cortical areas in the left hemisphere, to investigate whether any grouping could be made (Rolls et al., 2022c). It was found that three communities were formed, as follows: Group 1: STSva STSvp TE1a TGd PGi 10v 9m 10pp 47s 8Av 8BL 9a 9p. Group 2: TGv 44 45 47l SFL 55b. Group 3: A5 STGa STSda STSdp PSL STV TPOJ1. The connectivity of these three groups of language-related cortical regions (Rolls et al., 2022c) is described next.

6.10.1 A semantic system that includes the inferior bank of the superior temporal sulcus including object representations

The regions in Group 1 (STSva STSvp TE1a TGd PGi 10v 9m 10pp 47s 8Av 8BL 9a 9p), are considered first. The connectivities of regions in the ventral bank of the superior temporal sulcus, STSva and STSvp, are illustrated in Fig. 6.7, and of another region in this group, PGi, in Fig. 6.8. The regions involving cortical inferior regions of the superior temporal sulcus (STS) have effective connectivity with the adjacent inferior temporal visual cortex TE1a and temporal pole TG. The connected parietal PGi region has effective connectivity with inferior temporal visual cortex (TE) regions, with parietal PFm which also has connectivity with visual regions, with posterior cingulate cortex and parahippocampal TF memory-related regions, with the frontal pole, with the medial orbitofrontal cortex and related ventromedial prefrontal cortex reward-related regions, with the dorsolateral prefrontal cortex, and with 44 and 45 ('Broca's area') for output regions. PGi also receives inputs from PGs which includes visuo-motor as well as visual object system connectivity (Rolls et al., 2023a). This inferior STS semantic network is likely to be more involved in invariant object and face identity representations for static visual stimuli of the type represented in the inferior

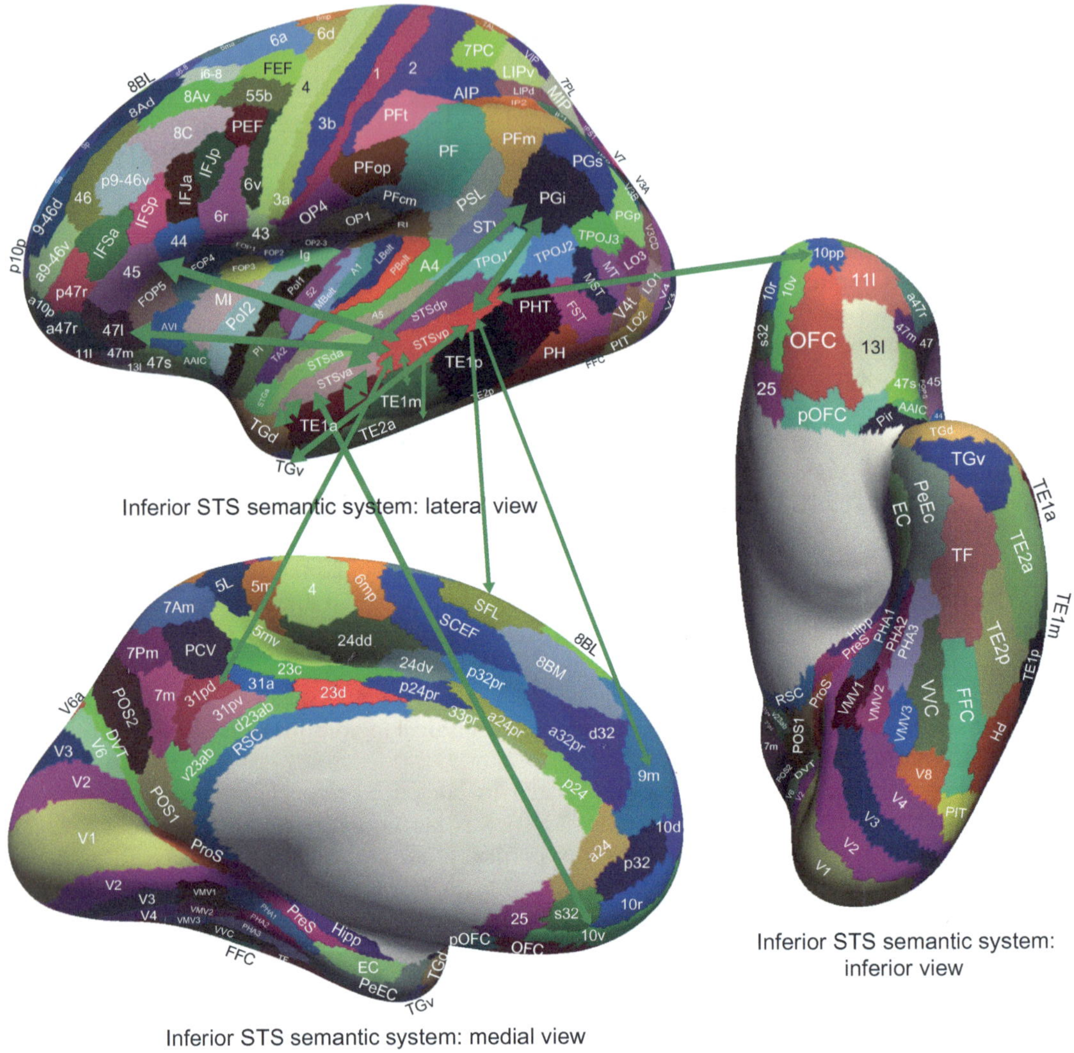

Fig. 6.7 Inferior bank of the Superior Temporal Sulcus (STS) cortex semantic system in STSva and STSvp: schematic overview based on effective connectivity. An output of the Ventrolateral Visual System for object and face information is to the cortex in the inferior parts of the Superior Temporal Sulcus, STSva and STSvp. The visual inputs to these inferior STS cortex regions are shown with green arrows, and come from TE1a, TE1m, TE2a, TGd, TGv, and PGi. Other connectivities are with the memory-related parts of the posterior cingulate cortex (31pd and 31pv); and from the vmPFC (10v and 10r). STSva and STSvp have connectivity directed towards Broca's area 44 and 45, and related areas (47s), and to the Superior Frontal Language region (SFL). (Data from Rolls E T, Deco G, Huang C-C and Feng J. 2023 Multiple cortical visual streams in humans. Cerebral Cortex 33: 3319-3349.)

temporal visual cortex, whereas the superior (/ dorsal) STS regions are more likely to be involved in responses to face expression and visual motion including that which can be combined with auditory stimuli such as the sight and sound of a vocalization (Baylis et al., 1987; Hasselmo et al., 1989b,a; Rolls, 2021a).

The connectivity of these Group 1 semantic regions with inferior temporal visual cortex TE regions provides for transform-invariant visual representations of objects and faces (Chapter 2) as shown by discoveries in macaques (Rolls, 2021c,a; Perrett, Rolls and Caan, 1982; Booth and Rolls, 1998; Arcaro and Livingstone, 2021; Freiwald, 2020; Lehky and Tanaka, 2016) with complementary evidence from human neuroimaging (Kanwisher et al., 1997; Finzi et al., 2021; Collins and Olson, 2014; Rolls et al., 2024b) to be incorporated into semantic representations provided by this network of semantic cortical regions.

The connectivity with inferior parietal cortex PGi and to a smaller extent PGs regions may allow information about the visual motion of objects, such as a head turning away, or the eyes closing, or the lips moving, to convey a social signal such as a greeting (lip-smack) or threat, or when vocalizing, to reach STS regions (Rolls et al., 2023a).

Effective connectivity of the Group 1 regions with the frontal pole (10pp, 9m, 9a, 9p) regions implicated in planning and sequencing (Shallice and Cipolotti, 2018; Shallice and Burgess, 1996; Gilbert and Burgess, 2008; Rolls et al., 2024a) and prospective as well as retrospective memory (Underwood et al., 2015) is likely to provide information about the temporal order / sequence of events, which is important in interpreting causal interactions in the world (Rolls, 2021i).

Similarly, the connectivity via the parahippocampal TF region with the hippocampal system (Fig. 6.8) implicated in episodic memory (Dere et al., 2008; Ekstrom and Ranganath, 2018; Rolls, 2018a, 2021a; Moscovitch et al., 2016) is likely to be important in building semantic information about the semantic meaning of events by enabling the memory of particular past episodic events to be taken into account (Rolls et al., 2025a) (Section 6.10.5). The connectivity of the Group 1 regions with the postero-ventral parts of the posterior cingulate cortex which are implicated in episodic memory (Vann et al., 2009; Leech and Smallwood, 2019; Leech and Sharp, 2014; Rolls et al., 2023g) by providing a route to the hippocampal system may in a similar way be important in enabling the Group 1 regions, especially those in the temporal lobe but also in the inferior parietal cortex (Davis et al., 2018; Papagno, 2018), to build semantic representations by taking particular previous episodes into account (Rolls et al., 2025a) (Section 6.10.5). Consistent with this, increasing activation of the angular gyrus and decreasing activation of the hippocampus occurs when episodic memories become incorporated into neocortical schemas (van der Linden et al., 2017).

The connectivity of the Group 1 regions with the orbitofrontal cortex (regions OFC and pOFC), pregenual anterior cingulate cortex (d32) and some ventromedial prefrontal cortex regions where reward value and emotion are represented (Grabenhorst and Rolls, 2011; Rolls et al., 2022b; Rolls, 2019e,c; Rolls et al., 2020b; Rolls, 2014a; Padoa-Schioppa and Conen, 2017; Cai and Padoa-Schioppa, 2012; Reber et al., 2017; Rolls, 2021d) provides another fundamental type of input, reward / emotion related, for building semantic representations about people and objects.

This ventral STS Group 1 semantic system also receives information from the top of the somatosensory hierarchy, parietal cortex region PF (Rolls et al., 2023a,c, 2022c) (Section 6.6). PF has effective connectivity directed towards language-related regions - the PeriSylvian language area (PSL), TPOJ2, STSvp (Rolls et al., 2022c) (and to prefrontal cortex areas involved in short-term memory related functions IFsa and 46 (Rolls et al., 2023a)). The diffusion tractography (Rolls et al., 2023a) further emphasizes connections of region PF with language-related areas 44, 45 and PSL (Rolls et al., 2023a). PF may on this connectivity evidence be the top of a somatosensory hierarchy which adds visual to somatosensory inputs to form semantic representations of felt objects, which can then gain access to language systems, as well as having outputs to superior parietal and intraparietal areas involved in performing actions such as reach and grasping for felt or seen objects. PF may thus build a multimodal, semantic, representation of felt objects. The PeriSylvian Language region (PSL) is very interesting, because it receives somatosensory inputs from the adjacent PF (Fig. 6.9), and has connections to STS semantic areas, so PSL may be a route for tactile inputs to become part of semantic object representations (Rolls et al., 2022c).

The extensive connectivity of the Group 1 regions with the lateral prefrontal cortex which is implicated in working memory and planning (Goldman-Rakic, 1996; Baddeley, 2021) is likely to

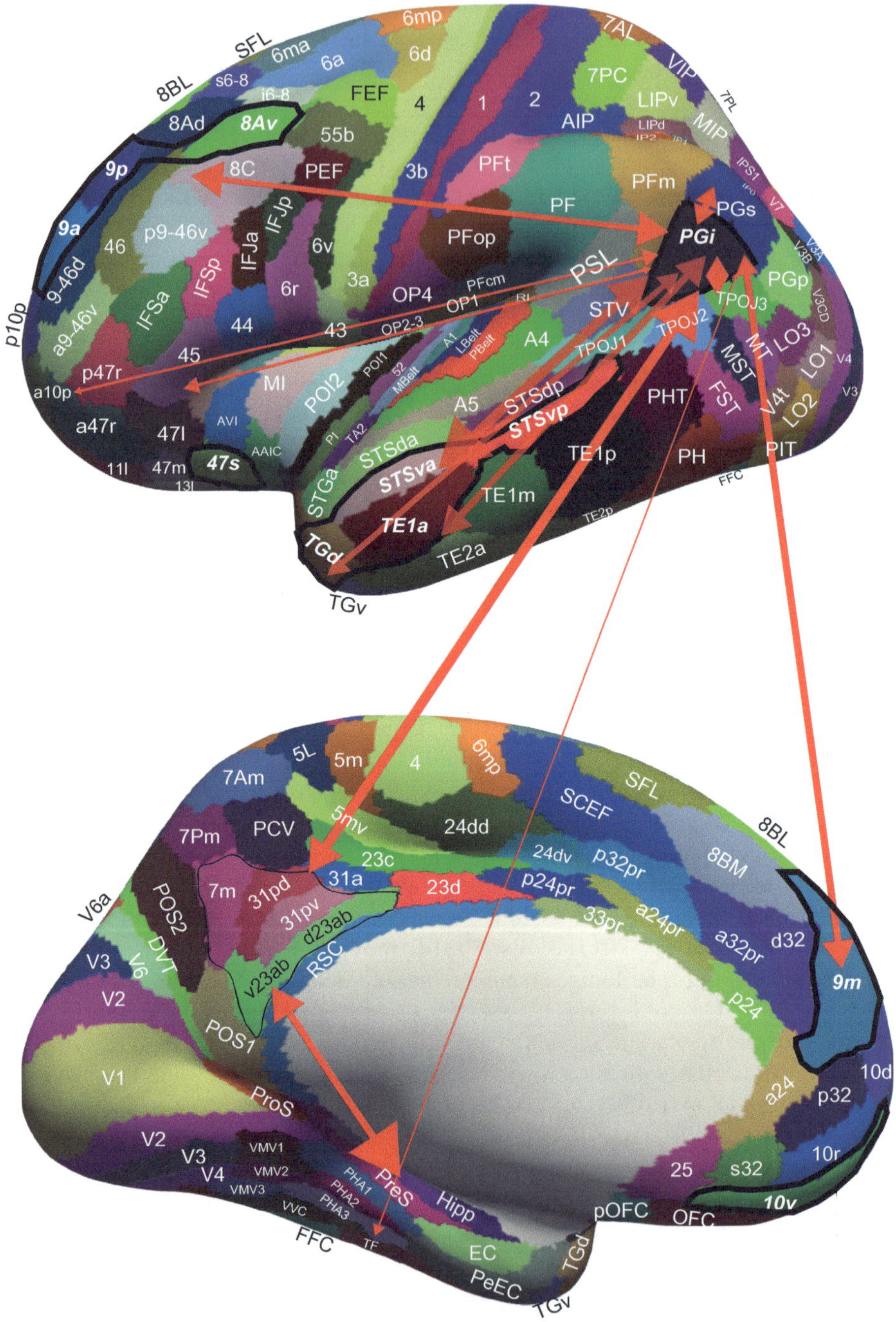

Fig. 6.8 Effective connectivity of region PGi chosen as an example of a Group 1 Semantic region with connectivity with the inferior STS stream, the posterior cingulate cortex, the frontal pole, and the dorsolateral prefrontal cortex. The Group 1 regions are indicated in bold italic font and are outlined in black (including PGi), and are: STSva STSvp TE1a TGd PGi 10v 9m 10pp 47s 8Av 8BL 9a 9p. The widths of the lines and the size of the arrowheads indicate the magnitude and direction of the effective connectivity. 7PM and IP1 connect to PGs which in turn connects to PGi, providing a route for PGi to receive visual motion information, and to connect that to areas such as STSva and STSvp. The thin black outline encloses the postero-ventral memory-related regions of the posterior cingulate cortex, which connect to the hippocampal system (Rolls et al., 2023g). The abbreviations are listed in Appendix A. (Data from Rolls E T, Deco G, Huang C-C and Feng J. 2022 The human language effective connectome. Neuroimage 258: 119352.)

enable semantic representations to be held online to facilitate planning, reasoning, and rearrangement of the order of components if a task fails (Rolls, 2020c).

Regions in the more inferior part of the STS and linked with inferior temporal cortex visual regions in Group 1 are implicated in 'semantic representations involved in speech, reading, and visual and supra-modal mental objects that are linked to word forms' (Hertrich et al., 2020). The temporo-parietal junction and angular gyrus (which include visual PG regions in HCP-MMP) regions in Group 1 are implicated in 'pragmatic processing, context integration, words with multiple meanings, perspective taking, and social cognition' (Hertrich et al., 2020), consistent with some of the points made above about the regions in the Group 1 network.

Overall, our evidence shows that there is a network of brain regions that includes regions in the ventral bank of the superior temporal sulcus that combines evidence from its strongly interconnected brain regions to build multimodal representations of objects, including visual and somatosensory information. The human connectivity described here shows that this is a remarkable system extending from the temporal pole TG regions through the inferior temporal visual cortex and the ventral bank of the superior temporal sulcus through to inferior parietal cortex region PGi as illustrated in Fig. 6.8, and which also has connectivity with some dorsal prefrontal cortex regions 9a, 9b and 9m (Fig. 6.8). This semantic system also has connectivity with the hippocampal memory system, in part through the posterior cingulate cortex postero-ventral 'what' memory-related regions (Fig. 6.8, (Rolls et al., 2023g)), and this connectivity is likely to be important in how episodic memory is used to help build semantic representations, as described in Section 6.10.5 (Rolls, 2022a; Rolls et al., 2025a).

6.10.2 A semantic system that includes the superior bank of the superior temporal sulcus including visual motion, auditory, somatosensory and social information

The regions in Group 3 (A5 STGa STSda STSdp PSL STV TPOJ1), are considered next, with their connectivities illustrated in Figs. 6.9 and 6.10.

In the superior (or dorsal) bank of the cortex in the Superior Temporal Sulcus (STS), regions STSda and STSdp were identified by a community analysis as part of a superior temporal lobe semantic system involved in language (Rolls et al., 2022c). Visual inputs to regions STSda and STSdp in this superior STS semantic system come from PGi (Fig. 6.10), and provide a route for moving visual stimuli / objects analyzed in the parietal cortex to reach STS regions (Rolls et al., 2023a). Visual inputs also reach STSdp from the Superior Temporal Visual (STV) region which receives from both MT and FFC (Fig. 6.10), and which could contribute to the neuronal activity in the cortex in the STS which has been shown to respond to moving heads, faces and objects in macaques (Hasselmo, Rolls, Baylis and Nalwa, 1989b; Hasselmo, Rolls and Baylis, 1989a), as described in Section 6.3. STSda and STSdp also have connectivity from STSva and STSvp which have strong connectivity with the Ventrolateral Visual Stream (Figs. 6.1 and 6.10). STSda and STSdp also receive auditory effective connectivity from A5 (Fig. 6.10). STSdp has connectivity directed towards Broca's area 44 and 45, and related areas (47l), and to the PeriSylvian Language region PSL and to the Superior Frontal Language region SFL (Rolls et al., 2022c).

It was proposed that this dorsal bank of the STS region is a third visual system that combines inputs from dorsal and ventral visual streams (Hasselmo et al., 1989b,a; Baylis et al., 1987; Rolls et al., 2023b), and this has received support (Pitcher and Ungerleider, 2021). Moreover, the same STS regions also contain neurons that respond to auditory stimuli including in some cases vocalization (Baylis, Rolls and Leonard, 1987). It is likely that neurons in these STS regions respond to combinations of auditory and synchronized visual stimuli such as vocalization and moving lips and are important in speech analysis by the viewer/hearer not only in noisy environments, but also to

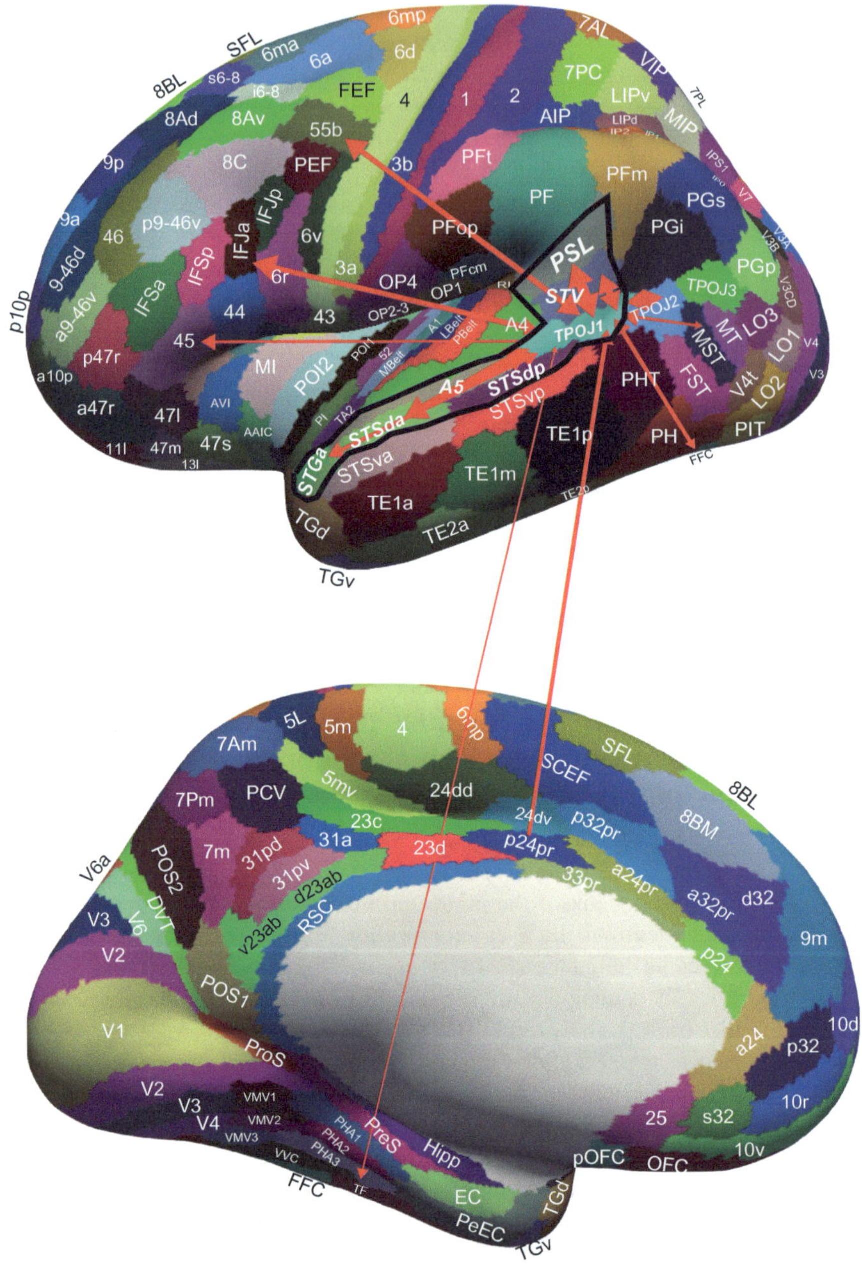

Fig. 6.9 Effective connectivity of region TPOJ1 chosen as an example of the typical effective connectivity of a Group 3 Semantic region with connectivity with a superior STS auditory-visual stream extending from STGA through STSda, A5, STSdp to TPOJ1. The Group 3 regions are indicated in bold italic font and are outlined in black, and are: A5 STGa STSda STSdp PSL STV TPOJ1. TPOJ1 also has effective connectivity with PSL, STV, TPOJ2, the inferior frontal gyrus IFJA and IFSp and 45, and with premotor area 55b and the midcingulate cortex p24pr. There is also effective connectivity with parahippocampal TF. The widths of the lines and the size of the arrowheads indicate the magnitude and direction of the effective connectivity. The abbreviations are listed in Appendix A. (Data from Rolls E T, Deco G, Huang C-C and Feng J. 2022 The human language effective connectome. Neuroimage 258: 119352.)

confirm who is speaking, and to learn associations between the sight of a person's face and the sound of their voice, which is part of what is required for the semantic representation of a person. Neurons of these types, including responsiveness to face expression, underlie the importance of this region in interpreting the social meaning of visual and auditory stimuli, which is a key part of semantic representations implemented in these STS regions, and that contribute to theory of mind, that is, of another individual's intentions and even thoughts. Consistent with this, individual STS cortex neurons have now been reported in macaques to respond to both visual face stimuli and to auditory stimuli (Khandhadia, Murphy, Romanski, Bizley and Leopold, 2021). In areas such as this, 'concept cells', which respond to multimodal inputs (Mormann, Dubois, Kornblith, Milosavljevic, Cerf, Ison, Tsuchiya, Kraskov, Quian Quiroga, Adolphs, Fried and Koch, 2011; Quian Quiroga, 2013), may be formed (Rolls, 2023d).

Region TPOJ1 has robust effective connectivity with auditory association cortex, including A5, A4 and to some extent LBelt, PBelt and TA2 (Fig. 6.9). The visual input to TPOJ1 is not from inferior temporal visual cortex (TE regions), but instead with some earlier visual cortical regions involved in motion processing (MST, MT, especially to STV and TPOJ1) and in face processing (the fusiform face cortex FFC and posterior inferior temporal cortex PIT). This is proposed to contribute to the responsiveness of neurons in this region, and in the dorsal bank of the STS, to moving heads and face movements important in social behavior as described in Section 6.3 (Rolls et al., 2023b, 2022c).

Another type of influence on these Group 3 regions comes from somatosensory cortical areas, as is especially evident in the functional connectivity (Rolls et al., 2022c). The functional connectivity of Group 3 regions is not only with somatosensory cortex 1-3b and 5, but also with frontal opercular and insular cortex, and there is some evidence for this too in the effective connectivity, including with parietal PF which can be considered as the top of the somatosensory hierarchy (Rolls et al., 2023a). This may contribute to semantic representations of body image and the sense of self (Rolls et al., 2023a).

This 'superior STS cortex semantic system' thus enables multimodal representations including visual, auditory, and probably also somatosensory via PF and earlier somatosensory cortex regions, to gain access to the language system (Rolls et al., 2022c). There is also a link via TF to the hippocampal memory system (Fig. 6.10), which is likely to be important in how episodic memory is used to help build semantic representations, as described in Section 6.10.5 (Rolls, 2022a; Rolls et al., 2025a).

Although the description of a region around the temporo-parieto-occipital junction as Wernicke's area involved in language comprehension is now much less clear than previously thought (Rolls, 2023c, 2026d), what is clear based on the evidence described by Rolls, Deco, Huang and Feng (2022c) is that regions in this general area of the brain in what are parts of the dorsal and ventral STS semantic streams are very strongly implicated now as having key roles in semantic representations, and with communicating with Broca's area as described in Section 6.10.4.

An anterior part of the superior temporal lobe (in Group 3) is identified as an auditory word form area (Hertrich et al., 2020; DeWitt and Rauschecker, 2013; Binder, 2017, 2015). A more posterior part of the superior temporal lobe in the superior temporal gyrus is involved in auditory-phonological processing (Hertrich et al., 2020), consistent with the auditory connectivity described here. Just inferior to that in posterior parts of the superior temporal gyrus, superior temporal sulcus, and middle temporal gyrus is a region implicated in 'syntax and semantics' (Hertrich et al., 2020). This is probably close to the temporo-parieto-occipital junction region (part of Group 3) the connectivity of which is shown in Fig. 6.9.

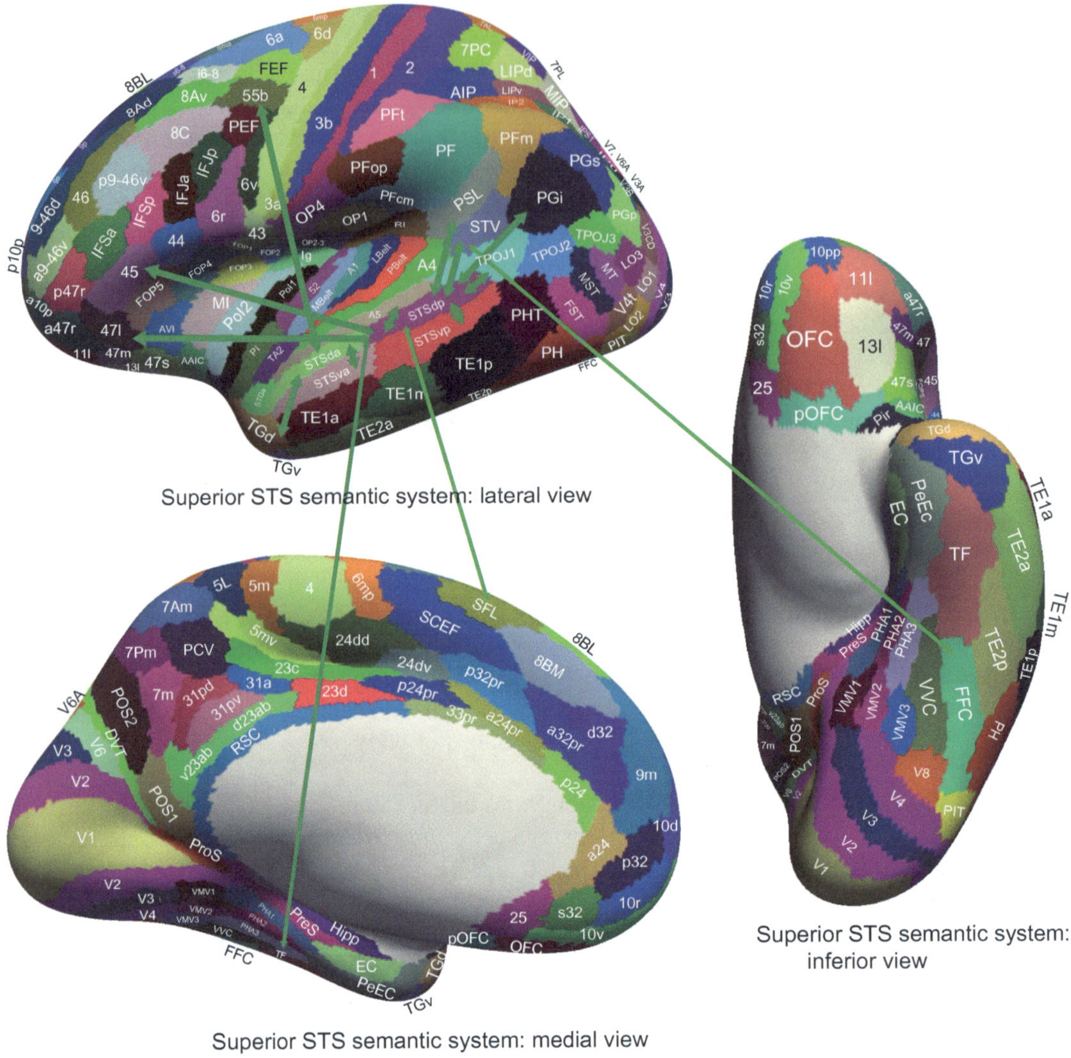

Fig. 6.10 Superior bank of the STS semantic system in STSda and STSdp: schematic overview. Visual inputs for moving visual stimuli / objects reach STSdp from PGi. Visual inputs also reach STSdp from the Superior Temporal Visual area which receives from FFC. STSda and STSdp also have connectivity from STSva and STSvp which have strong connectivity with the Ventrolateral Visual Stream (Fig. 6.1). STSda and STSdp also receive auditory effective connectivity from A5. STSdp has connectivity directed towards Broca's area 44 and 45, and related areas (47l), and to the PeriSylvian Language region PSL and to the Superior Frontal Language region SFL. (Data from Rolls E T, Deco G, Huang C-C and Feng J. 2023 Multiple cortical visual streams in humans. Cerebral Cortex 33: 3319-3349.)

6.10.3 Multimodal semantic representations

So far, the emphasis has been on visual and auditory semantic representations, partly because 'What' visual and auditory processing, which is primarily about the external world, is performed in the temporal lobes. However, some neurons in the cortex in the macaque superior temporal sulcus do respond to somatosensory stimuli (Baylis, Rolls and Leonard, 1987). Even the inferior parietal PGi system is at least dominated by visual system connectivity (Rolls et al., 2023a), and what seems to be missing so far in the PGi part of this system, is information about the somatosensory properties of objects, and how the objects can be manipulated, perhaps as tools, under somatosensory as well as visual control. Now of course any information from the somatosensory system is quite different in kind from what is represented in the temporal lobes, for the somatosensory system involves touch to the body or the position of the body, which is not a property that needs to be taken into account by the temporal lobe analysis of objects 'out there' in the world. The way is which somatosensory information is incorporated into semantic systems appears to be implemented using the Group 3 regions that have effective connectivity with somatosensory regions (frontal opercular FOP, insula and parietal PF). Consistent with this, the Group 3 regions such as STV receive inputs from posterior cingulate cortex region PCV (precuneus visual region) (Rolls et al., 2023g), and this may provide for visual inputs relating to visually-guided reaching to grasp for objects, to be incorporated into semantic representations of objects.

A key region for multimodal convergence is the set of temporo-parieto-occipital junction regions TPOJ1-3, which are highly connected with each other, and which between them have effective connectivities with all the major systems just described involved in language, including visual, auditory, STS, PF, and PGi (Rolls et al., 2023a). The implication is that in terms of connectivity the TPOJ2 and TPOJ3 regions both of which have connectivity with the Group 3 regions (Rolls et al., 2023a) contribute to bringing together semantic representations involving all of these types of representation in the brain. Further, the Group 3 region PSL, the Peri-Sylvian Language region, receives effective connectivity from PF which can be considered as the top of the somatosensory hierarchy (Rolls et al., 2023a). Another input to the TPOJ regions comes from the cortex just lateral to the fusiform face area, which is the visual word-form area (VWFA) (Yeatman and White, 2021; Dehaene et al., 2005; Dehaene and Cohen, 2011; Caffarra et al., 2021; Woolnough et al., 2021; Agrawal and Dehaene, 2024). Disruption of the connectivity of the VWFA is likely to account for problems such as alexia and dyslexia in which the visual analysis required for reading cannot reach the semantic TPOJ regions (Pritchard et al., 2018; Liuzzi et al., 2019). Indeed, the FFC has effective connectivity to TPOJ1 and TPOJ2 and to the superior temporal visual area (STV) which in turn connects strongly with the nearby TPOJ regions. Further, PHT, lateral to the FFC, projects to TPOJ2.

The reward value is also an important semantic property of people, objects, places, etc, and this also can be incorporated into semantic representations by inputs to these semantic systems, especially Group 1, from the orbitofrontal cortex, ventromedial prefrontal cortex, and anterior cingulate cortex (Rolls et al., 2022c).

The temporal pole TG regions of the cortex are also likely to form multimodal 'semantic' representations, but primarily based on visual and auditory properties of objects, which is what are primarily represented in the temporal lobes.

6.10.4 Broca's area and related regions (TGv 44 45 47l SFL 55b)

An interesting discovery is that the semantic systems just described in Groups 1 and 3 have effective connectivity directed towards Broca's area and related regions, which comprise Group 2 in this investigation (Rolls, Deco, Huang and Feng, 2022c). That supports the hypothesis that Broca's area (regions 44, 45 and 47l in HCP-MMP) and its related regions in Group 2 are involved in language output, in preparing for example to produce a stream of speech, and/or in syntax (Rolls, 2023c, 2026d; Rolls and Deco, 2015b).

Another interesting discovery made is that regions 44 and 45 are strongly connected to adjoining inferior frontal gyrus regions including IFSp and IFJa, and 47l and 47s (Rolls et al., 2022c), and that this whole system is expanded in the left hemisphere. It is proposed that this connectivity with inferior frontal gyrus regions provides 45 and 44 with extra cortical attractor networks, which could be very useful in the computations that may in involved in syntax, which require representations with different roles in sentences to be organised into an appropriate sequential stream, with hypotheses and a model on how linked attractor networks could be involved developed elsewhere (Rolls and Deco, 2015b; Rolls, 2023c, 2026d). A summary of the connectivities of these Group 2 regions is provided next.

The Group 2 language-related regions are indicated in bold italic font and are outlined in black in Fig. 6.11, and are: TGv 44 45 47l SFL 55b (Rolls et al., 2022c). Effective connectivity of region 45 is chosen as an example of the effective connectivity of a Group 2 region (Fig. 6.11). There is connectivity with the superior STS auditory-visual semantic stream (e.g. STSdp and STGa); with the inferior STS visual semantic stream (e.g. STSvp and TG); with the Peri Sylvian Language area and TPOJ1; and with nearby inferior frontal gyrus regions 44, IFSp, 47l and 47s. Probable output regions for region 45 include premotor 55b and the (supplementary premotor area) Superior Frontal Language region (SFL). There is some effective connectivity too with the frontal pole, and dorsolateral prefrontal 9a and 8BL.

The Group 2 regions and network are likely to be involved in speech processing (Rolls et al., 2022c). One region in Group 2 is 55b, which is situated in the premotor cortex sandwiched between the frontal eye fields FEF and PEF. 55b is likely to be an output region for 44 and 45, and it was proposed that it is involved in vocalization (Rolls et al., 2022c) given the importance of vocal cord control in speech production (Ekert et al., 2021). Consistent with this proposal, it has now been found that 55b (a 'dorsal precentral speech area') codes voice pitch during vocalization, and may be involved in the prosody of speech (Hickok et al., 2023). Similarly, the Superior Frontal Language area (SFL) also part of Group 2 may be a supplementary premotor cortical area involved in speech production.

The involvement of 47l and 47s in the inferior frontal gyrus regions linked by effective connectivity with language systems is found here in the context of left hemisphere functions. These orbital parts of the language system may be involved in retrieving articulatory plans from semantic stimuli (i.e. semantic-to-articulatory recoding) (Ekert et al., 2021). The left inferior frontal sulcus (IFS) regions found here to have connectivity with the Group 2 regions may be involved in similar functions (Ekert et al., 2021). In the right hemisphere the corresponding 47l and 47s regions may be able to function more as lateral orbitofrontal areas involved in non-reward, punishment, and emotion, and consistent with this, functional connectivities and activations related to emotion computations are more prominent in the right hemisphere (Cheng et al., 2016; Rolls et al., 2020b,a,e).

The connectivity of Broca's region (Rolls et al., 2022c) extends previous concepts about Broca's area (Clos et al., 2013; Amunts and Zilles, 2012), by showing that there is a whole group of interconnected areas in the frontal lobe of the left hemisphere that are implicated in language, not only regions 44 and 45 in the HCP-MMP atlas (which may not correspond exactly to BA44 and BA45), but also IFSp and IFJa, and 47l and 47s.

Similarly, the concept of a Wernicke's area involved in speech comprehension at the temporo-parietal junction is no longer accepted (DeWitt and Rauschecker, 2013; Binder, 2017), but the concept that I have proposed is that there is a semantic system that extends from the temporal pole to PGi (and PF for body including somatosensory representations (Rolls et al., 2023a)) in the parietal lobe, which in the temporal lobe has an inferior more visual and object-related part (Group 1), and a more superior auditory/visual motion/somatosensory related part (Group 3). Region 45 has more effective connectivity with (especially anterior) STS regions than does region 44 (Rolls et al., 2023f). Parts of the supramarginal part of the parietal lobe (which include PF regions) are implicated in auditory short-term memory and articulatory sequencing (Oberhuber et al., 2016). Parts of the angular gyrus of the parietal lobe (which include PG regions) are implicated in semantic processing involved for example in comprehension when reading (Seghier et al., 2010).

Broca's area, BA44 and BA45, is no longer thought to be essential for speech production in terms of articulation as shown by the analysis of effects of damage to it (Gajardo-Vidal et al., 2021), but

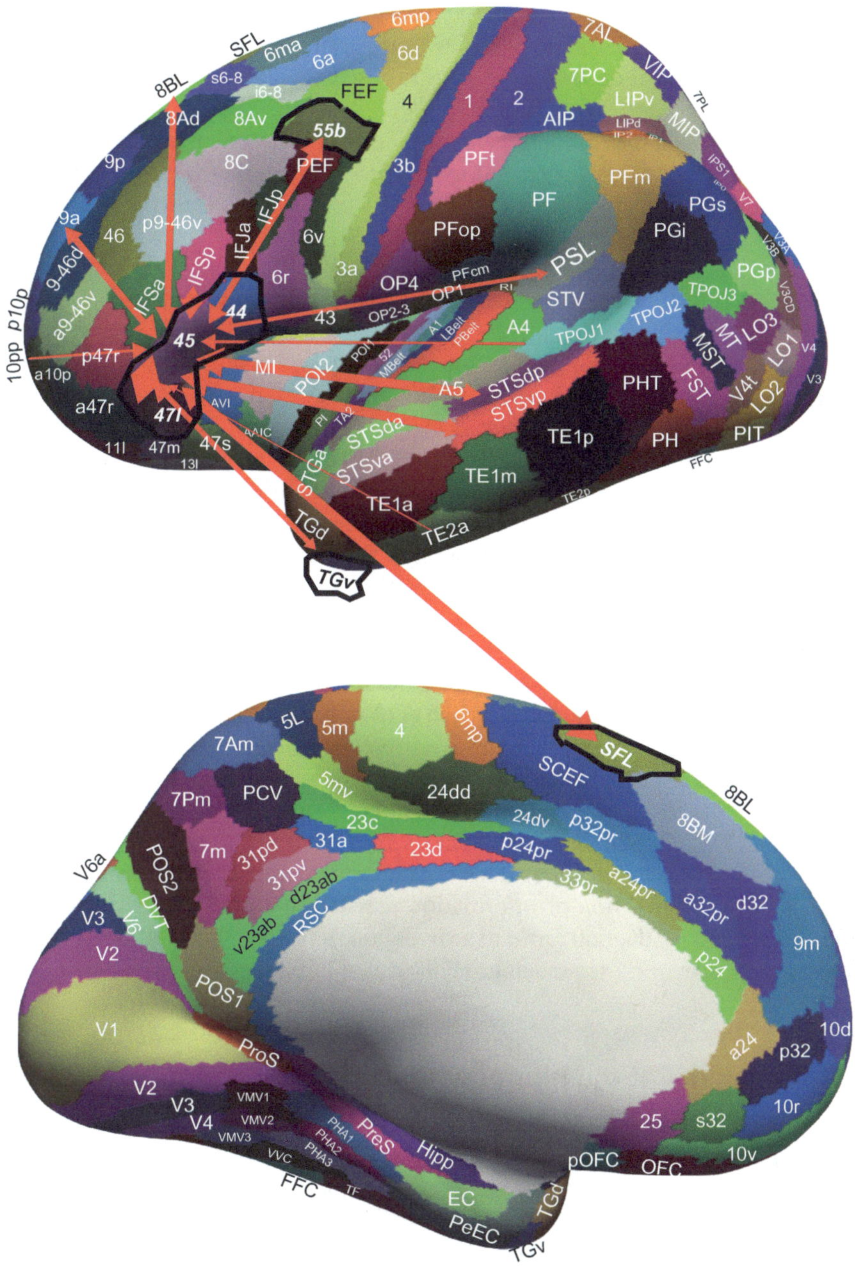

Fig. 6.11 Effective connectivity of region 45 chosen as an example of the effective connectivity of a Group 2 language-related region. The Group 2 regions are indicated in bold italic font and are outlined in black, and are: TGv 44 45 47l SFL 55b. There is connectivity with the superior STS auditory-visual stream (e.g. STSdp and STGa), with the inferior STS visual stream (e.g. STSvp and TG), with the Peri Sylvian Language area (PSL) and TPOJ1, and with nearby inferior frontal gyrus regions 44, IFSp, 47l and 47s. Probable output regions for 45 include premotor 55b and the (supplementary area) Superior Frontal Language region (SFL). There is some effective connectivity too with the frontal pole, dorsolateral prefrontal 9a and 8BL. The widths of the lines and the size of the arrowheads indicate the magnitude and direction of the effective connectivity. The abbreviations are listed in Appendix A. (Data from Rolls E T, Deco G, Huang C-C and Feng J. 2022 The human language effective connectome. Neuroimage 258: 119352.)

is implicated in some aspects of language, such as the timing of speech items (Long et al., 2016), and syntax as described next (Friederici et al., 2017; Flinker et al., 2015; Matchin and Hickok, 2020; Hickok, 2022).

The Group 2 network described here involving outputs from other language regions to regions 44, 45 and closely connected inferior frontal gyrus regions including IFja and IFSp, and extending into left orbitofrontal cortex regions 47s and 47l, is, I have suggested, involved in syntax (Rolls et al., 2022c), with the syntactic processing (Rolls and Deco, 2015b) being driven by semantic inputs from the Group 1 inferior STS temporal lobe visual object 'What' semantic network and the Group 3 superior STS temporal lobe visual motion / auditory / somatosensory semantic network.

Consistent with this, there is evidence from neuroimaging and related neuropsychological studies that implicates area 44 in syntax processing, manipulating and detecting structures above the word level (Hertrich et al., 2020; Friederici et al., 2017). In more detail, area 44 is seen as the key region for syntactic processing by implementing the 'merge' operation that is key for understanding syntax (Friederici, Chomsky, Berwick, Moro and Bolhuis, 2017). Merge is an operation that takes exactly two (syntactic) elements – call them x and y – and puts them together to form the unordered set $[x, y]$ (Friederici et al., 2017). The elements x and y can be word-like building blocks that are drawn from the lexicon or previously constructed phrases that are assembled into an unbounded array of hierarchically structured internal representations (phrases and, ultimately, sentences) (Friederici, Chomsky, Berwick, Moro and Bolhuis, 2017). Merge is a recursive operation, but as analyzed by Treves and Rolls (1994), recurrent processing in the brain could be implemented by an equivalent series of linked feedforward networks. Area 45 is described as a lexical-semantic core area linking phonological codes to lexical meanings (Hertrich et al., 2020; Friederici et al., 2017).

From a study using direct cortical recordings during vocal repetition of written and spoken words, it was found that Broca's area mediates a cascade of activation from sensory representations of words in temporal cortex to their corresponding articulatory gestures in motor cortex, but is quiescent during articulation (Flinker, Korzeniewska, Shestyuk, Franaszczuk, Dronkers, Knight and Crone, 2015). It was proposed that Broca's area does not participate in the production of individual words, but coordinates the transformation of information processing across large-scale cortical networks involved in spoken word production, prior to articulation. It has also been proposed that Broca's area is involved in lexical selection / sequencing (Hickok et al., 2023).

6.10.5 Connectivity of the hippocampal episodic memory system with the semantic system and its implications

Semantic representations are based on knowledge acquired over time about for example semantic categories such as cars and animals. As we have seen, the human anterior temporal lobe is implicated in the storage of semantic memories (Rolls, 2023c, 2026d; Hickok and Poeppel, 2015; Patterson et al., 2007; Kemmerer, 2015; Milton et al., 2021; Bonner and Price, 2013; Rolls et al., 2022c; Peelen and Caramazza, 2012). This contrasts with episodic memory, which is prototypically about an event at a particular time, about for example a particular car ('What') in a particular location ('Where'), and in which the hippocampus is implicated (Moscovitch et al., 2016; Squire and Wixted, 2011; Bennett and Stark, 2016; Rolls et al., 2024e; Rolls and Treves, 2024). The slow learning in the semantic system contrasts with the fast learning in the hippocampal episodic memory system (McClelland, McNaughton and Lampinen, 2020; McClelland, McNaughton and O'Reilly, 1995). Interestingly, there is evidence that the hippocampal episodic memory is implicated in learning new semantic representations, based for example on semantic learning impairments after hippocampal damage (Duff, Covington, Hilverman and Cohen, 2019).

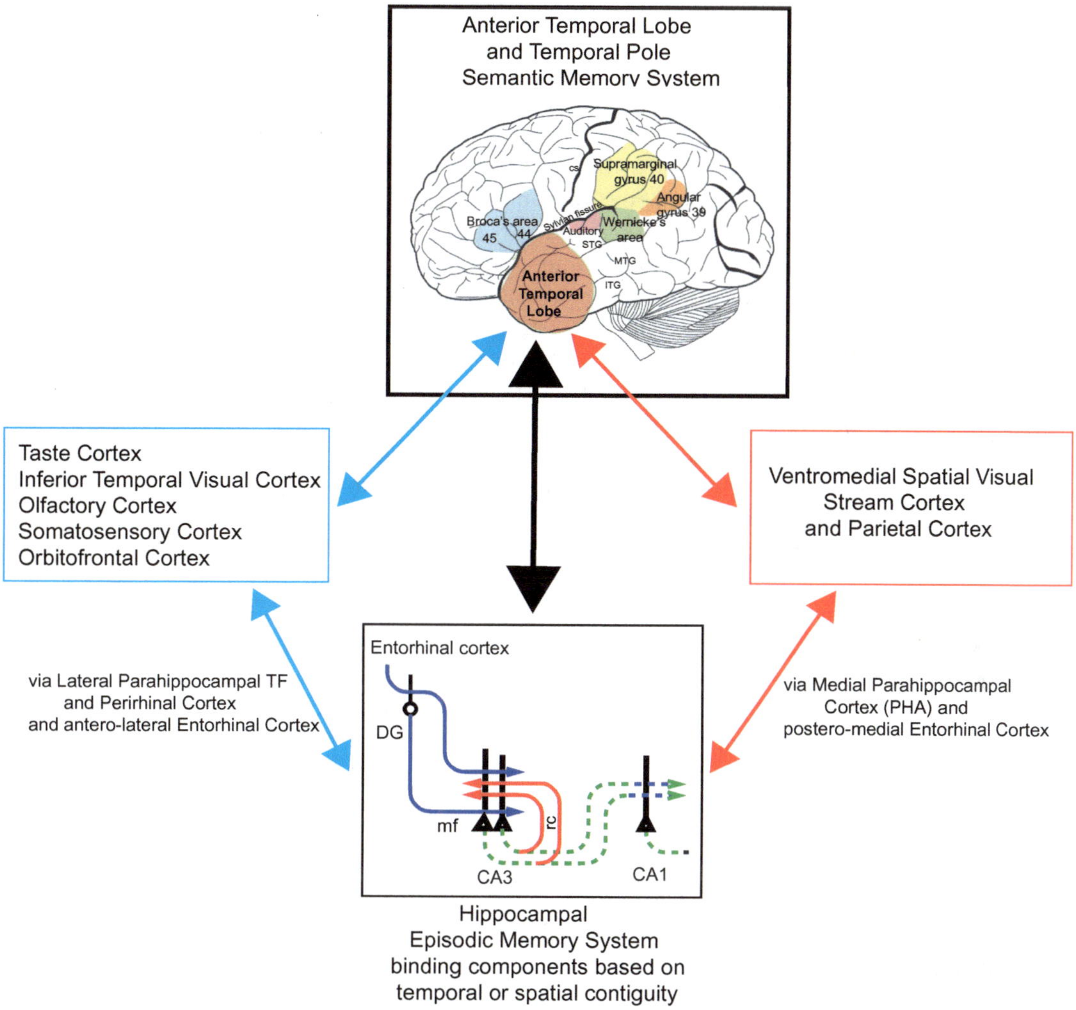

Fig. 6.12 Direct and hippocampal routes to the anterior temporal lobe semantic memory system. The anterior temporal lobe semantic memory system and the hippocampal episodic memory system both receive inputs from the 'What' cortical systems (blue, left, taste cortex, inferior temporal visual cortex, olfactory cortex, somatosensory cortex and reward value orbitofrontal cortex) and from the 'Where' cortical systems (red, right, ventromedial spatial stream cortex for spatial view, and parietal cortex). In addition, the hippocampal episodic memory system has connectivity with the anterior temporal lobe semantic system. It is proposed that because the hippocampal episodic memory system can bind together elements of an episodic memory in CA3 ('What', 'Where', When, and Reward), the anterior temporal lobe semantic memory system benefits from the hippocampal bindings of certain combinations of inputs from a number of similar episodic memories, and using slow learning can associate together the overlapping information from these episodic memories when they are recalled to help build a new semantic memory. The hippocampal binding could be based on spatial or on temporal contiguity. Some of the other cortical areas indicated in this Figure in color are involved in language, and are connected to the anterior temporal lobe semantic regions, which have connectivity that extends up through the temporo-parietal junction (labelled here Wernicke's area) to the inferior parietal cortex (Rolls et al., 2022c; Rolls, 2023c, 2026d). Further evidence for the relevant connectivity is cited in the text. On the view of the human brain: STG - superior temporal gyrus; MTG - middle temporal gyrus; ITG - inferior temporal gyrus; cs - central sulcus. On the hippocampal network architecture: DG - dentate gyrus; mf - mossy fibres; rc - CA3 recurrent collaterals which provide a single autoassociation / attractor network to bind current inputs together to form an episodic memory. Connectivity of the hippocampal episodic memory system via the lateral parahippocampal cortex region TF is shown for the human brain with many cortical regions in and connected to anterior temporal lobe semantic memory systems.

I have developed the hypothesis that one way in which the anterior temporal lobe could learn new semantic categories is when the items in the category occur close together in time, or are recalled close together in time, so that slow learning in attractor networks in the anterior temporal lobe could associate together the items in a category (Rolls et al., 2025a). This temporal association that is potentially useful for learning semantic memory could occur if a series of events occur close together in time, e.g. the items in a lecture; or the temporal association could occur if a series of related episodic memories was recalled in sequence, helping for example to build up a semantic representation based on episodes remembered from a movie. It is suggested that the hippocampus could contribute usefully to this process not only by being involved in the recall of recent episodic events, but also by binding together the components of those events (e.g. 'What', 'Where', and Reward value) and presenting those bound or 'chunked' events to the semantic system in the anterior temporal lobe (Rolls et al., 2025a). Learning from the chunked hippocampal representations is proposed as more efficient for learning new semantic representations in the anterior temporal lobe than starting with all the individual components that might characterize a semantic representation (Rolls et al., 2025a). The overall scheme is illustrated in Fig. 6.12.

I propose that the hippocampal system by performing chunking or binding based on hippocampal specialization for spatial and temporal processing can play an important role in forming new semantic categories based on spatial or temporal binding of components (Rolls et al., 2025a). The components that are bound together because they occur in the same part of space, or close together in time, need not themselves be spatial, but could include, objects, people, concepts, etc.

These concepts could be developed if we knew what the connectivity is between the hippocampal episodic memory system and the anterior temporal lobe semantic memory system. I therefore performed an investigation of the connectivity between the hippocampal episodic memory system and the anterior temporal lobe memory system in humans (Rolls et al., 2025a). The results for the main route with which the hippocampus communicates with the anterior temporal lobe, the lateral parahippocampal gyrus region TF, are shown in Fig. 6.13. This connectivity is largely with what may be termed the 'What' cortical systems concerned with representations of objects and faces (Rolls, 2023c, 2026d). (The connectivity of TF with the hippocampus is largely via the perirhinal cortex (PeEC) and then entorhinal cortex (Fig. 6.13).)

Fig. 6.13 shows first that there is connectivity of TF with the inferior temporal cortex largely unimodal visual cortical regions for invariant object and face representations, TE2p and TE1p.

Second, there is connectivity with the orbitofrontal cortex (47m), which provides a route for reward value to be incorporated into episodic memories (Rolls et al., 2022b).

Third, there is connectivity of TF with many regions in the anterior temporal lobe semantic network (Rolls et al., 2022c; Milton et al., 2021), including STSda, STSva, TE2a, and temporal pole TGd, all of which are multimodal (Rolls et al., 2022c).

Fourth, there is connectivity of TF with temporo-parieto-occipital regions TPOJ2 and TPOJ3 that are implicated in language (Rolls et al., 2022c; Milton et al., 2021), and that are part of what was referred as 'Wernicke's area' (Rolls, 2023c, 2026d).

Fifth, there is connectivity of TF with inferior parietal cortex region PGi, which is massively connected with anterior temporal lobe semantic regions (Rolls et al., 2022c, 2023a), and which may encode information about actions which is an important part of semantic representations (Rolls et al., 2022c, 2024b; Rolls, 2023c, 2026d; Rolls et al., 2023a).

Sixth, there is connectivity of TF with inferior frontal cortex regions IFSp and IFJa (Fig. 6.13), which are massively connected with Broca's area regions 44 and 45, and which are suggested to add extra attractor capacity to the proposed roles of regions 44 and 45 in language-based syntactic operations on semantic information (Rolls and Deco, 2015b; Rolls et al., 2022c; Rolls, 2023c, 2026d).

Fig. 6.14 shows the effective connectivity of the human medial parahippocampal cortex regions PHA1-3 involved in spatial scene representations (probably corresponding to TH in macaques) with

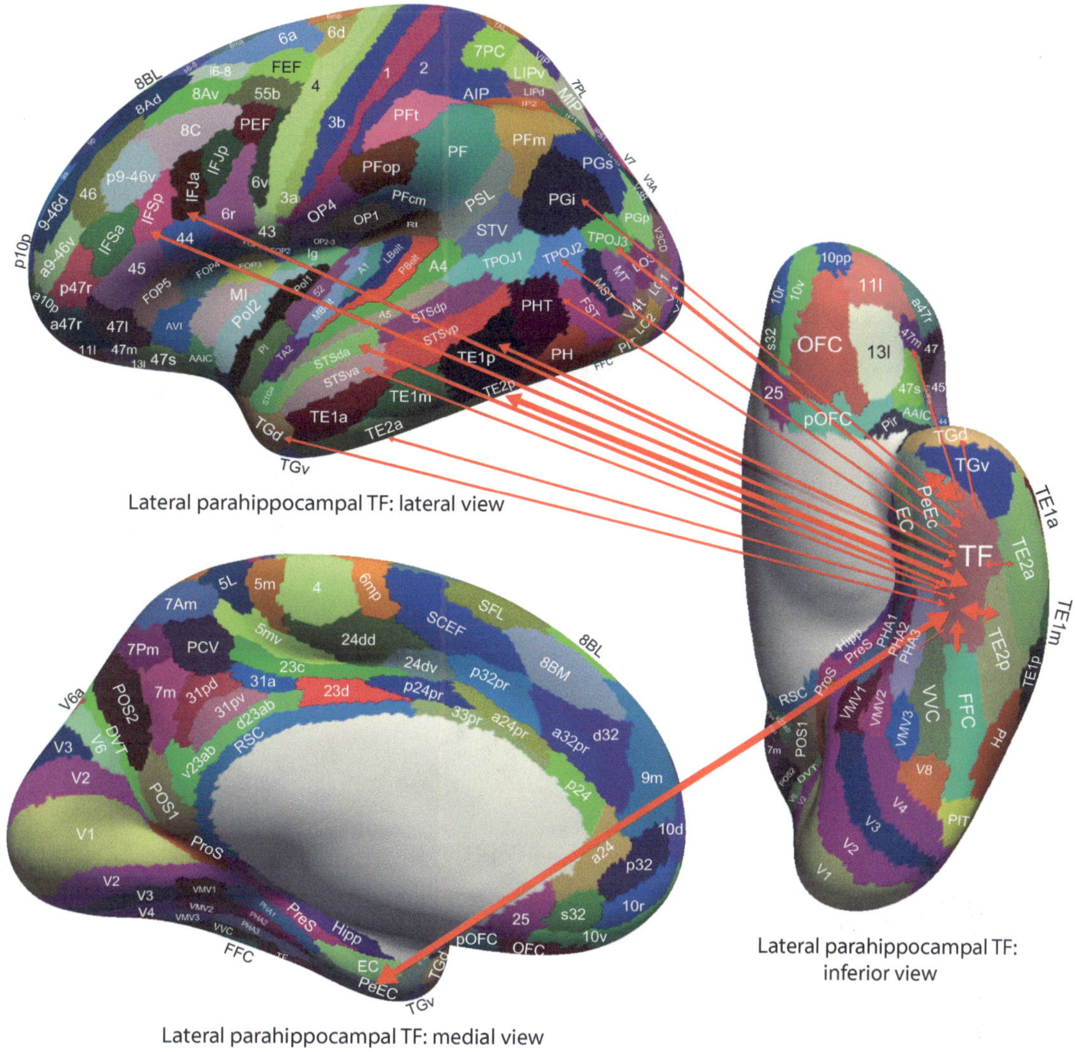

Fig. 6.13 Connectivity of the hippocampal episodic memory system includes connectivity with many cortical regions in and connected to anterior temporal lobe semantic memory systems. The connectivity of the lateral parahippocampal cortex region TF which includes connectivity to semantic systems related to 'What' representations is shown (see text). The effective connectivity between cortical regions in the Human Connectome Project Multimodal Parcellation atlas (HCP-MMP) (Glasser et al., 2016a; Huang et al., 2022) is shown. A list of the cortical regions in the HCP-MMP atlas is shown in Appendix 1.3. The width of the arrows represents the magnitude of the effective connectivity, and the size of the arrowheads the magnitude in each direction. The data are from effective connectivity matrices computed previously from resting state timeseries in 171 HCP participants imaged at 7T (Rolls et al., 2022c,a, 2023b), but this is the first cortical diagram for the connectivity of TF from this dataset. Connectivity of the hippocampal episodic memory system via the lateral parahippocampal cortex region TF is shown for the human brain with many cortical regions in and connected to anterior temporal lobe semantic memory systems. (Data from Rolls E T, Zhang C and Feng J. 2025 Slow semantic learning in the cerebral cortex, and its relation to the hippocampal episodic memory system. Cerebral Cortex 35: bhaf107.)

neocortical regions. This connectivity is largely with what may be termed the 'Where' cortical systems concerned with representations of spatial view, and actions in space (Rolls, 2024c, 2023c, 2026d). The connectivity of PHA1-3 with the hippocampus is largely via the entorhinal cortex (Fig. 6.14) (Rolls et al., 2022a; Huang et al., 2021).

First, the ventromedial cortical visual pathway for spatial scenes using spatial view cells provides

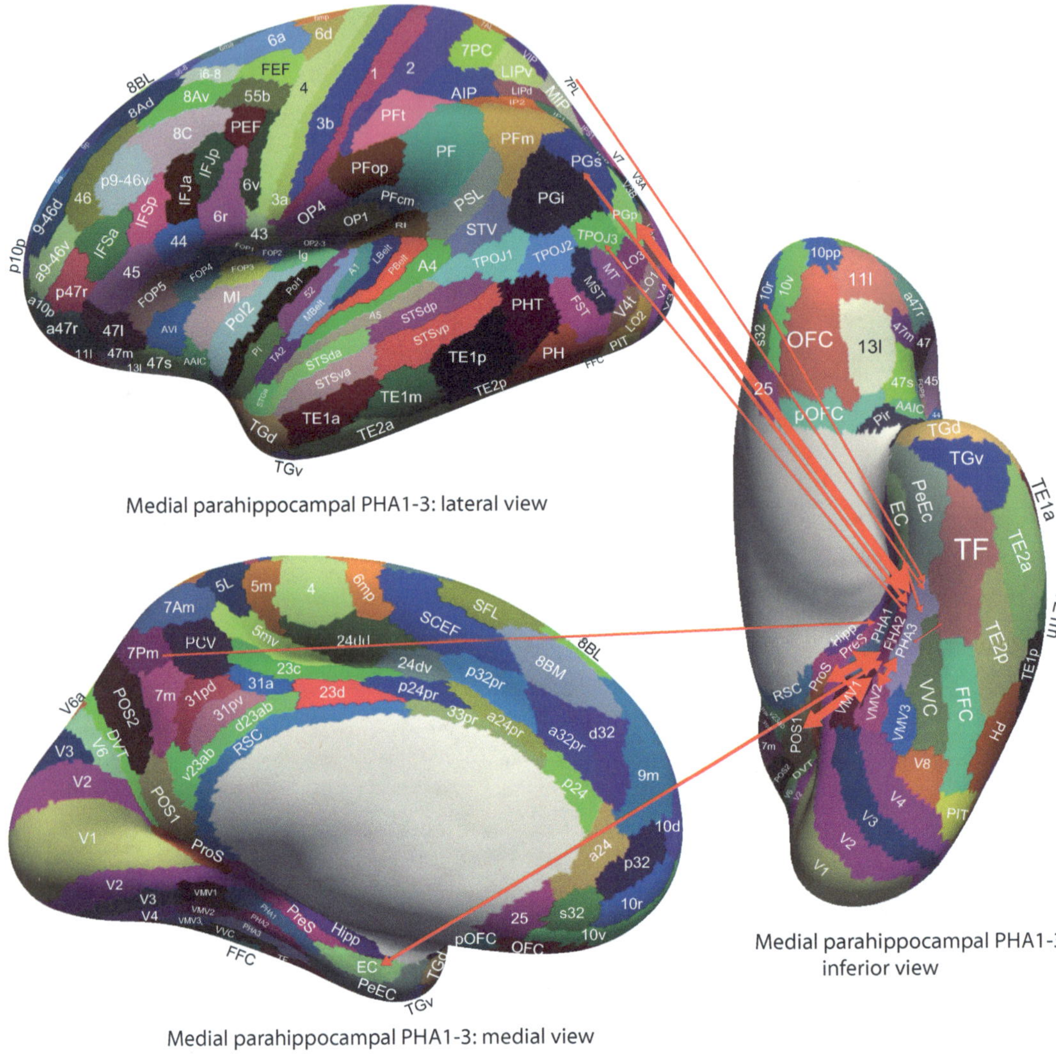

Fig. 6.14 Connectivity of the hippocampal episodic memory system includes connectivity with many cortical regions in and connected to anterior temporal lobe semantic memory systems. The connectivity of the medial parahippocampal cortex regions PHA1-3 which includes connectivity to semantic systems related to 'Where' representations is shown. The effective connectivity between cortical regions in the Human Connectome Project Multimodal Parcellation atlas (HCP-MMP) (Glasser et al., 2016a; Huang et al., 2022) is shown. A list of the cortical regions in the HCP-MMP atlas is shown in Appendix A. The width of the arrows represents the magnitude of the effective connectivity, and the size of the arrowheads the magnitude in each direction. The data are from effective connectivity matrices computed previously from resting state timeseries in 171 HCP participants imaged at 7T (Rolls et al., 2022c,a, 2023b), but this is the first time that a cortical diagram for the connectivity of the medial parahippocampal cortex regions PHA1-3 has been provided. (Data from Rolls E T, Zhang C and Feng J. 2025 Slow semantic learning in the cerebral cortex, and its relation to the hippocampal episodic memory system. Cerebral Cortex 35: bhaf107.)

inputs to PHA1-3 from regions that include VMV1-3, the ProStriate Cortex (ProS), and POS1 (Huang et al., 2021; Rolls et al., 2022a, 2023b; Rolls and Turova, 2025; Rolls et al., 2024c; Rolls, 2023a, 2025b).

Second, there is connectivity with TPOJ3, which is a semantic region in the temporo-parietal region (see above).

Third, there is connectivity with inferior parietal PGs (adjacent to and connected with PGi), which is connected with anterior temporal lobe semantic regions (Rolls et al., 2022c, 2023a), and which may encode information about actions which is an important part of semantic representations (Rolls et al., 2022c, 2024b; Rolls, 2023c, 2026d; Rolls et al., 2023a).

Fourth, there is connectivity from some area 7 parietal and related regions involved in visuo-motor actions in space (7Pm, 7Am, 7PL, IP0 and PCV) (Rolls et al., 2023a).

Fifth, there is some connectivity with reward-related ventromedial prefrontal cortex region 10r (Rolls et al., 2022b; Rolls, 2023b, 2022a).

The connectivity in Figs. 6.13 and 6.14 shows that the hippocampal episodic memory system does have remarkably extensive connectivity with cortical regions involved in semantic representations, and these two Figures do bring out this important connectivity in humans quite clearly for the first time (Rolls, Zhang and Feng, 2025a). In a check, significantly other more direct connectivity with semantic systems was not found from the perirhinal cortex and entorhinal cortex that are closer to the hippocampus, though more connectivity with the orbitofrontal cortex and anterior cingulate cortex reward-related cortical regions was found from these cortical regions (Rolls et al., 2022b).

The import of this research is that the hippocampal episodic memory system is highly connected with the semantic memory systems in the anterior temporal lobe, temporo-parietal junction, and the inferior parietal cortex, all of which are greatly developed in humans compared to macaques. This encourages more thinking about how the episodic memory system could contribute to the learning of new semantic representations. One proposal I am making is that the hippocampal episodic memory system binds together components of environmental events such as what happened, where, when, and its reward value, and that these bound structures are then easier structures to be associated together in networks in the semantic memory systems (Rolls et al., 2025a). I have proposed that attractor networks in these semantic systems could associate together these episodic building blocks if they occur close together in time, perhaps during sequential recall of related episodic memories (Rolls et al., 2025a) (Fig. 6.12). The hippocampal system would help this computation by providing bound chunks of information, rather than relying on the semantic attractor networks having to deal with large numbers of separate components to try to build a semantic representation.

(It is noted that the hippocampal process for binding described by Eichenbaum and colleagues (Sellami, Al Abed, Brayda-Bruno, Etchamendy, Valerio, Oule, Pantaleon, Lamothe, Potier, Bernard, Jabourian, Herry, Mons, Piazza, Eichenbaum and Marighetto, 2017) involved associations between temporally discontiguous events to form episodic memories in the hippocampus, whereas the processes described here are different and show how what may be temporally discontiguous hippocampal episodic memories when recalled together close in time can be used by slow semantic neocortical learning to help build new semantic neocortical representations.)

The proposal that attractor networks in neocortical regions build semantic representations raises some key issues (Rolls et al., 2025a). There is not a single attractor network for the semantics (only one is shown in Fig. 6.15 to help elucidate the concepts), but probably hundreds, in the cortical regions extending from the temporal pole through the anterior temporal cortex through the temporo-parietal junction to the inferior parietal cortex (Figs. 6.8 and 6.9) (Rolls et al., 2022c; Rolls, 2023c, 2026d). Each local cortical attractor network operates within a range of approximately 2 mm, corresponding to the high density of local recurrent excitatory collaterals between pyramidal cells (Rolls, 2023c, 2026d), so there is plenty of cortical area for very many separate local attractor networks in the anterior temporal lobe to inferior parietal cortex semantic regions (Rolls et al., 2022c; Rolls, 2023c, 2026d). Moreover, each local attractor network can potentially store up to in the order of 10,000 representations (though fewer if they are correlated as will be the case for semantic representations), given that there are in the order of 10,000 excitatory recurrent collateral synapses on each neuron,

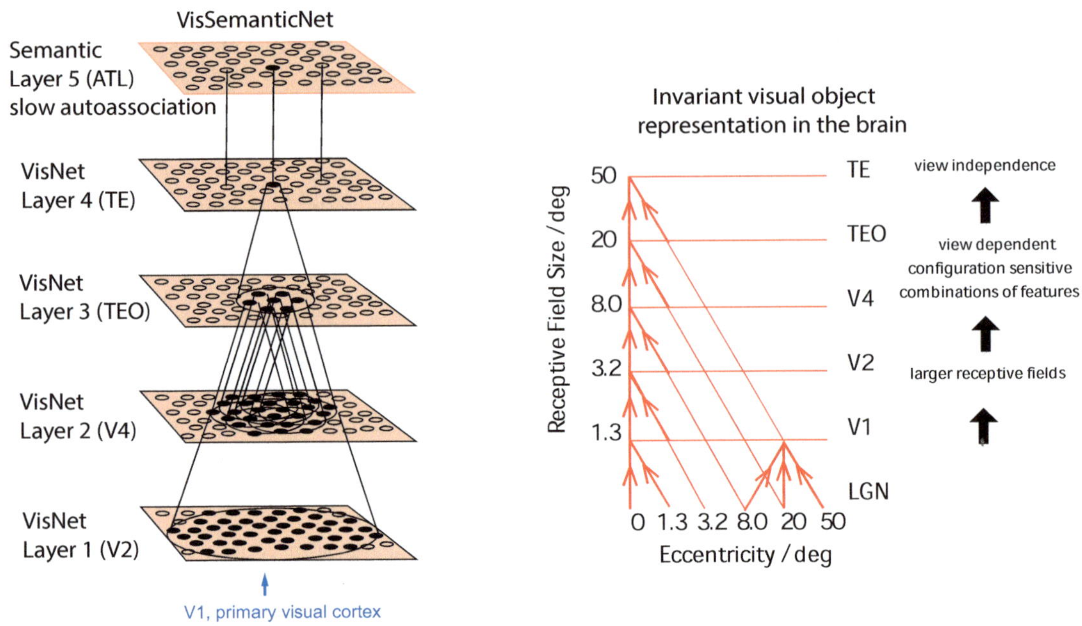

Fig. 6.15 A model, VisSemanticNet, for building semantic representations in the anterior temporal lobe (ATL), which is layer 5 in the model. The first 4 layers are VisNet (Rolls, 2012b, 2021c), which operate by convergence from layer to layer, competitive feedforward learning that utilizes a local associative trace learning rule with a sufficiently long time constant to capture different views of an object while it transforms. Layer 5 is an autoassociation network that utilizes a local associative trace learning rule with a sufficiently long time constant to capture the different objects in one semantic category while that semantic category is being processed. Part of the input to Layer 5 could also be from other sensory systems such as auditory, somatosensory, taste and olfactory, to contribute to building multimodal semantic representations. TEO - posterior inferior temporal visual cortex. TE - inferior temporal visual cortex. (Data from Rolls E T, Zhang C and Feng J. 2025 Slow semantic learning in the cerebral cortex, and its relation to the hippocampal episodic memory system. Cerebral Cortex 35: bhaf107.)

and a sparse representation (Rolls, 2023c, 2026d; Rolls and Treves, 2024; Treves and Rolls, 1991; Treves, 1991a).

In this context, I propose the concept that each of these semantic attractor networks will have different inputs from different parts of the cortex (visual object, visual face, visual scene, auditory, somatosensory, taste, olfactory, action ...), because of the different connectivity of each region in the anterior temporal lobe, as we have shown (Rolls et al., 2022c). This results, together with the probabilistic local connectivity typical of the cerebral cortex (Rolls, 2023c, 2026d), in somewhat different semantic representations in each differently connected local attractor network (Rolls et al., 2025a). This is computationally useful, for though each single attractor network may be able to perform moderately well at storing hierarchically organised representations (Parga and Virasoro, 1986), having the semantics set up in different attractor networks may help (Boboeva, Brasselet and Treves, 2018). For example, more general semantic concepts may be represented more anteriorly in the temporal lobe towards or in the temporal pole, being fed from more posterior semantic regions in the anterior temporal lobe that receive inputs from more unimodal sensory cortical processing streams (Rolls et al., 2022c). All of these semantic cortical areas in the anterior temporal lobe and inferior parietal cortex would need connectivity with cortical regions that may be involved in syntax such as regions in and near regions 45 and 44, and that is the case (Rolls et al., 2022c; Rolls, 2023c, 2026d).

To develop these concepts towards the level of how this type of semantic learning could be implemented in our brains, I proposed a model of slow semantic learning in the cerebral cortex, which is illustrated in Fig. 6.15 (Rolls et al., 2025a). The first four layers perform invariant visual object recognition just like VisNet described in Section 4.7 (Rolls, 2012b, 2021c). The fifth layer for the anterior temporal lobe semantic networks is one or many attractor networks that associate together the episodic memories each with chunks of information that represent a set of similar episodic memories that might be recalled close together in time, to help build a new semantic category. Later, in a different slow time window, other chunked items from another set of related episodic memories recalled closely together in time might be associated to form a different semantic category. The computational model was able to use this type of slow semantic learning to build new semantic categories in this way (Rolls et al., 2025a).

In this section (6.10), I have thus shown how we have made new discoveries about the connectivity of the human semantic memory systems, how these systems are connected to the human hippocampal episodic memory system, and how these are stimulating new concepts and ideas about how semantic memory systems in the human brain are built and how they operate.

6.11 Evaluation

In order to understand how the human brain works, it is essential to discover what each region computes, and what information is passed through connections to other brain regions that use the results of each other's computations.

The research described in this Chapter provides a framework for this understanding, not only by describing the connectivity between the 360 cortical regions in the most well-founded parcellation available for the human cerebral cortex, the Human Connectome Project Multimodal Parcellation, HCP-MMP (Glasser et al., 2016a), but also considers, wherever possible, the computations that take place in different cortical regions and connected cortical systems.

In this context, the pioneering research described here on both cortical connectivity and function provides I believe the type of framework that will be essential in future to understand the operation of the human brain. More detailed analyses of cortical function and connectivity are provided in Brain Computations and Connectivity (Rolls, 2023c) and **Brain Computations and Principles; and AI** (Rolls, 2026d).

It is important to consider why this type of approach to understanding the human brain and how its regions are connected is useful. The argument is that cortical computation is at least partly localized to local specialized cortical regions in order to minimize the wiring length in the brain, and therefore brain size (Rolls, 2016b, 2026d). (Brain size is a fundamental issue in land animals, and indeed in humans, babies are born immature so that the size of the head is not larger for childbirth.) Part of the fundamental design of the neocortex is to have localized recurrent collateral connections between pyramidal cells within a distance of 1-2 mm, which it is proposed form local attractor networks (Rolls, 2016b, 2026d). Another key principle of cortical design is to have cortical regions involved in different stages of a computation in a hierarchy close together in the brain, with key examples the hierarchically organised sensory systems (visual, taste, somatosensory, auditory etc), for this minimizes cortical connection lengths and thereby brain size (Rolls, 2016b, 2026d).

Understanding the connectivity as well as computations performed by different regions of the human brain is also of great importance for understanding human brain function, because the human brain has many cortical regions that are highly developed compared even to what is present and can be analyzed in macaques. Examples include the anterior temporal lobe and temporal pole semantic regions; the inferior frontal gyrus regions such as Broca's area involved in language; the temporo-parietal junction regions in Wernicke's area that are involved in language; dorsolateral prefrontal cortical regions involved in planning and executive function; the inferior parietal cortex involved in spatial functions; and the orbitofrontal cortex involved in emotion including emotional feelings (Rolls, 2023c, 2026d).

Despite these differences, where cortical regions do correspond in non-human primates and humans, being able to analyze activity at the neuronal level in non-human primates is important, for it is at the neuronal level that information is represented in the brain, and is transmitted between the computing elements of the brain, the neurons, so the neuron-level evidence is very important (Rolls, 2016b, 2023c, 2026d). The rodent brain is a less useful model for the systems-level organization of the human brain, because the computations performed in its brain regions are in many cases different from what is found in primates including humans (see Section 29.4 of Rolls (2026d)), and there may be less localisation of function in the rodent brain because of its smaller size, given the argument just presented (Rolls, 2026d).

Overall, the discoveries described in this Chapter provide a connectivity framework that is combined with evidence about what is computed in each cortical region to provide a foundational approach with many new concepts about how many cortical systems operate in the human brain (Rolls, 2026d).

7 Computational neuroscience, stochastic cortical attractor neurodynamics, psychiatric disorders, and aging

7.1 Introduction

Part of the aim of this book is to show how discoveries in systems-level neuroscience have been and can be made by focussing on what is computed in each brain region, and how it is computed. The brain is a set of computational systems, and given the aim just described, I have developed a strong research interest in computational neuroscience. My focus has been on biologically plausible neuronal networks, and I have set out in a tutorial way with sample code the principles of operation and properties of such networks, including pattern association networks, autoassociation networks, and competitive networks, in Appendix B of Brain Computations and Connectivity (Rolls, 2023c) and Appendix B of **Brain Computations and Principles; and AI** (Rolls, 2026d).

The dynamics, the operation in real time, of these biologically plausible networks is also very interesting, including how long it takes them to compute, which we have pioneered and which is remarkably fast (Appendix B, Rolls (2023c, 2026d)), and also including how the almost random spiking times of individual neurons for a given mean firing rate introduce stochasticity, noise or randomness, into the operation of these networks, and influence their stability (Rolls and Deco, 2010). Understanding the stochastic properties of these networks and their stability has in turn enabled us to make advances in understanding some mental disorders, and the cognitive and memory changes in aging. Some of these fundamental advances and discoveries are highlighted in this Chapter, with more details in Rolls (2023c, 2026d) and the original research publications.

In the course of the research described here, many discoveries have been made about how stochastic neurodynamics (Rolls and Deco, 2010) is important in brain function in health and disease.

The cerebral neocortex has highly developed local recurrent collateral connections allowing neurons within 2–3 mm to have a probability of approximately 0.1 of connecting with each other, and given associative synaptic modifiability, these can form local attractor networks (Appendix B of Rolls (2023c, 2026d) and Rolls (2016b)). These attractor networks provide a foundation for *what* is computed in many cortical systems, including short-term memory, decision-making (Section 3.12), language, episodic memory (Chapter 5), and visual perception by contributing to a trace of previous neuronal firing (Chapter 4). The stochastic dynamics of attractor or autoassociation networks is described in Section 3.12, and in more detail in Appendix B of Rolls (2023c, 2026d).

In this chapter I describe *how* the operation of attractor networks in the brain is influenced by noise in the brain produced by the almost random firing times of neurons for a given mean firing rate; how this can in fact be beneficial to the operation of the brain; and how the stability of these systems and how they are influenced by noise in the brain is relevant to understanding a number of mental disorders. These concepts can inform understanding of the brain that is relevant not only to its operation in health, but also in disease, and how it may be possible to ameliorate some of the effects found in these mental and other disorders, which is a key aim of the research described in this book. I start with an overview of stochastic dynamics in the brain (Rolls, 2023c, 2026d).

7.2 The noisy cortex: stochastic dynamics, decisions, and memory

The spiking of the neurons in the brain is almost random in time for a given mean firing rate, i.e. the spiking is approximately Poissonian, and this randomness introduces noise into the system, which makes the system behave stochastically. The effect that this stochasticity has on a neural system, and more generally the dynamics of the system, for example how long it takes to respond, and how long it remains in a given state, can be investigated using integrate-and-fire neuronal network simulations, which model how the currents through the different synaptic receptor-activated ion channels are integrated to produce the membrane potential of a neuron, which fires an action potential when the threshold is reached. The integrate-and-fire neuronal network approach is described in Section 3.12, and as used here dynamically models the membrane potential of point neurons, and the dynamics of the opening and closing of different receptor-activated ion channels. Each of the simulated neurons shows spiking that is very much like that of neurons recorded in the brain. From such integrate-and-fire models it is possible to make predictions not only about synaptic and neuronal activity, but also about the signals recorded in functional neuroimaging experiments, and the behavior of the system, in terms for example of its probabilistic choices, its reaction times, and its stability. This makes a close link with human behavior possible (Rolls, Grabenhorst and Deco, 2010a,b). The theory as to how networks operate stochastically (Rolls and Deco, 2010; Deco et al., 2013; Rolls, 2023c, 2026d) has implications throughout the cortex, some of which are described in this chapter.

First we consider the sources of the noise in the cortex (Section 7.2.1). Then we consider how the spiking-related noise influences the stability of short-term memory mechanisms in the cortex. How the spiking-related noise influences decision-making mechanisms in the cortex has been described in Section 3.12. Some advantages of noise in the brain, and how these play out in some different ways in the cortex, are described in Section 7.2.3. Fundamental approaches to stochastic dynamics in the cortex, including equations, are provided in Sections B.6 - B.8 of Rolls (2023c) and Rolls (2026d).

7.2.1 Reasons why the brain is inherently noisy and stochastic

Why is the spiking activity of neurons probabilistic, and what are the advantages that this may confer? The answer suggested (see Rolls (2016b); Rolls and Deco (2010); Rolls (2023c, 2026d)) is that the spiking activity is approximately Poisson-like (as if generated by a random process with a given mean rate, both in the brain and in the integrate-and-fire simulations I describe), because the neurons are held close to (just slightly below) their firing threshold, so that any incoming input can rapidly cause sufficient further depolarization to produce a spike (see Fig. 3.20). It is this ability to respond rapidly to an input, rather than having to charge up the cell membrane from the resting potential to the threshold, a slow process determined by the time constant of the neuron and influenced by that of the synapses, that enables neuronal networks in the brain, including attractor networks, to operate and retrieve information so rapidly, within about 2 time constants of the synapses, that is, within 15 ms (Treves, 1993; Rolls and Treves, 1998; Battaglia and Treves, 1998a; Rolls, 2016b; Panzeri, Rolls, Battaglia and Lavis, 2001; Rolls and Deco, 2010; Rolls, 2023c, 2026d). The spike trains are essentially Poisson-like because the cell potential hovers noisily close to the threshold for firing (Dayan and Abbott, 2001), the noise being generated in part by the Poisson-like firing of the other neurons in the network (Jackson, 2004). The noise and spontaneous firing help to ensure that when a stimulus arrives, there are always some neurons very close to threshold that respond rapidly, and then communicate their firing to other neurons through the modified synaptic weights, so that an attractor process can take place very rapidly (Battaglia and Treves, 1998a; Rolls, 2016b; Rolls and Deco, 2010; Rolls, 2023c, 2026d).

The implication of these concepts is that the operation of networks in the brain is inherently noisy because of the Poisson-like timing of the spikes of the neurons, which itself is related to the mechanisms that enable neurons to respond rapidly to their inputs (Rolls, 2016b, 2023c, 2026d). However, the consequence of the Poisson-like firing is that, even with quite large attractor networks of thousands of neurons with hundreds of neurons representing each pattern or memory, the network inevitably settles probabilistically to a given attractor state. This results in decision-making being probabilistic. Factors that influence the probabilistic behavior of the network include the strength of the inputs (with the difference in the inputs / the magnitude of the inputs being relevant to decision-making as described in Section 3.12); the depth and position of the basins of attraction, which if shallow or correlated with other basins will tend to slow the network; noise in the synapses due to probabilistic release of transmitters (Koch, 1999; Abbott and Regehr, 2004); and the mean firing rates of the neurons during the decision-making itself; the firing rate distribution (Webb, Rolls, Deco and Feng, 2011); and the dilution of the connectivity (Rolls and Webb, 2012). In terms of synaptic noise, if the probability of release, p, is constant, then the only effect is that the releases in response to a Poisson presynaptic train at rate r are described by a Poisson process with rate pr, rather than r. However, if the synapse has short-term depression or facilitation, the statistics change (Fuhrmann et al., 2002b; Abbott and Regehr, 2004). A combination of these factors contributes to the probabilistic behavior of the network becoming poorer and slower as the difference between the inputs divided by the base frequency is decreased (Deco and Rolls, 2006; Rolls et al., 2010a) (Section 3.12).

In fact, there are many sources of noise, both external to the brain and internal (Faisal, Selen and Wolpert, 2008). The sources of noise include the following, with citations to original findings provided by Faisal et al. (2008).

First, there is sensory noise, arising external to the brain. Thermodynamic noise is present for example in chemical sensing (including smell and gustation) because molecules arrive at the receptor at random rates owing to diffusion. Quantum noise can also be a factor, with photons arriving at the photoreceptor at a rate governed by a Poisson process.

Second, there is cellular or neuronal noise arising even in the absence of variation in synaptic input. This cellular noise arises from the stochastic opening and closing of voltage or ligand-gated ion channels, the fusing of synaptic vesicles, and the diffusion and binding of signalling molecules to the receptors. The channel noise becomes appreciable in cell bodies (where the number of channels and molecules involved becomes relatively small). This channel noise can affect the membrane potential of the cell body and the initiation of spikes.

Third, there is synaptic noise evident for example as spontaneous miniature postsynaptic currents (mPSCs) that are generated by the 'quantal' (discrete) nature of released neurotransmitter vesicles (Fatt and Katz, 1950; Manwani and Koch, 2001). This synaptic noise can be produced by processes that include spontaneous opening of intracellular Ca^{2+} stores, synaptic Ca^{2+}-channel noise, spontaneous triggering of the vesicle-release pathway, and spontaneous fusion of a vesicle with the membrane. In addition, at the synapses there is variation in the number of transmitter molecules in a released vesicle, and there are random diffusion processes of the molecules (Franks et al., 2003). In addition, there is randomness in whether an action potential releases a vesicle, with the probability sometimes being low, and being influenced by factors such as synaptic plasticity and synaptic adaptation (Faisal et al., 2008; Abbott and Regehr, 2004; Manwani and Koch, 2001).

The magnitude of the fluctuations caused by these factors is described by the standard deviation of the firing rates divided by the mean rate, and this can be used to measure the noise in the network. These effects are described in detail in the Appendix of **The Noisy Brain: Stochastic Dynamics as a Principle of Brain Function** (Rolls and Deco, 2010) (available for download at https://www.oxcns.org).

Although many approaches have seen noise as a problem in the brain that needs to be averaged out or resolved (Faisal, Selen and Wolpert, 2008), and this is certainly important in sensory systems, the thrust of **The Noisy Brain** (Rolls and Deco, 2010) is to show that in fact noise inherent in brain activity has a number of advantages by making the dynamics stochastic, which allows for many remarkable features of the brain, including creativity, probabilistic decision-making, stochastic resonance, unpredictability, conflict resolution, symmetry breaking, allocation to discrete categories, and many other important properties described later in this Chapter and by Rolls (2023c, 2026d).

The size of a network is an important factor in influencing how noisy the network is in relation to the statistical fluctuations related to the Poisson nature of the neuronal spiking. An infinite size network would have no noise, and its performance can be calculated analytically using mean field methods and checked with numerical simulations (Hopfield, 1982; Amit, 1989; Treves and Rolls, 1991; Treves, 1991a, 1990; Rolls, Treves, Foster and Perez-Vicente, 1997c; Rolls, 2012c, 2023c, 2026d). In terms of relevance to cortical function, attractor networks with even thousands of neurons show probabilistic behavior, showing that the spiking-related noise of neurons does influence cortical networks of biologically relevant size. For example, even when a fully connected recurrent attractor network has 4,000 neurons, the operation of the network is still probabilistic (Deco and Rolls, 2006). Under these conditions, the probabilistic spiking of the excitatory (pyramidal) cells in the recurrent collateral firing, rather than variability in the external inputs to the network, is what makes the major contribution to the noise in the network (Deco and Rolls, 2006). Thus, once the firing in the recurrent collaterals is spike-implemented by integrate-and-fire neurons, the probabilistic behavior seems inevitable, even up to quite large attractor network sizes (Rolls, 2023c, 2026d). We also showed that the graded nature of the firing rates of neurons (Webb, Rolls, Deco and Feng, 2011), and the dilution of connectivity within a network with fewer synapses per neuron than neurons in the network (Rolls and Webb, 2012), can influence the noise in the network (Rolls, 2023c, 2026d).

In Section 3.12 the concept of stability in attractor networks used for memory and decision-making is described (see Fig. 3.19). The network is stable if it has a low energy in the sense analyzed by Hopfield (1982), in which stability might arise if connected pairs of neurons have high firing rates and are connected by strong synapses (with an introduction provided by Rolls and Deco (2010) and Rolls (2023c, 2026d)). Although the diluted connectivity just described that is typical of the brain prevents the formal analysis from theoretical physics (Hopfield, 1982) being applied, nevertheless the same concepts of stability and calculation of the properties of attractor networks can still be shown to apply (see Treves (1991b); Treves and Rolls (1991); Rolls (2023c, 2026d)). The network that we have mainly investigated for stability (Rolls and Deco, 2010; Rolls, 2023c, 2026d) has parameters selected so that the spontaneous firing state is stable even when the memory retrieval or decision cues are being applied, for this produces more realistic decision time distributions because then the decisions are dominated by the noise, which may be different from earlier versions of memory and decision-making networks in which the spontaneous state is not stable when the cues are applied (Brunel and Wang, 2001; Wang, 2002). The concepts are set out in Chapter 21 of Rolls (2026d).

7.2.2 Stochastic dynamics and the stability of short-term memory

We next took this approach forward to investigate how noise influences the stability and the operation of attractor networks, which in turn leads towards models of some of the changes in the operation of networks during aging and in psychiatric disorders such as depression, schizophrenia, and obsessive-

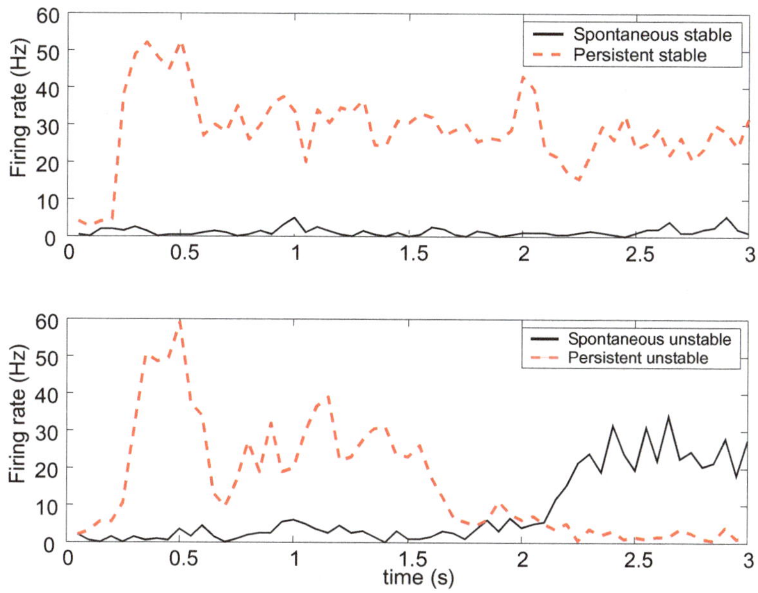

Fig. 7.1 The attractor network model for the stability of short-term memory. Example trials of the Integrate-and-Fire attractor network simulations of short-term memory. The average firing rate of all the neurons in the S1 population or pool is shown. Top. Normal operation. On a trial in which a recall stimulus was applied to S1 at 0–500 ms, firing continued normally until the end of the trial in the 'persistent' simulation condition. On a trial on which no recall stimulus was applied to S1, spontaneous firing continued until the end of the trial in the 'spontaneous' simulation condition. Bottom: Unstable operation. On this persistent condition trial, the firing decreased during the trial as the network fell out of the attractor because of the statistical fluctuations caused by the spiking dynamics. On the spontaneous condition trial, the firing increased during the trial because of the statistical fluctuations. In these simulations the network parameter was w_+=2.1. (Data from Rolls E T, Loh M and Deco G. 2008 An attractor hypothesis of obsessive-compulsive disorder. European Journal of Neuroscience 28: 782–793.)

compulsive disorder. I now introduce concepts on how noise produced by neuronal spiking, or noise from other sources, influences a single attractor network used for short-term memory, with illustrations drawn from Loh, Rolls and Deco (2007a) and Rolls, Loh and Deco (2008c).

In the spontaneous state simulations, no memory retrieval cue was applied, and we are interested in whether the network remains stably in the spontaneous firing state, or whether it is unstable and on some trials due to statistical fluctuations entered one of the attractor memory states, thus falsely retrieving a memory. Figure 7.1 (top) shows an example of a trial on which the network correctly stayed in the low spontaneous firing rate regime, and (bottom) another trial (labelled spontaneous unstable) in which statistical spiking-related fluctuations in the network caused it to enter a high activity state, moving into one of the attractors even without a stimulus.

In the persistent state simulations (in which the short-term memory was implemented by the continuing neuronal firing), a strong excitatory input was given to the S1 neuronal population between 0 and 500 ms. Two such trials are shown in Fig. 7.1. In Fig. 7.1 (top), the S1 neurons (correctly) keep firing at approximately 30 Hz after the retrieval cue is removed at 500 ms. However, due to statistical fluctuations in the network related to the spiking activity, on the trial labelled persistent unstable the high firing rate in the attractor for S1 was not stable, and the firing decreased back towards the spontaneous level, in the example shown starting after 1.5 s (Fig. 7.1 bottom). This trial illustrates a failure to maintain a stable short-term memory state.

These attractor network simulations (Loh et al., 2007a; Rolls et al., 2008c) illustrate how noise in the network can lead to the noisy (stochastic) memory recall, and the same applied to decision-making networks (Section 3.12.)

7.2.3 Stochastic dynamics in decision-making, and the evolutionary utility of probabilistic choice

The next stage in this research was to consider what the advantages might be of the stochastic dynamics of decision-making and memory recall (see further Rolls and Deco (2010); Rolls (2023c, 2026d)).

Probabilistic decision-making can be evolutionarily advantageous in that sometimes taking a decision that is not optimal based on previous history may provide information that is useful, and that may contribute to learning. By making the less favored choice sometimes, the organism can keep obtaining evidence on whether the environment is changing (for example on whether the probability of a reward for another choice has increased), and by doing this approximately according to the matching law minimizes the cost of the disadvantageous choices in obtaining information about the environment. The matching law is that the proportion of choices made approximately matches the outcome reward values that are available.

This probabilistic exploration of the environment is very important in trial-and-error learning, and indeed has been incorporated into a simple reinforcement algorithm in which noise is added to the system, and if this improves outcomes above the expected value, then changes are made to the synaptic weights in the correct direction (in the associative reward-penalty algorithm) (Sutton and Barto, 1981; Barto, 1985; Sutton and Barto, 2018; Rolls, 2016b, 2026d).

In perceptual learning, probabilistic exploratory behavior may be part of the mechanism by which perceptual representations can be shaped to have appropriate selectivities for the behavioral categorization being performed (Sigala and Logothetis, 2002; Szabo et al., 2006).

Another example is in food foraging, which probabilistically may reflect the outcomes (Krebs and Davies, 1991; Kacelnik and Brito e Abreu, 1998; Rolls et al., 2024a), and is a way to keep sampling the environment and exploring the space of possible choices. Foraging can be performed optimally in terms of the costs and benefits of staying, or of moving on to forage elsewhere. Some noise in the system may help the amount of foraging to be optimally adjusted.

Another sense in which probabilistic decision-making may be evolutionarily advantageous is with respect to detecting signals that are close to threshold in the process termed stochastic resonance (Rolls and Deco, 2010; Deco, Rolls, Albantakis and Romo, 2013).

Another area where the spiking-related noise in the decision-making process may be evolutionarily advantageous is in the generation of unpredictable behavior, which can be advantageous in a number of situations, for example when a prey is trying to escape from a predator, and perhaps in some social and economic situations in which organisms may not wish to reveal their intentions (Maynard Smith, 1982, 1984; Dawkins, 1995). Intrinsic indeterminacy may be essential for unpredictable behavior (Glimcher, 2005). For example, in interactive games like matching pennies or rock–paper–scissors, any trend that deviates from random choice by an agent could be exploited to his or her opponent's advantage.

Another computation in which stochastic fluctuations may influence human behavior is when choices are being made between the emotional system in the brain, and the rational system in the brain (Section 3.11.5; and Rolls (2025e, 2026d).

Another advantage of stochastic memory recall is that jumping slightly noisily in a semantic space may help generate new ideas, and indeed we discovered that the variability of functional connectivity over time, which may reflect noise in the brain, is correlated with human creativity (Sun, Liu, Rolls, Chen, Yao, Yang, Wei, Zhang, Zhang, Feng and Qiu, 2019).

Thus, far from being always a problem with neural computation, some noise in the brain, stochastic fluctuations, driven by the almost random timing of neuronal spikes for a given mean firing rate, can have some very advantageous effects, I have argued (Rolls, 2023c, 2026d).

7.3 Attractor dynamics and schizophrenia

7.3.1 Introduction

Schizophrenia is a major mental illness, which has a great impact on patients and their environment. We have developed a top-down approach to understanding schizophrenia based on the different types of symptoms, and relate them to instabilities in attractor neural networks (Rolls, 2005; Loh, Rolls and Deco, 2007a,b; Rolls, Loh, Deco and Winterer, 2008d; Rolls, 2016b; Deco, Rolls and Romo, 2009; Rolls and Deco, 2011b; Rolls, 2012a, 2014a, 2016b, 2021f, 2023c, 2026d, 2024b).

In our approach, we relate the three types of symptoms of schizophrenia to the dynamical systems attractor framework described earlier as follows (Loh, Rolls and Deco, 2007a,b; Rolls, Loh, Deco and Winterer, 2008d; Rolls, 2016b; Deco, Rolls and Romo, 2009; Rolls and Deco, 2011b; Rolls, 2012a, 2014a, 2023c, 2026d).

The **cognitive symptoms** of schizophrenia include distractibility, poor attention, and the dysexecutive syndrome (Liddle, 1987; Green, 1996; Mueser and McGurk, 2004; Owen et al., 2016). The core of the cognitive symptoms is a working memory deficit in which there is a difficulty in maintaining items in short-term memory (Goldman-Rakic, 1994, 1999; Glausier and Lewis, 2018; Dienel, Schoonover and Lewis, 2022; Smucny, Dienel, Lewis and Carter, 2022), which could directly or indirectly account for a wide range of the cognitive symptoms. We propose that these symptoms may be related to instabilities of persistent states in attractor neural networks, consistent with the body of theoretical research on network models of working memory (Durstewitz, Seamans and Sejnowski, 2000b). The neurons are firing at a lower rate, leading to shallower basins of attraction of the persistent states, and thus a difficulty in maintaining a stable short-term memory, normally the source of the bias in biased competition models of attention (Rolls and Deco, 2002; Deco and Rolls, 2005a; Rolls, 2008a, 2016b). The shallower basins of attraction would thus result in working memory deficits, poor attention, distractibility, and problems with executive function and action selection (Deco and Rolls, 2003, 2005a; Rolls, 2016b, 2023c, 2026d).

The **negative symptoms** refer to the flattening of affect and a reduction in emotion. Behavioral indicators are blunted affect, emotional and passive withdrawal, poor rapport, lack of spontaneity, motor retardation, and disturbance of volition (Liddle, 1987; Mueser and McGurk, 2004; Owen et al., 2016). The negative symptoms are a major source of difference between schizophrenic patients, and show a continuous gradation in magnitude between different people with schizophrenia (Rolls, Lu, Wan, Yan, Wang, Yang, Tan, Li, Chinese Schizophrenia Collaboration, Yu, Liddle, Palaniyappan, Zhang, Yue and Feng, 2017). Moreover, we have found in a large-scale study that the negative symptoms, as well as the positive and general symptoms, are reduced by treatment with antipsychotic drugs (Rolls et al., 2017). We propose that the negative symptoms are related to decreases in neuronal firing rates in the orbitofrontal cortex and/or anterior cingulate cortex (Rolls, 2014a, 2016b, 2023c, 2026d), where neuronal firing rates and activations in fMRI investigations are correlated with reward value and pleasure (Rolls, 2014a, 2006, 2007d; Rolls and Grabenhorst, 2008; Rolls, 2008e, 2019c,e, 2023c, 2026d) (Chapter 3). Consistent with this, imaging studies have identified a relationship between negative symptoms and prefrontal hypometabolism, i.e. a reduced activation of frontal areas (Wolkin et al., 1992; Aleman and Kahn, 2005), and there are reduced frontal and anterior cingulate cortex glutamate excitatory transmitter levels in schizophrenia (Howes et al., 2024).

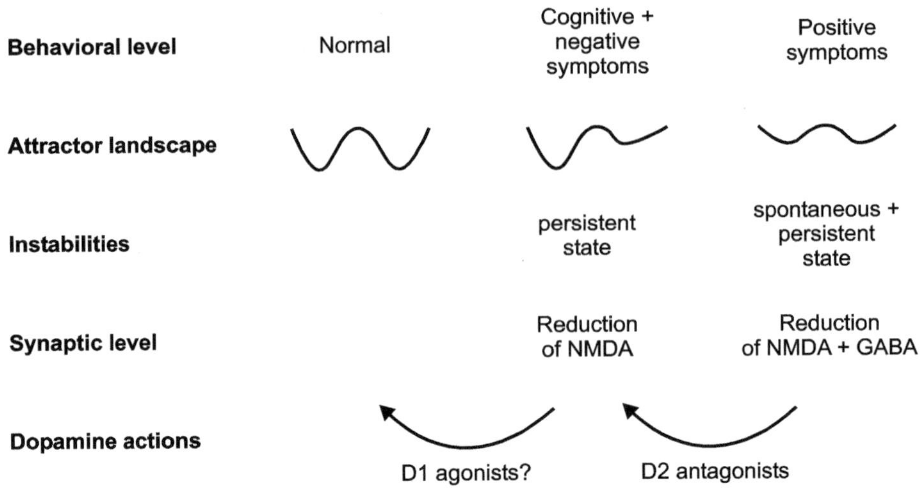

Fig. 7.2 Summary of attractor hypothesis of schizophrenic symptoms and simulation results (see text). The first basin (from the left) in each energy landscape is the spontaneous state, and the second basin is the persistent attractor state. The vertical axis of each landscape is the energy potential. (Data from Loh M, Rolls E T and Deco G. 2007 A dynamical systems hypothesis of schizophrenia. PLoS Computational Biology 3: e228.)

The **positive symptoms** of schizophrenia include bizarre (psychotic) trains of thoughts, hallucinations, and (paranoid) delusions (Liddle, 1987; Mueser and McGurk, 2004; Owen, Sawa and Mortensen, 2016; Rolls, Lu, Wan, Yan, Wang, Yang, Tan, Li, Chinese Schizophrenia Collaboration, Yu, Liddle, Palaniyappan, Zhang, Yue and Feng, 2017). We propose that these symptoms are related to shallow basins of attraction of both the spontaneous and persistent states in the temporal lobe semantic memory networks and to the statistical fluctuations caused by the probabilistic spiking of the neurons. This could result in activations arising spontaneously, and thoughts moving too freely round the energy landscape, loosely from thought to weakly associated thought, leading to bizarre thoughts and associations, which may eventually over time be associated together in semantic memory to lead to false beliefs and delusions. Consistent with this, neuroimaging studies suggest higher activation especially in areas of the temporal lobe (Weiss and Heckers, 1999; Shergill et al., 2000; Scheuerecker et al., 2008).

7.3.2 A model of attractor dynamics in schizophrenia

To further investigate our hypotheses, we performed simulations with the integrate-and-fire attractor network which includes channels activated by AMPA, NMDA and $GABA_A$ receptors modified from Brunel and Wang (2001). The main findings were as follows, and are summarized in Fig. 7.2 (Loh, Rolls and Deco, 2007a,b; Rolls, Loh, Deco and Winterer, 2008d).

A reduction of the NMDA receptor activated synaptic conductances (−NMDA) reduces the stability of the persistent state drastically. We hypothesized that this type of change might be related to the cognitive symptoms, since it shows a reduced stability of the working memory properties. The network was also more sensitive to a weak external stimulus, and we related this to increased distractibility. Consistent with this, there is a reduced spine density on excitatory neurons in the prefrontal cortex in schizophrenia (Glausier and Lewis, 2013; Konopaske et al., 2014; Glantz and Lewis, 2000; Dienel et al., 2022). If present also in the orbitofrontal and anterior cingulate cortex, this could reduce the firing rates of these cortical neurons, and thus produce the negative symptoms of schizophrenia. This simulation relates directly to the glutamate hypothesis of schizophrenia, and

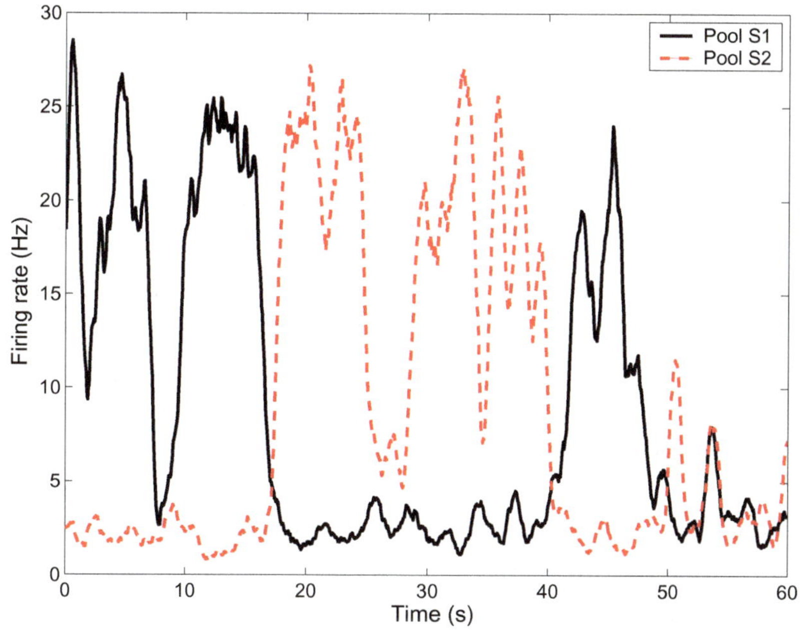

Fig. 7.3 Wandering between attractor states by virtue of statistical fluctuations caused by the randomness of the spiking activity. We simulated a single long trial (60 s) in the spontaneous test condition for the synaptic modification (−NMDA, −GABA). The two curves show the activity of the two selective pools over time smoothed with a 1 s sliding averaging window. The activity moves noisily between the attractor for the spontaneous state and the two persistent states S1 and S2. (Data from Loh M, Rolls E T and Deco G. 2007 A dynamical systems hypothesis of schizophrenia. PLoS Computational Biology 3: e228.)

particularly to hypofunction of the NMDA receptor (Uno and Coyle, 2019; Coyle, Ruzicka and Balu, 2020).

When both NMDA and GABA synaptic conductances are reduced, the stability of the high firing rate attractor state was reduced, and so was the stability of the spontaneous low firing rate state. We relate this pattern to the positive symptoms of schizophrenia, in which both the spontaneous and attractor states are shallow, and the system by the influence of statistical fluctuations wanders between the different (spontaneous and high firing rate attractor) states, as illustrated in Fig. 7.3.

The positive symptoms (Fig. 7.2, right column) of schizophrenia include delusions, hallucinations, thought disorder, and bizarre behavior. Examples of delusions are beliefs that others are trying to harm the person, impressions that others control the person's thoughts, and delusions of grandeur. Hallucinations are perceptual experiences, which are not shared by others, and are frequently auditory but can affect any sensory modality. These symptoms may be related to activity in the temporal lobes (Liddle, 1987; Epstein et al., 1999; Mueser and McGurk, 2004). The attractor framework approach taken here hypothesizes that the basins of attraction of both spontaneous and persistent states are shallow (Fig. 7.2). Due to the shallowness of the spontaneous state, the system can jump spontaneously up to a high activity state causing hallucinations to arise and leading to bizarre thoughts and associations. This might be the cause for the higher activations in schizophrenics in temporal lobe areas which are identified in neuroimaging experiments (Shergill et al., 2000; Scheuerecker et al., 2008).

We relate the positive symptoms to not only a reduction in NMDA conductance (cf. Coyle, Ruzicka and Balu (2020)), but especially also to a reduction in what we modelled as GABA conductance, but more generally to any reduction of inhibitory neuron efficacy that might lead to higher firing rates of cortical excitatory neurons (Loh, Rolls and Deco, 2007a,b; Rolls, Loh, Deco and Winterer, 2008d). This is consistent with the fact that the positive symptoms usually follow after the cognitive and negative symptoms and represent a qualitative worsening of the illness (Mueser and McGurk, 2004). Alterations in GABA receptors have been identified in schizophrenia (Lewis, 2014; Gonzalez-Burgos et al., 2010; Gonzalez-Burgos and Lewis, 2012; Glausier and Lewis, 2018), and the hippocampus has increased metabolism and activity in schizophrenia (Wegrzyn, Juckel and Faissner, 2022).

The evidence on GABA-mediated inhibition impairments in schizophrenia, and also of decreased spine density that would reduce excitatory transmission (Lewis, 2014; Gonzalez-Burgos et al., 2010; Gonzalez-Burgos and Lewis, 2012; MacDonald et al., 2017; Hoftman et al., 2017; Glausier and Lewis, 2018; Rolls, 2021f; Dienel et al., 2022), is an indication that the stability of cortical attractor networks is likely to be impaired in schizophrenia. The models described here have shown some of the effects that would be produced by altered levels of excitatory and inhibitory transmission on the stability of cortical circuitry, and how this might influence processes such as working memory and attention, and produce some of the symptoms of schizophrenia.

7.3.3 Reduced functional connectivity and stability of some cortical regions in schizophrenia

To test these hypotheses that some networks in the brain are less stable in schizophrenia, we moved ahead to measure whether the functional connectivity between some brain regions is lower in schizophrenia, and if so, what the implications for stability might be. As specified earlier, functional connectivity can be measured by the Pearson correlation between the BOLD signal for each pair of brain regions over a time period of several minutes. A higher correlation is interpreted as showing that the nodes (the brain regions) are more strongly connected, in that they are influencing each other's BOLD signals, or have a common input.

In one such investigation that accordingly we performed, the functional connectivity in a group of 123 patients with chronic schizophrenia compared to 136 matched healthy controls is shown in Fig. 7.4B (Rolls, Cheng and Feng, 2021a). The matrix shows the functional connectivity differences for pairs of brain areas from the automated anatomical labelling atlas 3 (AAL3) (Rolls, Huang, Lin, Feng and Joliot, 2020d). First, it is evident that many of the functional connectivities are significantly lower in schizophrenia. This is consistent with the hypothesis that the level of excitation between cortical areas is lower in schizophrenia, which is equivalent in the simulations described above to a reduction in the NMDA synaptic conductances. This is consistent with the disconnectivity hypothesis of schizophrenia (Friston and Frith, 1995).

Moving beyond the disconnectivity hypothesis, we hypothesized that the reduced functional connectivities evident in Fig. 7.4B might lead us to expect that there might be signs of less stability in the BOLD signal in schizophrenia, in that any basins of attraction might be weaker. This was shown to be the case, in that the temporal variability of the functional connectivities of many of the brain regions was higher in schizophrenia, as shown in Fig. 7.4A[6]. The higher temporal variability was

[6]The temporal variability measures how much the functional connectivity of a brain region k with other brain regions changes across time. The temporal variability of a given brain region was obtained as follows. We first segmented all BOLD signals into n non-overlapping windows with length l. The whole-brain FC network F_i (an $m * m$ matrix, with $m = 132$ nodes) in the ith time window was then constructed, with the Pearson correlation being the measure of Functional Connectivity. The Functional Connectivity profile of region k at time window i is denoted by $F_i(k, :)$ (shortened as $F_{i,k}$, which is an m-dimensional vector that represents all the functional connections of region k. The variability of a region k is defined as:

$$V_k = 1 - (\text{corrcoef}(F_{(i,k)}, F_{(j,k)})) \qquad i, j = 1, 2, 3, ..., n \quad i \neq j \tag{7.1}$$

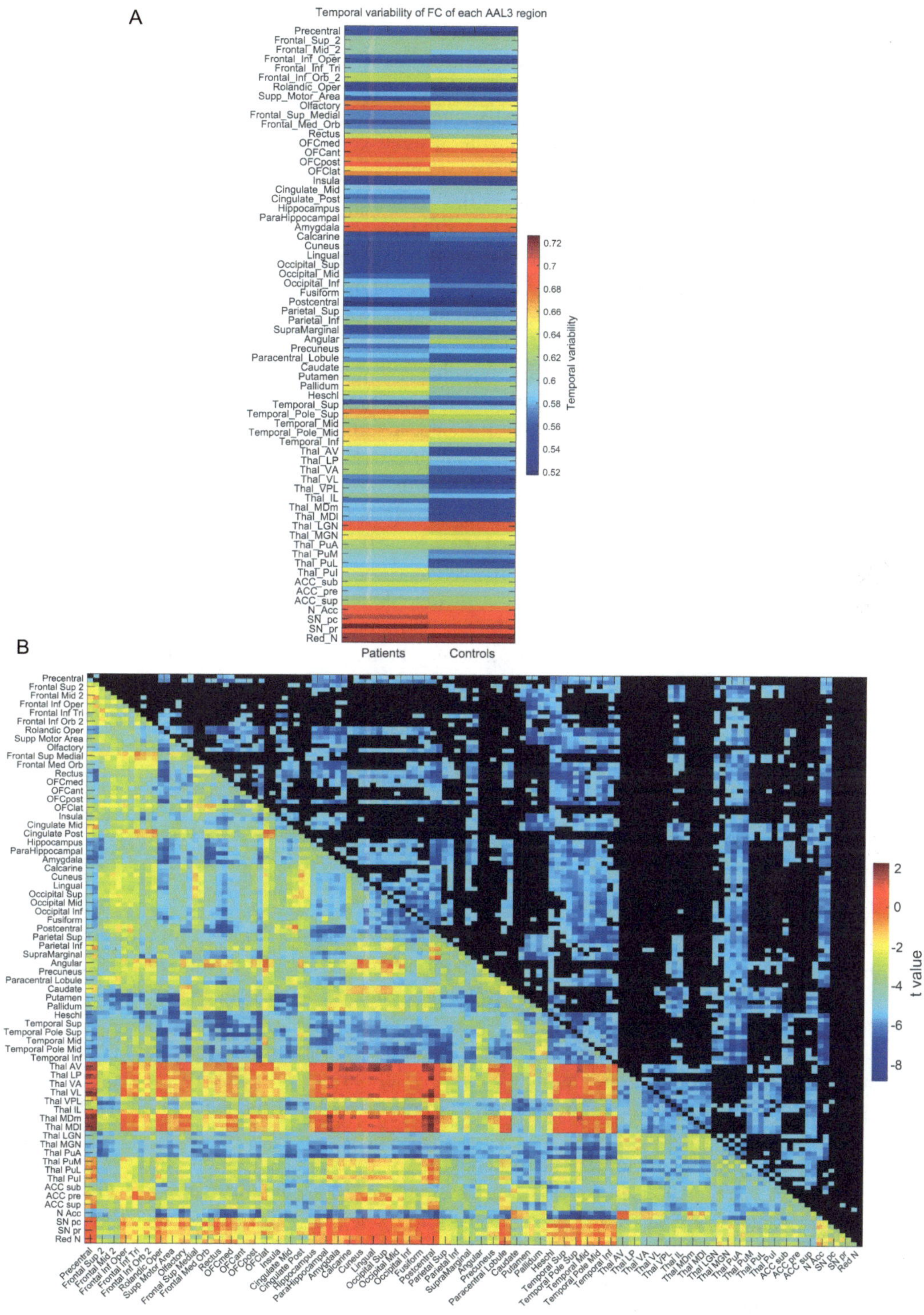

Fig. 7.4 A) The temporal variability of different AAL3 (automated anatomical labelling atlas 3 (Rolls et al., 2020d)) regions in the chronic schizophrenic and control groups. B) The functional connectivity of AAL3 regions for the chronic schizophrenic group minus controls. The lower left shows the t value for the difference in functional connectivity of patients - controls; the upper right shows the significance after Bonferroni correction. (Data from Rolls E T, Cheng W and Feng J. 2021 Brain dynamics: the temporal variability of connectivity, and differences in schizophrenia and ADHD. Translational Psychiatry 11: 70.)

especially clear for some early visual cortical areas (Inferior Occipital and Fusiform), the temporal lobe areas connected to these, and the orbitofrontal cortex. This is an indication of increased instability of these brain regions in schizophrenia (Rolls, Cheng and Feng, 2021a,b).

Very interestingly, this higher temporal variability reflecting instability of some early visual cortical areas, the temporal lobe areas connected to these, and the orbitofrontal cortex could be related to lower functional connectivities of especially these areas, as shown in Fig. 7.4B. Especially interesting was that the functional connectivities of the thalamic sensory visual relay, the lateral geniculate nucleus, and the thalamic sensory auditory relay, the medial geniculate nucleus, were lower in schizophrenia (Fig. 7.4B). This was in interesting contrast to the association thalamic nuclei, which had increased functional connectivity in schizophrenia. This finding was cross-validated in a different set of patients with first-episode schizophrenia, who had similar though somewhat smaller differences from controls (Rolls et al., 2021a).

These findings are consistent with the hypothesis that a factor in schizophrenia is a reduction in the connectivity and therefore excitability of some brain regions, which destabilizes attractor networks in these regions because the firing rates are insufficient to maintain the networks in a high firing rate state. In particular, we propose that in schizophrenia these differences bias processing away from external visual and auditory inputs, and towards internal cognitive processing in associative cortical areas such as the prefrontal and temporal cortical areas. We relate this to the tendency for people with schizophrenia to be disconnected from the world, and to be unable to maintain attention (Rolls, Cheng and Feng, 2021a).

Schizophrenia remains a difficult mental state to understand and treat, but the approach described here does provide an approach to understanding how the stochastic neurodynamics of cortical function may be different in schizophrenia, and how that may be helpful in better understanding and treating schizophrenia with its range of symptoms which this approach may help to understand better (see also Coyle et al. (2020); Dienel et al. (2022); Rolls (2024b); Rolls et al. (2017); Rolls (2021f); Rolls et al. (2020c); Scott and Tamminga (2018); Li et al. (2017); Pu et al. (2014); Rolls (2026d)).

7.4 Attractor dynamics and obsessive-compulsive disorder

7.4.1 Introduction

Obsessive-compulsive disorder (OCD) is a chronically debilitating disorder with a lifetime prevalence of 2–3% (Karno et al., 1988; Weissman et al., 1994; Pauls et al., 2014; Stein et al., 2016; Robbins et al., 2019; Fineberg and Robbins, 2021; Robbins et al., 2024). It is characterized by two sets of symptoms, obsessive and compulsive.

Obsessions are unwanted, intrusive, recurrent thoughts or impulses that are often concerned with themes of contamination and 'germs', checking household items in case of fire or burglary, order and symmetry of objects, or fears of harming oneself or others (Chamberlain, Solly, Hook, Vaghi and Robbins, 2021).

Compulsions are ritualistic, repetitive behaviors or mental acts carried out in relation to these obsessions e.g. washing, household safety checks, counting, rearrangement of objects in symmetrical arrays or constant checking of oneself and others to ensure no harm has occurred (Menzies et al., 2008; Stein et al., 2016; Robbins et al., 2019).

Patients with OCD experience the persistent intrusion of thoughts that they generally perceive as foreign and irrational but which cannot be dismissed. The anxiety associated with these unwanted and disturbing thoughts can be extremely intense; it is often described as a feeling that something is incomplete or wrong, or that terrible consequences will ensue if specific actions are not taken. Many patients engage in repetitive, compulsive behaviors that aim to discharge the anxieties associated with these obsessional thoughts. Severely affected patients can spend many hours each day in their obsessional thinking and resultant compulsive behaviors, leading to marked disability (Pittenger et al., 2006; Stein et al., 2016).

While patients with OCD and related disorders exhibit a wide variety of obsessions and compulsions, the symptoms tend to fall into specific clusters. Common patterns include obsessions of contamination, with accompanying cleaning compulsions; obsessions with symmetry or order, with accompanying ordering behaviors; obsessions of saving, with accompanying hoarding; somatic obsessions; aggressive obsessions with checking compulsions; and sexual and religious obsessions (Pittenger et al., 2006; Stein et al., 2016; van den Heuvel et al., 2016; Grant and Chamberlain, 2019; Robbins et al., 2019; Fineberg and Robbins, 2021). One approach to understanding compulsive behavior focuses on fronto-striatal systems implicated in goal-directed behavior and habits (Robbins et al., 2024).

In this section I describe a theory that we developed of how obsessive-compulsive disorders arise, and of the different symptoms. The theory (Rolls, Loh and Deco, 2008c; Rolls, 2012a) is based on the proposal that there is overstability of attractor neuronal networks in some cortical and related brain areas in obsessive-compulsive disorders.

7.4.2 A hypothesis about the increased stability of attractor networks and the symptoms of obsessive-compulsive disorder

We hypothesized (Rolls, Loh and Deco, 2008c) that cortical and related attractor networks become too stable in obsessive-compulsive disorder, so that once in an attractor state, the networks tend to remain there too long (Rolls, Loh and Deco, 2008c; Rolls, 2012a). The hypothesis is that the depths of the basins of attraction become deeper, and that this is what makes the attractor networks more stable. We further hypothesize that part of the mechanism for the increased depth of the basins of attraction is increased glutamatergic transmission, which increases the depth of the basins of attraction by increasing the firing rates of the neurons, and by increasing the effective value of the synaptic weights between the associatively modified synapses that define the attractor (as is made evident in Equation B.12 in Rolls (2023c)). The synaptic strength is effectively increased if more glutamate is released per action potential at the synapse, or if in other ways the currents injected into the neurons through the NMDA (N-methyl-d-aspartate) and/or AMPA synapses are larger. In addition, if NMDA receptor function is increased, this could also increase the stability of the system because of the temporal smoothing effect of the long time constant of the NMDA receptors (Wang, 1999).

This increased stability of cortical and related attractor networks, and the associated higher neuronal firing rates, could occur in different brain regions, and thereby produce different symptoms, as follows.

If these effects occurred in high order motor areas, the symptoms could include inability to move out of one motor pattern, resulting for example in repeated movements or actions. In parts of the cingulate cortex and dorsal medial prefrontal cortex, this could result in difficulty in switching between actions or strategies (Rushworth et al., 2007a,b, 2012; Rolls, 2019e,b), as the system would be locked into one action or strategy. If an action was locked into a high order motor area due to increased stability of an attractor network, then lower order motor areas might thereby not be able to escape easily what they implement, such as a sequence of movements, so that the sequence would be repeated.

A similar account, of becoming locked in one action and having difficulty in switching to another action, can be provided for response inhibition deficits, which have been found in OCD.

The response inhibition deficit has been found in tasks such as go/no-go and stop-signal reaction time (SSRT) which examine motor inhibitory processes, and also the Stroop task, a putative test of cognitive inhibition (Hartston and Swerdlow, 1999; Bannon et al., 2002; Penades et al., 2005; Bannon et al., 2006; Chamberlain et al., 2006, 2007; Penades et al., 2007; Robbins et al., 2019). For example, response inhibition deficits have been reported in OCD patients when performing the SSRT, which measures the time taken to internally suppress pre-potent motor responses (Chamberlain et al., 2006). Unaffected first-degree relatives of OCD patients are also impaired on this task compared with unrelated healthy controls, suggesting that response inhibition may be an endophenotype (or intermediate phenotype) for OCD (Chamberlain et al., 2007; Menzies et al., 2008).

If occurring in the lateral prefrontal cortex (including the dorsolateral and ventrolateral parts), the increased stability of attractor networks could produce symptoms that include a difficulty in shifting attention and in cognitive set shifting. These are in fact important symptoms that can be found in obsessive-compulsive disorder (Menzies, Chamberlain, Laird, Thelen, Sahakian and Bullmore, 2008; Chamberlain, Solly, Hook, Vaghi and Robbins, 2021). Two different forms of shifting have been investigated: affective set shifting, where the affective or reward value of a stimulus changes over time (e.g. a rewarded stimulus is no longer rewarded) (intradimensional or ID set shifting); and attentional set shifting, where the stimulus dimension (e.g. shapes or colors) to which the subject must attend is changed (extradimensional or ED set shifting). Deficits of attentional set shifting in OCD have been found in several neurocognitive studies using the CANTAB ID/ED set shifting task (Veale, Sahakian, Owen and Marks, 1996; Watkins, Sahakian, Robertson, Veale, Rogers, Pickard, Aitken and Robbins, 2005; Chamberlain, Fineberg, Blackwell, Robbins and Sahakian, 2006; Chamberlain, Fineberg, Menzies, Blackwell, Bullmore, Robbins and Sahakian, 2007; Robbins, Vaghi and Banca, 2019; Chamberlain, Solly, Hook, Vaghi and Robbins, 2021). This deficit is most consistently reported at the ED stage (in which the stimulus dimension, e.g. shape, color or number, alters and subjects have to inhibit their attention to this dimension and attend to a new, previously irrelevant dimension). The ED stage is analogous to the stage in the Wisconsin Card Sorting Task where a previously correct rule for card sorting is changed and the subject has to respond to the new rule (Berg, 1948; Chamberlain et al., 2021). This ED shift impairment in OCD patients is considered to reflect a lack of cognitive or attentional flexibility and may be related to the repetitive nature of OCD symptoms and behaviors(Chamberlain et al., 2021). Deficits in attentional set shifting are considered to be more dependent upon dorsolateral and ventrolateral prefrontal regions than the orbital prefrontal regions included in the orbitofronto-striatal model of OCD (Pantelis et al., 1999; Rogers et al., 2000; Nagahama et al., 2001; Hampshire and Owen, 2006; Robbins et al., 2019; Chamberlain et al., 2021), suggesting that cognitive deficits in OCD may not be underpinned exclusively by orbitofrontal cortex pathology. Indeed, intradimensional or affective set shifting may not be consistently impaired in OCD (Menzies et al., 2008; Chamberlain et al., 2021).

Planning may also be impaired in patients with OCD (Menzies et al., 2008), and this could arise because there is too much stability of attractor networks in the dorsolateral prefrontal cortex concerned with holding in mind the different short-term memory representations that encode the different steps of a plan (Rolls, 2016b, 2023c, 2026d). Indeed, there is evidence for dorsolateral prefrontal cortex (DLPFC) dysfunction in patients with OCD, in conjunction with impairment on a version of the Tower of London, a task often used to probe planning aspects of executive function (van den Heuvel et al., 2005). Impairment on the Tower of London task has also been demonstrated in healthy first-degree relatives of OCD patients (Delorme et al., 2007).

An increased firing rate of neurons in the orbitofrontal cortex, and anterior cingulate cortex, produced by hyperactivity of glutamatergic transmitter systems, would increase emotionality, which is frequently found in obsessive-compulsive disorder. Part of the increased anxiety found in obsessive-compulsive disorder could be related to an inability to complete tasks or actions in which one is locked.

But part of our unifying proposal is that part of the increased emotionality in OCD may be directly related to increased firing produced by the increased glutamatergic activity in brain areas such as the orbitofrontal and anterior cingulate cortex (Rolls et al., 2008c; Rolls, 2012a, 2024b) (cf. Robbins et al. (2024)). The orbitofrontal cortex and anterior cingulate cortex are involved in emotion, in that they are activated by primary and secondary reinforcers that produce affective states (Rolls, 2004b, 2014a, 2016b; Rolls and Grabenhorst, 2008; Rolls, 2019c,e, 2023b), and in that damage to these regions alters emotional behavior and emotional experience (Rolls et al., 1994a; Hornak et al., 1996, 2003, 2004; Berlin et al., 2004, 2005; Rolls, 2019c, 2021d) (Chapter 3). Indeed, negative emotions as well as positive emotions activate the orbitofrontal cortex, with the emotional states produced by negative events tending to be represented in the lateral orbitofrontal cortex and dorsal part of the anterior cingulate cortex (Kringelbach and Rolls, 2004; Rolls, 2014a, 2016b, 2009; Grabenhorst and Rolls, 2011; Rolls, 2019c; Rolls et al., 2020b; Zhang et al., 2024a; Rolls et al., 2023d) (Chapter 3). We may note that stimulus-reinforcer reversal tasks (also known as intra-dimensional shifts or affective reversal) are not generally impaired in patients with OCD (Menzies et al., 2008), and this is as predicted, for the machinery for the reversal including the detection of non-reward (Rolls, 2014a, 2019c) (Chapter 3) is present even if the system is hyperglutamatergic.

If the increased stability of attractor networks occurred in temporal lobe semantic memory networks, then this would result in a difficulty in moving from one thought to another, and possibly in stereotyped thoughts, which again may be a symptom of obsessive-compulsive disorder (Menzies et al., 2008; Robbins et al., 2019; Fineberg and Robbins, 2021; Robbins et al., 2024). The obsessional states are thus proposed to arise because cortical areas concerned with cognitive functions have states that become too stable. The compulsive states are proposed to arise partly in response to the obsessional states, but also partly because cortical areas concerned with actions have states that become too stable.

The theory provides a unifying computational account of both the obsessional and compulsive symptoms, in that both arise due to increased stability of cortical attractor networks, with the different symptoms related to overstability in different cortical areas (Rolls et al., 2008c; Rolls, 2012a, 2024b). The theory is also unifying in that a similar increase in glutamatergic activity in the orbitofrontal and anterior cingulate cortex could increase emotionality, as described earlier.

7.4.3 Glutamate and increased depth of the basins of attraction of attractor networks

To demonstrate how alterations of glutamate as a transmitter for the connections between the neurons may influence the stability of attractor networks, we (Rolls, Loh and Deco, 2008c) performed integrate-and-fire simulations of the stability of a short-term memory attractor network of the type described in Section 7.3. A feature of these simulations is that we simulated the currents produced by activation of NMDA and AMPA receptors in the recurrent collateral synapses, and took into account the effects of the spiking-related noise, which is an important factor in determining whether the attractor stays in a basin of attraction, or jumps over an energy barrier into another basin (Loh, Rolls and Deco, 2007a; Rolls, Loh, Deco and Winterer, 2008d; Deco, Rolls and Romo, 2009).

We found that increasing the NMDA receptor-activated synaptic currents by 3% or the AMPA by 10% increased the stability of the high firing rate short-term memory state of the network. Thinking about possible treatments, we showed that increasing the inhibition in the network by increasing the GABA receptor-mediated currents by 10% could ameliorate the overstability produced by increased glutamate-related effects. We also showed that the increased NMDA-receptor mediated effects produced less distractibility of the short-term memory state by another stimulus, which is a further sign of overstability (Rolls, Loh and Deco, 2008c).

7.4.4 Synthesis on obsessive-compulsive disorder

This provides a new approach to the symptoms of obsessive-compulsive disorder, for it deals with the symptoms in terms of overstability of attractor networks in the cerebral cortex (Rolls, Loh and Deco, 2008c; Rolls, 2012a, 2024b, 2026d). Consistent with this approach, hyperactivity of the lateral orbitofrontal cortex is found during symptom provocation, and this normalized over the course of behavioral therapy for OCD (Robbins et al., 2019). If the same generic change in activity and therefore in stability were produced in different cortical areas, then we have indicated how different symptoms might arise. Of course, if these changes were more evident in some areas than in others in different patients, this would help to account for the different symptoms in different patients. Having proposed a generic hypothesis for the disorder, we recognize of course that the exact symptoms that arise if stability in some systems is increased will be subject to the exact effects that these will have in an individual patient, who may react to these effects, and produce explanatory accounts for the effects, and ways to deal with them, that may be quite different from individual to individual.

This simulation evidence, that an increase of glutamatergic synaptic efficacy can increase the stability of attractor networks and thus potentially provide an account for some of the symptoms of obsessive-compulsive disorder, is consistent with evidence that glutamatergic function may be increased in some brain systems in obsessive-compulsive disorder (Rosenberg et al., 2000, 2001, 2004; Pittenger et al., 2006; Rolls, 2012a; Pittenger et al., 2011; Pittenger, 2015; Karthik et al., 2020). Consistent with this, agents with antiglutamatergic activity such as memantine and riluzole, which can decrease glutamate transmitter release, may be efficacious in obsessive-compulsive disorder (Pittenger et al., 2006; Bhattacharyya and Chakraborty, 2007; Pittenger et al., 2011; Pittenger, 2015, 2021).

Further evidence for a link between glutamate as a neurotransmitter and OCD comes from genetic studies. There is evidence for a significant association between the SLC glutamate transporter genes and OCD (Stewart et al., 2007; Pauls et al., 2014). These transporters are crucial in terminating the action of glutamate as an excitatory neurotransmitter and in maintaining extracellular glutamate concentrations within a normal range (Bhattacharyya and Chakraborty, 2007; Pauls et al., 2014). In addition, it has been postulated that N-methyl-d-aspartate (NMDA) receptors are involved in OCD, and specifically that polymorphisms in the 3′ untranslated region of GRIN2B (glutamate receptor, ionotropic, N-methyl-d-aspartate 2B) were associated with OCD in affected families (Arnold et al., 2004; Pittenger et al., 2011), and more recent evidence has also found some association to glutamate-related genes (Pauls et al., 2014). However, genetic studies of OCD have not yet produced robust conclusions (Fernandez, Leckman and Pittenger, 2018; Pittenger, 2021).

This research (Rolls, Loh and Deco, 2008c; Rolls, 2012a, 2024b) shows how the approach that we have adopted to develop computational neuroscience approaches to help understand the brain in health and disease (Rolls, 2026d) has direct application to help understand and potentially treat obsessive-compulsive disorder .

7.5 Depression and attractor dynamics

The stochastic neurodynamics approach that I have developed for understanding and treating depression involves increased sensitivity of the lateral orbitofrontal cortex to non-reward and increased functional connectivity with some other cortical regions; and reduced sensitivity and connectivity of the reward-related medial orbitofrontal cortex, both of which influence the stability of different attractor networks and that can account for different symptoms (Rolls, 2016d; Cheng et al., 2016; Rolls, 2017a, 2018b; Rolls et al., 2020b; Xie et al., 2021a; Zhang et al., 2024a). This approach to understanding depression is described in Section 3.13.

7.6 Attractor stochastic dynamics, aging, and memory

In Section 7.3 I considered how some cognitive symptoms such as poor short-term memory and attention could arise due to reduced depth in the basins of attraction of prefrontal cortical networks, and the effects of noise. The hypothesis is that the reduced depth in the basins of attraction would make short-term memory unstable, so that sometimes the continuing firing of neurons that implement short-term memory would cease, and the system under the influence of noise would fall back out of the short-term memory state into spontaneous firing.

Given that top-down attention requires a short-term memory to hold the object of attention in mind, and that this is the source of the top-down attentional bias that influences competition in other networks that are receiving incoming signals, then disruption of short-term memory is also predicted to impair the stability of attention.

These ideas are elaborated in Section 7.3, where the reduced depth of the basins of attraction in schizophrenia is related to down-regulation of NMDA receptors, or to factors that influence NMDA receptor generated ion channel currents such as dopamine D1 receptors.

Could similar processes, in which the stochasticity of the dynamics is increased because of a reduced depth in the basins of attraction, contribute to the changes in short-term memory and attention that are common in aging? Reduced short-term memory and less good attention are common in aging, as are impairments in episodic memory (Grady, 2008; Dahan et al., 2020). What changes in aging might contribute to a reduced depth in the basins of attraction?

Computational approaches to the memory problems in aging are considered next (Rolls and Deco, 2015a; Rolls, 2019f). Some of the changes in the aging brain include impaired synaptic plasticity, and impaired neurogenesis in the dentate gyrus of the hippocampal system (Dahan et al., 2020). The ways in which impaired synaptic plasticity could influence memory are described below in Section 7.6.3. Neurogenesis would if impaired reduce the ability of new items to generate separate patterns of firing in the CA3 neurons of the hippocampal episodic memory system, and thereby impair the formation of new episodic memories (see Chapter 5).

7.6.1 NMDA receptor hypofunction

One factor is that NMDA receptor functionality tends to decrease with aging (Kelly et al., 2006; Temido-Ferreira et al., 2019; Ploux et al., 2020; Orzylowski et al., 2021). This would act, as investigated in detail in Section 7.3, to reduce the depth of the basins of attraction, both by reducing the firing rate of the neurons in the active attractor, and effectively by decreasing the strength of the potentiated synaptic connections that support each attractor as the currents passing through these potentiated synapses would decrease (Rolls, 2023c, 2026d).

An example of the reductions of firing rate in an attractor network produced by even a small downregulation of the NMDA receptor activated ion channel conductances is shown in Fig. 7.5. The effect of these reductions would be to decrease the depth of the basins of attraction, both by reducing the firing rate, and by producing an effect similar to weakened synaptic strengths, as shown in Equation B.12 of Rolls (2026d).

The reduced depth in the basins of attraction could have a number of effects that are relevant to cognitive changes in aging.

First, the stability of short-term memory networks would be impaired, and it might be difficult to hold items in short-term memory for long, as the noise might push the network easily out of its shallow attractor.

Second, top-down attention would be impaired, in two ways. In one way, the short-term memory network holding the object of attention in mind would be less stable, so that the source of the top-

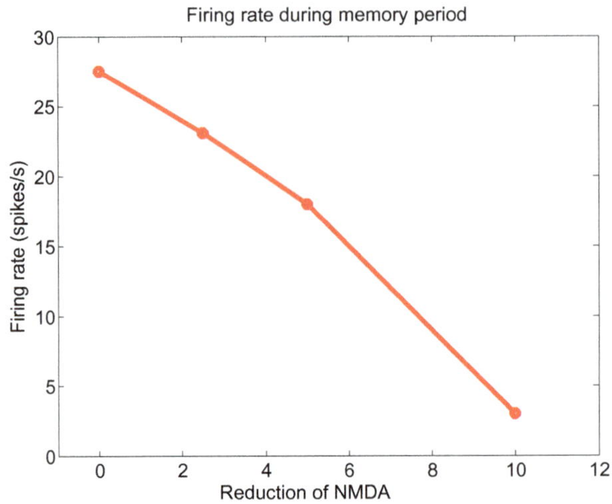

Fig. 7.5 Aging: reductions of NMDA receptor conductances reduce the firing rates (and therefore the stability) of short-term memory networks. Effects of reductions of NMDA receptor conductance on the firing rates of neurons in the short-term memory integrate-and-fire network during the short-term memory period of 4–5 s in the simulations. The firing rates are shown for correct trials only, that is when the spontaneous firing state was stable, and when the firing rate remained high in pool S1 (i.e. greater than 10 spikes/s more than in pool 2) until the end of the trial. (For the reduction of NMDA receptor activated synaptic conductance of 10%, there were no trials on which the memory was correctly maintained during this period, and the rates reflect a value close to the spontaneous firing rate before the memory retrieval cue was applied.) (Data from Rolls E T and Deco G. 2015 Stochastic cortical neurodynamics underlying the memory and cognitive changes in aging. Neurobiology of Learning and Memory 118: 150–161.)

down bias for the biased competition in other cortical areas might disappear. In a second way, and very interestingly, even when the short-term memory for attention is still in its persistent attractor state, it would be less effective as a source of the top-down bias, because the firing rates would be lower, as shown in Fig. 7.5.

Third, the recall of information from episodic memory systems in the temporal lobe (Rolls, 2016b; Dere et al., 2008; Rolls, 2008d; Rolls and Treves, 2024; Rolls et al., 2024d) (Chapter 5) would be impaired. This would arise because the positive feedback from the recurrent collateral synapses that helps the system to fall into a basin of attraction, representing in this case the recalled memory, would be less effective, and so the recall process would be more noisy overall.

Fourth, any reduction of the firing rate of the pyramidal cells caused by NMDA receptor hypofunction (Fig. 7.5) would itself be likely to impair new learning involving LTP.

In addition, if the NMDA receptor hypofunction were expressed not only in the prefrontal cortex where it would affect short-term memory, and in the temporal lobes where it would affect episodic memory, but also in the orbitofrontal cortex, then we would predict some reduction in emotion and motivation with aging, as these functions rely on the orbitofrontal cortex (see Rolls (2019c, 2023c, 2026d) and Chapter 3).

Although NMDA hypofunction may contribute to cognitive effects such as poor short-term memory and attention in aging and in schizophrenia, the two states are clearly very different. Part of the difference lies in the positive symptoms of schizophrenia (the psychotic symptoms, such as thought disorder, delusions, and hallucinations) which may be related to the additional downregulation of GABA in the temporal lobes, which would promote too little stability of the spontaneous firing rate state of temporal lobe attractor networks, so that the networks would have too great a tendency to enter states even in the absence of inputs, and to not be controlled normally by input signals (Loh, Rolls and Deco, 2007a,b; Rolls, Loh, Deco and Winterer, 2008d) (see Section 7.3). However, in relation to the cognitive symptoms of schizophrenia, there has always been the fact that schizophrenia is a condition that often has its onset in the late teens or twenties, and I suggest that there could be a link here to changes in NMDA and related receptor functions that are related to aging. In particular, short-term memory is at its peak when young, and it may be the case that by the late teens or early twenties NMDA and related receptor systems (including dopamine) may be less efficacious than when younger, so that the cognitive symptoms of schizophrenia are more likely to occur at this age than earlier.

Possible ways to minimize the effects of reduction in NMDA receptor efficacy with aging may be caffeine (Temido-Ferreira et al., 2019), and d-serine (Ploux et al., 2020).

7.6.2 Dopamine and norepinephrine

D1 receptor blockade in the prefrontal cortex can impair short-term memory (Sawaguchi and Goldman-Rakic, 1991, 1994; Goldman-Rakic, 1999; Castner et al., 2000; Wang et al., 2019; Cools and Arnsten, 2022). Part of the reason for this may be that D1 receptor blockade can decrease NMDA receptor activated ion channel conductances, among other effects (Seamans and Yang, 2004; Durstewitz et al., 1999, 2000a; Brunel and Wang, 2001; Durstewitz and Seamans, 2002) (see further Section 7.3). Thus part of the role of dopamine in the prefrontal cortex in short-term memory can be accounted for by a decreased depth in the basins of attraction of prefrontal attractor networks (Loh, Rolls and Deco, 2007a,b; Rolls, Loh, Deco and Winterer, 2008d). The decreased depth would be due to both the decreased firing rate of the neurons, and the reduced efficacy of the modified synapses as their ion channels would be less conductive (see Equation B.12 of Rolls (2023c)). The reduced depth of the basins of attraction can be thought of as decreasing the signal-to-noise ratio (Loh, Rolls and Deco, 2007b; Rolls, Loh, Deco and Winterer, 2008d). Given that dopaminergic function in the prefrontal cortex may decline with aging (Sikström, 2007; Cools and Arnsten, 2022), and in conditions in which there are cognitive impairments such as Parkinson's disease, the decrease in dopamine could contribute to the reduced short-term memory and attention in aging.

In attention deficit hyperactivity disorder (ADHD), in which there are attentional deficits including too much distractibility, catecholamine function more generally (dopamine and noradrenaline (i.e. norepinephrine)) may be reduced (Arnsten and Li, 2005), and I suggest that these reductions could produce less stability of short-term memory and thereby attentional states by reducing the depth of the basins of attraction. Indeed, treatment with guanfacine, a 2A-adrenoceptor agonist (i.e. working on receptors activated by norepinephrine) can improve cognitive including working memory functions implemented in the dorsolateral prefrontal cortex (Arnsten and Wang, 2016; Cools and Arnsten, 2022), and this may be helpful in healthy aging.

7.6.3 Impaired synaptic modification

Another factor that may contribute to the cognitive changes in aging is that long-lasting associative synaptic modification as assessed by long-term potentiation (LTP) is more difficult to achieve in older animals and decays more quickly (Barnes, 2003; Burke and Barnes, 2006; Kelly et al., 2006; Dahan et al., 2020; Orzylowski et al., 2021). This would tend to make the synaptic strengths that would support an attractor weaker, and weaken further over the course of time, and thus directly reduce the depth of the attractor basins. This would impact episodic memory, the memory for particular past episodes, such as where one was at breakfast on a particular day, who was present, and what was eaten (Rolls, 2016b, 2008d; Dere et al., 2008). The reduction of synaptic strength over time could also affect short-term memory, which requires that the synapses that support a short-term memory attractor be modified in the first place using LTP, before the attractor is used (Kesner and Rolls, 2001). Part of the mechanism involving the decrease in synaptic plasticity in aging may be a decrease in D-serine, which is a cofactor required for normal operation of the NMDA receptor in synaptic plasticity (Orzylowski et al., 2021).

In view of these changes, boosting glutamatergic transmission is being explored as a means of enhancing cognition and minimizing its decline in aging. Several classes of AMPA receptor potentiators have been described. These molecules bind to allosteric sites on AMPA receptors, slow desensitization, and thereby enhance signalling through the receptors. Some AMPA receptor potentiator agents have been explored in rodent models (Lynch and Gall, 2006; O'Neill and Dix, 2007; de Leon-Lopez et al., 2025). These treatments might increase the depth of the basins of attraction. Agents that activate the glycine or serine modulatory sites on the NMDA receptor (Coyle, 2006; Uno and Coyle, 2019; Ploux et al., 2020) would also be predicted to be useful in healthy aging, and there is some evidence that D-serine administration or agents that influence D-serine can reverse some of the memory problems in aging in humans and in animal models (Orzylowski et al., 2021).

Another factor is that Ca^{2+}-dependent processes affect Ca^{2+} signaling pathways and impair synaptic function in an aging-dependent manner, consistent with the Ca^{2+} hypothesis of brain aging and dementia (Kelly et al., 2006). In particular, an increase in Ca^{2+} conductance can occur in aged neurons, and CA1 pyramidal cells in the aged hippocampus have an increased density of L-type Ca^{2+} channels that might lead to disruptions in Ca^{2+} homeostasis, contributing to the plasticity deficits that occur during aging (Burke and Barnes, 2006; Orzylowski et al., 2021).

7.6.4 Cholinergic function and memory

7.6.4.1 Cholinergic function, memory, and aging

Another factor is acetylcholine. Acetylcholine in the neocortex has its origin largely in the cholinergic neurons in the basal magnocellular forebrain nuclei of Meynert (Mesulam, 1990; Zaborszky et al., 2018, 2008; Galvin et al., 2020). The correlation of clinical dementia ratings with the reductions in a number of cortical cholinergic markers such as choline acetyltransferase, muscarinic and nicotinic acetylcholine receptor binding, as well as levels of acetylcholine, suggested an association of cholinergic hypofunction with cognitive deficits, which led to the formulation of the cholinergic hypothesis of memory dysfunction in senescence and in Alzheimer's disease (Bartus, 2000; Schliebs and Arendt, 2006). Could the cholinergic system alter the function of the cerebral cortex in ways that can be illuminated by stochastic neurodynamics? The dorsolateral prefrontal cortex is a key brain area in working memory (Chapter 13 of Rolls (2026d)), and acetylcholine working mainly through nicotinic receptors is important in enabling neurons to maintain their firing in attractor states to implement working memory (Galvin, Arnsten and Wang, 2020; Cools and Arnsten, 2022). In addition, acetylcholine is important in hippocampal synaptic plasticity and memory (Hasselmo and Giocomo, 2006; Giocomo and Hasselmo, 2007; Hasselmo and Sarter, 2011; Newman et al., 2012; Zaborszky et al., 2018; Rolls, 2021a), as described in Chapter 5.

7.6.4.2 The responses of basal magnocellular forebrain nuclei cholinergic neurons

The cells in the basal magnocellular forebrain nuclei of Meynert lie just lateral to the lateral hypothalamus in the substantia innominata, and extend forward through the preoptic area into the diagonal band of Broca (Mesulam, 1990). These cells, many of which are cholinergic, project directly to the cerebral cortex (Divac, 1975; Kievit and Kuypers, 1975; Mesulam, 1990; Zaborszky et al., 2018; Galvin et al., 2020). These cells provide the major cholinergic input to the cerebral cortex, in that if they are lesioned the cortex is depleted of acetylcholine (Mesulam, 1990). Loss of these cells does occur in Alzheimer's disease, and there is consequently a reduction in cortical acetylcholine in this disease (Mesulam, 1990; Schliebs and Arendt, 2006). This loss of cortical acetylcholine may contribute to the memory loss in Alzheimer's disease, although it may not be the primary factor in the aetiology.

In order to investigate the role of the basal forebrain nuclei in memory, Aigner et al. (1991) made neurotoxic lesions of these nuclei in monkeys. Some impairments on a simple test of recognition memory, delayed non-match-to-sample, were found. Analysis of the effects of similar lesions in rats showed that performance on memory tasks was impaired, perhaps because of failure to attend properly (Muir et al., 1994). Damage to the cholinergic neurons in this region in monkeys with a selective neurotoxin was also shown to impair memory (Easton and Gaffan, 2000; Easton et al., 2002, 2020).

There are quite limited numbers of these basal forebrain neurons (in the order of thousands). Given that there are relatively few of these neurons, it is not likely that they carry the information to be stored in cortical memory circuits, for the number of different patterns that could be represented and stored is so small. (The number of different memory patterns composed of active neurons that could be stored is dependent in a leading way on the number of input connections on to each neuron in a pattern associator, see e.g. Rolls (2016b, 2023c, 2026d)). With these few neurons distributed throughout the cerebral cortex, the memory capacity of the whole system would be impractically small. This argument alone indicates that these cholinergic neurons are unlikely to carry the information to be stored in cortical memory systems. Instead, they could modulate storage in the cortex of information derived from what provides the numerically major input to cortical neurons, the glutamatergic terminals of other cortical neurons. This modulation may operate by setting thresholds for cortical cells to the appropriate value, or by more directly influencing the cascade of processes involved in long-term potentiation (Rolls, 2016b). There is indeed evidence that acetylcholine is necessary for cortical synaptic modifiability, as shown by studies in which depletion of acetylcholine and noradrenaline impaired cortical LTP/synaptic modifiability (Bear and Singer, 1986). However, non-specific effects of damage to the basal forebrain cholinergic neurons are also likely, with cortical neurons becoming much more sluggish in their responses, and showing much more adaptation, in the absence of cholinergic inputs (Markram and Tsodyks, 1996; Abbott, Varela, Sen and Nelson, 1997) (see later).

The question then arises of whether the basal forebrain cholinergic neurons tonically release acetylcholine, or whether they release it particularly in response to some external influence. To examine this, I decided to make recordings from basal forebrain neurons, at least some of which project to the cortex (see Rolls (2005)) and will have been the cholinergic neurons just described.

We discovered that some of these basal forebrain neurons respond to visual stimuli associated with rewards such as food (Rolls, 1975, 1981a,b, 1982, 1986a,b,c, 1990, 1993, 1999b; Rolls, Burton and Mora, 1976; Burton, Rolls and Mora, 1976; Mora, Rolls and Burton, 1976; Wilson and Rolls, 1990b,c), or with punishment (Rolls, Sanghera and Roper-Hall, 1979), that others respond to novel visual stimuli (Wilson and Rolls, 1990a), and that others respond to a range of visual stimuli. For example, in one set of recordings, one group of these neurons (1.5%) responded to novel visual stimuli while monkeys performed recognition or visual discrimination tasks (Wilson and Rolls, 1990a).

A complementary group of neurons more anteriorly responded to familiar visual stimuli in the same tasks (Rolls, Perrett, Caan and Wilson, 1982c; Wilson and Rolls, 1990a).

A third group of neurons (5.7%) responded to positively reinforcing visual stimuli in visual discrimination and in recognition memory tasks (Wilson and Rolls, 1990b,c).

In addition, a considerable proportion of these basal forebrain neurons (21.8%) responded to any visual stimuli shown in the tasks, and some (13.1%) responded to the tone cue that preceded the presentation of the visual stimuli in the task, and was provided to enable the monkey to alert to the visual stimuli (Wilson and Rolls, 1990a). None of these neurons responded to touch to the leg which induced arousal, so their responses did not simply reflect arousal.

Neurons in this region receive inputs from the amygdala (Mesulam, 1990; Amaral et al., 1992; Russchen et al., 1985) and orbitofrontal cortex (Rolls et al., 2022b; Rolls, 2022a), and it is probably via the amygdala and orbitofrontal cortex that the information described here reaches the basal forebrain neurons. For example, neurons with similar response properties have been found in the amygdala, and the amygdala appears to be involved in decoding visual stimuli that are associated with reinforcers, or are novel (Rolls, 1990, 1992b; Davis, 1992; Wilson and Rolls, 1993; Rolls, 2000c; LeDoux, 1995; Wilson and Rolls, 2005; Rolls, 2005, 2016b). Further, similar neurons responding to rewarding stimuli, aversive stimuli, and to novel stimuli are found in the orbitofrontal cortex (Thorpe, Rolls and Maddison, 1983; Rolls, Browning, Inoue and Hernadi, 2005a; Rolls, 2019a,c) (see Chapter 3).

7.6.4.3 Basal magnocellular forebrain nuclei cholinergic neurons as a cortical strobe to facilitate memory

On the basis of these findings, it is suggested that the normal physiological function of these basal forebrain neurons is to send a general activation signal to the cortex when certain classes of environmental stimulus occur. These stimuli are often stimuli to which behavioral activation is appropriate or required, such as positively or negatively reinforcing visual stimuli, or novel visual stimuli. The effect of the firing of these neurons on the cortex is excitatory, and in this way produces activation. This cortical activation may produce behavioral arousal, and may thus facilitate concentration and attention, which are both impaired in Alzheimer's disease. The reduced arousal and concentration may themselves contribute to the memory disorders. But the acetylcholine released from these basal magnocellular neurons may in addition be more directly necessary for memory formation, for Bear and Singer (1986) showed that long-term potentiation, used as an indicator of the synaptic modification which underlies learning, requires the presence in the cortex of acetylcholine as well as noradrenaline. For comparison, acetylcholine in the hippocampus makes it more likely that LTP will occur, probably through activation of an inositol phosphate second messenger cascade (Markram and Segal, 1992; Seigel and Auerbach, 1996; Hasselmo and Bower, 1993; Hasselmo et al., 1995; Hasselmo, 1999; Hasselmo and Sarter, 2011; Newman et al., 2012).

The adaptive value of the cortical strobe provided by the basal magnocellular neurons may thus be that it facilitates memory storage especially when significant (e.g. reinforcing) environmental stimuli are detected. This means that memory storage is likely to be conserved (new memories are less likely to be laid down) when significant environmental stimuli are not present. In that the basal forebrain projection spreads widely to many areas of the cerebral cortex, and in that there are relatively few basal forebrain neurons (in the order of thousands), the basal forebrain neurons do not determine the actual memories that are stored. Instead the actual memories stored are determined by the active subset of the thousands of cortical afferents on to a strongly activated cortical neuron (Treves and Rolls, 1994; Rolls and Treves, 1998; Rolls, 2016b, 2023c, 2026d). The basal forebrain magnocellular neurons would then according to this analysis when activated increase the probability that a memory would be stored. Impairment of the normal operation of the basal forebrain magnocellular neurons would be expected to interfere with normal memory by interfering with this function, and this interference could contribute in this way to the memory disorder in Alzheimer's disease.

An interesting new discovery is that in humans the cholinergic septal neurons that project to the hippocampus receive inputs in humans from the pregenual anterior cingulate cortex, which in turn receives from the orbitofrontal and ventromedial prefrontal cortex; and the basal forebrain cholinergic neurons that project to the neocortex receive from the medial orbitofrontal cortex (Rolls et al., 2022b). This places the human orbitofrontal cortex in a key position to influence memory consolidation, and indeed to be the origin of the cortical strobe that can increase memory consolidation (Rolls et al., 2022b; Rolls, 2022a). Consistent with this, damage to the ventromedial prefrontal cortex / pregenual anterior cingulate cortex impairs memory consolidation (Rolls, 2022a) (see Section 6.10.5). It is accordingly proposed that an additional factor in any decline in memory function in aging could relate to a less strong cortical strobe from the reward / punishment / novelty system in the orbitofrontal cortex, which might itself be less responsive to these emotion-provoking stimuli in some individuals as part of the normal aging process.

An important discovery is also that the main cortical region with effective connectivity to the basal forebrain region is region pOFC of the medial orbitofrontal cortex (Rolls, Deco, Huang and Feng, 2022b). An implication is that boosting this medial orbitofrontal cortex reward-related system in aging might help to reverse some effects of aging if they are related to declining cholinergic functioning in aging.

Thus one way in which impaired cholinergic neuron function is likely to impair memory is by reducing the depth of the basins of attraction of cortical networks, in that these networks store less strongly the representations that are needed for episodic memory and for short-term memory, thus making the recall of long-term episodic memories less reliable in the face of stochastic noise, and the maintenance of short-term memory less reliable in the face of stochastic noise. Such changes would thereby also impair attention.

7.6.4.4 Basal magnocellular forebrain nuclei cholinergic neurons reduce cortical adaptation

Another property of cortical neurons is that they tend to adapt with repeated input (Abbott, Varela, Sen and Nelson, 1997; Fuhrmann, Markram and Tsodyks, 2002a; Rolls, 2016b, 2023c, 2026d). However, this adaptation is most marked in slices, in which there is no acetylcholine. One effect of acetylcholine is to reduce this adaptation (Power and Sah, 2008). The mechanism is understood as follows. The afterpolarization (AHP) that follows the generation of a spike in a neuron is primarily mediated by two calcium-activated potassium currents, $I_{\rm AHP}$ and s$I_{\rm AHP}$ (Sah and Faber, 2002), which are activated by calcium influx during action potentials. The $I_{\rm AHP}$ current is mediated by small conductance calcium-activated potassium (SK) channels, and its time course primarily follows cytosolic calcium, rising rapidly after action potentials and decaying with a time constant of 50 to several hundred milliseconds (Sah and Faber, 2002). In contrast, the kinetics of the s$I_{\rm AHP}$ are slower, exhibiting a distinct rising phase and decaying with a time constant of 1–2 s (Sah, 1996). A variety of neuromodulators, including acetylcholine (ACh), noradrenaline, and glutamate acting via G-protein-coupled receptors, suppress the s$I_{\rm AHP}$ and thus reduce spike-frequency adaptation (Nicoll, 1988).

When recordings are made from single neurons operating in physiological conditions in the awake behaving monkey, peristimulus time histograms of inferior temporal cortex neurons to visual stimuli show only limited adaptation (see Chapter 4 and Rolls (2026d)). There is typically an onset of the neuronal response at 80–100 ms after the stimulus, followed within 50 ms by the highest firing rate. There is after that some reduction in the firing rate, but the firing rate is still typically more than half-maximal 500 ms later (see example in Fig. 4.13 (Tovee, Rolls, Treves and Bellis, 1993)). Thus under normal physiological conditions, firing rate adaptation can occur, but does not involve a major adaptation, even when cells are responding fast (at e.g. 100 spikes/s) to a visual stimulus. One of the factors that keeps the response relatively maintained may however be the presence of acetylcholine. The depletion of acetylcholine in aging and some disease states (Schliebs and Arendt, 2011; Cools and Arnsten, 2022) could lead to less sustained neuronal responses (i.e. more adaptation), and this may contribute to the symptoms found. In particular, the reduced firing rate that may occur as a function of time if acetylcholine is low would gradually over a few seconds reduce the depth of the basin of attraction, and thus destabilize short-term memory when noise is present, by reducing the firing rate component shown in Equation B.12 of Rolls (2023c). I suggest that such changes would thereby impair short-term memory, and thus also top-down attention.

This adaptation can be studied by including a time-varying intrinsic (potassium like) conductance in the cell membrane, as described in Section B.6.2 of Rolls (2026d). A specific implementation of the spike-frequency adaptation mechanism using Ca^{++}-activated K^{+} hyperpolarizing currents (Liu and Wang, 2001) is described (Rolls, 2023c, 2026d), and was also used by Deco and Rolls (2005c) in an investigation of sequence memory. (They also describe two other approaches to adaptation.)

On the basis of these hypotheses, it is predicted that enhancing cholinergic function will help to reduce the instability of attractor networks involved in short-term memory and attention that may occur in aging.

Part of the interest of this stochastic dynamics approach to aging (Rolls and Deco, 2015a; Rolls, 2019f) is that it provides a way to test combinations of pharmacological treatments, that may together help to minimize the cognitive symptoms of aging. Indeed, the approach facilitates the investigation of drug combinations that may together be effective in doses lower than when only one drug is given. Further, this approach may lead to predictions for effective treatments that need not necessarily restore the particular change in the brain that caused the symptoms, but may find alternative routes to restore the stability of the dynamics.

7.7 High blood pressure, reduced hippocampal functional connectivity, and impaired memory

In another investigation to help to relate the advances in the computational approaches described in this book to disorders of brain function, we discovered that in the general population (with 19,507 participants in the UK Biobank), there was a linear relation between higher blood pressure, reduced hippocampal functional connectivity with other brain regions, and impaired numeric and prospective memory (all $p < 10^{-10}$) (see Fig. 7.6) (Feng, Rolls, Cheng and Feng, 2020). Effects of this type are probably related to a number of neurovascular changes that occur in hypertension (Iadecola and Gottesman, 2019), but part of the interest of the new discovery is that effects are found throughout the range of systolic blood pressure on both decreased functional connectivity of the hippocampus and decreased memory performance.

These findings have implications for the treatment of hypertension, because the findings provide evidence about the effects of even small increases in blood pressure. It is also of interest and potential clinical relevance that these effects were more evident for systolic blood pressure than for diastolic blood pressure (Feng, Rolls, Cheng and Feng, 2020).

7.8 Human emotional and cognitive development, linked to structural differences in the brain

The differences in functional connectivity described in this Chapter that could produce different dynamics of different brain systems in some mental disorders could be partly related to genetic differences, and partly to the environment. To take this research forward, we have performed a number of very large-scale studies with thousands of participants, and have discovered that in children or adolescents, there may even be structural differences in the brain that relate to psychiatric and cognitive problems.

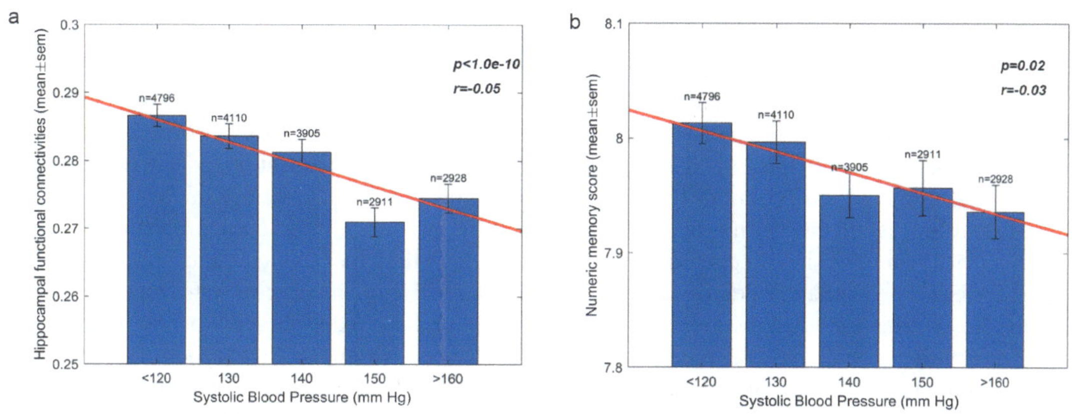

Fig. 7.6 The hippocampus, blood pressure, and memory. a. For 19,507 participants with neuroimaging data from the UK Biobank, there was a graded reduction in the functional connectivity of 88 hippocampal functional connectivity links and systolic blood pressure. The mean functional connectivity $\pm$ the sem is shown, together with the number of participants. The line shows the linear regression computed over all 19,507 participants, with age regressed out, and was significant at $p < 10^{-10}$. b. For the same participants there was a graded reduction in the numeric memory score with higher values of blood pressure, with the linear regression coefficient significant at $p < 10^{-10}$ and age regressed out. (Data from Feng R, Rolls E T, Cheng W and Feng J. 2020 Hypertension is associated with reduced hippocampal connectivity and impaired memory. EBioMedicine 61: 103082.)

In these investigations, we have discovered that a set of different factors can be associated with reduced area or volume of the orbitofrontal cortex, anterior cingulate and medial prefrontal cortex, the medial temporal lobe, and some related areas; and with behavioral and psychiatric problems including difficult behavior and depression.

In one investigation, this syndrome of altered structure of these brain regions and associated behavioral and psychiatric problems was associated with prolonged nausea and vomiting during pregnancy of the mothers (Wang, Rolls, Du, Du, Yang, Li, Li, Cheng and Feng, 2020).

In another investigation, family conflict was associated with a similar reduction of the area and volume of the same brain areas and similar behavioral, psychiatric and cognitive problems; and high parental monitoring (care) of the children with the opposite brain and behavioral differences (Gong, Rolls, Du, Feng and Cheng, 2021).

In a third investigation, reduced area and volume of similar brain areas in the children, and behavioral and psychiatric problems, were associated with a low age of the mother when the child was born (Du, Rolls, Gong, Cao, Vatansever, Zhang, Kang, Cheng and Feng, 2022).

Again, we do not know whether these syndromes are related to genetic or environmental effects, or to a combination, but these discoveries do pave the way for the better diagnosis of such syndromes, and for their prevention and treatment. Moreover, these investigations point again towards the importance of the orbitofrontal cortex, anterior cingulate and medial prefrontal cortex, the medial temporal lobe, and some related areas, in behavioral and psychiatric problems including difficult behavior and depression, and build on the foundation described in Chapter 3.

7.9 Evaluation

In this Chapter, I have described pioneering applications of stochastic neurodynamics to understanding major mental disorders including schizophrenia, obsessive-compulsive disorder, and depression, and the memory and cognitive changes in normal aging. There are implications not only for understanding, but also for treating, these states.

A key conceptual point arises given the importance of cortical attractor networks for understanding cortical function (Rolls, 2016b, 2023c, 2026d). The highly developed local excitatory and associatively modifiable recurrent collateral connections between nearby cortical pyramidal cells that can support attractor states are essentially the defining feature of cortex (Rolls, 2016b, 2023c, 2026d). The implementation of these attractor networks relies on excitatory connections between pyramidal cells, which produces what tend to be highly unstable systems. In this situation, the feedback (and feedforward) inhibition implemented by inhibitory interneurons must control these excitatory dynamics using precise balancing, and more than that, the balancing must result in suitable firing rates of the excitatory neurons for entering and maintaining an attractor state. As we have seen, if the firing rates are too low, then attractor networks used for example for working memory and attention will become unstable, fall out of the attractor state, and impair these types of cognitive and related function. If the firing rates are too high, then overstability may arise, which could relate to many types of disorder, as shown here and in more detail by Rolls (2023c, 2026d). Thus, setting up and maintaining cortical attractor networks at the correct 'working point' in terms of the amount of excitation and inhibition in each brain region is a major challenge for the brain, and can result in some disorders.

A further somewhat new concept is that synaptic or neuronal adaptation that produces some reduction of firing rate during the first 500 ms of a neuronal response (see example in Fig. 4.13) may be one way in which the cortex helps itself to respond quickly and fall rapidly into a basin of attraction, but then as the adaptation sets in, ensure that the network does not become dangerously overexcited. The adaptation will also help the network to fall out of the basin of attraction within a short time period of typically 1 to a few seconds, so that the brain does not become stuck with one behavior, but can be sensitive to other inputs, and move on to different behaviors. One example of these processes at work is in the hippocampo-cortical system for memory recall to the neocortex, where adaptation can help to prevent runaway excitation in the long loops shown in Fig. 5.14 (Rolls et al., 2024d).

In addition to this, the processes at work in evolution can be expected to select different working points in each person's brain for each of their networks. That is part of the process by which natural selection using variation can build individuals that may do well in different environments. Overall, natural selection may be searching for adaptive value of particular combinations of stability in different brain networks.

For example, setting the working points of attractor networks to be useful for creativity facilitated by stochastic jumps in the high-dimensional semantic space may be difficult to achieve. It may not work similarly in different individuals because of the variation that is involved in natural selection. Indeed, these evolutionary processes will make individuals different to each other, and that is fundamental to understanding personality (Rolls, 2014a, 2023c, 2026d).

More than that, the 'noise' in the brain may be influenced by factors such as the time of day, and each person may find that there are particular times of the day when they can be most creative, to provide an example. Evidence for this is that introverts tend to be more mentally active early in the day, and extraverts later in the day (Rolls, 2014a, 2023c, 2026d).

Thus understanding the stochastic dynamics of the brain has many implications for understanding the brain better in health and disease, and the research on stochastic dynamics that I have developed and described in this Chapter has helped to pioneer this key approach to understanding brain function in health and disorder. An important part of this pioneering approach has been to link these concepts about stochastic dynamics to empirical measures that influence the stability of attractor networks, including the functional connectivity between brain regions, the levels of neurotransmitters such as glutamate, acetylcholine, dopamine, and noradrenaline, and the receptors for these transmitters.

8 Biologically plausible computations in the brain, deep learning, and artificial intelligence

One of the key approaches that I have developed in the research described in this book is to complement the empirical discoveries in neuroscience described here on key brain systems with computational neuroscience approaches to understand not just what each brain system does, but how it performs the computations. Indeed, I see a key part of understanding the brain and human behavior as understanding the brain's computational systems, and how they interact. The approach that I take is to develop our computational understanding of real systems in the brain.

My computational neuroscience approach is thus somewhat different from some approaches in theoretical neuroscience (Dayan and Abbott, 2001; Hertz, Krogh and Palmer, 1991; Gerstner, Kistler, Naud and Paninski, 2014; Izhikevich, 2007; Ermentrout and Terman, 2010), which seek to understand the computations that can be performed by neural systems theoretically, but frequently without applying this to understand directly what computations are performed by different cortical regions, and how each cortical region performs its computations. Indeed, I think that the approach of the computational neuroscience research described in this book is rather pioneering in developing computational neuroscience theories and models of what many cortical systems in the brain compute, and how they compute it.

Given the importance of this computational neuroscience approach to understanding real neural systems in the brain, it then becomes of great interest to examine whether the principles of operation and computations performed by the brain (Rolls, 2016b, 2026d) are similar to how computations are performed in computers. This is especially timely now, for artificial intelligence (AI) approaches implemented on digital computers are being used by some as models of how the brain performs its computations, and this does need to be very carefully considered, as I show in this Chapter and in more detail elsewhere (Rolls, 2026d).

Given this background, in this Chapter I describe comparisons I have developed (Rolls, 2023c, 2026d) between computations in the brain and computations performed in computers (Section 8.1). This is intended to be helpful to those engineers, computer scientists, AI specialists and others interested in designing new computers that emulate aspects of brain function.

The comparison between computation in the brain and computation in computers in Section 8.1 is also intended to be of interest to neuroscientists to emphasize the differences in the implementations of computation in the brain and in digital computers, including digital computers when they are running artificial intelligence algorithms that involve deep learning in multilayer networks.

Part of the interest here is that many computer implementations of 'neural networks' implement very different algorithms to those implemented in the brain. This means that there is a lot left not only to learn about the brain, but also to learn from the brain. For example, Section 8.2 specifically compares deep learning in artificial neural networks to the type of self-organising learning implemented in the brain.

Section 8.3 considers the pioneering approach taken in Brain Computations and Connectivity (Rolls, 2023c), in **Brain Computations and Principles; and AI** (Rolls, 2026d), and in this book to biologically plausible computations performed by each of many brain regions. A key question is whether the current approach to biologically plausible brain computation using local synaptic learning rules is sufficient to understand brain computation, in contrast to the learning rules that rely on backpropagation of errors across multilayer networks to train synaptic weights as in artificial intelligence algorithms.

Finally, Section 8.4 makes it clear why the focus of this book and of Rolls (2023c, 2026d) is on computations in primate (and that very much includes human) brains, rather than on rodent (rat and mouse) brains. It is because the systems-level organization of primate including human brains is quite different from that in rodents. This must be borne in mind when considering how the human brain works in health and in disease.

8.1 Brain computation compared to computation on a digital computer

An important component of the approach taken in **Brain Computations and Connectivity** (Rolls, 2023c), in **Brain Computations and Principles; and AI** (Rolls, 2026d), and in this book is to describe *what* computations are performed by each brain region, and then *how* they may be performed. This enables the different levels of analysis in neuroscience to be brought together in a multilevel causal account (Rolls, 2026d, 2021g,i).

However, the types of computation that are performed by the brain, and the computational style, are very different from the types of computation performed by a digital computer, which performs specified logical / syntactic operations on exact data retrieved from memory, and then stores the exact result back in memory. To highlight some of the principles of brain computation described in this book, and to emphasize how the type of computation is very different from that performed by digital computers, the principles of computation by the brain are compared next with those of a digital computer, with a summary of some of the differences in Table 8.1.

Data addressing. An item of data is retrieved from the memory of a digital computer by providing the address of the data in memory, and then the data can be manipulated (moved, compared, added to the data at another address in the computer etc.) using typically a 32-bit or 64-bit binary word of data. Pointers to memory locations are thus used extensively. In contrast, in the cortex, the data are used as the access key (in for example a pattern associator, autoassociator, and competitive network, see Section 1.4 and Rolls (2023c, 2026d)), and the neurons with synaptic weights that match the data respond. Memory in the brain is thus **content-addressable**. In one time constant of the synapses / cell membranes, the brain has thus found the correct output. In contrast, in a digital computer a serial search is required, in which the data at every address must be retrieved and compared in turn to the test data to discover if there is a match.

The principle by which this content addressability and vector similarity computation is performed is illustrated by the dot product computation performed by simple models of neurons in the brain. Fig. 8.1 shows that the activation h_i of each neuron i is computed as

$$h_i = \sum_j x_j w_{ij} \tag{8.1}$$

where the sum is over the C input axons, indexed by j.

Because this inner or dot product is essentially a correlation (un-normalized) between the j input firings x and the j synaptic weights w_{ij} on neuron i, the activation and then the firing rate of the

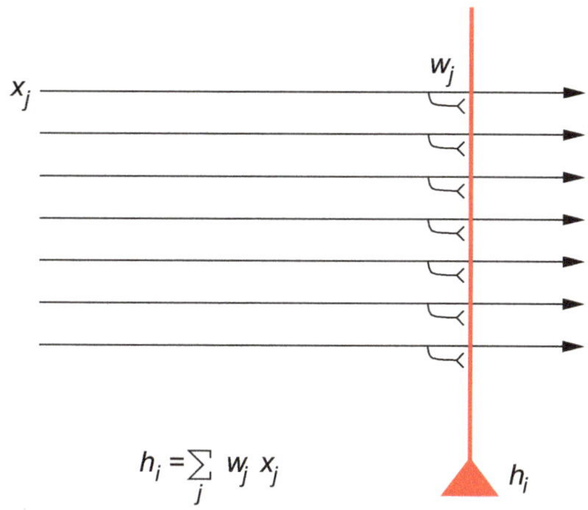

Fig. 8.1 A neuron i that computes a dot product of the input pattern of firing in axons x_j with its synaptic weight vector w_{ij} generalizes well to other patterns based on their similarity measured in terms of dot (inner) product or correlation, but shows no translation (or size, etc.) invariance.

neuron i reflects how well that neuron's synaptic weights match the input firings on the x inputs. Because this is a similarity computation, the firing of the neuron also generalises well to somewhat similar inputs.

Vector similarity vs logical operations. Cortical computation including that performed by associative memories and competitive networks operates by vector similarity – the dot (inner) product of the input and of the synaptic weight vector of each neuron are produced, and the neurons with the highest dot product will be most activated (Rolls, 2023c, 2026d) (Fig. 8.1). Even if an exact match is not found, some output is likely to result. In contrast, in a digital computer, logical operations (such as AND, OR, XOR) and exact mathematical operations (such as addition, subtraction, multiplication, and division) are computed. (There is no bit-wise similarity between the binary representations of 7 (0111) and 8 (1000).) The similarity computations performed by the brain may be very useful, in enabling similarities to be seen and parallels to be drawn, and this may be an interesting aspect of human creativity, realized for example in **Finnegans Wake** by James Joyce (1939) in which thoughts reached by associative thinking abound. However, the lateral thinking must be controlled, to prevent bizarre similarities being found, and this is argued to be related to the symptoms of schizophrenia in Section 7.3.

Fault tolerance. Because exact computations are performed in a digital computer, there is no in-built fault tolerance or graceful degradation. If one bit of a memory has a fault, the whole memory chip must be discarded. In contrast, the brain is naturally fault tolerant, because it uses vector similarity (by neurons using their input firing rate vectors and synaptic weight vectors) in its calculations, and linked to this, distributed representations (Fig. 8.1). This makes the brain robust developmentally with respect to 'missing synapses', and robust with respect to later losing some synapses or neurons (Rolls, 2023c, 2026d).

Word length. To enable the vector similarity comparison to have high capacity (for example memory capacity) the 'word length' in the brain is typically long, with between 10,000 and 50,000 synapses onto every neuron being common in cortical areas. (Remember that the leading term in the factor that determines the storage capacity of an associative memory is the number of synapses per neuron – see Rolls (2023c, 2026d).) In contrast, the word length in typical digital computers at 32 or 64

Table 8.1 Brain computation vs Digital computer computation

Brain Computation	Digital computer computation
1. Dot product similarity followed by a threshold	vs logical functions e.g. AND, NAND, OR
2. 10,000 inputs per neuron	vs several inputs to a logic gate
3. Content-addressable memory	vs data accessed by address
4. Fault tolerant: dot product	vs no inherent fault tolerance
5. Generalization and completion	vs only exact match in the hardware
6. Fast: parallel update of a neuron and of neurons in a network	vs serial processing with fast components
7. Approximate, low precision	vs high precision 64 bit
8. Stochastic spiking noise: probabilistic computation helps originality, creativity	vs noise free implemented by non-linearities
9. Dynamics are parallel e.g. an attractor network retrieves in 1–2 time constants of the synapses	vs inherently serial
10. Heuristics, e.g. Invariant object recognition analyzes only part of a scene.	vs attempt to understand a whole scene
11. Syntax: none inherent	vs syntactical operations on data at an address
12. The architecture adapts by learning to implement the computation.	vs fixed architecture with different software
13. Sparse distributed representation	vs binary encoding in a computer word
14. Mind vs brain	vs software vs hardware

bits is much shorter, though with the binary and exact encoding used this allows great precision in a digital computer.

Readability of the code. To comment further on the encoding: in the cortex, the code must not be too compact, so that it can be read by neuronally plausible dot product decoding and can use vector similarity to generalize to similar patterns (Rolls, 2016b), as shown in Rolls (2023c, 2026d). In contrast, the binary encoding used in a digital computer is optimally efficient, with one bit stored and retrievable for each binary memory location. However, the computer binary code cannot be read by neuronally plausible dot-product decoding (Fig. 8.1).

Precision. The precision of the components in a digital computer is that every modifiable memory location must store one bit accurately. In contrast, synapses in the brain need not work with exact precision, with for example typically less that one bit per synapse being usable in associative memories (Treves and Rolls, 1991; Rolls and Treves, 1998). The precision of the encoding of information in the firing rate of a neuron is likely to be a few bits – perhaps 3 – as judged by the standard deviation and firing rate range of individual cortical neurons (Section 4.6.2 and Rolls (2023c, 2026d)).

The speed of computation. This brings us to the speed of computation. In the brain, considerable information can be read in 20 ms from the firing rate of an individual neuron (e.g. 0.2 bits), leading to estimates of 10–30 bits/s for primate temporal cortex visual neurons (Rolls, Treves and Tovee, 1997d; Rolls and Tovee, 1994), and 2–3 bits/s for rat hippocampal cells (Skaggs, McNaughton, Gothard and Markus, 1993; Rolls, 2016b; Rolls and Treves, 2011) (Section 4.6.2). Though this is very slow compared to a digital computer, the brain does have the advantage that a single neuron receives spikes from thousands of individual neurons, and computes its output from all of these inputs within a period of approximately 10–20 ms (determined largely by the time constant of the synapses) (Rolls, 2016b, 2023c, 2026d). Moreover, each neuron, up to at least the order of tens of neurons, conveys independent information, as described in Section 4.6.2 and by Rolls (2023c, 2026d).

Parallel vs serial processing. Computation in a conventional digital computer is inherently serial, with a single central processing unit that must fetch the data from a memory address, manipulate the word of data, and store it again at a memory address. In contrast, brain computation is parallel in at least three senses.

First, an individual neuron in performing a dot product between its input firing rate vector and its synaptic weight vector does operate in an analog way to sum all the injected currents through the thousands of synapses to calculate the activation h_i, and fire if a threshold is reached, in a time in the order of the synaptic time constant. To implement this on a digital computer would take $2C$ operations (C multiply operations, and C add operations, where C is the number of synapses per neuron – see Equation 1.1 in (Rolls, 2023c, 2026d).

Second, each neuron in a single network (e.g. a small region of the cortex with of the order of hundreds of thousands of neurons) does this dot product computation in parallel, followed by interaction through the GABA inhibitory neurons, which again is fast. (It is in the order of the time constant of the synapses involved, operates in continuous time, and does not have to wait at all until the dot product operation of the pyramidal cells has been completed by all neurons given the spontaneous neuronal activity that allows some neurons to reflect their changed inputs rapidly by when the next spike occurs.) This interaction sets the threshold in associative and competitive networks, and helps to set the sparseness of the representation of the population of neurons.

Third, different brain regions operate in parallel. An example is that the ventral visual stream computes object representations, while simultaneously the dorsal visual stream computes (inter alia) the types of global motion described by Rolls and Stringer (2007) (Section 4.7.4), including for example a wheel rotating in the same direction as it traverses the visual field. Another example is that within a hierarchical system in the brain, every stage operates simultaneously, as a pipeline processor, with a good example being V1–V2–V4–IT (inferior temporal visual cortex), which can all operate simultaneously as the data are pipelined through (Chapter 4; Rolls (2023c, 2026d)).

We could refer to the computation that takes place in different modules, that is in networks that are relatively separate in terms of the number of connections between modules relative to those within modules, such as those in the dorsal and ventral visual streams, as being parallel computation. Within a single module or network, such as the CA3 region of the hippocampus, or inferior temporal visual cortex, we could refer to the computation as being *parallel distributed computation*, in that the closely connected neurons in the network all contribute to the result of the computation. For example, with distributed representations in an attractor network, the excitatory neurons interact with each other directly and through the inhibitory interneurons to retrieve and then maintain a stable pattern of neuronal firing in short-term memory (Section B.3 of Rolls (2026d)). In a competitive network involved in pattern categorization, all the neurons interact through the inhibitory interneurons to result in an active population of neurons that represents the best match between the input stimulus and what has been learned previously by the network, with neurons with a poor match being inhibited by neurons with a good match (Section B.4 of Rolls (2026d)). In a more complicated scenario with closely connected interacting modules, such as the prefrontal cortex and the inferior temporal cortex during top-down attention tasks and more generally forward and backward connections between adjacent cortical areas, we might also use the term parallel distributed computation, as the bottom-up and top-down interactions may be important in how the whole dynamical system of interconnected networks settles (see examples in Rolls (2023c, 2026d); Rolls et al. (2024d)).

Stochastic dynamics and probabilistic computation. Digital computers do not have noise to contend with as part of the computation, as they use binary logic levels, and perform exact computation. In contrast, brain computation is inherently noisy, and this gives it a non-exact, probabilistic, character. One of the sources of noise in the brain is the spiking activity of each neuron. Each neuron must transmit information by spikes, for an all-or-none spike carried along an axon ensures that the signal arrives faithfully, and is not subject to the uncertain cable transmission line losses of analog potentials. But once a neuron needs to spike, then it turns out to be important to have spontaneous activity, so that neurons do not all have to charge up from a hyperpolarized baseline whenever a new input is received. The fact that neurons are kept near threshold, with therefore some spontaneous spiking, is inherent to the rapid operation of for example autoassociative retrieval, as described in Section 3.12 and Chapter 7 of this book, and in Section B.3 of Rolls (2026d). But keeping the neurons close to threshold, and the spiking activity received from other neurons, results in spontaneous spike trains that are approximately Poisson, that is with spikes randomly timed for a given mean firing rate. The result of the interaction of all these randomly timed inputs is that in a network of finite size (i.e. with a limited number of neurons) there will be statistical fluctuations, that influence which memory is recalled, which decision is taken, etc. as described in Section 3.12 and Chapter 7. Thus brain computation is inherently noisy and probabilistic, and this has many advantages, as described in Section 3.12, Chapter 7, and by Rolls and Deco (2010); Rolls (2016b, 2023c, 2026d).

Syntax. Digital computers can perform arbitrary syntactical operations on operands, because they use pointers to address each of the different operands required (corresponding even for example to the subject, the verb, and the object of a sentence). In contrast, as data are not accessed in the brain by pointers that can point anywhere, but instead just having neurons firing to represent a data item, a real problem arises in specifying which of the firing neurons represent for example the subject, the verb, and the object, and distributed representations potentially make this even more difficult. The brain thus inherently finds syntactical operations difficult (as explained in Chapter 14 of Rolls (2026d)). We do not know how the brain implements the syntax required for language. But we do know that the firing of neurons conveys 'meaning' based on spatial location in the brain. For example, a neuron firing in V1 indicates that a bar or edge matching the filter characteristic of the neuron is present at a particular location in space. Another neuron in V1 encodes another feature at another position in space. A neuron in the inferior temporal visual cortex indicates (with other neurons helping to form a distributed representation) that a particular object or face is present in the visual scene. Perhaps the implementation of the syntax required for language that is implemented in the brain also utilizes the spatial location of the network in the cortex to help specify what syntactical role the representation should perform. This is a suggestion I make, as it is one way that the brain could deal with the implementation of the syntax required for language (Rolls, 2023c, 2026d).

Modifiable connectivity. The physical architecture (what is connected to what) of a digital computer is fixed. In contrast, the connectivity of the brain alters as a result of experience and learning, and indeed it is alterations in the strength of the synapses (which implement the connectivity) that underlie learning and memory. Indeed, self-organization in for example competitive networks has a strong influence on how the brain is matched to the statistics of the incoming signals from the world, and of the architecture that develops. In a digital computer, every connection must be specified. In contrast, in the brain there are far too few genes (of order 25,000) for the synaptic connections in the brain (of order 10^{15}, given approximately 10^{11} neurons each with in the order of 10^{4} synapses) for the genes to specify every connection[7]. The genes must therefore specify some much more general rules, such as that each CA3 neuron should make approximately 12,000 synapses with other CA3 neurons, and receive approximately 48 synapses from dentate granule cells (see Chapter 5). The actual connections made would then be made randomly within these constraints, and then strengthened or weakened as

[7]For comparison, a computer with 1 Gb of memory has approximately 10^{10} modifiable locations, and if it had a 100 Gb disk that would have approximately 10^{12} modifiable locations.

a result of self-organization based on for example conjunctive pre- and postsynaptic activity. Some of the rules that may be specified genetically have been suggested on the basis of a comparison of the architecture of different brain areas (Rolls and Stringer, 2000; Rolls, 2016b, 2023c, 2026d). Moreover, it has been shown that if these rules are selected by a genetic algorithm based on the fitness of the network that self-organizes and learns based on these rules, then architectures are built that solve different computational problems in one-layer networks, including pattern association learning, autoassociation memory, and competitive learning (Rolls and Stringer, 2000; Rolls, 2016b, 2023c, 2026d)). The architecture of the brain is thus interestingly adaptive, but guided in the long term by genetic selection of the building rules.

Logic. The learning rules that are implemented in the brain that are most widely accepted are associative, as exemplified by LTP and LTD. This, and the vector similarity operations implemented by neurons, set the stage for processes such as pattern association, autoassociation, and competitive learning to occur naturally, but not for logical operations such as XOR and NAND or arithmetic operations. Of course, the non-linearity inherent in the firing threshold of neurons is important in many of the properties of associative memories and competitive learning, as described by Rolls (2023c, 2026d), and indeed are how some of the non-linearities that can be seen with attention can arise (Deco and Rolls, 2005d).

Dynamical interaction between modules. Because the brain has populations of neurons that are simultaneously active (operating in parallel), but are interconnected, many properties arise naturally in dynamical neural systems, including the interactions that give rise to top-down attention (Chapter 22 and Sections 13.4 and 2.2 of Rolls (2026d)), the effects of mood on memory (Rolls and Stringer, 2001b; Rolls, 2014a) etc. Because simultaneous activity of different computational nodes does not occur in digital computers, these dynamical systems properties that arise from interacting subsystems do not occur naturally in digital computers, though they can be simulated.

The cortex has recurrent excitatory connections within a cortical area, and reciprocal, forward and feedback, connections between adjacent cortical areas in a hierarchy. The excitatory connections enable cortical activity to be maintained over short periods, making short-term memory an inherent property of the cortex. They also provide the autoassociative long-term memory with completion from a partial cue (given associative synaptic modifiability in these connections) (Rolls, 2023c, 2026d). However, completion on a digital computer is a difficult and serial process to identify a possible correct partial match.

Another comparison is that the short-term memory property of the cortex is part of what makes the cortex a dynamical interacting system, with for example what is in short-term memory in the prefrontal cortex acting to influence memory recall, perception, and even what decision is taken, in other networks, by top-down biased competition (Rolls, 2023c, 2026d).

There is a price that the brain pays for this positive feedback inherent in its recurrent cortical circuitry, which is that this circuitry is inherently unstable, and requires strong control by inhibitory interneurons to minimize the risk of epilepsy and other disorders (Chapter 7; Fig. 5.14; Rolls (2016b); Rolls et al. (2024d)).

Modular organization. Brain organization is modular, with many relatively independent modules each performing a different function, whereas digital computers typically have a single central processing unit connected to memory. The cortex has many localized modules with dense connectivity within a module, and then connections to a few other modules. The reasons for the modularity of the brain are considered by Rolls (2016b), and include the importance of minimizing the connection lengths between neurons that are performing a particular computation, in order to minimize brain size.

Hierarchical organization. As described by Rolls (2016b, 2023c, 2026d), many cortical systems are organized hierarchically. A major reason for this is that this enables the connectivity to be kept within the limits of which neurons appear capable (up to 50,000 synapses per neuron), yet also enables global computation (such as the presence of a particular object anywhere in the visual field) to be achieved, as exemplified by VisNet, a model of invariant visual object recognition (see Fig. 4.18 and Chapter 4). Another important reason is that hierarchical organization simplifies the learning that is required at each stage and enables it to be a local operation, in contrast to backpropagation of error networks where similar problems could in principle be solved in a two-layer network (with one hidden layer), but would require training with a non-local learning rule (Section 4.6.2) as well as potentially neurons with very large numbers of connections. Another feature of cortical organization is that the number of regions or levels in any hierarchy in the brain is not more than 4 or 5, because each region requires 20–30 ms of computation and transmission time. (This of course contrasts with the hundreds of successive layers being explored with artificial neural networks, as described in Section B.14 of Rolls (2026d), which is completely biologically implausible in terms of response times.)

8.2 A comparison of brain computation with learning in artificial deep learning networks using error backpropagation

Learning in deep networks with error backpropagation is described by LeCun, Bengio and Hinton (2015); Lillicrap, Santoro, Marris, Akerman and Hinton (2020); Bishop and Bishop (2024), and in Section B.12 of Rolls (2023c, 2026d). Attempts to compare the operation of hierarchical convolutional deep neural networks to the operation of the ventral visual cortical stream (Yamins and DiCarlo, 2016; Rajalingham, Issa, Bashivan, Kar, Schmidt and DiCarlo, 2018; Macpherson, Churchland, Sejnowski, DiCarlo, Kamitani, Takahashi and Hikida, 2021) are evaluated in Section 4.7.6 and by Zhang, Rolls and Feng (2026a). There is increasing interest in how artificial intelligence systems including large language models (Paass and Giesselbach, 2023; Bishop and Bishop, 2024) compare to computations performed by the brain (Song, Lukasiewicz, Xu and Bogacz, 2020; Macpherson, Churchland, Sejnowski, DiCarlo, Kamitani, Takahashi and Hikida, 2021; Sejnowski, 2020, 2023; Song, Millidge, Salvatori, Lukasiewicz, Xu and Bogacz, 2024; Rolls, 2024a, 2026d). Here the biological plausibility and advantages and disadvantages of this deep learning compared with brain computation are considered. We also consider whether deep learning does take place in the brain, for prima facie it is not biologically plausible.

A first problem in terms of biological plausibility of deep networks trained by error backpropagation is that these are typically supervised systems, with a separate teacher for each output neuron informing it how each output neuron should be responding for a given input pattern at the bottom layer of the network (LeCun et al., 2015; Lillicrap et al., 2020). There is no architecture that looks remotely like this in the cerebral cortex, with a separate teacher for each neuron. The only part of the brain where there is what might be a teacher for each neuron is in the cerebellum, where there is a single climbing fibre for each Purkinje cell, and a network with no hidden layers (Chapter 17 of Rolls (2023c, 2026d)). Progress is being made in unsupervised training of deep convolutional neural networks (Zhuang, Yan, Nayebi, Schrimpf, Frank, DiCarlo and Yamins, 2021) using for example principles utilized in autoencoders in which higher layers aim to reconstruct the signals in the input layer and update the connections based on the differences (the errors) (Lillicrap et al., 2020), but these still implement convolution and deep learning using non-local learning rules including backpropagation of error to optimize representations, so are still biologically implausible.

But the problem is deeper than that there is no teacher for each output neuron (and the other problems that appear to make deep learning biologically implausible described below and elsewhere (Rolls, 2026d)). A bigger problem for backpropagation of error networks is that somehow what the output of every neuron should be needs to be specified by a teacher, that is, backprop is a supervised learning system in that error correction is performed. The great beauty of the brain in contrast is that it is a largely self-organizing system, without a teacher telling each neuron what to do, as is made evident throughout Rolls (2023c, 2026d) and in this book, with a simple example of such a system a hierarchically organised set of competitive networks with convergence from stage to stage, as described in Chapter 4. Such self-organizing networks learn what categories to form based on the statistics of the inputs arriving from the world. They often use special adaptations to learn those statistics in a useful way, such as the trace short-term memory learning rule described in Chapter 4 that enables invariant visual object and face representations to be learned because over short time periods it is likely to be the same object that is being viewed. And they may benefit from guidance (mild supervision) provided by top-down signals from higher layers, as described in Section B.4.5 of Rolls (2023c, 2026d).

In essence, a deep network trained by backpropagation of error tries to achieve an optimal mapping between input patterns and output patterns of neural activity. (A pattern of neuronal activity refers throughout this book to the firing rates of a population of neurons, that is to a vector of firing rates of a set of neurons, such as all the x_j's in Fig. 8.1, that might provide the inputs to a neuron.) The number of neurons in each deep layer must be carefully adjusted to ensure that the system does not become just a look-up table, yet has sufficient neurons to enable the mapping and generalization required, and the number of layers is often adjusted to be 100 or more to help to minimize overall errors (Bishop and Bishop, 2024). But a problem that arises is that what is being represented at each intermediate layer is very difficult to understand (Plebe and Grasso, 2019; Sejnowski, 2020), and is likely to yield very little useful information to help understand the brain. There may be a lesson here for AI and machine learning, as follows:

At each stage of cortical processing, some output is always taken and directed to structures such as the striatum and basal ganglia for use in behavioral output (see Fig. 8.3). That is a key aspect of cortical design (Rolls, 2016b, 2023c, 2026d). It may reflect an interesting way in which the cortex evolves, by adding new regions each of which can produce useful output. That is not true of deep backprop networks.

An important point is that the brain does not solve a predetermined mapping problem from input to output, as in deep learning. Instead, the general architecture of the brain is specified by a relatively small number of genes that appear to specify what type of neuron connects to what other type to generate the architecture within and between cortical areas (Section B.20 of Rolls (2023c) and Chapter 32 of Rolls (2026d)). The brain then builds the details of its synaptic connectivities using self-organizing learning based on the environment in which it finds itself. This view has been developed by comparing the connectivity of different cortical areas, and then allowing genetic algorithms to select different combinations of the connectivity rules to build neural networks in computer simulations that can successfully solve different types of computational problem for which different architectures are needed (Rolls and Stringer, 2000; Rolls, 2016b, 2023c, 2026d).

Another key feature of cortical design is that with its self-organizing principles rather than being taught what to compute, the cortex is able to solve complex problems with just a few layers, e.g. four, as described in Chapter 4 and by Rolls (2016b, 2026d). The cortex is constrained to a few layers in the hierarchy, because each layer takes about 15 ms to compute including recurrent attractor operations, and the total compute time needs to be kept to less than 100 ms for biological utility. More layers could be tried in artificial self-organizing neural networks.

I hope that those interested in brain-inspired intelligence (Lu et al., 2024) will look carefully at what the computational problems are that are solved by different parts of the brain, and will not only learn from our current understanding of how the brain performs its computations, but will also suggest new ways in which those computations may be performed. In this context, **Brain Computations and Principles; and AI** (Rolls, 2026d) does set out principles of the organization of the brain many of which underlie its computational abilities and style.

There are other factors that make deep networks trained by backprop biologically implausible. Given that the signed error for a hidden neuron in an error backpropagation network is calculated by propagating backwards information based on the errors of all the output neurons to which a hidden neuron is connected, and all the relevant synaptic weights, and the activations of the output neurons to define the part of the activation function on which they are operating, it is implausible to suppose that the correct information to provide the appropriate signed error for each hidden neuron is propagated backwards between real neurons. A hidden neuron would have to 'know', or receive information about, the errors of all the neurons to which it is connected, and its synaptic weights to them, and their current activations (Section B.12 of Rolls (2023c, 2026d)). If there were more than one hidden layer, this would be even more difficult.

To expand on the difficulties: first, there would have to be a mechanism in the brain for providing an appropriate error signal to each output neuron in the network, though it may be possible in some approaches to relax this constraint (Zhuang et al., 2021). With the possible exception of the cerebellum, an architecture where a separate error signal could be provided for each output neuron is difficult to identify in the brain. Second, as noted in Chapter 5 and by Rolls (2016b, 2026d), the backprojection pathways that are present in the cortex seem suited to perform recall (Chapter 5) and to implement top-down attention (Rolls, 2023c, 2026d)), and this might make it difficult for them also to have the correct strength to carry the correct error signal.

A problem with the backpropagation of error approach in a biological context is thus that in order to achieve their competence, backpropagation networks use what is almost certainly a learning rule that is much more powerful than those that could be implemented biologically, and achieve their excellent performance by performing the mapping though a minimal number of hidden neurons (Lillicrap, Santoro, Marris, Akerman and Hinton, 2020; LeCun, Bengio and Hinton, 2015; Bishop and Bishop, 2024). In contrast, real neuronal networks in the brain probably use much less powerful learning rules, in which errors are not propagated backwards, and at the same time have very large numbers of hidden neurons, without the bottleneck that helps to provide backpropagation networks with their good performance. A consequence of these differences between backpropagation and biologically plausible networks may be that the way in which biological networks solve difficult problems may be rather different from the way in which backpropagation networks find mappings. Thus the algorithms used in artificial intelligence systems may not always be excellent guides to how biologically plausible networks may perform on similar problems. Part of the challenge for future work is to discover how more biologically plausible networks than backpropagation networks can solve comparably hard problems, and then to examine the properties of these networks, as a perhaps more accurate guide to brain computation (Lillicrap, Santoro, Marris, Akerman and Hinton, 2020; Rolls, 2026d).

As stated above, it is a major challenge for brain research to discover whether there are algorithms that will solve comparably difficult problems to backpropagation, but with a local learning rule (Lillicrap et al., 2020; Rolls, 2026d)). Such algorithms may be expected to require many more hidden neurons than backpropagation networks, in that the brain does not appear to use information bottlenecks to help it solve difficult problems. The issue here is that much of the power of backpropagation algorithms arises because there is a minimal number of hidden neurons to perform the required mapping using typically a final one-layer delta-rule network. Useful generalization arises in such networks because with a minimal number of hidden neurons, the net sets the representation they provide to enable appropriate generalization. The danger with more hidden neurons is that the network becomes a look-up table, with one hidden neuron for every required output, and generalization when the inputs vary becomes poor. The challenge is to find a more biologically plausible type of network that operates with large numbers of neurons, and yet that still provides useful generalization. An example of such an approach, VisNet, is described in Chapter 4 (Rolls, 2021c; Zhang et al., 2026a; Rolls, 2026d).

Another key difference between artificial deep learning approaches and real networks in humans and other primates is in the training. For deep learning with supervision, a very large number of exemplars, for example visual images, is presented each with a label to state its category. In contrast, primates including humans experience a continuous world in which first one object is encountered for typically many seconds as it is held in the hand and viewed from different angles, or the individual moves round the world for example by crawling and sees the world gradually change as obstacles are moved round or bumped into or grasped. The latter type of real-world training allows temporal and spatial continuities about objects in the world to be learned, and that is the style of training used with VisNet (Chapter 4) in which continuities about visually transforming objects are used to help learn about the properties of objects and how they transform in the real world. With such different styles of training, it is likely to be difficult to use artificial deep learning networks trained by thousands of different labelled exemplars to understand how the real primate including human brain learns about the rules of perception, visual object recognition, language, etc. It is not clear what would be learned for example if a primate or human brain was trained only on thousands of labelled exemplars, rather than on the continuous input from the world in which the properties of objects and space can be learned as they continuously transform. It is not clear that whatever such an individual learned would be at all like what is learned and is present in a real brain trained in a real natural environment. An interesting implication is that training AI systems in ways more similar to the ways in which real biological systems learn from the natural environment may be useful (Rolls, 2026d). AI may benefit from knowledge of how real brains learn in their natural environments.

Another key problem for AI networks such as deep convolutional neural networks trained by error backpropagation to predict the category into which an image exemplar should be placed, is that the training used results in local image features being learned that can be used to perform the categorisation required by the training, whereas human perception typically uses the shape of the whole object (Bowers et al., 2022; Rolls, 2026d). The implication is that although artificial deep neural network models may learn to predict how to categorize a set of visual stimuli, they may do so in low-level ways not used by humans, and thus may not reveal how human perception works (Bowers et al., 2022; Rolls, 2026d). In this context, the VisNet architecture described in Chapter 4 is as far as I know the most biologically plausible approach to how computations in the ventral visual system for invariant object and face recognition are performed using unsupervised 'slow', memory trace-based, learning (Rolls, 2021c; Zhang et al., 2026a; Rolls, 2026d).

Another problem with deep networks trained with backpropagation of error, is that if something new is learned later, that may disrupt what has already been learned (Song, Millidge, Salvatori, Lukasiewicz, Xu and Bogacz, 2024). For a biological system, that would be a real problem, even if with a lot more training the correct mappings could then be achieved. In that context, a different type of learning algorithm has been proposed, 'prospective configuration' learning, in which the network first infers the pattern of neural activity that should result from learning, and then the synaptic weights are modified to consolidate the change in neural activity (Song et al., 2024). However this training algorithm still depends on some type of error correction to adjust the synaptic weights towards those that would produce the correct firing rates, and how biologically plausible that is remains to be made clear, especially for unsupervised learning which seems so typical of learning by the brain.

What is described here and elsewhere (Rolls, 2016b, 2021c,a, 2026d) may it is hoped be useful for developing better artificial neural networks and artificial intelligence. For example, deep hierarchical convolutional neural networks are typically trained on very large numbers of single training image exemplars (snapshots) of the classes to be learned, and can fail if a few pixels are altered, implying that they learn pixel-level representations. It is proposed that training such networks with different transforms of objects with statistics of how objects transform in short time periods in the real world as in trace rule learning (Chapter 4) would much better enable transform-invariant shape-based representations to be learned, leading to much more powerful performance.

Deep learning by error backpropagation is so powerful as a training algorithm to learn a mapping that it can be used to produce neuronal activity in one of its layers that resembles that found in the brain (Yamins and DiCarlo, 2016; Rajalingham et al., 2018; Macpherson et al., 2021; Zhuang et al., 2021; Sejnowski, 2020, 2023; Rolls, 2024a). But, for the reasons provided here (including above and in Section 4.7.6), it is not clear from these artificial networks whether we learn much useful about how the brain performs its computations, as the architecture, training algorithms, and training procedures are typically so different.

Limitations of current understanding of deep learning networks and how different they are from the computations performed by the brain, and how difficult it would be to implement a backprop equivalent in the brain perhaps with autoencoders without raising many further difficulties, have been highlighted by others (Plebe and Grasso, 2019; Sejnowski, 2020; Lillicrap, Santoro, Marris, Akerman and Hinton, 2020; Bowers, Malhotra, Dujmovic, Llera Montero, Tsvetkov, Biscione, Puebla, Adolfi, Hummel, Heaton, Evans, Mitchell and Blything, 2022). The mild guidance of competitive learning by cortico-cortical backprojections described in Section B.4.5 of Rolls (2023c, 2026d) may provide a brain-feasible and at least biologically plausible approach.

Many further differences between computations in the brain and those performed by current AI systems are described and evaluated in **Brain Computations and Principles; and AI** (Rolls, 2026d). The evidence available now is that brain computations are performed very differently to those in current AI systems; that current AI systems can not be used as a guide to how the brain performs its computations; and that some future developments in AI will be made by understanding better how the brain performs its computations (Rolls, 2026d).

8.3 Biologically plausible computation in the brain: a grand unifying theory?

In an interesting and perceptive review in the journal Brain of **Brain Computations: What and How** (Rolls, 2021a), Manohar (2022) raised the question of whether the book provides 'a grand unifying theory' of brain function with the computational neuroscience approach taken in that book. My response to that is that the approach taken in **Brain Computations: What and How** (Rolls, 2021a), in **Brain Computations and Connectivity** (Rolls, 2023c), and in **Brain Computations and Principles; and AI** (Rolls, 2026d) is indeed aiming to provide exactly that, a comprehensive theory of the operation of the brain, and is pioneering in that it forges and provides the foundation of a biologically plausible computational approach to brain function.

The approach taken by Rolls (2023c, 2026d) and here is a biologically plausible approach to how the brain computes, rather than invoking the artificial intelligence approach of deep learning to try to account for brain computation. The approach to how the brain computes in this book is biologically plausible in that it utilizes only local synaptic modification rules to set up the synaptic connections between neurons, and primarily unsupervised learning, yet considers how far this can take us in understanding how the brain operates. This approach is quite different to the deep learning approach used now in much machine learning and AI, which typically backpropagates an error back down through the network to correct every synaptic weight in all preceding layers of the network (LeCun et al., 2015; Lillicrap et al., 2020), but that is not easily consistent with cortical connectivity and function, as described in Section 8.2. Moreover, the learning in the brain is largely unsupervised, whereas deep learning approaches typically require a supervisor for each output neuron, which in turn requires what must be computed for each output neuron to be specified by the designer of the deep network. The point is developed (in Section 8.2) that machine learning might benefit from understanding some of the unsupervised approaches used by the brain to solve complex problems, including navigation, attention, semantics, and even syntax. The title of **Brain Computations and Connectivity** (Rolls, 2023c), refers to the concept that the actual connectivity at the local level of neurons and synapses that is found in the brain, and between brain regions, is important in understanding what and how the brain computes.

In Brain Computations and Connectivity (Rolls, 2023c) and in **Brain Computations and Principles; and AI** (Rolls, 2026d), I provided much new evidence that is important to this biologically plausible computational neuroscience approach in that the connectivity studies of the 360 cortical areas defined in the Human Connectome Project Multimodal Parcellation atlas (HCP-MMP) (Glasser et al., 2016a; Huang et al., 2022) that are described in *Brain Computations and Principles; and AI* (Huang et al., 2021; Ma et al., 2022; Rolls et al., 2022a,b, 2023g,a, 2022c, 2023b,f,c,d,e, 2024c,a) provide a framework to help understand better what computation is being performed in each of these brain regions, and how the results of that computation are fed primarily to a limited set of further cortical regions to which strong effective connectivity is directed.

We can then ask whether the biologically plausible networks described in these books (Rolls, 2016b, 2021a, 2023c, 2026d) are sufficient to account for brain computations.

A pattern association network (Section 1.4 and Fig. 1.3; and Rolls (2026d) Section B.2) is excellent for recalling a stimulus or event with which a stimulus has been associated. A simple example is an association between a previously neutral visual stimulus, such as the sight of an object, with the taste of a food reward, to produce an expected reward value representation, that can be used to guide goal-directed actions to obtain the taste reward outcome. This type of learning is key in emotional learning (Chapter 3) (Rolls, 2023b). This type of pattern association is also fundamental to how the backprojection pathways from the hippocampus to the neocortex in the brain are used to recall episodic memories to the neocortex (Section 9.3.8 of Rolls (2026d) and Rolls and Treves (2024); Rolls et al. (2024d)), and also to how backprojection pathways are used to implement top-down attention (Section 2.12 of Rolls (2026d)).

An autoassociation network (Section 1.4 and Fig. 1.3; and Rolls (2026d) Section B.3) performs *the* key type of cortical computation, and is implemented by the highly developed excitatory local recurrent collateral connections between nearby pyramidal cells in the neocortex (Section 1.4.5 of Rolls (2026d)), or throughout the CA3 neurons of the hippocampus (Chapter 5). One key computational property is that the self-re-excitation of any one of many different sets of neurons with strong connections between them can maintain the firing of that set of neurons to implement short-term memory, and to provide a foundation for working memory and planning (Chapter 13 of Rolls (2026d)).

This computational ability of autoassociation networks to maintain activity is useful not only for what we might think of as short-term memory functions, but for many other functions. An important example in sensory cortices is to maintain activity for sufficiently long, in the order of one to several seconds, for the statistics of the world to be able to guide the learning of representations that are useful because they capture the properties of stimuli that all can be produced by the same object or stimulus. A key example is the trace learning rule in VisNet to help learn invariant representations (Section 4.7). A key computational property that is provided here is that this is relatively unsupervised learning: there is no teacher for each output neuron, and no need to prespecify what categories should be formed, which is a limitation of brain computation that is used in typical supervised deep learning artificial neural networks. The point here is that the cortical recurrent collaterals can help to maintain neuronal activity in a population of neurons for sufficiently long to enable invariant representations to be formed based on the statistics of the input from the world. The same principle is useful in coordinate transforms learned by gain modulation ((Rolls, 2020b) and Section 3.5 of Rolls (2026d)), and in the learning of global motion in the dorsal visual system (Section 4.7.4).

Another key property of autoassociation networks is that the whole of the representation, for example a whole memory, can be retrieved from any one part. This is key in the mechanisms of episodic memory (Chapter 5). It is emphasized that this is not a property of a pattern association network.

A key feature of pattern association and autoassociation networks in the brain is that they have a high capacity, which is in the order of the number of associatively modifiable synapses on any one neuron in a network with sparse representations (Sections B.2 and B.3 of Rolls (2026d)). A typical cortical neuron might have approximately 10,000 associatively modifiable synapses, so the storage capacity of a single local neocortical network which might be approximately 1–2 mm in diameter (and containing 100,000 neurons given the diluted connectivity that is a feature of the cortex) could store in the order of 10,000 memories, which is the size of a person's typical working vocabulary. (And that may not be just a co-incidence.)

However a key property of the neocortex is that it learns new representations: it can categorize incoming stimuli for example as one type of object, or another. The key type of biologically plausible network for this type of learning is a competitive network (Section 1.4 and Fig. 1.3; and Rolls (2026d) Section B.4). It can usefully allocate inputs that regularly occur together to form a category defined by their combination or conjunction. Competitive networks with sparse representations can learn to categorize inputs with a fairly high capacity. Moreover, competitive networks implement a type of unsupervised learning. Improvements to the learning rules used in competitive networks as described in Sections 4.7.3.1 and 4.7.3.2 and by Zhang, Rolls and Feng (2026a) make competitive networks even more biologically plausible.

A very important property of competitive networks is that they can be combined with other synapses on the same neurons that implement short-term memory using attractor dynamics, and with another set of synaptic inputs that learn by pattern association to initiate memory recall later, or to implement top-down attention. This has been tested computationally: all three types of computation can be performed using these three types of computation by populations of such single neurons, and this is proposed to be *the* key feature of the design of the neocortex (Rolls, 2021e).

The architecture that was simulated to prove the operation of this computational combination is shown in Fig. 8.2 (Rolls, 2021e), and is a part of the neocortical architecture illustrated in Fig. 8.3. The computational model combines competitive learning using the forward inputs from the preceding cortical region, with associatively modifiable recurrent collateral synapses to form an autoassociation attractor network to maintain neuronal firing for short periods, and with pattern association learning to implement recall or top-down attention using cortico-cortical backprojections. It was shown that during training, the forward inputs could be utilized for competitive learning, and at the same time the recurrent collateral connections could be trained to set up an attractor network between the set of neurons that had learned to recognize a pattern by competitive learning. Also during training, the top-down backprojections that were active could be associated with the pyramidal cells that responded to any one category.

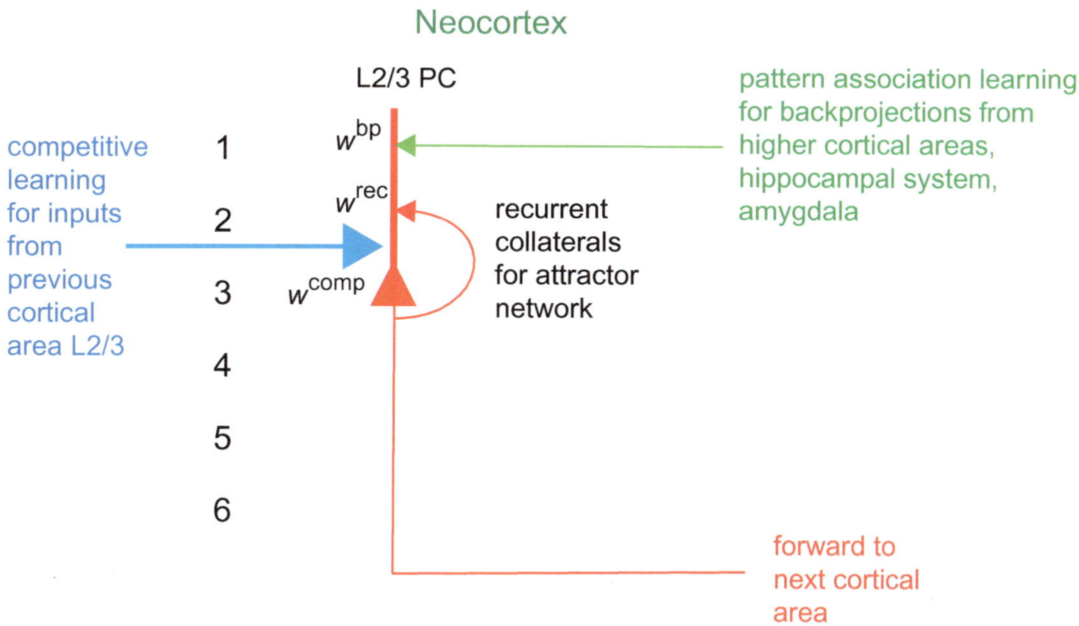

Fig. 8.2 The neocortical model that combines competitive learning using the forward inputs from the preceding cortical region, with associatively modifiable recurrent collateral synapses to form an autoassociation attractor network to maintain neuronal firing for short periods, and with pattern association learning to implement recall or top-down attention using cortico-cortical backprojections. L2/3 PC is a layer 2/3 pyramidal cell. The thick line above the pyramidal cell body is the dendrite, receiving inputs from the previous cortical area and operating by competitive learning using the synapses w^{comp}; from the recurrent collaterals that operate as an attractor network using synapses w^{rec}; and back-projections from higher cortical areas for memory recall and top-down attention that operate by pattern association learning using the synapses w^{bp}. 1-6 : the layers of the neocortex. (Data from Rolls E T. 2021 The connections of neocortical pyramidal cells can implement the learning of new categories, attractor memory, and top-down recall and attention. Brain Structure and Function 226: 2523-2536.)

During the recall stage when there was no forward input from the preceding cortical stage, the top-down backprojection inputs could retrieve the correct set of pyramidal cells that had been active in association with the top-down recall cue, and the firing of the pyramidal cells could be maintained in short-term memory by the attractor synapses. The whole model was useful not only in showing that this could all be achieved in a single population of pyramidal cells with these three types of input, but also in showing suitable ranges for the strengths of the three types of synapse on each neocortical pyramidal cell (Fig. 8.2) (Rolls, 2021e).

Whether competitive networks are sufficiently powerful, when combined with pattern association and autoassociation properties on the same neurons, to account for the new categories that can be learned, typically in hierarchical networks, remains to be determined. This is a key frontier for future research: if competitive networks, constructed with the addition of pattern and autoassociators in the neocortex, are not sufficiently powerful to account for the learning of new categories and the formation of semantic networks, then what alternatives are there? Potts attractor networks, which are combinations of attractor networks, may have high storage capacity (Song, Yao and Treves, 2014; Boboeva, Brasselet and Treves, 2018; Ryom, Stendardi, Ciaramelli and Treves, 2023; Rolls and Treves, 2024), but it is not clear how the new representations that are needed could be learned by them. So the area of how biologically plausible networks can learn new representations if competitive networks are not sufficiently powerful is of great interest for the future.

Another type of learning that is probably implemented in the brain is reinforcement learning

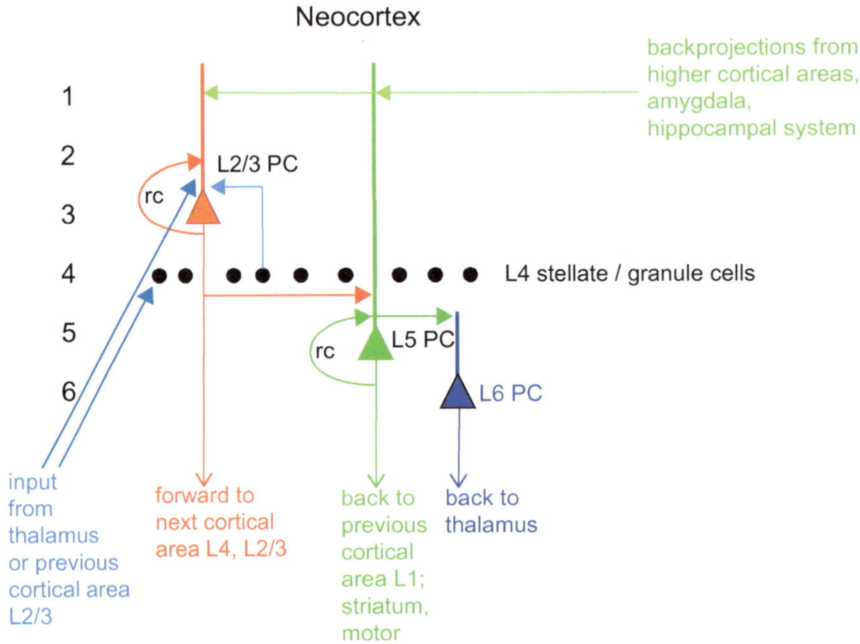

Fig. 8.3 Functional canonical microcircuit of the neocortex (see text). The cortical layers are numbered 1–6, with layer 1 at the surface of the cortex, and layer 6 adjacent to the white matter consisting of axonal connections to other brain areas. Recurrent collateral connections (rc) are shown as a loop back to a particular population of cells, of which just one neuron is shown in this Figure. In primates the feedforward projection neurons are concentrated in L3B; and the main feedback projection neurons are in L6 and Lower L5 (L5B), but some L2 and L3A neurons do send backprojections especially for adjacent regions in a hierarchy (Markov et al., 2014). Some L6 cortico-thalamic neurons send a projection to L4 (see text). (Data from Rolls E T. 2016 Cerebral Cortex: Principles of Operation. Oxford University Press: Oxford.)

(Sutton and Barto, 2018) using dopamine as a single reward prediction error signal (Schultz, 2016c,b, 2017). One of its main uses in the brain is for stimulus-response learning of rather fixed habits using the dopamine pathway that projects to the basal ganglia (Schultz, 2016b) (Chapter 16 of Rolls (2026d), and that is one of the relatively simple but relatively slow types of learning performed by the brain, and may even predate the neocortex in evolution.

Another type of learning may be some perceptron type of learning with a separate teacher for each neuron (Section B.11 of Rolls (2026d)) in one part of the brain, the cerebellum, where the teacher may be the climbing fibre which forces a pyramidal cell to learn what is present on its parallel fibre inputs (Chapter 17 of Rolls (2026d)). However, it is not clear how the teacher for each Purkinje cell knows what to teach that Purkinje cells about how to respond to its parallel fibre inputs (Rolls, 2023c, 2026d).

Novelty detection is relatively easy to set up in neural networks, and is found in for example the orbitofrontal cortex (Rolls, Browning, Inoue and Hernadi, 2005a). The opposite, familiarity detection, which might help to represent one's friends, one's belongings, and one's territory, and indeed to reflect ownership, appears to be represented in the perirhinal cortex (Holscher, Rolls and Xiang, 2003), and again can be easily learned in neuronal networks (Rolls, Franco and Stringer, 2005b).

The approach taken in this book and in other books (Rolls, 2016b, 2021a, 2023c, 2026d) may thus not be a 'a grand unifying theory' (Manohar, 2022) of how the brain works. But it is a pioneering approach to how biologically plausible computation may be performed in very many parts of our brains to implement very many of the wonderful things that our brains can do. It is an open issue at present about whether different principles are needed, and if so, how they are implemented in biologically plausible networks in our brains.

These books (Rolls, 2016b, 2021a, 2023c, 2026d) are intended to pioneer a foundation for future developments, by setting out what we know about what computations are performed in many different brain regions, and how biologically plausible networks could perform the computations in many brain regions; and leave open the question of whether the computations that are performed might need some more powerful but biologically plausible algorithms than those described in them and here.

These books are also pioneering in that they go elsewhere than mainly theoretical studies of neural networks (Dayan and Abbott, 2001; Gerstner, Kistler, Naud and Paninski, 2014; Izhikevich, 2007; Ermentrout and Terman, 2010; Wang, 2025), and approach instead what particular networks may be implemented in different brain regions, and how these networks or modules are interconnected to enable processes such as episodic memory, invariant visual object recognition, and emotion to be implemented in the real brain of primates including humans.

A feature of my previous books and this book is that they highlight many of the fundamental conceptual and empirical discoveries that have been made to help understand cortical function and computation. Highlighting what the fundamental discoveries were, and how they were made, is didactically important to help foster future fundamental advances, which is an aim of this book.

8.4 Primates including humans have a different systems-level organization of many brain systems compared to rodents

Primates including humans have a different systems-level organization of many brain systems compared to rodents (rats and mice), as set out in detail in Section 19.10 of Rolls (2023c) and in Section 29.4 of Rolls (2026d). This is evident for at least the following systems: the visual system; the taste system; the olfactory system; the somatosensory system; the auditory system; the hippocampal system for memory and navigation; the orbitofrontal cortex and anterior cingulate cortex for reward value and emotion; the visuo-motor system; and of course the language systems (Rolls, 2023c, 2026d).

This makes rodents in many cases not a good model for the systems-level analysis of the computations performed in many human (and other primate) brain systems. The focus of this book and Rolls (2023c, 2026d) is therefore on primates including humans, with evidence frequently provided by neurophysiological discoveries in primates, that are important for understanding the computations being performed, and that are closely related to corresponding processing in humans as shown by neuroimaging investigations and the effects of brain damage in humans. The interest here is to understand the human brain in health and disease, including providing an understanding and a foundation that are relevant to treatment in humans.

In addition, neuronal responses in mice to impending motor action, and reward delivery and consumption, are widespread across many brain regions (International Brain et al., 2025), in contrast to the much greater selectivity of neuron types found in different cortical regions in primates including humans that are described throughout this book, again providing evidence that the rodent brain provides a poor model of the systems-level organization of the primate including human brain.

The investigation performed on hundreds of thousands of neurons in mice (International Brain et al., 2025) also raises an interesting strategic point in neuroscience research. The point is to compare what can be learned from studying hundreds of thousands of neurons in a fixed single task or set of tasks, which is tempting with the availability of electrodes with multiple recording sites, compared to focussing on a single neuron at a time, and performing extensive tests while measuring its responses to unravel what makes that neuron respond and what it encodes. In almost all of the primate neurophysiology studies described in this book, the focus was on understanding exactly what the effective stimuli are for each individual neuron, and what information is represented, and how it is represented by groups of such neurons, and that allowed many of the discoveries and advances described in this book to be made. This approach to analyze exactly what is encoded by single neurons by detailed investigations on each neuron with many different hypotheses tested on each neuron was indeed essential when pioneering discoveries were being made when little was already known about each brain region (Rolls, 2026d). I believe that this way of advancing our understanding of the brain needs to be kept in mind in future, for that is how many discoveries are made.

I also believe that in addition to performing multiple regression analyses on large populations of neurons, it is also important to use information theory to measure what is encoded by populations of neurons, for information theory provides a way to quantitatively assess encoding using a scale in which information is additive and usefully quantitative in that sense, as described in this book and more fully in Rolls (2026d).

8.5 Reflections on discoveries in neuroscience

I have the following reflections on the discoveries that I have made described in this book, which may be of interest to other scientists and to historians of science.

To understand a brain computation, a set of integrated interdisciplinary discoveries is needed, including empirical discoveries at the neuronal level about what is represented in connected brain regions, and computational neuroscience theories about how the computations are performed. That approach with a set of integrated interdisciplinary discoveries to understand the operation of brain systems is described here and in **Brain Computations and Principles; and AI** (Rolls, 2026d) and elsewhere (Rolls, 2025a), and contrasts with the single advance that is sometimes recognized by awards in neuroscience.

The discoveries that I have made have often been pioneering and at the forefront of scientific investigations (e.g. of reward neurons in the orbitofrontal cortex, e.g. of face cells, e.g. of spatial view cells in primates, e.g. of theories of hippocampal function, of invariant visual object recognition, and of emotion). Being at the forefront is very exciting, and because one is ahead of others, one can chart one's own path of how to move forward with further discoveries to improve understanding.

But at the same time, when at the forefront, when one submits a paper to a scientific journal, the editors may not immediately respond to move a paper forwards. They will not know much or anything about the new concepts, and some of the discoveries may seem surprising to them. So they may be conservative, and accept instead papers in the following waves of scientific investigation, which are moving ahead with steady progress (and in recent years by emphasis on the use of the latest techniques), without being able to assess what are conceptual advances in a field.

That in turn may make the authors of pioneering research not promote their new ideas too much when the ideas are very new, for it can take some time for the scientific establishment to get used to the new ideas, and, as it were, for reviewers to have had time to see for themselves by running their own investigations. At these later stages, that follow the conceptual discoveries, a flood of new papers can come somewhat easily, but without necessarily much new conceptual content.

I think that other investigators may be interested in these reflections on what it feels like to be at the forefront of scientific discoveries.

Appendix 1 Human Cerebral Cortex Parcellation

A.1 The Human Connectome Project Multimodal Parcellation of the Human Cerebral Cortex: Introduction

The connectivity between regions of the human cerebral cortex described in this book and in our connectivity papers (Rolls et al., 2022a; Ma et al., 2022; Rolls et al., 2022b,c, 2023b,f,c,g,a,d, 2024a, 2023e, 2024c; Rolls and Turova, 2025) uses the Human Connectome Project Multimodal Parcellation (HCP-MMP) atlas (Glasser, Coalson, Robinson, Hacker, Harwell, Yacoub, Ugurbil, Andersson, Beckmann, Jenkinson, Smith and Van Essen, 2016a). A brief description of why this atlas is used, its cortical regions and how they are separated into divisions, and the abbreviations, are described for convenience in this Appendix.

The Human Connectome Project Multimodal Parcellation atlas (HCP-MMP) provides a very useful parcellation of human cerebral cortical regions (Glasser et al., 2016b,a). The atlas is multimodal in that each region is defined by a combination of criteria: architecture (T1w/T2w myelin content and cortical thickness maps which can be identified with MRI); resting-state functional connectivity; and task-based fMRI activation in seven tasks (Glasser et al., 2016a). The HCP-MMP v1.0 atlas includes 180 cortical regions in each hemisphere, based on analyses in 210 participants. The regions defined in this atlas include many regions that are of interest because there is evidence on their functions. These regions include early visual cortex regions including V1, V2, V3, V4; ventral visual system regions such as FFC, PIT, V8, TE1p and TE2p; dorsal visual system motion regions MT, MST and the connected intraparietal LIP and VIP etc as shown in Figs. 1.1 and 1.2. The boundaries of each cortical region are defined using this HCP-MMP approach based on a combination of the multimodal evidence obtained in all of these ways. This potentially makes this atlas able to discriminate more reliably between different cortical regions with potentially different functions and connectivity than any other atlas based on evidence from a single modality such as only functional connectivity.

Given the usefulness of the parcellation provided in the HCP-MMP atlas (Glasser et al., 2016a), we have provided an extended version of the atlas, HCPex, that extends it with 66 subcortical areas, provides it in volumetric form useful for many types of neuroimaging software (e.g. SPM (Friston et al., 2006) https://www.fil.ion.ucl.ac.uk/spm/ and FSL (Jenkinson et al., 2012; Smith et al., 2004) https://fsl.fmrib.ox.ac.uk), and by providing a reordered version with an option to revert to the original order (Huang, Rolls, Feng and Lin, 2022). The HCPex version of the HCP-MMP atlas (Huang, Rolls, Feng and Lin, 2022) is available at https://www.oxcns.org. The order of cortical regions defined in HCPex and shown in Tables A.1 – A.6 is used throughout this book. This atlas helps in the analysis of brain structure and function, for many of the cortical areas defined in the HCP-MMP / HCPex atlas are known about functionally, many can be related to corresponding regions in macaques in which much detailed neurophysiology is available, and the atlas helps new investigations of the brain to be related to the known functions of those anatomically defined brain regions.

The surface-based registration such as that provided by the HCP-MMP leads to better spatial localization of cortical areas than volumetric methods (Coalson et al., 2018; Glasser et al., 2016b) such as those commonly used and made available in neuroimaging software such as SPM and FSL. However, given that much neuroimaging analysis is performed with software such as SPM and FSL, and that volumetric analysis enables subcortical areas to be easily included straightforwardly in the atlas which is an aim of HCPex, we have provided HCPex which is in volumetric space (Huang, Rolls, Feng and Lin, 2022). At the same time, where better localization is required than may be provided with HCPex, we appreciate and recommend the use of the surface-based version HCP-MMP v1.0 (Glasser et al., 2016a), for the reasons set out by Coalson et al. (2018). In fact, for most of the effective connectivity analyses described in this book, the cortical connectivity was measured with the surface-based parcellation, and the connectivity with subcortical areas was measured with HCPex.

Given these advantages of the HCP-MMP human brain parcellation for helping to delineate different regions of the human cerebral cortex, we have built on it as follows. We have measured the connectivity between the different cortical regions in this parcellation, to help provide a foundation for understanding how different human cortical regions are connected, to provide a foundation for understanding how information is processed through each of our brain systems. We analysed the direct connections between all of these cortical regions with diffusion tractography measures (Huang, Rolls, Hsu, Feng and Lin, 2021), and used this connection evidence to advance understanding of many different cortical systems. (Further information about these methods is provided in Chapter 6 which describes our advances in understanding the connectivity of the human brain). In addition, I decided to measure the effective connectivity between all these 360 cortical regions in the HCP-MMP atlas, where effective connectivity estimates the effects of each pair of brain regions on each other using time delays in the fMRI BOLD signal (Rolls, Deco, Huang and Feng, 2022a). This is important for providing evidence about the directionality of the flow of information between cortical regions (see Chapter 6), and hence about cortical computation. In addition, we also measure the functional connectivity between all 360 cortical regions in the HCP-MMP atlas, to provide evidence of the interactions between cortical regions. We are the only research group that has used all three measures to advance understanding of most of the information processing streams in the human brain (Rolls et al., 2022a; Ma et al., 2022; Rolls et al., 2022b,c, 2023b,f,c,g,a,d, 2024a, 2023e, 2024c; Rolls and Turova, 2025), as described in Chapter 6 and Sections 3.10, 4.2, 5.2.10 and 5.2.11.

The cortical regions defined in the HCP-MMP / HCPex atlas are shown in Figs. 1.1 and 1.2, with a list of the cortical regions in Tables A.1–A.6 and of the subcortical regions in HCPex in Table A.7. Coronal slices to further help visualization of the locations in the brain of these cortical regions are provided by Huang, Rolls, Feng and Lin (2022). Descriptions of many of these cortical regions including their connectivity and functions are provided in the different Chapters of this book and more fully by Rolls (2023c, 2026d). Short summaries of some of the key Divisions in the HCP-MMP and HCPex atlases (Glasser et al., 2016a; Huang et al., 2022) are provided next with reference to Tables A.1–A.6. The Divisions in the HCP-MMP atlas (Glasser et al., 2016a) are sets of cortical regions grouped together based on their connectivity and functions by the Human Connectome Project (Glasser et al., 2016a).

Table A.1 Cortical regions in the HCP-MMP atlas in the order used in the HCPex atlas

ID (L, R)	Region	Region Long Name	Cortical Division
1, 181	V1	Primary Visual Cortex	Primary Visual
2, 182	V2	Second Visual Area	Early Visual
3, 183	V3	Third Visual Area	Early Visual
4, 184	V4	Fourth Visual Area	Early Visual
5, 185	IPS1	IntraParietal Sulcus Area 1	Dorsal Stream Visual
6, 186	V3A Area	V3A	Dorsal Stream Visual
7, 187	V3B Area	V3B	Dorsal Stream Visual
8, 188	V6	Sixth Visual Area	Dorsal Stream Visual
9, 189	V6A	Area V6A	Dorsal Stream Visual
10, 190	V7	Seventh Visual Area	Dorsal Stream Visual
11, 191	FFC	Fusiform Face Complex	Ventral Stream Visual
12, 192	PIT	Posterior InferoTemporal complex	Ventral Stream Visual
13, 193	V8	Eighth Visual Area	Ventral Stream Visual
14, 194	VMV1	VentroMedial Visual Area 1	Ventral Stream Visual
15, 195	VMV2	VentroMedial Visual Area 2	Ventral Stream Visual
16, 196	VMV3	VentroMedial Visual Area 3	Ventral Stream Visual
17, 197	VVC	Ventral Visual Complex	Ventral Stream Visual
18, 198	FST	Area FST	MT+ Complex
19, 199	LO1	Area Lateral Occipital 1	MT+ Complex
20, 200	LO2	Area Lateral Occipital 2	MT+ Complex
21, 201	LO3	Area Lateral Occipital 3	MT+ Complex
22, 202	MST	Medial Superior Temporal Area	MT+ Complex
23, 203	MT	Middle Temporal Area	MT+ Complex
24, 204	PH	Area PH	MT+ Complex
25, 205	V3CD	Area V3CD	MT+ Complex
26, 206	V4t	Area V4t	MT+ Complex

A.2 Visual processing regions

The main visual processing cortical regions are in the Primary Visual, Early Visual, Dorsal Visual Stream, Ventral Visual Stream, and MT+ Complex Divisions in Table A.1, and in the Lateral Temporal Division in Table A.3, and their connectivity and functions are described in Chapters 2 and 3 of Rolls (2023c, 2026d), and by Rolls et al. (2023b); Rolls (2024c), and in Chapters 4 and 5 of this book.

A.3 Somatosensory and premotor cortex regions

The main somatosensory and motor cortical regions are in the SomatoSensory, ParaCentral and MidCingulate, Premotor, and Posterior Opercular Divisions in Table A.2, and in the Insula / Frontal Opercular Division in Table A.3, and their connectivity and functions are described in Chapters 6 and 15 of Rolls (2026d), and by Rolls et al. (2023c), and in Section 6.6 in this book. Regions 23c, 24dd and 24dv are the midcingulate premotor cortex described by Vogt (2016) and Rolls et al. (2023g).

A.4 Auditory cortex regions

The main auditory cortical regions are in the Early Auditory, and Auditory Association Divisions in Table A.2, and their connectivity and functions are described in Chapter 7 of Rolls (2026d) and by Rolls, Rauschecker, Deco, Huang and Feng (2023f), and in Section 6.7 of this book. However some of the regions in the 'Auditory Association' Division are in fact multimodal, include visual and auditory responsiveness to social stimuli, and are also involved in semantic representations for language, as described in Chapters 8 and 2 and Section 14.5 of Rolls (2023c), and by Rolls et al. (2022c), Rolls et al. (2023b), and Pitcher and Ungerleider (2021).

A.5 Hippocampal, medial temporal lobe, and related posterior cingulate cortex regions

The hippocampal memory system is in the Medial Temporal Division in Table A.3, with its closely related regions in the Posterior Cingulate Division shown in Table A.4. Their connectivity and functions are described in Chapters 9 and 10 of Rolls (2026d), and by Rolls et al. (2022a, 2023g,a); Rolls (2023a, 2022a); Rolls and Treves (2024); Rolls (2025a), and in Chapter 5 of this book.

A.6 Language-related regions

The human language-related cortical regions are in the Temporo-Parietal-Occipital TPO Division in Table A.3, and regions 44, 45 and 47l (Broca's area) in Table A.6. Their connectivity and functions are described in Chapter 14 of Rolls (2026d), and by Rolls et al. (2022c), and in Section 6.10 of this book.

Table A.2 Cortical regions in the HCP-MMP atlas in the order used in the HCPex atlas

ID (L, R)	Region	Region Long Name	Cortical Division
27, 207	1	Area 1	SomaSens Motor
28, 208	2	Area 2	SomaSens Motor
29, 209	3a	Area 3a	SomaSens Motor
30, 210	3b	Primary Sensory Cortex	SomaSens Motor
31, 211	4	Primary Motor Cortex	SomaSens Motor
32, 212	23c	Area 23c	ParaCentral MidCing
33, 213	24dd	Dorsal Area 24d	ParaCentral MidCing
34, 214	24dv	Ventral Area 24d	ParaCentral MidCing
35, 215	5L	Area 5L	ParaCentral MidCing
36, 216	5m	Area 5m	ParaCentral MidCing
37, 217	5mv	Area 5m ventral	ParaCentral MidCing
38, 218	6ma	Area 6m anterior	ParaCentral MidCing
39, 219	6mp	Area 6mp	ParaCentral MidCing
40, 220	SCEF	Supplementary and Cingulate Eye Field	ParaCentral MidCing
41, 221	55b	Area 55b	Premotor
42, 222	6a	Area 6 anterior	Premotor
43, 223	6d	Dorsal area 6	Premotor
44, 224	6r	Rostral Area 6	Premotor
45, 225	6v	Ventral Area 6	Premotor
46, 226	FEF	Frontal Eye Fields	Premotor
47, 227	PEF	Premotor Eye Field	Premotor
48, 228	43	Area 43	Posterior Opercular
49, 229	FOP1	Frontal Opercular Area 1	Posterior Opercular
50, 230	OP1	Area OP1-SII	Posterior Opercular
51, 231	OP2-3	Area OP2-3-VS	Posterior Opercular
52, 232	OP4	Area OP4-PV	Posterior Opercular
53, 233	52	Area 52	Early Auditory
54, 234	A1	Primary Auditory Cortex	Early Auditory
55, 235	LBelt	Lateral Belt Complex	Early Auditory
56, 236	MBelt	Medial Belt Complex	Early Auditory
57, 237	PBelt	ParaBelt Complex	Early Auditory
58, 238	PFcm	Area PFcm	Early Auditory
59, 239	RI	RetroInsular Cortex	Early Auditory
60, 240	A4	Auditory 4 Complex	Auditory Association
61, 241	A5	Auditory 5 Complex	Auditory Association
62, 242	STGa	Area STGa	Auditory Association
63, 243	STSda	Area STSd anterior	Auditory Association
64, 244	STSdp	Area STSd posterior	Auditory Association
65, 245	STSva	Area STSv anterior	Auditory Association
66, 246	STSvp	Area STSv posterior	Auditory Association
67, 247	TA2	Area TA2	Auditory Association

Table A.3 Cortical regions in the HCP-MMP atlas in the order used in the HCPex atlas

ID (L, R)	Region	Region Long Name	Cortical Division
68, 248	AAIC	Anterior Agranular Insula Complex	Insula FrontalOperc
69, 249	AVI	Anterior Ventral Insular Area	Insula FrontalOperc
70, 250	FOP2	Frontal Opercular Area 2	Insula FrontalOperc
71, 251	FOP3	Frontal Opercular Area 3	Insula FrontalOperc
72, 252	FOP4	Frontal Opercular Area 4	Insula FrontalOperc
73, 253	FOP5	Area Frontal Opercular 5	Insula FrontalOperc
74, 254	Ig	Insular Granular Complex	Insula FrontalOperc
75, 255	MI	Middle Insular Area	Insula FrontalOperc
76, 256	PI	Para-Insular Area	Insula FrontalOperc
77, 257	Pir	Piriform Cortex	Insula FrontalOperc
78, 258	PoI1	Area Posterior Insular 1	Insula FrontalOperc
79, 259	PoI2	Posterior Insular Area 2	Insula FrontalOperc
80, 260	H	Hippocampus	Medial Temporal
81, 261	PreS	PreSubiculum	Medial Temporal
82, 262	EC	Entorhinal Cortex	Medial Temporal
83, 263	PeEc	Perirhinal Ectorhinal Cortex	Medial Temporal
84, 264	TF	Area TF	Medial Temporal
85, 265	PHA1	ParaHippocampal Area 1	Medial Temporal
86, 266	PHA2	ParaHippocampal Area 2	Medial Temporal
87, 267	PHA3	ParaHippocampal Area 3	Medial Temporal
88, 268	PHT	Area PHT	Lateral Temporal
89, 269	TE1a	Area TE1 anterior	Lateral Temporal
90, 270	TE1m	Area TE1 Middle	Lateral Temporal
91, 271	TE1p	Area TE1 posterior	Lateral Temporal
92, 272	TE2a	Area TE2 anterior	Lateral Temporal
93, 273	TE2p	Area TE2 posterior	Lateral Temporal
94, 274	TGd	Area TG dorsal	Lateral Temporal
95, 275	TGv	Area TG Ventral	Lateral Temporal
96, 276	PSL	PeriSylvian Language Area	TPO
97, 277	STV	Superior Temporal Visual Area	TPO
98, 278	TPOJ1	Area TemporoParietoOccipital Junction 1	TPO
99, 279	TPOJ2	Area TemporoParietoOccipital Junction 2	TPO
100, 280	TPOJ3	Area TemporoParietoOccipital Junction 3	TPO

Table A.4 Cortical regions in the HCP-MMP atlas in the order used in the HCPex atlas

ID (L, R)	Region	Region Long Name	Cortical Division
101, 281 7AL	Lateral Area 7A	Superior Parietal	
102, 282	7Am	Medial Area 7A	Superior Parietal
103, 283	7PC	Area 7PC	Superior Parietal
104, 284	7PI	Lateral Area 7P	Superior Parietal
105, 285	7Pm	Medial Area 7P	Superior Parietal
106, 286	AIP	Anterior IntraParietal Area	Superior Parietal
107, 287	LIPd	Area Lateral IntraParietal dorsal	Superior Parietal
108, 288	LIPv	Area Lateral IntraParietal ventral	Superior Parietal
109, 289	MIP	Medial IntraParietal Area	Superior Parietal
110, 290	VIP	Ventral IntraParietal Complex	Superior Parietal
111, 291	IP0	Area IntraParietal 0	Inferior Parietal
112, 292	IP1	Area IntraParietal 1	Inferior Parietal
113, 293	IP2	Area IntraParietal 2	Inferior Parietal
114, 294	PF	Area PF Complex	Inferior Parietal
115, 295	PFm	Area PFm Complex	Inferior Parietal
116, 296	PFop	Area PF Opercular	Inferior Parietal
117, 297	PFt	Area PFt	Inferior Parietal
118, 298	PGi	Area PGi	Inferior Parietal
119, 299	PGp	Area PGp	Inferior Parietal
120, 300	PGs	Area PGs	Inferior Parietal
121, 301	23d	Area 23d	Posterior Cingulate
122, 302	31a	Area 31a	Posterior Cingulate
123, 303	31pd	Area 31pd	Posterior Cingulate
124, 304	31pv	Area 31p ventral	Posterior Cingulate
125, 305	7m	Area 7m	Posterior Cingulate
126, 306	d23ab	Area dorsal 23 a+b	Posterior Cingulate
127, 307	DVT	Dorsal Transitional Visual Area	Posterior Cingulate
128, 308	PCV	PreCuneus Visual Area	Posterior Cingulate
129, 309	POS1	Parieto-Occipital Sulcus Area 1	Posterior Cingulate
130, 310	POS2	Parieto-Occipital Sulcus Area 2	Posterior Cingulate
131, 311	ProS	ProStriate Area	Posterior Cingulate
132, 312	RSC	RetroSplenial Complex	Posterior Cingulate
133, 313	v23ab	Area ventral 23 a+b	Posterior Cingulate

A.7 Parietal cortex regions

The parietal cortex regions involved in visuo-spatial and related multimodal functions are in the Superior Parietal and Inferior Parietal Divisions in Table A.4, and their connectivity and functions are described in Chapter 10 of Rolls (2026d), and by Rolls et al. (2023a), and in Section 6.5 of this book.

Table A.5 Cortical regions in the HCP-MMP atlas in the order used in the HCPex atlas

ID (L, R)	Region	Region Long Name	Cortical Division
134, 314	10r	Area 10r	AntCing MedPFC
135, 315	10v	Area 10v	AntCing MedPFC
136, 316	25	Area 25	AntCing MedPFC
137, 317	33pr	Area 33 prime	AntCing MedPFC
138, 318	8BM	Area 8BM	AntCing MedPFC
139, 319	9m	Area 9 Middle	AntCing MedPFC
140, 320	a24	Area a24	AntCing MedPFC
141, 321	a24pr	Anterior 24 prime	AntCing MedPFC
142, 322	a32pr	Area anterior 32 prime	AntCing MedPFC
143, 323	d32	Area dorsal 32	AntCing MedPFC
144, 324	p24	Area posterior 24	AntCing MedPFC
145, 325	p24pr	Area Posterior 24 prime	AntCing MedPFC
146, 326	p32	Area p32	AntCing MedPFC
147, 327	p32pr	Area p32 prime	AntCing MedPFC
148, 328	pOFC	Posterior OFC Complex	AntCing MedPFC
149, 329	s32	Area s32	AntCing MedPFC
150, 330	10d	Area 10d	OrbPolaFrontal
151, 331	10pp	Polar 10p	OrbPolarFrontal
152, 332	11l	Area 11l	OrbPolarFrontal
153, 333	13l	Area 13l	OrbPolarFrontal
154, 334	47m	Area 47m	OrbPolarFrontal
155, 335	47s	Area 47s	OrbPolarFrontal
156, 336	a10p	Area anterior 10p	OrbPolarFrontal
157, 337	OFC	Orbital Frontal Complex	OrbPolarFrontal
158, 338	p10p	Area posterior 10p	OrbPolarFrontal

A.8 Orbitofrontal cortex and related anterior cingulate cortex regions

The orbitofrontal cortex and ventromedial prefrontal cortex regions involved in reward valuation and emotion, and the anterior cingulate cortex involved in linking reward to action in action-outcome learning, are in the Orbitofrontal and Polar Frontal, and Anterior Cingulate and Medial Prefrontal Cortex Divisions in Table A.5, and their connectivity and functions are described in Chapters 11 and 12 of Rolls (2026d), and by Rolls et al. (2022b, 2023d); Rolls (2023b), and in Chapter 3 of this book.

Table A.6 Cortical regions in the HCP-MMP atlas in the order used in the HCPex atlas

ID (L, R)	Region	Region Long Name	Cortical Division
159, 339	44	Area 44	Inferior Frontal
160, 340	45	Area 45	Inferior Frontal
161, 341	47l	Area 47l (47 lateral)	Inferior Frontal
162, 342	a47r	Area anterior 47r	Inferior Frontal
163, 343	IFJa	Area IFJa	Inferior Frontal
164, 344	IFJp	Area IFJp	Inferior Frontal
165, 345	IFSa	Area IFSa	Inferior Frontal
166, 346	IFSp	Area IFSp	Inferior Frontal
167, 347	p47r	Area posterior 47r	Inferior Frontal
168, 348	46	Area 46	Dorsolateral Prefrontal
169, 349	8Ad	Area 8Ad	Dorsolateral Prefrontal
170, 350	8Av	Area 8Av	Dorsolateral Prefrontal
171, 351	8BL	Area 8B Lateral	Dorsolateral Prefrontal
172, 352	8C	Area 8C	Dorsolateral Prefrontal
173, 353	9-46d	Area 9-46d	Dorsolateral Prefrontal
174, 354	9a	Area 9 anterior	Dorsolateral Prefrontal
175, 355	9p	Area 9 Posterior	Dorsolateral Prefrontal
176, 356	a9-46v	Area anterior 9-46v	Dorsolateral Prefrontal
177, 357	i6-8	Inferior 6-8 Transitional Area	Dorsolateral Prefrontal
178, 358	p9-46v	Area posterior 9-46v	Dorsolateral Prefrontal
179, 359	s6-8	Superior 6-8 Transitional Area	Dorsolateral Prefrontal
180, 360	SFL	Superior Frontal	
Language Area	Dorsolateral Prefrontal		

A.9 Prefrontal cortex regions

The prefrontal cortex cortical regions involved in short-term memory, working memory, planning, and attention are in the Inferior Frontal, and Dorsolateral Prefrontal Cortex Divisions in Table A.6, and their connectivity and functions are described in Chapter 13 of Rolls (2026d), and by Rolls et al. (2023c) and Rolls et al. (2024a).

Table A.7 Subcortical regions in the HCPex atlas

ID (L, R)	Region	Region Long Name	Cortical Division
361, 394	AV	Thalamus: Anteroventral Nucleus	
362, 395	CeM	Thalamus: Central medial	
363, 396	CL	Thalamus: Central lateral	
364, 397	CM	Thalamus: Centralmedian	
365, 398	LD	Thalamus: Laterodorsal	
366, 399	LGN	Thalamus: Lateral Geniculate	
367, 400	LP	Thalamus: Lateral Posterior	
368, 401	L-Sg	Thalamus: Limitans Suprageniculate	
369, 402	MDl	Thalamus: Mediodorsolateral parvocellular	
370, 403	MDm	Thalamus: Mediodorsomedial magnocellular	
371, 404	MGN	Thalamus: Medial Geniculate	
372, 405	MV(Re)	Thalamus: Reuniens	
373, 406	Pf	Thalamus: Parafascicular	
374, 407	PuA	Thalamus: Pulvinar anterior	
375, 408	PuI	Thalamus: Pulvinar inferior	
376, 409	PuL	Thalamus: Pulvinar lateral	
377, 410	PuM	Thalamus: Pulvinar medial	
378, 411	VA	Thalamus: Ventral Anterior	
379, 412	VLa	Thalamus: Ventral Lateral Anterior	
380, 413	VLp	Thalamus: Ventral Lateral Posterior	
381, 414	VPL	Thalamus: Ventral posterolateral	
382, 415	Putam	Putamen	
383, 416	Caud	Caudate	
384, 417	NAc	Nucleus Accumbens	
385, 418	Gpe	Globus pallidus externalis	
386, 419	Gpi	Globus pallidus internalis	
387, 420	Amyg	Amygdala	
388, 421	SNpc	Substantia nigra pars compacta	
389, 422	SNpr	Substantia nigra pars reticulata	
390, 423	VTA	Ventral tegmental area	
391, 424	MB	Mammillary bodies	
392, 425	Septum	Septal nuclei	
393, 426	Nb	Nucleus basalis	

A.10 Subcortical regions defined in the HCPex atlas

The subcortical regions involved in many functions that are in the HCPex atlas (Huang, Rolls, Feng and Lin, 2022) are listed in Table A.7, and their connectivity and functions are described in Chapter 16 of Rolls (2026d) and elsewhere (Rolls et al., 2023d, 2022b; Rolls, 2022a).

Bibliography

Abbott LF and Regehr WG (2004). Synaptic computation. *Nature* 431: 796–803.

Abbott LF, Rolls ET, and Tovee MJ (1996). Representational capacity of face coding in monkeys. *Cereb Cortex* 6: 498–505.

Abbott LF, Varela JA, Sen K, and Nelson SB (1997). Synaptic depression and cortical gain control. *Science* 275: 220–224.

Abdallah CG, Adams TG, Kelmendi B, Esterlis I, Sanacora G, and Krystal JH (2016). Ketamine's mechanism of action: a path to rapid-acting antidepressants. *Depression and Anxiety* 33: 689–697.

Adolphs R, Gosselin F, Buchanan TW, Tranel D, Schyns P, and Damasio AR (2005). A mechanism for impaired fear recognition after amygdala damage. *Nature* 433: 68–72.

Afraz A, Yamins DL, and DiCarlo JJ (2014). Neural mechanisms underlying visual object recognition. *Cold Spring Harb Symp Quant Biol* 79: 99–107.

Aggelopoulos NC and Rolls ET (2005). Natural scene perception: inferior temporal cortex neurons encode the positions of different objects in the scene. *European Journal of Neuroscience* 22: 2903–2916.

Aggelopoulos NC, Franco L, and Rolls ET (2005). Object perception in natural scenes: encoding by inferior temporal cortex simultaneously recorded neurons. *J Neurophysiol* 93: 1342–57.

Agrawal A and Dehaene S (2024). Cracking the neural code for word recognition in convolutional neural networks. *PLoS Comput Biol* 20: e1012430.

Aigner TG, Mitchell SJ, Aggleton JP, DeLong MR, Struble RG, Price DL, Wenk GL, Pettigrew KD, and Mishkin M (1991). Transient impairment of recognition memory following ibotenic acid lesions of the basal forebrain in macaques. *Experimental Brain Research* 86: 18–26.

Aleksandrova LR, Phillips AG, and Wang YT (2017). Antidepressant effects of ketamine and the roles of ampa glutamate receptors and other mechanisms beyond nmda receptor antagonism. *J Psychiatry Neurosci* 42: 222–229.

Aleman A and Kahn RS (2005). Strange feelings: do amygdala abnormalities dysregulate the emotional brain in schizophrenia? *Progress in Neurobiology* 77: 283–298.

Amaral DG, Ishizuka N, and Claiborne B (1990). Neurons, numbers and the hippocampal network. *Prog Brain Res* 83: 1–11.

Amaral DG, Price JL, Pitkanen A, and Carmichael ST (1992). Anatomical organization of the primate amygdaloid complex. In Aggleton JP, editor, *The Amygdala*, chap. 1, 1–66. Wiley-Liss, New York.

Amir O, Biederman I, and Hayworth KJ (2011). The neural basis for shape preferences. *Vision Res* 51: 2198–206.

Amit DJ (1989). *Modeling Brain Function.* Cambridge University Press, Cambridge.

Amunts K and Zilles K (2012). Architecture and organizational principles of broca's region. *Trends Cogn Sci* 16: 418–26.

Andersen P, Morris RGM, Amaral DG, Bliss TVP, and OKeefe J (2007). *The Hippocampus Book.* Oxford University Press, Oxford and New York.

Andersen RA (1995). Encoding of intention and spatial location in the posterior parietal cortex. *Cereb Cortex* 5: 457–69.

Arcaro MJ and Livingstone MS (2021). On the relationship between maps and domains in inferotemporal cortex. *Nat Rev Neurosci* 22: 573–583.

Arnold PD, Rosenberg DR, Mundo E, Tharmalingam S, Kennedy JL, and Richter MA (2004). Association of a glutamate (NMDA) subunit receptor gene (GRIN2B) with obsessive-compulsive disorder: a preliminary study. *Psychopharmacology (Berl)* 174: 530–538.

Arnsten AF and Li BM (2005). Neurobiology of executive functions: catecholamine influences on prefrontal cortical functions. *Biological Psychiatry* 57: 1377–1384.

Arnsten AF and Wang M (2016). Targeting prefrontal cortical systems for drug development: potential therapies for cognitive disorders. *Annu Rev Pharmacol Toxicol* 56: 339–60.

Aron AR, Robbins TW, and Poldrack RA (2014). Inhibition and the right inferior frontal cortex: one decade on. *Trends Cogn Sci* 18: 177–85.

Auger SD and Maguire EA (2013). Assessing the mechanism of response in the retrosplenial cortex of good and poor navigators. *Cortex* 49: 2904–2913.

Avila E, Lakshminarasimhan KJ, DeAngelis GC, and Angelaki DE (2019). Visual and vestibular selectivity for self-motion in macaque posterior parietal area 7a. *Cereb Cortex* 29: 3932–3947.

Baddeley AD (2021). Developing the concept of working memory: the role of neuropsychology. *Arch Clin Neuropsychol* 36: 861–873.

Baddeley RJ, Abbott LF, Booth MJA, Sengpiel F, Freeman T, Wakeman EA, and Rolls ET (1997). Responses of neurons in primary and inferior temporal visual cortices to natural scenes. *Proceedings of the Royal Society of London B* 264: 1775–1783.

Ballard DH (1993). Subsymbolic modelling of hand-eye co-ordination. In Broadbent DE, editor, *The Simulation of Human Intelligence*, chap. 3, 71–102. Blackwell, Oxford.

Ballesta S, Shi W, Conen KE, and Padoa-Schioppa C (2020). Values encoded in orbitofrontal cortex are causally related to economic choices. *Nature* 588: 450–453.

Ballesta S, Shi W, and Padoa-Schioppa C (2022). Orbitofrontal cortex contributes to the comparison of values underlying economic choices. *Nat Commun* 13: 4405.

Bannon S, Gonsalvez CJ, Croft RJ, and Boyce PM (2002). Response inhibition deficits in obsessive-compulsive disorder. *Psychiatry Research* 110: 165–174.

Bannon S, Gonsalvez CJ, Croft RJ, and Boyce PM (2006). Executive functions in obsessive-compulsive disorder: state or trait deficits? *Aust N Z Journal of Psychiatry* 40: 1031–1038.

Baraduc P, Duhamel JR, and Wirth S (2019). Schema cells in the macaque hippocampus. *Science* 363: 635–639.

Barat E, Wirth S, and Duhamel JR (2018). Face cells in orbitofrontal cortex represent social categories. *Proc Natl Acad Sci U S A* 115: E11158–E11167.

Barlow HB (1972). Single units and sensation: a neuron doctrine for perceptual psychology. *Perception* 1: 371–394.

Barnes CA (2003). Long-term potentiation and the ageing brain. *Philosophical Transactions of the Royal Society of London B* 358: 765–772.

Barto AG (1985). Learning by statistical cooperation of self-interested neuron-like computing elements. *Human Neurobiology* 4: 229–256.

Bartus RT (2000). On neurodegenerative diseases, models, and treatment strategies: lessons learned and lessons forgotten a generation following the cholinergic hypothesis. *Experimental Neurology* 163: 495–529.

Battaglia F and Treves A (1998a). Stable and rapid recurrent processing in realistic autoassociative memories. *Neural Computation* 10: 431–450.

Battaglia FP and Treves A (1998b). Attractor neural networks storing multiple space representations: a model for hippocampal place fields. *Physical Review E* 58: 7738–7753.

Battista A, Padoa-Schioppa C, and Wang XJ (2025). A neural circuit framework for economic choice: From building blocks of valuation to compositionality in multitasking. *bioRxiv* .

Battistella G, Borghesani V, Henry M, Shwe W, Lauricella M, Miller Z, Deleon J, Miller BL, Dronkers N, Brambati SM, Seeley WW, Mandelli ML, and Gorno-Tempini ML (2020). Task-free functional language networks: reproducibility and clinical application. *J Neurosci* 40: 1311–1320.

Baylis GC, Rolls ET, and Leonard CM (1985). Selectivity between faces in the responses of a population of neurons in the cortex in the superior temporal sulcus of the monkey. *Brain Research* 342: 91–102.

Baylis GC, Rolls ET, and Leonard CM (1987). Functional subdivisions of the temporal lobe neocortex. *Journal of Neuroscience* 7: 330–342.

Baylis LL, Rolls ET, and Baylis GC (1995). Afferent connections of the caudolateral orbitofrontal cortex taste area of the primate. *Neuroscience* 64: 801–12.

Bayne T, Seth AK, Massimini M, Shepherd J, Cleeremans A, Fleming SM, Malach R, Mattingley JB, Menon DK, Owen AM, Peters MAK, Razi A, and Mudrik L (2024). Tests for consciousness in humans and beyond. *Trends Cogn Sci* 28: 454–466.

Bear MF and Singer W (1986). Modulation of visual cortical plasticity by acetylcholine and noradrenaline. *Nature* 320: 172–176.

Beck AT (2008). The evolution of the cognitive model of depression and its neurobiological correlates. *American Journal of Psychiatry* 165: 969–977.

Becker NB, Jesus SN, Joao K, Viseu JN, and Martins RIS (2017). Depression and sleep quality in older adults: a meta-analysis. *Psychol Health Med* 22: 889–895.

Belin P (2019). The vocal brain: core and extended cerebral networks for voice processing. In Fruhholz S and Belin P, editors, *Oxford Handbook of Voice Perception*, 37–60. Oxford University Press, Oxford.

Bengio Y, Goodfellow I, and Courville A (2017). *Deep Learning*. MIT Press, Boston, MA.

Bennett IJ and Stark CE (2016). Mnemonic discrimination relates to perforant path integrity: An ultra-high resolution diffusion tensor imaging study. *Neurobiol Learn Mem* 129: 107–112.

Berg E (1948). A simple objective technique for measuring flexibility in thinking. *Journal of General Psychology* 39: 15–22.

Berlin H and Rolls ET (2004). Time perception, impulsivity, emotionality, and personality in self-harming borderline personality disorder patients. *Journal of Personality Disorders* 18: 358–378.

Berlin H, Rolls ET, and Kischka U (2004). Impulsivity, time perception, emotion, and reinforcement sensitivity in patients with orbitofrontal cortex lesions. *Brain* 127: 1108–1126.

Berlin H, Rolls ET, and Iversen SD (2005). Borderline personality disorder, impulsivity and the orbitofrontal cortex. *American Journal of Psychiatry* 162: 2360–2373.

Berridge KC, Robinson TE, and Aldridge JW (2009). Dissecting components of reward: 'liking', 'wanting', and learning. *Curr Opin Pharmacol* 9: 65–73.

Bhattacharyya S and Chakraborty K (2007). Glutamatergic dysfunction–newer targets for anti-obsessional drugs. *Recent Patents CNS Drug Discovery* 2: 47–55.

Bicanski A and Burgess N (2018). A neural-level model of spatial memory and imagery. *Elife* 7: e33752.

Bicanski A and Burgess N (2020). Neuronal vector coding in spatial cognition. *Nat Rev Neurosci* 21: 453–470.

Biederman I (1987). Recognition-by-components: A theory of human image understanding. *Psychological Review* 94: 115–147.

Biederman I and Gerhardstein PC (1993). Recognizing depth-rotated objects: Evidence and conditions for 3D viewpoint invariance. *Journal of Experimental Psychology: Human Perception and Performance* 20: 80.

Binder JR (2015). The wernicke area: Modern evidence and a reinterpretation. *Neurology* 85: 2170–5.

Binder JR (2017). Current controversies on wernicke's area and its role in language. *Curr Neurol Neurosci Rep* 17: 58.

Bishop CM and Bishop H (2024). *Deep Learning: Foundations and Concepts*. Springer Nature, Switzerland.

Block N (2005). Two neural correlates of consciousness. *Trends Cogn Sci* 9: 46–52.

Boboeva V, Brasselet R, and Treves A (2018). The capacity for correlated semantic memories in the cortex. *Entropy (Basel)* 20: 824.

Bonner MF and Price AR (2013). Where is the anterior temporal lobe and what does it do? *J Neurosci* 33: 4213–5.

Bonnici HM and Maguire EA (2018). Two years later - revisiting autobiographical memory representations in vmpfc and hippocampus. *Neuropsychologia* 110: 159–169.

Booth DA (1985). Food-conditioned eating preferences and aversions with interoceptive elements: learned appetites and satieties. *Annals of the New York Academy of Sciences* 443: 22–37.

Booth MCA and Rolls ET (1998). View-invariant representations of familiar objects by neurons in the inferior temporal visual cortex. *Cerebral Cortex* 8: 510–523.

Bornkessel-Schlesewsky I, Schlesewsky M, Small SL, and Rauschecker JP (2015). Neurobiological roots of language in primate audition: common computational properties. *Trends Cogn Sci* 19: 142–50.

Bouchard KE, Mesgarani N, Johnson K, and Chang EF (2013). Functional organization of human sensorimotor cortex for speech articulation. *Nature* 495: 327–32.

Boussaoud D, Desimone R, and Ungerleider LG (1991). Visual topography of area TEO in the macaque. *Journal of Computational Neurology* 306: 554–575.

Bowers JS, Malhotra G, Dujmovic M, Llera Montero M, Tsvetkov C, Biscione V, Puebla G, Adolfi F, Hummel JE, Heaton RF, Evans BD, Mitchell J, and Blything R (2022). Deep problems with neural network models of human vision. *Behav Brain Sci* 46: e385.

Brain WR (1960). *Clinical Neurology*. Oxford University Press, Oxford.

Brandon MP, Bogaard AR, Libby CP, Connerney MA, Gupta K, and Hasselmo ME (2011). Reduction of theta rhythm dissociates grid cell spatial periodicity from directional tuning. *Science* 332: 595–9.

Brown TI, Carr VA, LaRocque KF, Favila SE, Gordon AM, Bowles B, Bailenson JN, and Wagner AD (2016). Prospective representation of navigational goals in the human hippocampus. *Science* 352: 1323–6.

Bruce C, Desimone R, and Gross CG (1981). Visual properties of neurons in a polysensory area in superior temporal sulcus of the macaque. *J Neurophysiol* 46: 369–84.

Brunel N and Wang XJ (2001). Effects of neuromodulation in a cortical network model of object working memory dominated by recurrent inhibition. *Journal of Computational Neuroscience* 11: 63–85.

Bubb EJ, Kinnavane L, and Aggleton JP (2017). Hippocampal - diencephalic - cingulate networks for memory and emotion: An anatomical guide. *Brain Neurosci Adv* 1: 1–20.

Buckley MJ (2005). The role of the perirhinal cortex and hippocampus in learning, memory, and perception. *Q J Exp Psychol B* 58: 246–68.

Buckley MJ and Gaffan D (2000). The hippocampus, perirhinal cortex, and memory in the monkey. In Bolhuis JJ, editor, *Brain, Perception, and Memory: Advances in Cognitive Neuroscience*, 279–298. Oxford University Press, Oxford.

Buckner RL and DiNicola LM (2019). The brain's default network: updated anatomy, physiology and evolving insights. *Nat Rev Neurosci* 20: 593–608.

Buffalo EA (2025). Visual exploration and the primate hippocampal formation. *Hippocampus* 35: e23673.

Burgess N (2008). Spatial cognition and the brain. *Ann N Y Acad Sci* 1124: 77–97.

Burgess N and O'Keefe J (1996). Neuronal computations underlying the firing of place cells and their role in navigation. *Hippocampus* 6: 749–62.

Burgess N, Jackson A, Hartley T, and O'Keefe J (2000). Predictions derived from modelling the hippocampal role in navigation. *Biol Cybern* 83: 301–312.

Burke SN and Barnes CA (2006). Neural plasticity in the ageing brain. *Nature Reviews Neuroscience* 7: 30–40.

Burton MJ, Rolls ET, and Mora F (1976). Effects of hunger on the responses of neurones in the lateral hypothalamus to the sight and taste of food. *Experimental Neurology,* 51: 668–677.

Butter C (1969). Perseveration in extinction and in discrimination reversal tasks following selective prefrontal ablations in macaca mulatta. *Physiology and Behavior* 4: 163–171.

Cadieu CF, Hong H, Yamins DL, Pinto N, Ardila D, Solomon EA, Majaj NJ, and DiCarlo JJ (2014). Deep neural networks rival the representation of primate it cortex for core visual object recognition. *PLoS Comput Biol* 10: e1003963.

Caffarra S, Karipidis I, Yablonski M, and Yeatman JD (2021). Anatomy and physiology of word-selective visual cortex: from visual features to lexical processing. *Brain Struct Funct* 226: 3051–3065.

Cahusac PMB, Miyashita Y, and Rolls ET (1989). Responses of hippocampal formation neurons in the monkey related to delayed spatial response and object-place memory tasks. *Behavioural Brain Research* 33: 229–240.

Cahusac PMB, Rolls ET, Miyashita Y, and Niki H (1993). Modification of the responses of hippocampal neurons in the monkey during the learning of a conditional spatial response task. *Hippocampus* 3: 29–42.

Cai X and Padoa-Schioppa C (2012). Neuronal encoding of subjective value in dorsal and ventral anterior cingulate cortex. *Journal of Neuroscience* 32: 3791–3808.

Cai X and Padoa-Schioppa C (2019). Neuronal evidence for good-based economic decisions under variable action costs. *Nat Commun* 10: 393.

Camille N, Tsuchida A, and Fellows LK (2011). Double dissociation of stimulus-value and action-value learning in humans with orbitofrontal or anterior cingulate cortex damage. *J Neurosci* 31: 15048–52.

Campbell FW (1980). The physics of visual perception. *Philos Trans R Soc Lond B Biol Sci* 290: 5–9.

Carmichael ST and Price JL (1996). Connectional networks within the orbital and medial prefrontal cortex of macaque monkeys. *Journal of Comparative Neurology* 371: 179–207.

Carruthers P (2019). *Human and Animal Minds*. Oxford University Press, Oxford.

Castner SA, Williams GV, and Goldman-Rakic PS (2000). Reversal of antipsychotic-induced working memory deficits by short-term dopamine D1 receptor stimulation. *Science* 287: 2020–2022.

Catani M and Thiebaut de Schotten M (2008). A diffusion tensor imaging tractography atlas for virtual in vivo dissections. *Cortex* 44: 1105–32.

Cavanna AE and Trimble MR (2006). The precuneus: a review of its functional anatomy and behavioural correlates. *Brain* 129: 564–83.

Cavina-Pratesi C, Monaco S, Fattori P, Galletti C, McAdam TD, Quinlan DJ, Goodale MA, and Culham JC (2010). Functional magnetic resonance imaging reveals the neural substrates of arm transport and grip formation in reach-to-grasp actions in humans. *J Neurosci* 30: 10306–23.

Chadwick MJ, Hassabis D, Weiskopf N, and Maguire EA (2010). Decoding individual episodic memory traces in the human hippocampus. *Curr Biol* 20: 544–7.

Chadwick MJ, Mullally SL, and Maguire EA (2013). The hippocampus extrapolates beyond the view in scenes: an fmri study of boundary extension. *Cortex* 49: 2067–79.

Chamberlain SR, Fineberg NA, Blackwell AD, Robbins TW, and Sahakian BJ (2006). Motor inhibition and cognitive flexibility in obsessive-compulsive disorder and trichotillomania. *American Journal of Psychiatry* 163: 1282–1284.

Chamberlain SR, Fineberg NA, Menzies LA, Blackwell AD, Bullmore ET, Robbins TW, and Sahakian BJ (2007). Impaired cognitive flexibility and motor inhibition in unaffected first-degree relatives of patients with obsessive-compulsive disorder. *American Journal of Psychiatry* 164: 335–338.

Chamberlain SR, Solly JE, Hook RW, Vaghi MM, and Robbins TW (2021). Cognitive inflexibility in ocd and related disorders. *Curr Top Behav Neurosci* 49: 125–145.

Chang EF, Raygor KP, and Berger MS (2015). Contemporary model of language organization: an overview for neurosurgeons. *J Neurosurg* 122: 250–61.

Chang L and Tsao DY (2017). The code for facial identity in the primate brain. *Cell* 169: 1013–1028 e14.

Chau BK, Sallet J, Papageorgiou GK, Noonan MP, Bell AH, Walton ME, and Rushworth MF (2015). Contrasting roles for orbitofrontal cortex and amygdala in credit assignment and learning in macaques. *Neuron* 87: 1106–1118.

Chen H and Naya Y (2020). Forward processing of object-location association from the ventral stream to medial temporal lobe in nonhuman primates. *Cereb Cortex* 30: 1260–1271.

Chen M, Li B, Guang J, Wei L, Wu S, Liu Y, and Zhang M (2016). Two subdivisions of macaque lip process visual-oculomotor information differently. *Proc Natl Acad Sci U S A* 113: E6263–E6270.

Cheng W, Rolls ET, Gu H, Zhang J, and Feng J (2015). Autism: reduced functional connectivity between cortical areas involved in face expression, theory of mind, and the sense of self. *Brain* 138: 1382–1393.

Cheng W, Rolls ET, Qiu J, Liu W, Tang Y, Huang CC, Wang X, Zhang J, Lin W, Zheng L, Pu J, Tsai SJ, Yang AC, Lin CP, Wang F, Xie P, and Feng J (2016). Medial reward and lateral non-reward orbitofrontal cortex circuits change in opposite directions in depression. *Brain* 139: 3296–3309.

Cheng W, Rolls ET, Zhang J, Sheng W, Ma L, Wan L, Luo Q, and Feng J (2017). Functional connectivity decreases in autism in emotion, self, and face circuits identified by knowledge-based enrichment analysis. *Neuroimage* 148: 169–178.

Cheng W, Rolls ET, Qiu J, Xie X, Lyu W, Li Y, Huang CC, Yang AC, Tsai SJ, Lyu F, Zhuang K, Lin CP, Xie P, and Feng J (2018a). Functional connectivity of the human amygdala in health and in depression. *Soc Cogn Affect Neurosci* 13: 557–568.

Cheng W, Rolls ET, Qiu J, Xie X, Wei D, Huang CC, Yang AC, Tsai SJ, Li Q, Meng J, Lin CP, Xie P, and Feng J (2018b). Increased functional connectivity of the posterior cingulate cortex with the lateral orbitofrontal cortex in depression. *Transl Psychiatry* 8: 90.

Cheng W, Rolls ET, Qiu J, Yang D, Ruan H, Wei D, Zhao L, Meng J, Xie P, and Feng J (2018c). Functional connectivity of the precuneus in unmedicated patients with depression. *Biol Psychiatry Cogn Neurosci Neuroimaging* 3: 1040–1049.

Cheng W, Rolls ET, Ruan H, and Feng J (2018d). Functional connectivities in the brain that mediate the association between depressive problems and sleep quality. *JAMA Psychiatry* 75: 1052–1061.

Cheng W, Rolls ET, Gong W, Du J, Zhang J, Zhang XY, Li F, and Feng J (2021). Sleep duration, brain structure, and psychiatric and cognitive problems in children. *Mol Psychiatry* 26: 3992–4003.

Chettih SN, Mackevicius EL, Hale S, and Aronov D (2024). Barcoding of episodic memories in the hippocampus of a food-caching bird. *Cell* 187: 1922–1935 e20.

Clos M, Amunts K, Laird AR, Fox PT, and Eickhoff SB (2013). Tackling the multifunctional nature of broca's region meta-analytically: co-activation-based parcellation of area 44. *Neuroimage* 83: 174–88.

Coalson TS, Van Essen DC, and Glasser MF (2018). The impact of traditional neuroimaging methods on the spatial localization of cortical areas. *Proc Natl Acad Sci U S A* 115: E6356–E6365.

Colby CL, Duhamel JR, and Goldberg ME (1993). Ventral intraparietal area of the macaque: anatomic location and visual response properties. *J Neurophysiol* 69: 902–14.

Colby CL, Duhamel JR, and Goldberg ME (1996). Visual, presaccadic, and cognitive activation of single neurons in monkey lateral intraparietal area. *J Neurophysiol* 76: 2841–52.

Collins JA and Olson IR (2014). Beyond the ffa: The role of the ventral anterior temporal lobes in face processing. *Neuropsychologia* 61: 65–79.

Connolly JD, Andersen RA, and Goodale MA (2003). Fmri evidence for a parietal reach region in the human brain. *Exp Brain Res* 153: 140–5.

Cools R and Arnsten AFT (2022). Neuromodulation of prefrontal cortex cognitive function in primates: the powerful roles of monoamines and acetylcholine. *Neuropsychopharmacology* 47: 309–328.

Cowey A and Rolls ET (1975). Human cortical magnification factor and its relation to visual acuity. *Experimental Brain Research* 21: 447–454.

Coyle JT (2006). Glutamate and schizophrenia: beyond the dopamine hypothesis. *Cellular and Molecular Neurobiology* 26: 365–384.

Coyle JT, Ruzicka WB, and Balu DT (2020). Fifty years of research on schizophrenia: the ascendance of the glutamatergic synapse. *Am J Psychiatry* 177: 1119–1128.

Craig AD (2009). How do you feel–now? the anterior insula and human awareness. *Nat Rev Neurosci* 10: 59–70.

Craig AD (2011). Significance of the insula for the evolution of human awareness of feelings from the body. *Ann N Y Acad Sci* 1225: 72–82.

Critchley H, Daly E, Phillips M, Brammer M, Bullmore E, Williams S, Van Amelsvoort T, Robertson D, David A, and Murphy D (2000). Explicit and implicit neural mechanisms for processing of social information from facial expressions: a functional magnetic resonance imaging study. *Hum Brain Mapp* 9: 93–105.

Critchley HD and Harrison NA (2013). Visceral influences on brain and behavior. *Neuron* 77: 624–38.

Critchley HD and Rolls ET (1996a). Hunger and satiety modify the responses of olfactory and visual neurons in the primate orbitofrontal cortex. *J Neurophysiol* 75: 1673–86.

Critchley HD and Rolls ET (1996b). Olfactory neuronal responses in the primate orbitofrontal cortex: analysis in an olfactory discrimination task. *J Neurophysiol* 75: 1659–72.

Cromwell HC, Tremblay L, and Schultz W (2018). Neural encoding of choice during a delayed response task in primate striatum and orbitofrontal cortex. *Exp Brain Res* 236: 1679–1688.

Culham JC, Danckert SL, DeSouza JF, Gati JS, Menon RS, and Goodale MA (2003). Visually guided grasping produces fmri activation in dorsal but not ventral stream brain areas. *Exp Brain Res* 153: 180–9.

Culham JC, Cavina-Pratesi C, and Singhal A (2006). The role of parietal cortex in visuomotor control: what have we learned from neuroimaging? *Neuropsychologia* 44: 2668–84.

Cunningham PM, Roe LS, Keller KL, and Rolls BJ (2023). Variety and portion size combine to increase food intake at single-course and multi-course meals. *Appetite* 191: 107089.

Dahan L, Rampon C, and Florian C (2020). Age-related memory decline, dysfunction of the hippocampus and therapeutic opportunities. *Prog Neuropsychopharmacol Biol Psychiatry* 102: 109943.

Dalley JW and Robbins TW (2017). Fractionating impulsivity: neuropsychiatric implications. *Nat Rev Neurosci* 18: 158–171.

Damasio A, Damasio H, and Tranel D (2013). Persistence of feelings and sentience after bilateral damage of the insula. *Cereb Cortex* 23: 833–46.

Darwin C (1859). *The Origin of Species*. John Murray [reprinted (1982) by Penguin Books Ltd], London.

Davis M (1992). The role of the amygdala in conditioned fear. In Aggleton JP, editor, *The Amygdala*, chap. 9, 255–306. Wiley-Liss, New York.

Davis M (2011). Nmda receptors and fear extinction: implications for cognitive behavioral therapy. *Dialogues Clin Neurosci* 13: 463–74.

Davis SW, Wing EA, and Cabeza R (2018). Contributions of the ventral parietal cortex to declarative memory. *Handb Clin Neurol* 151: 525–553.

Dawkins MS (1995). *Unravelling Animal Behaviour*. Longman, Harlow, 2nd edn.

Dawkins MS (2026). *Who is Conscious? A Guide to the Minds of Animals*. Oxford University Press, Oxford.

Dawkins R (1976). *The Selfish Gene*. Oxford University Press, Oxford.

Dawkins R (1982). *The Extended Phenotype: The Gene as the Unit of Selection*. W.H. Freeman, San Francisco.

Dawkins R (1986). *The Blind Watchmaker*. Longman, Harlow.

Dawkins R (1989). *The Selfish Gene*. Oxford University Press, Oxford, 2nd edn.

Dayan P and Abbott LF (2001). *Theoretical Neuroscience*. MIT Press, Cambridge, MA.

De Araujo IE and Rolls ET (2004). Representation in the human brain of food texture and oral fat. *J Neurosci* 24: 3086–93.

De Araujo IE, Kringelbach ML, Rolls ET, and McGlone F (2003a). Human cortical responses to water in the mouth, and the effects of thirst. *J Neurophysiol* 90: 1865–76.

De Araujo IE, Rolls ET, Kringelbach ML, McGlone F, and Phillips N (2003b). Taste-olfactory convergence, and the representation of the pleasantness of flavour, in the human brain. *Eur J Neurosci* 18: 2059–68.

De Araujo IE, Rolls ET, Velazco MI, Margot C, and Cayeux I (2005). Cognitive modulation of olfactory processing. *Neuron* 46: 671–9.

De Araujo IE, Schatzker M, and Small DM (2020). Rethinking food reward. *Annu Rev Psychol* 71: 139–164.

De Araujo IET, Rolls ET, and Stringer SM (2001). A view model which accounts for the spatial fields of hippocampal primate spatial view cells and rat place cells. *Hippocampus* 11: 699–706.

de Haan EHF and Dijkerman HC (2020). Somatosensation in the brain: a theoretical re-evaluation and a new model. *Trends Cogn Sci* 24: 529–541.

de Leon-Lopez CAM, Carretero-Rey M, and Khan ZU (2025). Ampa receptors in synaptic plasticity, memory function, and brain diseases. *Cell Mol Neurobiol* 45: 14.

De Luca F, McCormick C, Ciaramelli E, and Maguire EA (2019). Scene processing following damage to the ventromedial prefrontal cortex. *Neuroreport* 30: 828–833.

Dean HL and Platt ML (2006). Allocentric spatial referencing of neuronal activity in macaque posterior cingulate cortex. *J Neurosci* 26: 1117–27.

Deco G and Rolls ET (2002a). A neurodynamical theory of visual attention: comparisons with fmri- and single-neuron data. *International Conference on Artificial Neural Networks* LNCS 2415: 3–8.

Deco G and Rolls ET (2002b). Object-based visual neglect: a computational hypothesis. *European Journal of Neuroscience* 16: 1994–2000.

Deco G and Rolls ET (2003). Attention and working memory: a dynamical model of neuronal activity in the prefrontal cortex. *Eur J Neurosci* 18: 2374–90.

Deco G and Rolls ET (2004). A neurodynamical cortical model of visual attention and invariant object recognition. *Vision Res* 44: 621–42.

Deco G and Rolls ET (2005a). Attention, short-term memory, and action selection: a unifying theory. *Progress in Neurobiology* 76: 236–256.

Deco G and Rolls ET (2005b). Synaptic and spiking dynamics underlying reward reversal in the orbitofrontal cortex. *Cereb Cortex* 15: 15–30.

Deco G and Rolls ET (2005c). Sequential memory: a putative neural and synaptic dynamical mechanism. *Journal of Cognitive Neuroscience* 17: 294–307.

Deco G and Rolls ET (2005d). Neurodynamics of biased competition and co-operation for attention: a model with spiking neurons. *Journal of Neurophysiology* 94: 295–313.

Deco G and Rolls ET (2006). A neurophysiological model of decision-making and Weber's law. *European Journal of Neuroscience* 24: 901–916.

Deco G, Rolls ET, and Horwitz B (2004). "what" and "where" in visual working memory: a computational neurodynamical perspective for integrating fmri and single-neuron data. *Journal of Cognitive Neuroscience* 16: 683–701.

Deco G, Rolls ET, and Romo R (2009). Stochastic dynamics as a principle of brain function. *Progress in Neurobiology* 88: 1–16.

Deco G, Rolls ET, and Romo R (2010). Synaptic dynamics and decision-making. *Proceedings of the National Academy of Sciences* 107: 7545–7549.

Deco G, Rolls ET, Albantakis L, and Romo R (2013). Brain mechanisms for perceptual and reward-related decision-making. *Prog Neurobiol* 103: 194–213.

Deco G, Cruzat J, Cabral J, Tagliazucchi E, Laufs H, Logothetis NK, and Kringelbach ML (2019). Awakening: Predicting external stimulation to force transitions between different brain states. *Proceedings of the National Academy of Sciences* 116: 18088–18097.

Dede AJ, Frascino JC, Wixted JT, and Squire LR (2016). Learning and remembering real-world events after medial temporal lobe damage. *Proc Natl Acad Sci U S A* 113: 13480–13485.

Deen B, Koldewyn K, Kanwisher N, and Saxe R (2015). Functional organization of social perception and cognition in the superior temporal sulcus. *Cereb Cortex* 25: 4596–4609.

Dehaene S (2014). *Consciousness and the Brain.* Penguin, New York.

Dehaene S and Cohen L (2011). The unique role of the visual word form area in reading. *Trends Cogn Sci* 15: 254–62.

Dehaene S, Cohen L, Sigman M, and Vinckier F (2005). The neural code for written words: a proposal. *Trends Cogn Sci* 9: 335–41.

Dehaene S, Lau H, and Kouider S (2017). What is consciousness, and could machines have it? *Science* 358: 486–492.

Delgado MR, Jou RL, and Phelps EA (2011). Neural systems underlying aversive conditioning in humans with primary and secondary reinforcers. *Front Neurosci* 5: 71.

Delhaye BP, Long KH, and Bensmaia SJ (2018). Neural basis of touch and proprioception in primate cortex. *Compr Physiol* 8: 1575–1602.

Delorme R, Gousse V, Roy I, Trandafir A, Mathieu F, Mouren-Simeoni MC, Betancur C, and Leboyer M (2007). Shared executive dysfunctions in unaffected relatives of patients with autism and obsessive-compulsive disorder. *European Psychiatry* 22: 32–38.

Deng WL, Rolls ET, Ji X, Robbins TW, Banaschewski T, Bokde ALW, Bromberg U, Buechel C, Desrivieres S, Conrod P, Flor H, Frouin V, Gallinat J, Garavan H, Gowland P, Heinz A, Ittermann B, Martinot JL, Lemaitre H, Nees F, Papadoulos Orfanos D, Poustka L, Smolka MN, Walter H, Whelan R, Schumann G, and Feng J (2017). Separate neural systems for behavioral change and for emotional responses to failure during behavioral inhibition. *Human Brain Mapping* 38: 3527–3537.

Dennett D (1991). *Consciousness Explained.* Penguin, London.

Dere E, Easton A, Nadel L, and Huston JP (2008). *Handbook of Episodic Memory.* Handbook of Behavioral Neuroscience. Elsevier, Amsterdam.

Descartes R (1644). *The Philosophical Writings of Descartes (3 volumes, 1984-1991).* Cambridge University Press, Cambridge.

Desimone R (1991). Face-selective cells in the temporal cortex of monkeys. *J Cogn Neurosci* 3: 1–8.

Desimone R and Duncan J (1995). Neural mechanisms of selective visual attention. *Annual Review of Neuroscience* 18: 193–222.

Desimone R and Ungerleider L (1989). Neural mechanisms of visual processing in monkeys. In Boller F and Grafman J, editors, *Handbook of Neuropsychology*, vol. 2, 267–289. Elsevier, New York.

Desimone R, Albright TD, Gross CG, and Bruce C (1984). Stimulus-selective properties of inferior temporal neurons in the macaque. *J Neurosci* 4: 2051–62.

DeWitt I and Rauschecker JP (2013). Wernicke's area revisited: parallel streams and word processing. *Brain Lang* 127: 181–91.

DiCarlo JJ and Cox DD (2007). Untangling invariant object recognition. *Trends Cogn Sci* 11: 333–41.

DiCarlo JJ and Maunsell JHR (2000). Form representation in monkey inferotemporal cortex is virtually unaltered by free viewing. *Nature Neuroscience* 3: 814–821.

Dickinson A (2016). Instrumental conditioning revisited: Updating dual-process theory. *Associative Learning and Cognition* 51: 177–195.

Dienel SJ, Schoonover KE, and Lewis DA (2022). Cognitive dysfunction and prefrontal cortical circuit alterations in schizophrenia: developmental trajectories. *Biol Psychiatry* 92: 450–459.

Ding N, Melloni L, Zhang H, Tian X, and Poeppel D (2016). Cortical tracking of hierarchical linguistic structures in connected speech. *Nat Neurosci* 19: 158–64.

DiNicola LM, Braga RM, and Buckner RL (2020). Parallel distributed networks dissociate episodic and social functions within the individual. *J Neurophysiol* 123: 1144–1179.

Divac I (1975). Magnocellular nuclei of the basal forebrain project to neocortex, brain stem, and olfactory bulb. Review of some functional correlates. *Brain Research* 93: 385–398.

Dolan RJ, Fink GR, Rolls ET, Booth M, Holmes A, Frackowiak RSJ, and Friston KJ (1997). How the brain learns to see objects and faces in an impoverished context. *Nature* 389: 596–599.

Donahue CJ, Sotiropoulos SN, Jbabdi S, Hernandez-Fernandez M, Behrens TE, Dyrby TB, Coalson T, Kennedy H, Knoblauch K, Van Essen DC, and Glasser MF (2016). Using diffusion tractography to predict cortical connection strength and distance: a quantitative comparison with tracers in the monkey. *J Neurosci* 36: 6758–70.

Drevets WC (2007). Orbitofrontal cortex function and structure in depression. *Annals of the New York Academy of Sciences* 1121: 499–527.

Du J, Rolls ET, Cheng W, Li Y, Gong W, Qiu J, and Feng J (2020). Functional connectivity of the orbitofrontal cortex, anterior cingulate cortex, and inferior frontal gyrus in humans. *Cortex* 123: 185–199.

Du J, Rolls ET, Gong W, Cao M, Vatansever D, Zhang J, Kang J, Cheng W, and Feng J (2022). Association between parental age, brain structure, and behavioral and cognitive problems in children. *Mol Psychiatry* 27: 967–975.

Duff MC, Covington NV, Hilverman C, and Cohen NJ (2019). Semantic memory and the hippocampus: revisiting, reaffirming, and extending the reach of their critical relationship. *Front Hum Neurosci* 13: 471.

Duman RS and Aghajanian GK (2012). Synaptic dysfunction in depression: potential therapeutic targets. *Science* 338: 68–72.

Dunlop BW, Rajendra JK, Craighead WE, Kelley ME, McGrath CL, Choi KS, Kinkead B, Nemeroff CB, and Mayberg HS (2017). Functional connectivity of the subcallosal cingulate cortex and differential outcomes to treatment with cognitive-behavioral therapy or antidepressant medication for major depressive disorder. *Am J Psychiatry* 174: 533–545.

Durstewitz D and Seamans JK (2002). The computational role of dopamine D1 receptors in working memory. *Neural Networks* 15: 561–572.

Durstewitz D, Kelc M, and Gunturkun O (1999). A neurocomputational theory of the dopaminergic modulation of working memory functions. *Journal of Neuroscience* 19: 2807–2822.

Durstewitz D, Seamans JK, and Sejnowski TJ (2000a). Dopamine-mediated stabilization of delay-period activity in a network model of prefrontal cortex. *Journal of Neurophysiology* 83: 1733–1750.

Durstewitz D, Seamans JK, and Sejnowski TJ (2000b). Neurocomputational models of working memory. *Nature Neuroscience* 3 Suppl: 1184–1191.

Easton A and Gaffan D (2000). Amygdala and the memory of reward: the importance of fibres of passage from the basal forebrain. In Aggleton JP, editor, *The Amygdala: a Functional Analysis*, chap. 17, 569–586. Oxford University Press, Oxford, 2nd edn.

Easton A, Ridley RM, Baker HF, and Gaffan D (2002). Unilateral lesions of the cholinergic basal forebrain and fornix in one hemisphere and inferior temporal cortex in the opposite hemisphere produce severe learning impairments in rhesus monkeys. *Cerebral Cortex* 12: 729–736.

Easton A, Barros M, and Lever C (2020). Acetylcholine and spontaneous recognition memory in rodents and primates. *Curr Top Behav Neurosci* 45: 29–45.

Edvardsen V, Bicanski A, and Burgess N (2020). Navigating with grid and place cells in cluttered environments. *Hippocampus* 30: 220–232.

Eichenbaum H (2014). Time cells in the hippocampus: a new dimension for mapping memories. *Nat Rev Neurosci* 15: 732–44.

Ekert JO, Lorca-Puls DL, Gajardo-Vidal A, Crinion JT, Hope TMH, Green DW, and Price CJ (2021). A functional dissociation of the left frontal regions that contribute to single word production tasks. *Neuroimage* 245: 118734.

Ekstrom AD and Ranganath C (2018). Space, time, and episodic memory: The hippocampus is all over the cognitive map. *Hippocampus* 28: 680–687.

Ekstrom AD, Kahana MJ, Caplan JB, Fields TA, Isham EA, Newman EL, and Fried I (2003). Cellular networks underlying human spatial navigation. *Nature* 425: 184–8.

Elias GJB, Germann J, Boutet A, Loh A, Li B, Pancholi A, Beyn ME, Naheed A, Bennett N, Pinto J, Bhat V, Giacobbe P, Woodside DB, Kennedy SH, and Lozano AM (2022a). 3t mri of rapid brain activity changes driven by subcallosal cingulate deep brain stimulation. *Brain* 145: 2214–2226.

Elias GJB, Germann J, Boutet A, Pancholi A, Beyn ME, Bhatia K, Neudorfer C, Loh A, Rizvi SJ, Bhat V, Giacobbe P, Woodside DB, Kennedy SH, and Lozano AM (2022b). Structuro-functional surrogates of response to subcallosal cingulate deep brain stimulation for depression. *Brain* 145: 362–377.

Elliffe MC, Rolls ET, and Stringer SM (2002). Invariant recognition of feature combinations in the visual system. *Biol Cybern* 86: 59–71.

Epstein J, Stern E, and Silbersweig D (1999). Mesolimbic activity associated with psychosis in schizophrenia. Symptom-specific PET studies. *Annals of the New York Academy of Sciences* 877: 562–574.

Epstein R (2005). The cortical basis of visual scene processing. *Visual Cognition* 12: 954–978.

Epstein R and Kanwisher N (1998). A cortical representation of the local visual environment. *Nature* 392: 598–601.

Epstein RA (2008). Parahippocampal and retrosplenial contributions to human spatial navigation. *Trends Cogn Sci* 12: 388–96.

Epstein RA and Baker CI (2019). Scene perception in the human brain. *Annu Rev Vis Sci* 5: 373–397.

Epstein RA and Julian JB (2013). Scene areas in humans and macaques. *Neuron* 79: 615–7.

Ermentrout B and Terman DH (2010). *Mathematical Foundations of Neuroscience*. Springer, New York.

Eshel N and Roiser JP (2010). Reward and punishment processing in depression. *Biological Psychiatry* 68: 118–124.

Faisal A, Selen L, and Wolpert D (2008). Noise in the nervous system. *Nature Reviews Neuroscience* 9: 292–303.

Fatt P and Katz B (1950). Some observations on biological noise. *Nature* 166: 597–598.

Feigenbaum JD and Morris RG (2004). Allocentric versus egocentric spatial memory after unilateral temporal lobectomy in humans. *Neuropsychology* 18: 462–72.

Feigenbaum JD and Rolls ET (1991). Allocentric and egocentric spatial information processing in the hippocampal formation of the behaving primate. *Psychobiology* 19: 21–40.

Feinstein JS, Adolphs R, Damasio A, and Tranel D (2011). The human amygdala and the induction and experience of fear. *Curr Biol* 21: 34–8.

Feldman JA (1985). Four frames suffice: a provisional model of vision and space. *Behavioural Brain Sciences* 8: 265–289.

Fellows LK (2011). Orbitofrontal contributions to value-based decision making: evidence from humans with frontal lobe damage. *Ann N Y Acad Sci* 1239: 51–8.

Fellows LK and Farah MJ (2003). Ventromedial frontal cortex mediates affective shifting in humans: evidence from a reversal learning paradigm. *Brain* 126: 1830–7.

Fellows LK and Farah MJ (2005). Different underlying impairments in decision-making following ventromedial and dorsolateral frontal lobe damage in humans. *Cerebral Cortex* 15: 58–63.

Feng R, Rolls ET, Cheng W, and Feng J (2020). Hypertension is associated with reduced hippocampal connectivity and impaired memory. *EBioMedicine* 61: 103082.

Fernandez TV, Leckman JF, and Pittenger C (2018). Genetic susceptibility in obsessive-compulsive disorder. *Handb Clin Neurol* 148: 767–781.

Filimon F, Nelson JD, Huang RS, and Sereno MI (2009). Multiple parietal reach regions in humans: cortical representations for visual and proprioceptive feedback during on-line reaching. *J Neurosci* 29: 2961–71.

Fineberg NA and Robbins TW (2021). *The Neurobiology and Treatment of OCD*, vol. 49 of *Current Topics in Behavioral Neurosciences*. Springer Nature, Switzerland.

Finzi D, Gomez J, Nordt M, Rezai AA, Poltoratski S, and Grill-Spector K (2021). Differential spatial computations in ventral and lateral face-selective regions are scaffolded by structural connections. *Nat Commun* 12: 2278.

Flinker A, Korzeniewska A, Shestyuk AY, Franaszczuk PJ, Dronkers NF, Knight RT, and Crone NE (2015). Redefining the role of broca's area in speech. *Proc Natl Acad Sci U S A* 112: 2871–5.

Földiák P (1991). Learning invariance from transformation sequences. *Neural Computation* 3: 193–199.

Forgeard MJ, Haigh EA, Beck AT, Davidson RJ, Henn FA, Maier SF, Mayberg HS, and Seligman ME (2011). Beyond depression: towards a process-based approach to research, diagnosis, and treatment. *Clinical Psychology (New York)* 18: 275–299.

Francis S, Rolls ET, Bowtell R, McGlone F, O'Doherty J, Browning A, Clare S, and Smith E (1999). The representation of pleasant touch in the brain and its relationship with taste and olfactory areas. *Neuroreport* 10: 453–9.

Franco L, Rolls ET, Aggelopoulos NC, and Treves A (2004). The use of decoding to analyze the contribution to the information of the correlations between the firing of simultaneously recorded neurons. *Experimental Brain Research* 155: 370–384.

Franco L, Rolls ET, Aggelopoulos NC, and Jerez JM (2007). Neuronal selectivity, population sparseness, and ergodicity in the inferior temporal visual cortex. *Biological Cybernetics* 96: 547–560.

Franks KM, Stevens CF, and Sejnowski TJ (2003). Independent sources of quantal variability at single glutamatergic synapses. *Journal of Neuroscience* 23: 3186–3195.

Franzius M, Sprekeler H, and Wiskott L (2007). Slowness and sparseness lead to place, head-direction, and spatial-view cells. *PLoS Comput Biol* 3: e166.

Freedman DJ (2015). Learning-dependent plasticity of visual encoding in inferior temporal cortex. *J Vis* 15: 1420.

Freiwald WA (2020). The neural mechanisms of face processing: cells, areas, networks, and models. *Curr Opin Neurobiol* 60: 184–191.

Freiwald WA, Tsao DY, and Livingstone MS (2009). A face feature space in the macaque temporal lobe. *Nat Neurosci* 12: 1187–96.

Freton M, Lemogne C, Bergouignan L, Delaveau P, Lehericy S, and Fossati P (2014). The eye of the self: precuneus volume and visual perspective during autobiographical memory retrieval. *Brain Struct Funct* 219: 959–68.

Fried I, Rutishauser U, Cerf M, and Kreiman G (2014). *Single Neuron Studies of the Human Brain: Probing Cognition*. MIT Press, Cambridge, MA.

Friederici AD, Chomsky N, Berwick RC, Moro A, and Bolhuis JJ (2017). Language, mind and brain. *Nat Hum Behav* 1: 713–722.

Friston KJ and Frith CD (1995). Schizophrenia: a disconnection syndrome? *Clin Neurosci* 3: 89–97.

Friston KJ, Buechel C, Fink GR, Morris J, Rolls E, and Dolan RJ (1997). Psychophysiological and modulatory interactions in neuroimaging. *Neuroimage* 6: 218–29.

Friston KJ, Ashburner JT, Kiebel SJ, Nichols TE, and Penny WD (2006). *Statistical Parametric Mapping: the analysis of functional brain images*. Academic Press.

Fuhrmann G, Markram H, and Tsodyks M (2002a). Spike frequency adaptation and neocortical rhythms. *Journal of Neurophysiology* 88: 761–770.

Fuhrmann G, Segev I, Markram H, and Tsodyks M (2002b). Coding of temporal information by activity-dependent synapses. *Journal of Neurophysiology* 87: 140–148.

Funahashi S, Bruce CJ, and Goldman-Rakic PS (1989). Mnemonic coding of visual space in monkey dorsolateral prefrontal cortex. *Journal of Neurophysiology* 61: 331–349.

Gaffan D (1987). Amnesia, personal memory and the hippocampus: experimental neuropsychological studies in monkeys. In Stahl S, Iversen S, and Goodman E, editors, *Cognitive Neurochemistry*, 46–56. Oxford University Press, Oxford.

Gage FH (2025). Adult neurogenesis in the human dentate gyrus. *Hippocampus* 35: e23655.

Gajardo-Vidal A, Lorca-Puls DL, Team P, Warner H, Pshdary B, Crinion JT, Leff AP, Hope TMH, Geva S, Seghier ML, Green DW, Bowman H, and Price CJ (2021). Damage to broca's area does not contribute to long-term speech production outcome after stroke. *Brain* 144: 817–832.

Gallant JL, Connor CE, and Van-Essen DC (1998). Neural activity in areas V1, V2 and V4 during free viewing of natural scenes compared to controlled viewing. *NeuroReport* 9: 85–90.

Galletti C and Fattori P (2018). The dorsal visual stream revisited: Stable circuits or dynamic pathways? *Cortex* 98: 203–217.

Gallistel CR, Rolls E, and Greene D (1969). Neuron function inferred from behavioral and electrophysiological estimates of refractory period. *Science* 166: 1028–1030.

Gallivan JP, McLean A, and Culham JC (2011). Neuroimaging reveals enhanced activation in a reach-selective brain area for objects located within participants' typical hand workspaces. *Neuropsychologia* 49: 3710–21.

Galvin VC, Arnsten AFT, and Wang M (2020). Involvement of nicotinic receptors in working memory function. *Curr Top Behav Neurosci* 45: 89–99.

Gamberini M, Passarelli L, Fattori P, and Galletti C (2020). Structural connectivity and functional properties of the macaque superior parietal lobule. *Brain Struct Funct* 225: 1349–1367.

Gardner EP (2008). Dorsal and ventral streams in the sense of touch. In *The Senses: A Comprehensive Reference*, vol. 6, 233–258. Academic Press, New York.

Ge T, Feng J, Grabenhorst F, and Rolls ET (2012). Componential granger causality, and its application to identifying the source and mechanisms of the top-down biased activation that controls attention to affective vs sensory processing. *Neuroimage* 59: 1846–1858.

Geesaman BJ and Andersen RA (1996). The analysis of complex motion patterns by form/cue invariant MSTd neurons. *Journal of Neuroscience* 16: 4716–4732.

Georges-François P, Rolls ET, and Robertson RG (1999). Spatial view cells in the primate hippocampus: allocentric view not head direction or eye position or place. *Cerebral Cortex* 9: 197–212.

Georgopoulos AP, Schwartz AB, and Kettner RE (1986). Neuronal population coding of movement direction. *Science* 233: 1416–9.

Georgopoulos AP, Lurito JT, Petrides M, Schwartz AB, and Massey JT (1989). Mental rotation of the neuronal population vector. *Science* 243: 234–6.

Gerstner W, Kistler WM, Naud R, and Paninski L (2014). *Neuronal dynamics: From single neurons to networks and models of cognition*. Cambridge University Press, Cambridge.

Ghasemi M, Phillips C, Fahimi A, McNerney MW, and Salehi A (2017). Mechanisms of action and clinical efficacy of nmda receptor modulators in mood disorders. *Neurosci Biobehav Rev* 80: 555–572.

Ghosh VE and Gilboa A (2014). What is a memory schema? a historical perspective on current neuroscience literature. *Neuropsychologia* 53: 104–14.

Ghosh VE, Moscovitch M, Melo Colella B, and Gilboa A (2014). Schema representation in patients with ventromedial pfc lesions. *J Neurosci* 34: 12057–70.

Gibbs J, Maddison SP, and Rolls ET (1981). Satiety role of the small intestine examined in sham-feeding rhesus monkeys. *Journal of Comparative and Physiological Psychology* 95: 1003–15.

Gilbert SJ and Burgess PW (2008). Executive function. *Curr Biol* 18: R110–4.

Gilboa A and Marlatte H (2017). Neurobiology of schemas and schema-mediated memory. *Trends Cogn Sci* 21: 618–631.

Giocomo LM and Hasselmo ME (2007). Neuromodulation by glutamate and acetylcholine can change circuit dynamics by regulating the relative influence of afferent input and excitatory feedback. *Mol Neurobiol* 36: 184–200.

Giocomo LM, Moser MB, and Moser EI (2011). Computational models of grid cells. *Neuron* 71: 589–603.

Glantz LA and Lewis DA (2000). Decreased dendritic spine density on prefrontal cortical pyramidal neurons in schizophrenia. *Arch Gen Psychiatry* 57: 65–73.

Glasser MF, Coalson TS, Robinson EC, Hacker CD, Harwell J, Yacoub E, Ugurbil K, Andersson J, Beckmann CF, Jenkinson M, Smith SM, and Van Essen DC (2016a). A multi-modal parcellation of human cerebral cortex. *Nature* 536: 171–8.

Glasser MF, Smith SM, Marcus DS, Andersson JL, Auerbach EJ, Behrens TE, Coalson TS, Harms MP, Jenkinson M, Moeller S, Robinson EC, Sotiropoulos SN, Xu J, Yacoub E, Ugurbil K, and Van Essen DC (2016b). The human connectome project's neuroimaging approach. *Nat Neurosci* 19: 1175–87.

Glausier JR and Lewis DA (2013). Dendritic spine pathology in schizophrenia. *Neuroscience* 251: 90–107.

Glausier JR and Lewis DA (2018). Mapping pathologic circuitry in schizophrenia. *Handb Clin Neurol* 150: 389–417.

Glimcher PW (2005). Indeterminacy in brain and behavior. *Annu Rev Psychol* 56: 25–56.

Glimcher PW and Fehr E (2013). *Neuroeconomics: Decision-Making and the Brain.* Academic Press, New York, 2nd edn.

Gnadt JW and Andersen RA (1988). Memory related motor planning activity in posterior parietal cortex of macaque. *Exp Brain Res* 70: 216–20.

Gold PW (2015). The organization of the stress system and its dysregulation in depressive illness. *Molecular Psychiatry* 20: 32–47.

Goldman-Rakic P (1994). Working memory dysfunction in schizophrenia. *Journal of Neuropsychology and Clinical Neuroscience* 6: 348–357.

Goldman-Rakic PS (1996). The prefrontal landscape: implications of functional architecture for understanding human mentation and the central executive. *Philos Trans R Soc Lond B Biol Sci* 351: 1445–53.

Goldman-Rakic PS (1999). The physiological approach: functional architecture of working memory and disordered cognition in schizophrenia. *Biological Psychiatry* 46: 650–661.

Gong W, Rolls ET, Du J, Feng J, and Cheng W (2021). Brain structure is linked to the association between family environment and behavioral problems in children in the abcd study. *Nature Communications* 12: 3769.

Gonzalez-Burgos G and Lewis DA (2012). Nmda receptor hypofunction, parvalbumin-positive neurons, and cortical gamma oscillations in schizophrenia. *Schizophr Bull* 38: 950–7.

Gonzalez-Burgos G, Hashimoto T, and Lewis DA (2010). Alterations of cortical gaba neurons and network oscillations in schizophrenia. *Curr Psychiatry Rep* 12: 335–44.

Goodale MA and Milner AD (1992). Separate visual pathways for perception and action. *Trends Neurosci* 15: 20–5.

Gotlib IH and Hammen CL (2009). *Handbook of Depression.* Guilford Press, New York.

Grabenhorst F and Rolls ET (2008). Selective attention to affective value alters how the brain processes taste stimuli. *European Journal of Neuroscience* 27: 723–729.

Grabenhorst F and Rolls ET (2010). Attentional modulation of affective vs sensory processing: functional connectivity and a top-down biased activation theory of selective attention. *Journal of Neurophysiology* 104: 1649–1660.

Grabenhorst F and Rolls ET (2011). Value, pleasure, and choice in the ventral prefrontal cortex. *Trends in Cognitive Sciences* 15: 56–67.

Grabenhorst F, Rolls ET, Margot C, da Silva MAAP, and Velazco MI (2007). How pleasant and unpleasant stimuli combine in different brain regions: odor mixtures. *Journal of Neuroscience* 27: 13532–13540.

Grabenhorst F, Rolls ET, and Bilderbeck A (2008a). How cognition modulates affective responses to taste and flavor: top down influences on the orbitofrontal and pregenual cingulate cortices. *Cerebral Cortex* 18: 1549–1559.

Grabenhorst F, Rolls ET, and Parris BA (2008b). From affective value to decision-making in the prefrontal cortex. *European Journal of Neuroscience* 28: 1930–1939.

Grabenhorst F, D'Souza AA, Parris BA, Rolls ET, and Passingham RE (2010a). A common neural scale for the subjective pleasantness of different primary rewards. *Neuroimage* 51: 1265–74.

Grabenhorst F, Rolls ET, Parris BA, and D'Souza A (2010b). How the brain represents the reward value of fat in the mouth. *Cerebral Cortex* 20: 1082–1091.

Grady CL (2008). Cognitive neuroscience of aging. *Annals of the New York Academy of Sciences* 1124: 127–144.

Grant JE and Chamberlain SR (2019). Obsessive compulsive personality traits: Understanding the chain of pathogenesis from health to disease. *J Psychiatr Res* 116: 69–73.

Grattan LE and Glimcher PW (2014). Absence of spatial tuning in the orbitofrontal cortex. *PLoS One* 9: e112750.

Gray J (1975). *Elements of a Two-Process Theory of Learning.* Academic Press, London.

Grayson J (1961). *Nerves, Brain and Man.* Scientific Book Club, London.

Graziano MSA, Andersen RA, and Snowden RJ (1994). Tuning of MST neurons to spiral motions. *Journal of Neuroscience* 14: 54–67.

Green MF (1996). What are the functional consequences of neurocognitive deficits in schizophrenia? *American Journal of Psychiatry* 153: 321–330.

Grimaldi P, Saleem KS, and Tsao D (2016). Anatomical connections of the functionally defined "face patches" in the macaque monkey. *Neuron* 90: 1325–1342.

Gross CG, Bender DB, and Rocha-Miranda CE (1969). Visual receptive fields of neurons in inferotemporal cortex of the monkey. *Science* 166: 1303–6.

Gross CG, Rocha-Miranda CE, and Bender DB (1972). Visual properties of neurons in inferotemporal cortex of the macaque. *J Neurophysiol* 35: 96–111.

Gross CG, Desimone R, Albright TD, and Schwartz EL (1985). Inferior temporal cortex and pattern recognition. *Experimental Brain Research* 11: 179–201.

Grossberg S (1976). Adaptive pattern classification and universal recoding: Ii. feedback, expectation, olfaction, illusions. *Biol Cybern* 23: 187–202.

Guerguiev J, Lillicrap TP, and Richards BA (2017). Towards deep learning with segregated dendrites. *Elife* 6: e22901.

Hamilton JP, Chen MC, and Gotlib IH (2013). Neural systems approaches to understanding major depressive disorder: an intrinsic functional organization perspective. *Neurobiology of Disease* 52: 4–11.

Hamilton WD (1964). The genetical evolution of social behaviour. *Journal of Theoretical Biology* 7: 1–52.

Hamilton WD (1996). *Narrow Roads of Gene Land.* W. H. Freeman, New York.

Hampshire A and Owen AM (2006). Fractionating attentional control using event-related fMRI. *Cerebral Cortex* 16: 1679–1689.

Harel J, Koch C, and Perona P (2006a). Graph-based visual saliency. *Advances in Neural Information Processing Systems (NIPS)* 19: 545–552.

Harel J, Koch C, and Perona P (2006b). A saliency implementation in MATLAB http://www.klab.caltech.edu/ harel/share/gbvs.php.

Harmer CJ and Cowen PJ (2013). 'it's the way that you look at it'–a cognitive neuropsychological account of ssri action in depression. *Philosophical Transactions of the Royal Society of London B Biological Sciences* 368: 20120407.

Hartley T, Lever C, Burgess N, and O'Keefe J (2014). Space in the brain: how the hippocampal formation supports spatial cognition. *Philos Trans R Soc Lond B Biol Sci* 369: 20120510.

Hartston HJ and Swerdlow NR (1999). Visuospatial priming and Stroop performance in patients with obsessive compulsive disorder. *Neuropsychology* 13: 447–457.

Hassabis D, Chu C, Rees G, Weiskopf N, Molyneux PD, and Maguire EA (2009). Decoding neuronal ensembles in the human hippocampus. *Curr Biol* 19: 546–54.

Hasselmo ME (1999). Neuromodulation: acetylcholine and memory consolidation. *Trends Cogn Sci* 3: 351–359.

Hasselmo ME and Bower JM (1993). Acetylcholine and memory. *Trends in Neurosciences* 16: 218–222.

Hasselmo ME and Giocomo LM (2006). Cholinergic modulation of cortical function. *J Mol Neurosci* 30: 133–135.

Hasselmo ME and McGaughy J (2004). High acetylcholine levels set circuit dynamics for attention and encoding and low acetylcholine levels set dynamics for consolidation. *Prog Brain Res* 145: 207–31.

Hasselmo ME and Nadel L (2025). Scientific histories of hippocampal research: Introduction to the special issue. *Hippocampus* 35: e23680.

Hasselmo ME and Sarter M (2011). Modes and models of forebrain cholinergic neuromodulation of cognition. *Neuropsychopharmacology* 36: 52–73.

Hasselmo ME and Wyble BP (1997). Free recall and recognition in a network model of the hippocampus: simulating effects of scopolamine on human memory function. *Behav Brain Res* 89: 1–34.

Hasselmo ME, Rolls ET, and Baylis GC (1989a). The role of expression and identity in the face-selective responses of neurons in the temporal visual cortex of the monkey. *Behav Brain Res* 32: 203–18.

Hasselmo ME, Rolls ET, Baylis GC, and Nalwa V (1989b). Object-centred encoding by face-selective neurons in the cortex in the superior temporal sulcus of the monkey. *Experimental Brain Research* 75: 417–429.

Hasselmo ME, Schnell E, and Barkai E (1995). Learning and recall at excitatory recurrent synapses and cholinergic modulation in hippocampal region CA3. *Journal of Neuroscience* 15: 5249–5262.

Hebb D (1949). *The Organization of Behavior: A Neuropsychological Theory.* Wiley, New York.

Heberlein AS, Padon AA, Gillihan SJ, Farah MJ, and Fellows LK (2008). Ventromedial frontal lobe plays a critical role in facial emotion recognition. *J Cogn Neurosci* 20: 721–33.

Hertrich I, Dietrich S, and Ackermann H (2020). The margins of the language network in the brain. *Frontiers in Communication* 5: 519955.

Hertz J, Krogh A, and Palmer RG (1991). *Introduction to the Theory of Neural Computation.* Addison-Wesley, Redwood City, CA.

Hesse JK and Tsao DY (2020). The macaque face patch system: a turtle's underbelly for the brain. *Nat Rev Neurosci* 21: 695–716.

Hetherington MM (1996). Sensory-specific satiety and its importance in meal termination. *Neurosci Biobehav Rev* 20: 113–7.

Hickok G (2022). The dual stream model of speech and language processing. *Handb Clin Neurol* 185: 57–69.

Hickok G and Poeppel D (2007). The cortical organization of speech processing. *Nat Rev Neurosci* 8: 393–402.

Hickok G and Poeppel D (2015). Neural basis of speech perception. *Handb Clin Neurol* 129: 149–60.

Hickok G, Venezia J, and Teghipco A (2023). Beyond broca: neural architecture and evolution of a dual motor speech coordination system. *Brain* 146: 1775–1790.

Hinton GE and Anderson JA (1981). *Parallel Models of Associative Memory*. Erlbaum, Hillsdale, N.J.

Hoftman GD, Datta D, and Lewis DA (2017). Layer 3 excitatory and inhibitory circuitry in the prefrontal cortex: developmental trajectories and alterations in schizophrenia. *Biol Psychiatry* 81: 862–873.

Holscher C, Rolls ET, and Xiang J (2003). Perirhinal cortex neuronal activity related to long-term familiarity memory in the macaque. *Eur J Neurosci* 18: 2037–46.

Holtzheimer PE, Husain MM, Lisanby SH, Taylor SF, Whitworth LA, McClintock S, Slavin KV, Berman J, McKhann GM, Patil PG, Rittberg BR, Abosch A, Pandurangi AK, Holloway KL, Lam RW, Honey CR, Neimat JS, Henderson JM, DeBattista C, Rothschild AJ, Pilitsis JG, Espinoza RT, Petrides G, Mogilner AY, Matthews K, Peichel D, Gross RE, Hamani C, Lozano AM, and Mayberg HS (2017). Subcallosal cingulate deep brain stimulation for treatment-resistant depression: a multisite, randomised, sham-controlled trial. *Lancet Psychiatry* 4: 839–849.

Hopfield JJ (1982). Neural networks and physical systems with emergent collective computational abilities. *Proc Natl Acad Sci U S A* 79: 2554–8.

Hornak J, Rolls ET, and Wade D (1996). Face and voice expression identification in patients with emotional and behavioural changes following ventral frontal lobe damage. *Neuropsychologia* 34: 247–261.

Hornak J, Bramham J, Rolls ET, Morris RG, O'Doherty J, Bullock PR, and Polkey CE (2003). Changes in emotion after circumscribed surgical lesions of the orbitofrontal and cingulate cortices. *Brain* 126: 1691–1712.

Hornak J, O'Doherty J, Bramham J, Rolls ET, Morris RG, Bullock PR, and Polkey CE (2004). Reward-related reversal learning after surgical excisions in orbitofrontal and dorsolateral prefrontal cortex in humans. *Journal of Cognitive Neuroscience* 16: 463–478.

Howard MW and Eichenbaum H (2015). Time and space in the hippocampus. *Brain Res* 1621: 345–54.

Howes OD, Bukala BR, and Beck K (2024). Schizophrenia: from neurochemistry to circuits, symptoms and treatments. *Nat Rev Neurol* 20: 22–35.

Hsu CCH, Rolls ET, Huang CC, Chong ST, Lo CYZ, Feng J, and Lin CP (2020). Connections of the human orbitofrontal cortex and inferior frontal gyrus. *Cerebral Cortex* 30: 5830–5843.

Huang CC, Rolls ET, Hsu CC, Feng J, and Lin CP (2021). Extensive cortical connectivity of the human hippocampal memory system: beyond the "what" and "where" dual-stream model. *Cerebral Cortex* 31: 4652–4669.

Huang CC, Rolls ET, Feng J, and Lin CP (2022). An extended human connectome project multimodal parcellation atlas of the human cortex and subcortical areas. *Brain Struct Funct* 227: 763–778.

Hubel DH and Wiesel TN (1962). Receptive fields, binocular interaction and functional architecture in the cat's visual cortex. *J Physiol* 160: 106–154.

Hubel DH and Wiesel TN (1968). Receptive fields and functional architecture of macaque monkey striate cortex. *Journal of Physiology* 195: 215–243.

Hubel DH and Wiesel TN (1977). Ferrier lecture. functional architecture of macaque monkey visual cortex. *Proc R Soc Lond B Biol Sci* 198: 1–59.

Iadarola ND, Niciu MJ, Richards EM, Vande Voort JL, Ballard ED, Lundin NB, Nugent AC, Machado-Vieira R, and Zarate J C A (2015). Ketamine and other n-methyl-d-aspartate receptor antagonists in the treatment of depression: a perspective review. *Therapetic Advances in Chronic Disease* 6: 97–114.

Iadecola C and Gottesman RF (2019). Neurovascular and cognitive dysfunction in hypertension. *Circ Res* 124: 1025–1044.

Insabato A, Pannunzi M, Rolls ET, and Deco G (2010). Confidence-related decision-making. *Journal of Neurophysiology* 104: 539–547.

International Brain L, Angelaki D, Benson B, Benson J, Birman D, Bonacchi N, Bougrova K, Bruijns SA, Carandini M, Catarino JA, Chapuis GA, Churchland AK, Dan Y, Davatolhagh F, Dayan P, DeWitt EE, Engel TA, Fabbri M, Faulkner M, Fiete IR, Findling C, Freitas-Silva L, Gercek B, Harris KD, Hausser M, Hofer SB, Hu F, Hubert F, Huntenburg JM, Khanal A, Krasniak CS, Langdon C, Langfield C, Lau PYP, Mainen ZF, Meijer GT, Miska NJ, Mrsic-Flogel TD, Noel JP, Nylund K, Pan-Vazquez A, Paninski L, Pouget A, Rossant C, Roth N, Schaeffer R, Schartner M, Shi Y, Socha KZ, Steinmetz NA, Svoboda K, Urai AE, Wells MJ, West SJ, Whiteway MR, Winter O, and Witten IB (2025). A brain-wide map of neural activity during complex behaviour. *Nature* 645: 177–191.

Ison MJ, Quian Quiroga R, and Fried I (2015). Rapid encoding of new memories by individual neurons in the human brain. *Neuron* 87: 220–30.

Israel I, Chapuis N, Glasauer S, Charade O, and Berthoz A (1993). Estimation of passive horizontal linear whole-body displacement in humans. *J Neurophysiol* 70: 1270–3.

Itskov PM, Vinnik E, and Diamond ME (2011). Hippocampal representation of touch-guided behavior in rats: persistent and independent traces of stimulus and reward location. *PLoS One* 6: e16462.

Itti L and Koch C (2000). A saliency-based search mechanism for overt and covert shifts of visual attention. *Vision Research* 40: 1489–1506.

Iversen SD and Mishkin M (1970). Perseverative interference in monkeys following selective lesions of the inferior prefrontal convexity. *Exp Brain Res* 11: 376–86.

Izhikevich EM (2007). *Dynamical systems in neuroscience*. MIT press.

Izquierdo A (2017). Functional heterogeneity within rat orbitofrontal cortex in reward learning and decision making. *J Neurosci* 37: 10529–10540.

Jackson BS (2004). Including long-range dependence in integrate-and-fire models of the high interspike-interval variability of cortical neurons. *Neural Computation* 16: 2125–2195.

Jellema T and Perrett DI (2003). Cells in monkey sts responsive to articulated body motions and consequent static posture: a case of implied motion? *Neuropsychologia* 41: 1728–37.

Jenkinson M, Beckmann CF, Behrens TE, Woolrich MW, and Smith SM (2012). Fsl. *Neuroimage* 62: 782–90.

Jeurissen B, Descoteaux M, Mori S, and Leemans A (2019). Diffusion mri fiber tractography of the brain. *NMR Biomed* 32: e3785.

Jia X, Hong H, and DiCarlo JJ (2021). Unsupervised changes in core object recognition behavior are predicted by neural plasticity in inferior temporal cortex. *Elife* 10: 60830.

Johansen-Berg H, Gutman DA, Behrens TE, Matthews PM, Rushworth MF, Katz E, Lozano AM, and Mayberg HS (2008). Anatomical connectivity of the subgenual cingulate region targeted with deep brain stimulation for treatment-resistant depression. *Cereb Cortex* 18: 1374–83.

Johnson SC, Baxter LC, Wilder LS, Pipe JG, Heiserman JE, and Prigatano GP (2002). Neural correlates of self-reflection. *Brain* 125: 1808–14.

Johnston JN, Kadriu B, Kraus C, Henter ID, and Zarate J C A (2024). Ketamine in neuropsychiatric disorders: an update. *Neuropsychopharmacology* 49: 23–40.

Joyce J (1939). *Finnegans Wake*. Faber and Faber, London.

Kacelnik A and Brito e Abreu F (1998). Risky choice and Weber's Law. *Journal of Theoretical Biology* 194: 289–298.

Kadohisa M, Rolls ET, and Verhagen JV (2005). Neuronal representations of stimuli in the mouth: the primate insular taste cortex, orbitofrontal cortex, and amygdala. *Chemical Senses* 30: 401419.

Kamps FS, Julian JB, Kubilius J, Kanwisher N, and Dilks DD (2016). The occipital place area represents the local elements of scenes. *Neuroimage* 132: 417–424.

Kanwisher N, McDermott J, and Chun MM (1997). The fusiform face area: a module in human extrastriate cortex specialized for face perception. *J Neurosci* 17: 4302–11.

Kar K and DiCarlo JJ (2024). The quest for an integrated set of neural mechanisms underlying object recognition in primates. *Annu Rev Vis Sci* 10: 91–121.

Karno M, Golding JM, Sorenson SB, and Burnam MA (1988). The epidemiology of obsessive-compulsive disorder in five US communities. *Archives of General Psychiatry* 45: 1094–1099.

Karthik S, Sharma LP, and Narayanaswamy JC (2020). Investigating the role of glutamate in obsessive-compulsive disorder: current perspectives. *Neuropsychiatr Dis Treat* 16: 1003–1013.

Kastner S, Chen Q, Jeong SK, and Mruczek REB (2017). A brief comparative review of primate posterior parietal cortex: A novel hypothesis on the human toolmaker. *Neuropsychologia* 105: 123–134.

Kayaert G, Biederman I, Op de Beeck HP, and Vogels R (2005a). Tuning for shape dimensions in macaque inferior temporal cortex. *Eur J Neurosci* 22: 212–24.

Kayaert G, Biederman I, and Vogels R (2005b). Representation of regular and irregular shapes in macaque inferotemporal cortex. *Cerebral Cortex* 15: 1308–1321.

Kelly KM, Nadon NL, Morrison JH, Thibault O, Barnes CA, and Blalock EM (2006). The neurobiology of aging. *Epilepsy Research* 68, Supplement 1: S5–S20.

Kemmerer D (2015). *Cognitive Neuroscience of Language*. Psychology Press, New York.

Kennedy DP and Adolphs R (2011). Reprint of: Impaired fixation to eyes following amygdala damage arises from abnormal bottom-up attention. *Neuropsychologia* 49: 589–95.

Kesner RP and Rolls ET (2001). Role of long term synaptic modification in short term memory. *Hippocampus* 11: 240–250.

Kesner RP and Rolls ET (2015). A computational theory of hippocampal function, and tests of the theory: new developments. *Neuroscience and Biobehavioral Reviews* 48: 92–147.

Kesner RP, Hunsaker MR, and Ziegler W (2011). The role of the dorsal and ventral hippocampus in olfactory working memory. *Neurobiol Learn Mem* 96: 361–6.

Khandhadia AP, Murphy AP, Romanski LM, Bizley JK, and Leopold DA (2021). Audiovisual integration in macaque face patch neurons. *Curr Biol* 31: 1826–1835 e3.

Kievit J and Kuypers HGJM (1975). Subcortical afferents to the frontal lobe in the rhesus monkey studied by means of retrograde horseradish peroxidase transport. *Brain Research* 85: 261–266.

Kim JG and Biederman I (2012). Greater sensitivity to nonaccidental than metric changes in the relations between simple shapes in the lateral occipital cortex. *Neuroimage* 63: 1818–26.

Kim JJ and Fanselow MS (1992). Modality-specific retrograde amnesia of fear. *Science* 256: 675–7.

Kircher TT, Senior C, Phillips ML, Benson PJ, Bullmore ET, Brammer M, Simmons A, Williams SC, Bartels M, and David AS (2000). Towards a functional neuroanatomy of self processing: effects of faces and words. *Brain Res Cogn Brain Res* 10: 133–44.

Kircher TT, Brammer M, Bullmore E, Simmons A, Bartels M, and David AS (2002). The neural correlates of intentional and incidental self processing. *Neuropsychologia* 40: 683–92.

Klein-Flugge MC, Jensen DEA, Takagi Y, Priestley L, Verhagen L, Smith SM, and Rushworth MFS (2022). Relationship between nuclei-specific amygdala connectivity and mental health dimensions in humans. *Nat Hum Behav* 6: 1705–1722.

Knudsen EB and Wallis JD (2022). Taking stock of value in the orbitofrontal cortex. *Nat Rev Neurosci* 23: 428–438.

Kobayashi Y and Amaral DG (2007). Macaque monkey retrosplenial cortex: Iii. cortical efferents. *J Comp Neurol* 502: 810–33.

Koch C (1999). *Biophysics of Computation.* Oxford University Press, Oxford.

Kohonen T (1977). *Associative Memory: A System Theoretical Approach.* Springer, New York.

Kohonen T (1984). *Self-Organization and Associative Memory.* Springer-Verlag, Berlin.

Kohonen T, Oja E, and Lehtio P (1981). Storage and processing of information in distributed memory systems. In Hinton GE and Anderson JA, editors, *Parallel Models of Associative Memory*, 129–167. Lawrence Erlbaum, Hillsdale.

Kolibius LD, Roux F, Parish G, Ter Wal M, Van Der Plas M, Chelvarajah R, Sawlani V, Rollings DT, Lang JD, Gollwitzer S, Walther K, Hopfengartner R, Kreiselmeyer G, Hamer H, Staresina BP, Wimber M, Bowman H, and Hanslmayr S (2023). Hippocampal neurons code individual episodic memories in humans. *Nat Hum Behav* 7: 1968–1979.

Konopaske GT, Lange N, Coyle JT, and Benes FM (2014). Prefrontal cortical dendritic spine pathology in schizophrenia and bipolar disorder. *JAMA Psychiatry* 71: 1323–31.

Kraus BJ, Robinson RJ, White JA, Eichenbaum H, and Hasselmo ME (2013). Hippocampal "time cells": time versus path integration. *Neuron* 78: 1090–1101.

Kravitz DJ, Saleem KS, Baker CI, Ungerleider LG, and Mishkin M (2013). The ventral visual pathway: an expanded neural framework for the processing of object quality. *Trends Cogn Sci* 17: 26–49.

Krebs JR and Davies NB (1991). *Behavioural Ecology.* Blackwell, Oxford, 3rd edn.

Kringelbach ML and Rolls ET (2003). Neural correlates of rapid reversal learning in a simple model of human social interaction. *Neuroimage* 20: 1371–83.

Kringelbach ML and Rolls ET (2004). The functional neuroanatomy of the human orbitofrontal cortex: evidence from neuroimaging and neuropsychology. *Progress in Neurobiology* 72: 341–372.

Kringelbach ML, O'Doherty J, Rolls ET, and Andrews C (2003). Activation of the human orbitofrontal cortex to a liquid food stimulus is correlated with its subjective pleasantness. *Cerebral Cortex* 13: 1064–1071.

Kringelbach ML, Perl YS, Tagliazucchi E, and Deco G (2023). Toward naturalistic neuroscience: Mechanisms underlying the flattening of brain hierarchy in movie-watching compared to rest and task. *Sci Adv* 9: eade6049.

Kropff E, Carmichael JE, Moser MB, and Moser EI (2015). Speed cells in the medial entorhinal cortex. *Nature* 523: 419–24.

Kubie JL (2025). How ideas about context and remapping eveloped in brooklyn. *Hippocampus* 35: e23671.

Lally N, Nugent AC, Luckenbaugh DA, Niciu MJ, Roiser JP, and Zarate J C A (2015). Neural correlates of change in major depressive disorder anhedonia following open-label ketamine. *Journal of Psychopharmacology* 29: 596–607.

Lau H (2022). *In Consciousness We Trust: the cognitive neuroscience of conscious experience.* Oxford University Press, Oxford.

Lau H and Rosenthal D (2011). Empirical support for higher-order theories of conscious awareness. *Trends Cogn Sci* 15: 365–73.

Leaver AM and Rauschecker JP (2016). Functional topography of human auditory cortex. *J Neurosci* 36: 1416–28.

LeCun Y, Kavukcuoglu K, and Farabet C (2010). Convolutional networks and applications in vision. *IEEE International Symposium on Circuits and Systems* 253–256.

LeCun Y, Bengio Y, and Hinton G (2015). Deep learning. *Nature* 521: 436–44.

LeDoux J (2012). Rethinking the emotional brain. *Neuron* 73: 653–76.

LeDoux J, Brown R, Pine D, and Hofmann S (2018). Know thyself: well-being and subjective experience. *Cerebrum* 2018: https://www.ncbi.nlm.nih.gov/pmc/articles/PMC6353121/.

LeDoux JE (1995). Emotion: clues from the brain. *Annu Rev Psychol* 46: 209–35.

LeDoux JE (1996). *The Emotional Brain.* Simon and Schuster, New York.

LeDoux JE (2000). Emotion circuits in the brain. *Annual Review of Neuroscience* 23: 155–184.

LeDoux JE (2020). Thoughtful feelings. *Curr Biol* 30: R619–R623.

LeDoux JE and Brown R (2017). A higher-order theory of emotional consciousness. *Proc Natl Acad Sci U S A* 114: E2016–E2025.

LeDoux JE and Hofmann SG (2018). The subjective experience of emotion: a fearful view. *Curr Opin Behav Sci* 19: 67–72.

LeDoux JE and Pine DS (2016). Using neuroscience to help understand fear and anxiety: a two-system framework. *Am J Psychiatry* 173: 1083–1093.

LeDoux JE, Iwata J, Cicchetti P, and Reis DJ (1988). Different projections of the central amygdaloid nucleus mediate autonomic and behavioral correlates of conditioned fear. *J Neurosci* 8: 2517–29.

Leech R and Sharp DJ (2014). The role of the posterior cingulate cortex in cognition and disease. *Brain* 137: 12–32.

Leech R and Smallwood J (2019). The posterior cingulate cortex: Insights from structure and function. *Handb Clin Neurol* 166: 73–85.

Lehky SR and Tanaka K (2016). Neural representation for object recognition in inferotemporal cortex. *Curr Opin Neurobiol* 37: 23–35.

Leonard CM, Rolls ET, Wilson FAW, and Baylis GC (1985). Neurons in the amygdala of the monkey with responses selective for faces. *Behavioural Brain Research* 15: 159–176.

Lerch JP, van der Kouwe AJ, Raznahan A, Paus T, Johansen-Berg H, Miller KL, Smith SM, Fischl B, and Sotiropoulos SN (2017). Studying neuroanatomy using mri. *Nat Neurosci* 20: 314–326.

Leutgeb JK, Leutgeb S, Treves A, Meyer R, Barnes CA, McNaughton BL, Moser MB, and Moser EI (2005a). Progressive transformation of hippocampal neuronal representations in "morphed" environments. *Neuron* 48: 345–58.

Leutgeb S, Leutgeb JK, Barnes CA, Moser EI, McNaughton BL, and Moser MB (2005b). Independent codes for spatial and episodic memory in hippocampal neuronal ensembles. *Science* 309: 619–23.

Lewis DA (2014). Inhibitory neurons in human cortical circuits: substrate for cognitive dysfunction in schizophrenia. *Curr Opin Neurobiol* 26: 22–6.

Li N and DiCarlo JJ (2010). Unsupervised natural visual experience rapidly reshapes size-invariant object representation in inferior temporal cortex. *Neuron* 67: 1062–75.

Li N and DiCarlo JJ (2012). Neuronal learning of invariant object representation in the ventral visual stream is not dependent on reward. *J Neurosci* 32: 6611–20.

Li T, Wang Q, Zhang J, Rolls ET, Yang W, Palaniyappan L, Zhang L, Cheng W, Yao Y, Liu Z, Gong X, Luo Q, Tang Y, Crow TJ, Broome MR, Xu K, Li C, Wang J, Liu Z, Lu G, Wang F, and Feng J (2017). Brain-wide analysis of functional connectivity in first-episode and chronic stages of schizophrenia. *Schizophr Bull* 43: 436–448.

Liddle PF (1987). The symptoms of chronic schizophrenia: a re-examination of the positive-negative dichotomy. *British Journal of Psychiatry* 151: 145–151.

Lillicrap TP, Santoro A, Marris L, Akerman CJ, and Hinton G (2020). Backpropagation and the brain. *Nat Rev Neurosci* 21: 335–346.

Liu J, Harris A, and Kanwisher N (2010). Perception of face parts and face configurations: an fmri study. *J Cogn Neurosci* 22: 203–11.

Liu Y and Wang XJ (2001). Spike-frequency adaptation of a generalized leaky integrate-and-fire model neuron. *Journal of Computational Neuroscience* 10: 25–45.

Liuzzi AG, Bruffaerts R, and Vandenberghe R (2019). The medial temporal written word processing system. *Cortex* 119: 287–300.

Loh M, Rolls ET, and Deco G (2007a). A dynamical systems hypothesis of schizophrenia. *PLoS Computational Biology* 3: e228. doi:10.1371/journal.pcbi.0030228.

Loh M, Rolls ET, and Deco G (2007b). Statistical fluctuations in attractor networks related to schizophrenia. *Pharmacopsychiatry* 40: S78–84.

Long MA, Katlowitz KA, Svirsky MA, Clary RC, Byun TM, Majaj N, Oya H, Howard r M A, and Greenlee JDW (2016). Functional segregation of cortical regions underlying speech timing and articulation. *Neuron* 89: 1187–1193.

Lowe D (1985). *Perceptual Organization and Visual Recognition.* Kluwer, Boston.

Lozano AM, Giacobbe P, Hamani C, Rizvi SJ, Kennedy SH, Kolivakis TT, Debonnel G, Sadikot AF, Lam RW, Howard AK, Ilcewicz-Klimek M, Honey CR, and Mayberg HS (2012). A multicenter pilot study of subcallosal cingulate area deep brain stimulation for treatment-resistant depression. *J Neurosurg* 116: 315–22.

Lu W, Du X, Wang J, Zeng L, Ye L, Xiang S, Zheng Q, Zhang J, Xu N, Feng J, and Consortium DTB (2024). Simulation and assimilation of the digital human brain. *Nat Comput Sci* 4: 890–898.

Lujan JL, Chaturvedi A, Choi KS, Holtzheimer PE, Gross RE, Mayberg HS, and McIntyre CC (2013). Tractography-activation models applied to subcallosal cingulate deep brain stimulation. *Brain Stimul* 6: 737–9.

Luo Q, Ge T, Grabenhorst F, Feng J, and Rolls ET (2013). Attention-dependent modulation of cortical taste circuits revealed by granger causality with signal-dependent noise. *PLoS Computational Biology* 9: e1003265.

Lynch G and Gall CM (2006). AMPAkines and the threefold path to cognitive enhancement. *Trends in Neuroscience* 29: 554–562.

Ma Q, Rolls ET, Huang CC, Cheng W, and Feng J (2022). Extensive cortical functional connectivity of the human hippocampal memory system. *Cortex* 147: 83–101.

Ma Y (2015). Neuropsychological mechanism underlying antidepressant effect: a systematic meta-analysis. *Molecular Psychiatry* 20: 311–319.

MacDonald CJ, Lepage KQ, Eden UT, and Eichenbaum H (2011). Hippocampal "time cells" bridge the gap in memory for discontiguous events. *Neuron* 71: 737–49.

MacDonald ML, Alhassan J, Newman JT, Richard M, Gu H, Kelly RM, Sampson AR, Fish KN, Penzes P, Wills ZP, Lewis DA, and Sweet RA (2017). Selective loss of smaller spines in schizophrenia. *Am J Psychiatry* 174: 586–594.

Macpherson T, Churchland A, Sejnowski T, DiCarlo J, Kamitani Y, Takahashi H, and Hikida T (2021). Natural and artificial intelligence: A brief introduction to the interplay between ai and neuroscience research. *Neural Netw* 144: 603–613.

Maguire EA (2014). Memory consolidation in humans: new evidence and opportunities. *Exp Physiol* 99: 471–86.

Maier SF and Seligman ME (2016). Learned helplessness at fifty: Insights from neuroscience. *Psychol Rev* 123: 349–67.

Maier-Hein KH, Neher PF, Houde JC, Cote MA, Garyfallidis E, Zhong J, Chamberland M, Yeh FC, Lin YC, Ji Q, Reddick WE, Glass JO, Chen DQ, Feng Y, Gao C, Wu Y, Ma J, He R, Li Q, Westin CF, Deslauriers-Gauthier S, Gonzalez JOO, Paquette M, St-Jean S, Girard G, Rheault F, Sidhu J, Tax CMW, Guo F, Mesri HY, David S, Froeling M, Heemskerk AM, Leemans A, Bore A, Pinsard B, Bedetti C, Desrosiers M, Brambati S, Doyon J, Sarica A, Vasta R, Cerasa A, Quattrone A, Yeatman J, Khan AR, Hodges W, Alexander S, Romascano D, Barakovic M, Auria A, Esteban O, Lemkaddem A, Thiran JP, Cetingul HE, Odry BL, Mailhe B, Nadar MS, Pizzagalli F, Prasad G, Villalon-Reina JE, Galvis J, Thompson PM, Requejo FS, Laguna PL, Lacerda LM, Barrett R, Dell'Acqua F, Catani M, Petit L, Caruyer E, Daducci A, Dyrby TB, Holland-Letz T, Hilgetag CC, Stieltjes B, and Descoteaux M (2017). The challenge of mapping the human connectome based on diffusion tractography. *Nat Commun* 8: 1349.

Maltbie EA, Kaundinya GS, and Howell LL (2017). Ketamine and pharmacological imaging: use of functional magnetic resonance imaging to evaluate mechanisms of action. *Behav Pharmacol* 28: 610–622.

Manohar SG (2022). Computational neuroscience: a grand unifying theory? *Brain* 145: 1189–1190.

Manwani A and Koch C (2001). Detecting and estimating signals over noisy and unreliable synapses: information-theoretic analysis. *Neural Computation* 13: 1–33.

Mao D, Avila E, Caziot B, Laurens J, Dickman JD, and Angelaki DE (2021). Spatial modulation of hippocampal activity in freely moving macaques. *Neuron* 109: 3521–3534 e6.

Maravita A and Romano D (2018). The parietal lobe and tool use. *Handb Clin Neurol* 151: 481–498.

Markov NT, Vezoli J, Chameau P, Falchier A, Quilodran R, Huissoud C, Lamy C, Misery P, Giroud P, Ullman S, Barone P, Dehay C, Knoblauch K, and Kennedy H (2014). Anatomy of hierarchy: feedforward and feedback pathways in macaque visual cortex. *J Comp Neurol* 522: 225–59.

Markram H and Segal M (1992). The inositol 1,4,5-triphosphate pathway mediates cholinergic potentiation of rat hippocampal neuronal responses to NMDA. *Journal of Physiology* 447: 513–533.

Markram H and Tsodyks M (1996). Redistribution of synaptic efficacy between neocortical pyramidal neurons. *Nature* 382: 807–810.

Marr D (1971). Simple memory: a theory for archicortex. *Phil. Trans. Roy. Soc. Lond. B* 262: 23–81.

Martinez-Garcia M, Rolls ET, Deco G, and Romo R (2011). Neural and computational mechanisms of postponed decisions. *Proc Natl Acad Sci U S A* 108: 11626–31.

Matchin W and Hickok G (2020). The cortical organization of syntax. *Cereb Cortex* 30: 1481–1498.

Mayberg HS, Lozano AM, Voon V, McNeely HE, Seminowicz D, Hamani C, Schwalb JM, and Kennedy SH (2005). Deep brain stimulation for treatment-resistant depression. *Neuron* 45: 651–60.

Mayberg HS, Riva-Posse P, and Crowell AL (2016). Deep brain stimulation for depression: keeping an eye on a moving target. *JAMA Psychiatry* 73: 439–40.

Maynard Smith J (1982). *Evolution and the Theory of Games*. Cambridge University Press, Cambridge.

Maynard Smith J (1984). Game theory and the evolution of behaviour. *Behavioral and Brain Sciences* 7: 95–125.

McCabe C, Rolls ET, Bilderbeck A, and McGlone F (2008). Cognitive influences on the affective representation of touch and the sight of touch in the human brain. *Social, Cognitive and Affective Neuroscience* 3: 97–108.

McClelland JL, McNaughton BL, and O'Reilly RC (1995). Why there are complementary learning systems in the hippocampus and neocortex: insights from the successes and failures of connectionist models of learning and memory. *Psychological Review* 102: 419–457.

McClelland JL, McNaughton BL, and Lampinen AK (2020). Integration of new information in memory: new insights from a complementary learning systems perspective. *Philos Trans R Soc Lond B Biol Sci* 375: 20190637.

McCormick C, Ciaramelli E, De Luca F, and Maguire EA (2018). Comparing and contrasting the cognitive effects of hippocampal and ventromedial prefrontal cortex damage: a review of human lesion studies. *Neuroscience* 374: 295–318.

McNaughton BL (2025). Neuronal 'ensemble' recording and the search for the cell assembly: a personal history. *Hippocampus* 35: e23669.

McNaughton BL and Morris RGM (1987). Hippocampal synaptic enhancement and information storage within a distributed memory system. *Trends in Neurosciences* 10: 408–415.

McNaughton BL, Barnes CA, and O'Keefe J (1983). The contributions of position, direction, and velocity to single unit activity in the hippocampus of freely-moving rats. *Exp Brain Res* 52: 41–9.

McNaughton BL, Chen LL, and Markus EJ (1991). "dead reckoning," landmark learning, and the sense of direction: a neurophysiological and computational hypothesis. *J Cogn Neurosci* 3: 190–202.

McNaughton BL, Barnes CA, Gerrard JL, Gothard K, Jung MW, Knierim JJ, Kudrimoti H, Qin Y, Skaggs WE, Suster M, and Weaver KL (1996). Deciphering the hippocampal polyglot: the hippocampus as a path integration system. *J Exp Biol* 199: 173–85.

McNaughton BL, Battaglia FP, Jensen O, Moser EI, and Moser MB (2006). Path integration and the neural basis of the 'cognitive map'. *Nat Rev Neurosci* 7: 663–78.

Menzies L, Chamberlain SR, Laird AR, Thelen SM, Sahakian BJ, and Bullmore ET (2008). Integrating evidence from neuroimaging and neuropsychological studies of obsessive-compulsive disorder: The orbitofronto-striatal model revisited. *Neuroscience and Biobehavioral Reviews* 32: 525–549.

Mesulam MM (1990). Human brain cholinergic pathways. *Prog Brain Res* 84: 231–41.

Miller JF, Neufang M, Solway A, Brandt A, Trippel M, Mader I, Hefft S, Merkow M, Polyn SM, Jacobs J, Kahana MJ, and Schulze-Bonhage A (2013). Neural activity in human hippocampal formation reveals the spatial context of retrieved memories. *Science* 342: 1111–4.

Milner AD and Goodale MA (1995). *The Visual Brain in Action*. Oxford University Press, Oxford.

Milton CK, Dhanaraj V, Young IM, Taylor HM, Nicholas PJ, Briggs RG, Bai MY, Fonseka RD, Hormovas J, Lin YH, Tanglay O, Conner AK, Glenn CA, Teo C, Doyen S, and Sughrue ME (2021). Parcellation-based anatomic model of the semantic network. *Brain Behav* 11: e02065.

Mineault PJ, Khawaja FA, Butts DA, and Pack CC (2012). Hierarchical processing of complex motion along the primate dorsal visual pathway. *Proc Natl Acad Sci U S A* 109: E972–80.

Mishkin M (1979). Analogous neural models for tactual and visual learning. *Neuropsychologia* 17: 139–51.

Mishkin M (1982). A memory system in the monkey. *Philos Trans R Soc Lond B Biol Sci* 298: 83–95.

Mishkin M, Ungerleider LG, and Macko KA (1983). Object vision and spatial vision: two cortical pathways. *Trends Neurosci* 6: 414–417.

Miyashita Y, Rolls ET, Cahusac PM, Niki H, and Feigenbaum JD (1989). Activity of hippocampal formation neurons in the monkey related to a conditional spatial response task. *J Neurophysiol* 61: 669–78.

Mkrtchian A, Evans JW, Kraus C, Yuan P, Kadriu B, Nugent AC, Roiser JP, and Zarate J C A (2021). Ketamine modulates fronto-striatal circuitry in depressed and healthy individuals. *Mol Psychiatry* 26: 3292–3301.

Monaco S, Cavina-Pratesi C, Sedda A, Fattori P, Galletti C, and Culham JC (2011). Functional magnetic resonance adaptation reveals the involvement of the dorsomedial stream in hand orientation for grasping. *J Neurophysiol* 106: 2248–63.

Mora F, Rolls ET, and Burton MJ (1976). Modulation during learning of the responses of neurones in the lateral hypothalamus to the sight of food. *Experimental Neurology,* 53: 508–519.

Mora F, Avrith DB, and Rolls ET (1980). An electrophysiological and behavioural study of self-stimulation in the orbitofrontal cortex of the rhesus monkey. *Brain Research Bulletin* 5: 111–115.

Mormann F, Dubois J, Kornblith S, Milosavljevic M, Cerf M, Ison M, Tsuchiya N, Kraskov A, Quian Quiroga R, Adolphs R, Fried I, and Koch C (2011). A category-specific response to animals in the right human amygdala. *Nat Neurosci* 14: 1247–9.

Morris R, Amaral DG, Bliss T, Duff K, and O'Keefe J (2025). *The Hippocampus Book*. Oxford University Press, Oxford, 2nd edn.

Morris RG, Garrud P, Rawlins JN, and O'Keefe J (1982). Place navigation impaired in rats with hippocampal lesions. *Nature* 297: 681–3.

Moscovitch M, Cabeza R, Winocur G, and Nadel L (2016). Episodic memory and beyond: the hippocampus and neocortex in transformation. *Annu Rev Psychol* 67: 105–34.

Moser EI, Moser MB, and Roudi Y (2014). Network mechanisms of grid cells. *Philos Trans R Soc Lond B Biol Sci* 369: 20120511.

Moser EI, Moser MB, and McNaughton BL (2017). Spatial representation in the hippocampal formation: a history. *Nat Neurosci* 20: 1448–1464.

Moser MB, Rowland DC, and Moser EI (2015). Place cells, grid cells, and memory. *Cold Spring Harb Perspect Biol* 7: a021808.

Movshon JA, Adelson EH, Gizzi MS, and Newsome WT (1985). The analysis of moving visual patterns. In Chagas C, Gattass R, and Gross CG, editors, *Pattern Recognition Mechanisms*, 117–151. Springer-Verlag, New York.

Mudrik L, Boly M, Dehaene S, Fleming SM, Lamme V, Seth A, and Melloni L (2025). Unpacking the complexities of consciousness: Theories and reflections. *Neurosci Biobehav Rev* 170: 106053.

Mueser KT and McGurk SR (2004). Schizophrenia. *Lancet* 363: 2063–2072.

Muir JL, Everitt BJ, and Robbins TW (1994). AMPA-induced excitotoxic lesions of the basal forebrain: a significant role for the cortical cholinergic system in attentional function. *Journal of Neuroscience* 14: 2313–2326.

Munuera J and Duhamel JR (2020). The role of the posterior parietal cortex in saccadic error processing. *Brain Struct Funct* 225: 763–784.

Murray EA and Rudebeck PH (2018). Specializations for reward-guided decision-making in the primate ventral prefrontal cortex. *Nat Rev Neurosci* 19: 404–417.

Nagahama Y, Okada T, Katsumi Y, Hayashi T, Yamauchi H, Oyanagi C, Konishi J, Fukuyama H, and Shibasaki H (2001). Dissociable mechanisms of attentional control within the human prefrontal cortex. *Cerebral Cortex* 11: 85–92.

Nakazawa K, Quirk MC, Chitwood RA, Watanabe M, Yeckel MF, Sun LD, Kato A, Carr CA, Johnston D, Wilson MA, and Tonegawa S (2002). Requirement for hippocampal CA3 NMDA receptors in associative memory recall. *Science* 297: 211–218.

Nakazawa K, Sun LD, Quirk MC, Rondi-Reig L, Wilson MA, and Tonegawa S (2003). Hippocampal CA3 NMDA receptors are crucial for memory acquisition of one-time experience. *Neuron* 38: 305–315.

Nakazawa K, McHugh TJ, Wilson MA, and Tonegawa S (2004). NMDA receptors, place cells and hippocampal spatial memory. *Nature Reviews Neuroscience* 5: 361–372.

Natu VS, Arcaro MJ, Barnett MA, Gomez J, Livingstone M, Grill-Spector K, and Weiner KS (2021). Sulcal depth in the medial ventral temporal cortex predicts the location of a place-selective region in macaques, children, and adults. *Cereb Cortex* 31: 48–61.

Newman EL, Gupta K, Climer JR, Monaghan CK, and Hasselmo ME (2012). Cholinergic modulation of cognitive processing: insights drawn from computational models. *Front Behav Neurosci* 6: 24.

Newsome WT, Britten KH, and Movshon JA (1989). Neuronal correlates of a perceptual decision. *Nature* 341: 52–54.

Nicoll RA (1988). The coupling of neurotransmitter receptors to ion channels in the brain. *Science* 241: 545–551.

Nowicka A and Ringo JL (2000). Eye position-sensitive units in hippocampal formation and in inferotemporal cortex of the macaque monkey. *Eur J Neurosci* 12: 751–9.

Nugent AC, Milham MP, Bain EE, Mah L, Cannon DM, Marrett S, Zarate CA, Pine DS, Price JL, and Drevets WC (2006). Cortical abnormalities in bipolar disorder investigated with mri and voxel-based morphometry. *Neuroimage* 30: 485–497.

Nutt DJ, Peill JM, Weiss B, Godfrey K, Carhart-Harris RL, and Erritzoe D (2024). Psilocybin and other classic psychedelics in depression. *Curr Top Behav Neurosci* 66: 149–174.

Oberhuber M, Hope TMH, Seghier ML, Parker Jones O, Prejawa S, Green DW, and Price CJ (2016). Four functionally distinct regions in the left supramarginal gyrus support word processing. *Cereb Cortex* 26: 4212–4226.

O'Doherty J, Rolls ET, Francis S, Bowtell R, McGlone F, Kobal G, Renner B, and Ahne G (2000). Sensory-specific satiety related olfactory activation of the human orbitofrontal cortex. *NeuroReport* 11: 893–897.

O'Doherty J, Kringelbach ML, Rolls ET, Hornak J, and Andrews C (2001a). Abstract reward and punishment representations in the human orbitofrontal cortex. *Nature Neuroscience* 4: 95–102.

O'Doherty J, Rolls ET, Francis S, Bowtell R, and McGlone F (2001b). Representation of pleasant and aversive taste in the human brain. *J Neurophysiol* 85: 1315–21.

O'Keefe J (1979). A review of the hippocampal place cells. *Progress in Neurobiology* 13: 419–439.

O'Keefe J (1990). A computational theory of the hippocampal cognitive map. *Prog Brain Res* 83: 301–12.

O'Keefe J and Dostrovsky J (1971). The hippocampus as a spatial map: preliminary evidence from unit activity in the freely moving rat. *Brain Research* 34: 171–175.

O'Keefe J and Krupic J (2021). Do hippocampal pyramidal cells respond to nonspatial stimuli? *Physiol Rev* 101: 1427–1456.

O'Keefe J and Nadel L (1978). *The Hippocampus as a Cognitive Map*. Clarendon Press, Oxford.

O'Keefe J, Burgess N, Donnett JG, Jeffery KJ, and Maguire EA (1998). Place cells, navigational accuracy, and the human hippocampus. *Philos Trans R Soc Lond B Biol Sci* 353: 1333–40.

Olds J (1977). *Drives and Reinforcements: Behavioral Studies of Hypothalamic Functions*. Raven Press, New York.

Olds J and Milner P (1954). Positive reinforcement produced by electrical stimulation of septal area and other regions of the rat brain. *Journal of Comparative and Physiological Psychology* 47: 419–427.

Olds J and Olds M (1965). Drives, rewards, and the brain. In Barron F and Dement WC, editors, *New Directions in Psychology*, vol. 2. Holt, Rinehart and Winston, New York.

O'Mara SM, Rolls ET, Berthoz A, and Kesner RP (1994). Neurons responding to whole-body motion in the primate hippocampus. *Journal of Neuroscience* 14: 6511–6523.

O'Neill MJ and Dix S (2007). AMPA receptor potentiators as cognitive enhancers. *IDrugs* 10: 185–192.

Orban GA, Sepe A, and Bonini L (2021). Parietal maps of visual signals for bodily action planning. *Brain Struct Funct* 226: 2967–2988.

O'Reilly RC and Munakata Y (2000). *Computational Explorations in Cognitive Neuroscience*. MIT Press, Cambridge, MA.

Orzylowski M, Fujiwara E, Mousseau DD, and Baker GB (2021). An overview of the involvement of d-serine in cognitive impairment in normal aging and dementia. *Front Psychiatry* 12: 754032.

Owen MJ, Sawa A, and Mortensen PB (2016). Schizophrenia. *Lancet* 388: 86–97.

Paass G and Giesselbach S (2023). *Foundation Models for Natural Language Processing*. Springer, Switzerland.

Padoa-Schioppa C (2007). Orbitofrontal cortex and the computation of economic value. *Ann N Y Acad Sci* 1121: 232–53.

Padoa-Schioppa C (2011). Neurobiology of economic choice: a good-based model. *Annual Review of Neuroscience* 34: 333–359.

Padoa-Schioppa C and Assad JA (2006). Neurons in the orbitofrontal cortex encode economic value. *Nature* 441: 223–6.

Padoa-Schioppa C and Assad JA (2008). The representation of economic value in the orbitofrontal cortex is invariant for changes of menu. *Nat Neurosci* 11: 95–102.

Padoa-Schioppa C and Conen KE (2017). Orbitofrontal cortex: A neural circuit for economic decisions. *Neuron* 96: 736–754.

Pantelis C, Barber FZ, Barnes TR, Nelson HE, Owen AM, and Robbins TW (1999). Comparison of set-shifting ability in patients with chronic schizophrenia and frontal lobe damage. *Schizophrenia Research* 37: 251–270.

Panzeri S, Schultz SR, Treves A, and Rolls ET (1999a). Correlations and the encoding of information in the nervous system. *Proceedings of the Royal Society of London B* 266: 1001–1012.

Panzeri S, Treves A, Schultz S, and Rolls ET (1999b). On decoding the responses of a population of neurons from short time windows. *Neural Computation* 11: 1553–1577.

Panzeri S, Rolls ET, Battaglia F, and Lavis R (2001). Speed of feedforward and recurrent processing in multilayer networks of integrate-and-fire neurons. *Network: Computation in Neural Systems* 12: 423–440.

Panzeri S, Moroni M, Safaai H, and Harvey CD (2022). The structures and functions of correlations in neural population codes. *Nat Rev Neurosci* 23: 551–567.

Papagno C (2018). Memory deficits. *Handbook of Clinical Neurology, Parietal Cortex* 151: 377–393.

Parga N and Virasoro MA (1986). The ultrametric organization of memories in a neural network. *Journal de Physique* 47: 1857–1864.

Parkinson JK, Murray EA, and Mishkin M (1988). A selective mnemonic role for the hippocampus in monkeys: memory for the location of objects. *Journal of Neuroscience* 8: 4059–4167.

Passarelli L, Gamberini M, and Fattori P (2021). The superior parietal lobule of primates: a sensory-motor hub for interaction with the environment. *J Integr Neurosci* 20: 157–171.

Passingham RE (2021). *Understanding the Prefrontal Cortex: selective advantage, connectivity and neural operations*. Oxford University Press, Oxford.

Passingham RE and Wise SP (2012). *The Neurobiology of the Prefrontal Cortex*. Oxford University Press, Oxford.

Pastor-Bernier A, Stasiak A, and Schultz W (2021). Reward-specific satiety affects subjective value signals in orbitofrontal cortex during multicomponent economic choice. *Proc Natl Acad Sci U S A* 118.

Patterson K, Nestor PJ, and Rogers TT (2007). Where do you know what you know? the representation of semantic knowledge in the human brain. *Nat Rev Neurosci* 8: 976–87.

Pauls DL, Abramovitch A, Rauch SL, and Geller DA (2014). Obsessive-compulsive disorder: an integrative genetic and neurobiological perspective. *Nat Rev Neurosci* 15: 410–24.

Peelen MV and Caramazza A (2012). Conceptual object representations in human anterior temporal cortex. *J Neurosci* 32: 15728–36.

Penades R, Catalan R, Andres S, Salamero M, and Gasto C (2005). Executive function and nonverbal memory in obsessive-compulsive disorder. *Psychiatry Research* 133: 81–90.

Penades R, Catalan R, Rubia K, Andres S, Salamero M, and Gasto C (2007). Impaired response inhibition in obsessive compulsive disorder. *European Psychiatry* 22: 404–410.

Perrett DI and Rolls ET (1983). Neural mechanisms underlying the visual analysis of faces. In Ewert JP, Capranica R, and Ingle D, editors, *Advances in Vertebrate Neuroethology*, 543–566. Plenum Press, New York.

Perrett DI, Rolls ET, and Caan W (1979). Temporal lobe cells of the monkey with visual responses selective for faces. *Neuroscience Letters* S3: S358.

Perrett DI, Rolls ET, and Caan W (1982). Visual neurons responsive to faces in the monkey temporal cortex. *Experimental Brain Research* 47: 329–342.

Perrett DI, Smith PA, Mistlin AJ, Chitty AJ, Head AS, Potter DD, Broennimann R, Milner AD, and Jeeves MA (1985a). Visual analysis of body movements by neurones in the temporal cortex of the macaque monkey: a preliminary report. *Behav Brain Res* 16: 153–70.

Perrett DI, Smith PA, Potter DD, Mistlin AJ, Head AS, Milner AD, and Jeeves MA (1985b). Visual cells in the temporal cortex sensitive to face view and gaze direction. *Proceedings of the Royal Society of London B* 223: 293–317.

Perrett DI, Mistlin AJ, and Chitty AJ (1987). Visual neurons responsive to faces. *Trends in Neurosciences* 10: 358–364.

Perry G, Rolls ET, and Stringer SM (2006). Spatial vs temporal continuity in view invariant visual object recognition learning. *Vision Res* 46: 3994–4006.

Perry G, Rolls ET, and Stringer SM (2010). Continuous transformation learning of translation invariant representations. *Exp Brain Res* 204: 255–70.

Pert CB (1997). *Molecules of Emotion*. Simon and Schuster, London.

Pitcher D and Ungerleider LG (2021). Evidence for a third visual pathway specialized for social perception. *Trends Cogn Sci* 25: 100–110.

Pitcher D, Ianni G, and Ungerleider LG (2019). A functional dissociation of face-, body- and scene-selective brain areas based on their response to moving and static stimuli. *Sci Rep* 9: 8242.

Pittenger C (2015). Glutamatergic agents for ocd and related disorders. *Curr Treat Options Psychiatry* 2: 271–283.

Pittenger C (2021). Pharmacotherapeutic strategies and new targets in ocd. *Curr Top Behav Neurosci* 49: 331–384.

Pittenger C, Krystal JH, and Coric V (2006). Glutamate-modulating drugs as novel pharmacotherapeutic agents in the treatment of obsessive-compulsive disorder. *NeuroRx* 3: 69–81.

Pittenger C, Bloch MH, and Williams K (2011). Glutamate abnormalities in obsessive compulsive disorder: Neurobiology, pathophysiology, and treatment. *Pharmacology and Therapeutics* 132: 314–332.

Pitzalis S, Sereno MI, Committeri G, Fattori P, Galati G, Tosoni A, and Galletti C (2013). The human homologue of macaque area v6a. *Neuroimage* 82: 517–30.

Pitzalis S, Fattori P, and Galletti C (2015). The human cortical areas v6 and v6a. *Vis Neurosci* 32: E007.

Piza DB, Corrigan BW, Gulli RA, Do Carmo S, Cuello AC, Muller L, and Martinez-Trujillo J (2024). Primacy of vision shapes behavioral strategies and neural substrates of spatial navigation in marmoset hippocampus. *Nat Commun* 15: 4053.

Plebe A and Grasso G (2019). The unbearable shallow understanding of deep learning. *Minds and Machines* 29: 515–553.

Ploux E, Freret T, and Billard JM (2020). d-serine in physiological and pathological brain aging. *Biochim Biophys Acta Proteins Proteom* 1869: 140542.

Ponce-Alvarez A and Deco G (2024). The hopf whole-brain model and its linear approximation. *Sci Rep* 14: 2615.

Power JM and Sah P (2008). Competition between calcium-activated K+ channels determines cholinergic action on firing properties of basolateral amygdala projection neurons. *Journal of Neuroscience* 28: 3209–3220.

Prado J, Clavagnier S, Otzenberger H, Scheiber C, Kennedy H, and Perenin MT (2005). Two cortical systems for reaching in central and peripheral vision. *Neuron* 48: 849–58.

Price JL (2006). Connections of orbital cortex. In Zald DH and Rauch SL, editors, *The Orbitofrontal Cortex*, chap. 3, 39–55. Oxford University Press, Oxford.

Price JL and Drevets WC (2012). Neural circuits underlying the pathophysiology of mood disorders. *Trends Cogn Sci* 16: 61–71.

Pritchard SC, Coltheart M, Marinus E, and Castles A (2018). A computational model of the self-teaching hypothesis based on the dual-route cascaded model of reading. *Cogn Sci* 42: 722–770.

Pryce CR, Azzinnari D, Spinelli S, Seifritz E, Tegethoff M, and Meinlschmidt G (2011). Helplessness: a systematic translational review of theory and evidence for its relevance to understanding and treating depression. *Pharmacol Ther* 132: 242–267.

Pu WD, Rolls ET, Guo S, Liu H, Yu Y, Xue Z, Feng J, and Liu Z (2014). Altered functional connectivity links in neuroleptic-naïve and neuroleptic-treated patients with schizophrenia, and their relation to symptoms including volition. *Neuroimage Clinical* 6: 463–474.

Quadt L, Critchley H, and Nagai Y (2022). Cognition, emotion, and the central autonomic network. *Auton Neurosci* 238: 102948.

Quian Quiroga R (2013). Gnostic cells in the 21st century. *Acta Neurobiol Exp (Wars)* 73: 463–71.

Quirk GJ, Armony JL, Repa JC, Li XF, and LeDoux JE (1996). Emotional memory: a search for sites of plasticity. *Cold Spring Harb Symp Quant Biol* 61: 247–57.

Rajalingham R, Issa EB, Bashivan P, Kar K, Schmidt K, and DiCarlo JJ (2018). Large-scale, high-resolution comparison of the core visual object recognition behavior of humans, monkeys, and state-of-the-art deep artificial neural networks. *J Neurosci* 38: 7255–7269.

Rao VR, Sellers KK, Wallace DL, Lee MB, Bijanzadeh M, Sani OG, Yang Y, Shanechi MM, Dawes HE, and Chang EF (2018). Direct electrical stimulation of lateral orbitofrontal cortex acutely improves mood in individuals with symptoms of depression. *Curr Biol* 28: 3893–3902 e4.

Rauschecker JP (2012). Ventral and dorsal streams in the evolution of speech and language. *Front Evol Neurosci* 4: 7.

Rauschecker JP (2015). Auditory cortex. *Reference Module in Neuroscience and Biobehavioral Psychology. Brain Mapping* 2: 299–304.

Reber J, Feinstein JS, O'Doherty JP, Liljeholm M, Adolphs R, and Tranel D (2017). Selective impairment of goal-directed decision-making following lesions to the human ventromedial prefrontal cortex. *Brain* 140: 1743–1756.

Renoult L, Irish M, Moscovitch M, and Rugg MD (2019). From knowing to remembering: the semantic-episodic distinction. *Trends Cogn Sci* 23: 1041–1057.

Rhee TG, Shim SR, Forester BP, Nierenberg AA, McIntyre RS, Papakostas GI, Krystal JH, Sanacora G, and Wilkinson ST (2022). Efficacy and safety of ketamine vs electroconvulsive therapy among patients with major depressive episode: A systematic review and meta-analysis. *JAMA Psychiatry* 79: 1162–1172.

Ridley M (1993). *The Red Queen: Sex and the Evolution of Human Nature*. Penguin, London.

Riesenhuber M and Poggio T (1999). Hierarchical models of object recognition in cortex. *Nat Neurosci* 2: 1019–25.

Ringo JL, Sobotka S, Diltz MD, and Bunce CM (1994). Eye movements modulate activity in hippocampal, parahippocampal, and inferotemporal neurons. *J Neurophysiol* 71: 1285–8.

Riva-Posse P, Choi KS, Holtzheimer PE, Crowell AL, Garlow SJ, Rajendra JK, McIntyre CC, Gross RE, and Mayberg HS (2018). A connectomic approach for subcallosal cingulate deep brain stimulation surgery: prospective targeting in treatment-resistant depression. *Mol Psychiatry* 23: 843–849.

Riva-Posse P, Crowell AL, Wright K, Waters AC, Choi K, Garlow SJ, Holtzheimer PE, Gross RE, and Mayberg HS (2020). Rapid antidepressant effects of deep brain stimulation and their relation to surgical protocol. *Biol Psychiatry* 88: e37–e39.

Rizzolatti G and Sinigaglia C (2016). The mirror mechanism: a basic principle of brain function. *Nat Rev Neurosci* 17: 757–765.

Robbins TW, Vaghi MM, and Banca P (2019). Obsessive-compulsive disorder: puzzles and prospects. *Neuron* 102: 27–47.

Robbins TW, Banca P, and Belin D (2024). From compulsivity to compulsion: the neural basis of compulsive disorders. *Nat Rev Neurosci* 25: 313–333.

Robertson RG, Rolls ET, and Georges-François P (1998). Spatial view cells in the primate hippocampus: Effects of removal of view details. *Journal of Neurophysiology* 79: 1145–1156.

Robinson L and Rolls ET (2015). Invariant visual object recognition: biologically plausible approaches. *Biol Cybern* 109: 505–35.

Robinson TE and Berridge KC (2025). The incentive-sensitization theory of addiction 30 years on. *Annu Rev Psychol* 76: 29–58.

Rockland KS, Saleem KS, and Tanaka K (1994). Divergent feedback connections from areas v4 and teo in the macaque. *Visual Neuroscience* 11: 579–600.

Rogers RD, Andrews TC, Grasby PM, Brooks DJ, and Robbins TW (2000). Contrasting cortical and subcortical activations produced by attentional-set shifting and reversal learning in humans. *Journal of Cognitive Neuroscience* 12: 142–162.

Rogers Flattery CN, Rosen RF, Farberg AS, Dooyema JM, Hof PR, Sherwood CC, Walker LC, and Preuss TM (2020). Quantification of neurons in the hippocampal formation of chimpanzees: comparison to rhesus monkeys and humans. *Brain Struct Funct* 225: 2521–2531.

Rolls BJ (2016a). Creativity needs some serendipity: Reflections on a career in ingestive behavior. *Physiol Behav* 162: 186–95.

Rolls BJ and Rolls ET (1981). The control of drinking. *British Medical Bulletin* 37: 127–130.

Rolls BJ, Wood RJ, and Rolls ET (1980a). Thirst: the initiation, maintenance, and termination of drinking. *Progress in Psychobiology and Physiological Psychology* 9: 263–321.

Rolls BJ, Wood RJ, Rolls ET, Lind H, Lind R, and Ledingham J (1980b). Thirst following water deprivation in humans. *American Journal of Physiology* 239: R476–R482.

Rolls BJ, Rolls ET, Rowe EA, and Sweeney K (1981a). Sensory specific satiety in man. *Physiology and Behavior* 27: 137–142.

Rolls BJ, Rowe EA, Rolls ET, Kingston B, Megson A, and Gunary R (1981b). Variety in a meal enhances food intake in man. *Physiology and Behavior* 26: 215–221.

Rolls BJ, Rowe EA, and Rolls ET (1982a). How flavour and appearance affect human feeding. *Proceedings of the Nutrition Society,* 41: 109–117.

Rolls BJ, Rowe EA, and Rolls ET (1982b). How sensory properties of foods affect human feeding behavior. *Physiology and Behavior* 29: 409–417.

Rolls ET (1971). Involvement of brainstem units in medial forebrain bundle self-stimulation. *Physiology and Behavior* 7: 297–310.

Rolls ET (1974). The neural basis of brain-stimulation reward. *Progress in Neurobiology* 3: 71–160.

Rolls ET (1975). *The Brain and Reward*. Pergamon Press, Oxford.

Rolls ET (1976). The neurophysiological basis of brain-stimulation reward. In Wauquier A and Rolls ET, editors, *Brain-Stimulation Reward*, 65–87. North Holland, Amsterdam.

Rolls ET (1979). Effects of electrical stimulation of the brain on behaviour. *Psychology Surveys* 2: 151–169.

Rolls ET (1981a). Processing beyond the inferior temporal visual cortex related to feeding, learning, and striatal function. In Katsuki Y, Norgren R, and Sato M, editors, *Brain Mechanisms of Sensation*, chap. 16, 241–269. Wiley, New York.

Rolls ET (1981b). Central nervous mechanisms related to feeding and appetite. *British Medical Bulletin* 37: 131–134.

Rolls ET (1981c). Responses of amygdaloid neurons in the primate. In Ben-Ari Y, editor, *The Amygdaloid Complex*, 383–393. Elsevier, Amsterdam.

Rolls ET (1982). Neuronal mechanisms underlying the formation and disconnection of associations between visual stimuli and reinforcement in primates. In Woody CD, editor, *Conditioning: Representation of Involved Neural Functions*, 363–373. Plenum, New York.

Rolls ET (1984). Neurons in the cortex of the temporal lobe and in the amygdala of the monkey with responses selective for faces. *Human Neurobiology* 3: 209–222.

Rolls ET (1986a). A theory of emotion, and its application to understanding the neural basis of emotion. In Oomura Y, editor, *Emotions. Neural and Chemical Control*, 325–344. Japan Scientific Societies Press; and Karger, Tokyo; and Basel.

Rolls ET (1986b). Neural systems involved in emotion in primates. In Plutchik R and Kellerman H, editors, *Emotion: Theory, Research, and Experience*, vol. 3: Biological Foundations of Emotion, chap. 5, 125–143. Academic Press, New York.

Rolls ET (1986c). Neuronal activity related to the control of feeding. In Ritter R, Ritter S, and Barnes C, editors, *Feeding Behavior: Neural and Humoral Controls*, chap. 6, 163–190. Academic Press, New York.

Rolls ET (1987). Information representation, processing and storage in the brain: analysis at the single neuron level. In Changeux JP and Konishi M, editors, *The Neural and Molecular Bases of Learning*, 503–540. Wiley, Chichester.

Rolls ET (1989a). Functions of neuronal networks in the hippocampus and neocortex in memory. In Byrne JH and Berry WO, editors, *Neural Models of Plasticity: Experimental and Theoretical Approaches*, 240–265. Academic Press, San Diego.

Rolls ET (1989b). Functions of neuronal networks in the hippocampus and cerebral cortex in memory. In Cotterill RMJ, editor, *Models of Brain Function*, 15–33. Cambridge University Press, Cambridge.

Rolls ET (1989c). The representation and storage of information in neuronal networks in the primate cerebral cortex and hippocampus. In Durbin R, Miall C, and Mitchison G, editors, *The Computing Neuron*, 125–159. Addison-Wesley, Wokingham, England.

Rolls ET (1989d). Parallel distributed processing in the brain: Implications of the functional architecture of neuronal networks in the hippocampus. In Morris RGM, editor, *Parallel Distributed Processing: Implications for Psychology and Neurobiology*, 286–308. Clarendon Press, Oxford.

Rolls ET (1990). A theory of emotion, and its application to understanding the neural basis of emotion. *Cognition and Emotion* 4: 161–190.

Rolls ET (1991). Neural organisation of higher visual functions. *Current Opinion in Neurobiology* 1: 274–278.

Rolls ET (1992a). Neurophysiological mechanisms underlying face processing within and beyond the temporal cortical visual areas. *Philosophical Transactions of the Royal Society of London B* 335: 11–21.

Rolls ET (1992b). Neurophysiology and functions of the primate amygdala. In Aggleton JP, editor, *The Amygdala*, chap. 5, 143–165. Wiley-Liss, New York.

Rolls ET (1993). The neural control of feeding in primates. In Booth D, editor, *Neurophysiology of Ingestion*, chap. 9, 137–169. Pergamon, Oxford.

Rolls ET (1997a). Consciousness in neural networks? *Neural Networks* 10: 1227–1240.

Rolls ET (1997b). Brain mechanisms of vision, memory, and consciousness. In Ito M, Miyashita Y, and Rolls E, editors, *Cognition, Computation, and Consciousness*, 81–120. Oxford University Press, Oxford.

Rolls ET (1999a). Spatial view cells and the representation of place in the primate hippocampus. *Hippocampus* 9: 467–480.

Rolls ET (1999b). *The Brain and Emotion*. Oxford University Press, Oxford.

Rolls ET (1999c). The functions of the orbitofrontal cortex. *Neurocase* 5: 301–312.

Rolls ET (2000a). Precis of the brain and emotion. *Behav Brain Sci* 23: 177–91; discussion 192–233.

Rolls ET (2000b). Functions of the primate temporal lobe cortical visual areas in invariant visual object and face recognition. *Neuron* 27: 205–18.

Rolls ET (2000c). Neurophysiology and functions of the primate amygdala, and the neural basis of emotion. In Aggleton JP, editor, *The Amygdala: Second Edition. A Functional Analysis*, chap. 13, 447–478. Oxford University Press, Oxford.

Rolls ET (2003). Consciousness absent and present: a neurophysiological exploration. *Progress in Brain Research* 144: 95–106.

Rolls ET (2004a). A higher order syntactic thought (host) theory of consciousness. In Gennaro RJ, editor, *Higher-Order Theories of Consciousness: An Anthology*, 137–172. John Benjamins, Amsterdam.

Rolls ET (2004b). The functions of the orbitofrontal cortex. *Brain and Cognition* 55: 11–29.

Rolls ET (2005). *Emotion Explained*. Oxford University Press, Oxford.

Rolls ET (2006). The neurophysiology and functions of the orbitofrontal cortex. In Zald DH and Rauch SL, editors, *The Orbitofrontal Cortex*, chap. 5, 95–124. Oxford University Press, Oxford.

Rolls ET (2007a). The affective neuroscience of consciousness: higher order linguistic thoughts, dual routes to emotion and action, and consciousness. In Zelazo P, Moscovitch M, and Thompson E, editors, *Cambridge Handbook of Consciousness*, chap. 29, 831–859. Cambridge University Press, Cambridge.

Rolls ET (2007b). The representation of information about faces in the temporal and frontal lobes. *Neuropsychologia* 45: 125–143.

Rolls ET (2007c). A computational neuroscience approach to consciousness. *Neural Networks* 20: 962–82.

Rolls ET (2007d). Sensory processing in the brain related to the control of food intake. *Proceedings of the Nutrition Society* 66: 96–112.

Rolls ET (2007e). Memory systems: multiple systems in the brain and their interactions. In Roediger HL, Dudai Y, and Fitzpatrick SM, editors, *Science of Memory: Concepts*, chap. 59, 345–351. Oxford University Press, New York.

Rolls ET (2008a). *Memory, Attention, and Decision-Making: A Unifying Computational Neuroscience Approach*. Oxford University Press, Oxford.

Rolls ET (2008b). Emotion, higher order syntactic thoughts, and consciousness. In Weiskrantz L and Davies MK, editors, *Frontiers of Consciousness*, chap. 4, 131–167. Oxford University Press, Oxford.

Rolls ET (2008c). Face processing in different brain areas, and critical band masking. *J Neuropsychol* 2: 325–60.

Rolls ET (2008d). The primate hippocampus and episodic memory. In Dere E, Easton A, Nadel L, and Huston JP, editors, *Handbook of Episodic Memory*, chap. 4.2, 417–438. Elsevier, Amsterdam.

Rolls ET (2008e). Functions of the orbitofrontal and pregenual cingulate cortex in taste, olfaction, appetite and emotion. *Acta Physiologica Hungarica* 95: 131–164.

Rolls ET (2009). The anterior and midcingulate cortices and reward. In Vogt B, editor, *Cingulate Neurobiology and Disease*, chap. 8, 191–206. Oxford University Press, Oxford.

Rolls ET (2010a). Noise in the brain, decision-making, determinism, free will, and consciousness. In EPerry, DCollerton, FLebeau, and Ashton H, editors, *New Horizons in the Neuroscience of Consciousness*, 113–120. John Benjamins, Amsterdam.

Rolls ET (2010b). Neural representation of fat texture in the mouth. In Montmayeur JP and Coutre lJ, editors, *Fat Detection: Taste, Texture, and Postingestive Effects*, chap. 8, 197–223. CRC Press, Boca Raton, FL.

Rolls ET (2011a). Consciousness, decision-making, and neural computation. In VCutsuridis, AHussain, and JGTaylor, editors, *Perception-Action Cycle: Models, Algorithms and Systems*, chap. 9, 287–333. Springer, Berlin.

Rolls ET (2011b). Face neurons. In Calder AJ, Rhodes G, Johnson MH, and Haxby JV, editors, *The Oxford Handbook of Face Perception*, chap. 4, 51–75. Oxford University Press, Oxford.

Rolls ET (2012a). Glutamate, obsessive-compulsive disorder, schizophrenia, and the stability of cortical attractor neuronal networks. *Pharmacol Biochem Behav* 100: 736–51.

Rolls ET (2012b). Invariant visual object and face recognition: neural and computational bases, and a model, visnet. *Front Comput Neurosci* 6: 35.

Rolls ET (2012c). Advantages of dilution in the connectivity of attractor networks in the brain. *Biologically Inspired Cognitive Architectures* 1: 44–54.

Rolls ET (2012d). *Neuroculture: On the Implications of Brain Science*. Oxford University Press, Oxford.

Rolls ET (2013a). What are emotional states, and why do we have them? *Emotion Review* 5: 241–247.

Rolls ET (2013b). A biased activation theory of the cognitive and attentional modulation of emotion. *Frontiers in Human Neuroscience* 7: 74.

Rolls ET (2013c). The mechanisms for pattern completion and pattern separation in the hippocampus. *Front Syst Neurosci* 7: 74.

Rolls ET (2014a). *Emotion and Decision-Making Explained*. Oxford University Press, Oxford.

Rolls ET (2014b). Emotion and decision-making explained: Précis. *Cortex* 59: 185–193.

Rolls ET (2015a). Taste, olfactory, and food reward value processing in the brain. *Progress in Neurobiology* 127-128: 64–90.

Rolls ET (2015b). The neuronal representation of information in the human brain. Review. *Brain* 138: 3459–3462.

Rolls ET (2016b). *Cerebral Cortex: Principles of Operation*. Oxford University Press, Oxford.

Rolls ET (2016c). Functions of the anterior insula in taste, autonomic, and related functions. *Brain and Cognition* 110: 4–19.

Rolls ET (2016d). A non-reward attractor theory of depression. *Neurosci Biobehav Rev* 68: 47–58.

Rolls ET (2016e). Pattern separation, completion, and categorisation in the hippocampus and neocortex. *Neurobiology of Learning and Memory* 129: 4–28.

Rolls ET (2016f). Reward systems in the brain and nutrition. *Annual Review of Nutrition* 36: 435–470.

Rolls ET (2016g). Motivation explained: Ultimate and proximate accounts of hunger and appetite. *Advances in Motivation Science* 3: 187–249.

Rolls ET (2017a). The roles of the orbitofrontal cortex via the habenula in non-reward and depression, and in the responses of serotonin and dopamine neurons. *Neurosci Biobehav Rev* 75: 331–334.

Rolls ET (2017b). Cortical coding. *Language, Cognition and Neuroscience* 32: 316–329.

Rolls ET (2018a). The storage and recall of memories in the hippocampo-cortical system. *Cell and Tissue Research* 373: 577–604.

Rolls ET (2018b). *The Brain, Emotion, and Depression*. Oxford University Press, Oxford.

Rolls ET (2019a). The orbitofrontal cortex and emotion in health and disease, including depression. *Neuropsychologia* 128: 14–43.

Rolls ET (2019b). The cingulate cortex and limbic systems for action, emotion, and memory. In Vogt BA, editor, *Handbook of Clinical Neurology: Cingulate Cortex*, vol. 166, chap. 2, 23–37. Elsevier, Oxford.

Rolls ET (2019c). *The Orbitofrontal Cortex*. Oxford University Press, Oxford.

Rolls ET (2019d). Emotion and reasoning in human decision-making. *Economics: The Open-Access, Open-Assessment E-Journal* 13: http://dx.doi.org/10.5018/economics–ejournal.ja.2019–39.

Rolls ET (2019e). The cingulate cortex and limbic systems for emotion, action, and memory. *Brain Struct Funct* 224: 3001–3018.

Rolls ET (2019f). Attractor network dynamics, transmitters, and the memory and cognitive changes in aging. In Heilman KM and Nadeau SE, editors, *Cognitive Changes and the Aging Brain*, chap. 14, 203–225. Cambridge University Press, Cambridge.

Rolls ET (2020a). The texture and taste of food in the brain. *Journal of Texture Studies* 51: 23–44.

Rolls ET (2020b). Spatial coordinate transforms linking the allocentric hippocampal and egocentric parietal primate brain systems for memory, action in space, and navigation. *Hippocampus* 30: 332–353.

Rolls ET (2020c). Neural computations underlying phenomenal consciousness: a higher order syntactic thought theory. *Frontiers in Psychology (Consciousness Research)* 11: 655.

Rolls ET (2021a). *Brain Computations: What and How*. Oxford University Press, Oxford.

Rolls ET (2021b). Neurons including hippocampal spatial view cells, and navigation in primates including humans. *Hippocampus* 31: 593–611.

Rolls ET (2021c). Learning invariant object and spatial view representations in the brain using slow unsupervised learning. *Frontiers in Computational Neuroscience* 15: 686239.

Rolls ET (2021d). The neuroscience of emotional disorders. In Heilman KM and Nadeau SE, editors, *Handbook of Clinical Neurology: Disorders of Emotion in Neurologic Disease*, vol. 183, chap. 1, 1–26. Elsevier, Oxford.

Rolls ET (2021e). The connections of neocortical pyramidal cells can implement the learning of new categories, attractor memory, and top-down recall and attention. *Brain Structure and Function* 226: 2523–2536.

Rolls ET (2021f). Attractor cortical neurodynamics, schizophrenia, and depression. *Translational Psychiatry* 11: 215.

Rolls ET (2021g). A neuroscience levels of explanation approach to the mind and the brain. *Front Comput Neurosci* 15: 649679.

Rolls ET (2021h). On pattern separation in the primate including human hippocampus. *Trends in Cognitive Sciences* 25: 920–922.

Rolls ET (2021i). Mind causality: A computational neuroscience approach. *Front Comput Neurosci* 15: 706505.

Rolls ET (2022a). The hippocampus, ventromedial prefrontal cortex, and episodic and semantic memory. *Prog Neurobiol* 217: 102334.

Rolls ET (2022b). *The Brain, Emotion, and Depression*. East China Normal University Press, Shanghai, China, chinese edn.

Rolls ET (2023a). Hippocampal spatial view cells for memory and navigation, and their underlying connectivity in humans. *Hippocampus* 33: 533–572.

Rolls ET (2023b). Emotion, motivation, decision-making, the orbitofrontal cortex, anterior cingulate cortex, and the amygdala. *Brain Struct Funct* 228: 1201–1257.

Rolls ET (2023c). *Brain Computations and Connectivity*. Oxford University Press, Open Access, Oxford.

Rolls ET (2023d). Hippocampal spatial view cells, place cells, and concept cells: View representations. *Hippocampus* 33: 667–687.

Rolls ET (2024a). The memory systems of the human brain and generative artificial intelligence. *Heliyon* 10: e31965.

Rolls ET (2024b). Cortical neurodynamics, schizophrenia, depression, and obsessive-compulsive disorder. In Mishara AL, Moskalewicz M, Schwartz MA, Kranjec A, and Corlett PR, editors, *Phenomenological Neuropsychiatry: How Patient Experience Bridges the Clinic with Clinical Neuroscience*, chap. 11, 119–143. Springer, Switzerland.

Rolls ET (2024c). Two what, two where, visual cortical streams in humans. *Neurosci Biobehav Rev* 160: 105650.

Rolls ET (2025a). Hippocampal discoveries: spatial view cells, connectivity, and computations for memory and navigation, in primates including humans. *Hippocampus* 35: e23666.

Rolls ET (2025b). A theory and model of scene representations with hippocampal spatial view cells. *Hippocampus* 35: e70013.

Rolls ET (2025c). Speed cells in the primate hippocampus and navigation. In Mirdal G, editor, *Franchir les Frontieres pour Décrypter le Cerveau*, 161–165. Odile Jacob, Paris.

Rolls ET (2025d). The orbitofrontal cortex. *Oxford Research Encyclopedia of Neuroscience* https://doi.org/10.1093/acrefore/9780190264086.013.560.

Rolls ET (2025e). Emotion, motivation, reasoning, and how their brain systems are related. *Brain Sciences* 15: 507.

Rolls ET (2026a). From brain mechanisms of emotion and decision-making to neuroeconomics. In Teschl M and Kirman A, editors, *The State of Mind in Economics*. Cambridge University Press, Cambridge.

Rolls ET (2026b). Taste, olfactory, oral texture and visual processing of food reward in the brain. In Oria RB, Malva JO, and Netto CA, editors, *Nutritional Neuroscience*, chap. 2, 29–42. Academic Press, London.

Rolls ET (2026c). Hippocampal revolutions. *Neuroscience and Biobehavioral Reviews* 180: 106492.

Rolls ET (2026d). *Brain Computations and Principles; and AI.* Oxford University Press, Open Access, Oxford.

Rolls ET and Baylis GC (1986). Size and contrast have only small effects on the responses to faces of neurons in the cortex of the superior temporal sulcus of the monkey. *Experimental Brain Research* 65: 38–48.

Rolls ET and Baylis LL (1994). Gustatory, olfactory, and visual convergence within the primate orbitofrontal cortex. *Journal of Neuroscience* 14: 5437–5452.

Rolls ET and Cooper SJ (1973). Activation of neurones in the prefrontal cortex by brain-stimulation reward in the rat. *Brain Research* 60: 351–368.

Rolls ET and Cooper SJ (1974). Connection between the prefrontal cortex and pontine brain-stimulation reward sites in the rat. *Experimental Neurology* 42: 687–699.

Rolls ET and Cowey A (1970). Topography of the retina and striate cortex and its relationship to visual acuity in rhesus monkeys and squirrel monkeys. *Exp Brain Res* 10: 298–310.

Rolls ET and de Waal AWL (1985). Long-term sensory-specific satiety: evidence from an ethiopian refugee camp. *Physiology and Behavior* 34: 1017–1020.

Rolls ET and Deco G (2002). *Computational Neuroscience of Vision.* Oxford University Press, Oxford.

Rolls ET and Deco G (2006). Attention in natural scenes: neurophysiological and computational bases. *Neural Networks* 19: 1383–1394.

Rolls ET and Deco G (2010). *The Noisy Brain: Stochastic Dynamics as a Principle of Brain Function.* Oxford University Press, Oxford.

Rolls ET and Deco G (2011a). Prediction of decisions from noise in the brain before the evidence is provided. *Frontiers in Neuroscience* 5: 33.

Rolls ET and Deco G (2011b). A computational neuroscience approach to schizophrenia and its onset. *Neuroscience and Biobehavioral Reviews* 35: 1644–1653.

Rolls ET and Deco G (2011c). Prediction of decisions from noise in the brain before the evidence is provided. *Frontiers in Neuroscience* 5: 33.

Rolls ET and Deco G (2015a). Stochastic cortical neurodynamics underlying the memory and cognitive changes in aging. *Neurobiol Learn Mem* 118: 150–61.

Rolls ET and Deco G (2015b). Networks for memory, perception, and decision-making, and beyond to how the syntax for language might be implemented in the brain. *Brain Research* 1621: 316–334.

Rolls ET and Deco G (2016). Non-reward neural mechanisms in the orbitofrontal cortex. *Cortex* 83: 27–38.

Rolls ET and Grabenhorst F (2008). The orbitofrontal cortex and beyond: from affect to decision-making. *Prog Neurobiol* 86: 216–44.

Rolls ET and Mills P (2019). The generation of time in the hippocampal memory system. *Cell Rep* 28: 1649–1658 e6.

Rolls ET and Mills WPC (2018). Non-accidental properties, metric invariance, and encoding by neurons in a model of ventral stream visual object recognition, visnet. *Neurobiology of Learning and Memory* 152: 20–31.

Rolls ET and Milward T (2000). A model of invariant object recognition in the visual system: learning rules, activation functions, lateral inhibition, and information-based performance measures. *Neural Comput* 12: 2547–72.

Rolls ET and O'Mara SM (1995). View-responsive neurons in the primate hippocampal complex. *Hippocampus* 5: 409–24.

Rolls ET and Rolls BJ (1982). Brain mechanisms involved in feeding. In Barker L, editor, *Psychobiology of Human Food Selection*, 33–62. AVI Publishing Company, Westport, Connecticut.

Rolls ET and Rolls JH (1997). Olfactory sensory-specific satiety in humans. *Physiology and Behavior* 61: 461–473.

Rolls ET and Stringer SM (2000). On the design of neural networks in the brain by genetic evolution. *Prog Neurobiol* 61: 557–79.

Rolls ET and Stringer SM (2001a). Invariant object recognition in the visual system with error correction and temporal difference learning. *Network* 12: 111–29.

Rolls ET and Stringer SM (2001b). A model of the interaction between mood and memory. *Network: Computation in Neural Systems* 12: 89–109.

Rolls ET and Stringer SM (2006). Invariant visual object recognition: a model, with lighting invariance. *J Physiol Paris* 100: 43–62.

Rolls ET and Stringer SM (2007). Invariant global motion recognition in the dorsal visual system: a unifying theory. *Neural Computation* 19: 139–169.

Rolls ET and Tovee MJ (1994). Processing speed in the cerebral cortex and the neurophysiology of visual masking. *Proceedings of the Royal Society of London B* 257: 9–15.

Rolls ET and Tovee MJ (1995a). Sparseness of the neuronal representation of stimuli in the primate temporal visual cortex. *J Neurophysiol* 73: 713–26.

Rolls ET and Tovee MJ (1995b). The responses of single neurons in the temporal visual cortical areas of the macaque when more than one stimulus is present in the visual field. *Experimental Brain Research* 103: 409–420.

Rolls ET and Treves A (1990). The relative advantages of sparse versus distributed encoding for associative neuronal networks in the brain. *Network* 1: 407–421.

Rolls ET and Treves A (1994). Neural networks in the brain involved in memory and recall. *Prog Brain Res* 102: 335–41.

Rolls ET and Treves A (1998). *Neural Networks and Brain Function*. Oxford University Press, Oxford.

Rolls ET and Treves A (2011). The neuronal encoding of information in the brain. *Progress in Neurobiology* 95: 448–490.

Rolls ET and Treves A (2024). A theory of hippocampal function: New developments. *Prog Neurobiol* 238: 102636.

Rolls ET and Turova TS (2025). Visual cortical networks for 'what' and 'where' to the human hippocampus revealed with dynamical graphs. *Cerebral Cortex* 35: bhaf106.

Rolls ET and Webb TJ (2012). Cortical attractor network dynamics with diluted connectivity. *Brain Research* 1434: 212–225.

Rolls ET and Webb TJ (2014). Finding and recognising objects in natural scenes: complementary computations in the dorsal and ventral visual systems. *Frontiers in Computational Neuroscience* 8: 85.

Rolls ET and Williams GV (1987). Neuronal activity in the ventral striatum of the primate. In Carpenter MB and Jayamaran A, editors, *The Basal Ganglia II – Structure and Function – Current Concepts*, 349–356. Plenum, New York.

Rolls ET and Wirth S (2018). Spatial representations in the primate hippocampus, and their functions in memory and navigation. *Prog Neurobiol* 171: 90–113.

Rolls ET and Wirth S (2023). Hippocampal system neurons encoding views in different species: Introduction to the special issue of hippocampus 2023. *Hippocampus* 33: 445–447.

Rolls ET and Xiang JZ (2005). Reward-spatial view representations and learning in the hippocampus. *Journal of Neuroscience* 25: 6167 6174.

Rolls ET and Xiang JZ (2006). Spatial view cells in the primate hippocampus, and memory recall. *Reviews in the Neurosciences* 17: 175–200.

Rolls ET, Burton MJ, and Mora F (1976). Hypothalamic neuronal responses associated with the sight of food. *Brain Res* 111: 53–66.

Rolls ET, Judge SJ, and Sanghera MK (1977). Activity of neurones in the inferotemporal cortex of the alert monkey. *Brain Research* 130: 229–238.

Rolls ET, Sanghera MK, and Roper-Hall A (1979). The latency of activation of neurons in the lateral hypthalamus and substantia innominata during feeding in the monkey. *Brain Research* 164: 121–135.

Rolls ET, Burton MJ, and Mora F (1980c). Neurophysiological analysis of brain-stimulation reward in the monkey. *Brain Research* 194: 339–357.

Rolls ET, Perrett DI, and Thorpe SJ (1980d). The influence of motivation on the responses of neurons in the posterior parietal association cortex. *The Behavioral and Brain Sciences* 3: 514–515.

Rolls ET, Perrett DI, Caan AW, and Wilson FAW (1982c). Neuronal responses related to visual recognition. *Brain* 105: 611–646.

Rolls ET, Rolls BJ, and Rowe EA (1983). Sensory-specific and motivation-specific satiety for the sight and taste of food and water in man. *Physiology and Behavior* 30: 185–192.

Rolls ET, Baylis GC, and Leonard CM (1985). Role of low and high spatial frequencies in the face-selective responses of neurons in the cortex in the superior temporal sulcus in the monkey. *Vision Research* 25: 1021–35.

Rolls ET, Murzi E, Yaxley S, Thorpe SJ, and Simpson SJ (1986). Sensory-specific satiety: food-specific reduction in responsiveness of ventral forebrain neurons after feeding in the monkey. *Brain Res* 368: 79–86.

Rolls ET, Baylis GC, and Hasselmo ME (1987). The responses of neurons in the cortex in the superior temporal sulcus of the monkey to band-pass spatial frequency filtered faces. *Vision Research* 27: 311–326.

Rolls ET, Baylis GC, Hasselmo M, and Nalwa V (1989a). The representation of information in the temporal lobe visual cortical areas of macaque monkeys. In Kulikowski J, Dickinson C, and Murray I, editors, *Seeing Contour and Colour*. Pergamon, Oxford.

Rolls ET, Miyashita Y, Cahusac PMB, Kesner RP, Niki H, Feigenbaum J, and Bach L (1989b). Hippocampal neurons in the monkey with activity related to the place in which a stimulus is shown. *Journal of Neuroscience* 9: 1835–1845.

Rolls ET, Sienkiewicz ZJ, and Yaxley S (1989c). Hunger modulates the responses to gustatory stimuli of single neurons in the caudolateral orbitofrontal cortex of the macaque monkey. *Eur J Neurosci* 1: 53–60.

Rolls ET, Yaxley S, and Sienkiewicz ZJ (1990). Gustatory responses of single neurons in the caudolateral orbitofrontal cortex of the macaque monkey. *J Neurophysiol* 64: 1055–66.

Rolls ET, Cahusac PMB, Feigenbaum JD, and Miyashita Y (1993). Responses of single neurons in the hippocampus of the macaque related to recognition memory. *Experimental Brain Research* 93: 299–306.

Rolls ET, Hornak J, Wade D, and McGrath J (1994a). Emotion-related learning in patients with social and emotional changes associated with frontal lobe damage. *Journal of Neurology, Neurosurgery and Psychiatry* 57: 1518–1524.

Rolls ET, Tovee MJ, Purcell DG, Stewart AL, and Azzopardi P (1994b). The responses of neurons in the temporal cortex of primates, and face identification and detection. *Experimental Brain Research* 101: 474–484.

Rolls ET, Critchley HD, Mason R, and Wakeman EA (1996a). Orbitofrontal cortex neurons: role in olfactory and visual association learning. *Journal of Neurophysiology* 75: 1970–1981.

Rolls ET, Critchley HD, and Treves A (1996b). The representation of olfactory information in the primate orbitofrontal cortex. *Journal of Neurophysiology* 75: 1982–1996.

Rolls ET, Francis S, Bowtell R, Browning D, Clare S, Smith E, and McGlone F (1997a). Taste and olfactory activation of the orbitofrontal cortex. *NeuroImage* 5: S199.

Rolls ET, Robertson RG, and Georges-François P (1997b). Spatial view cells in the primate hippocampus. *European Journal of Neuroscience* 9: 1789–1794.

Rolls ET, Treves A, Foster D, and Perez-Vicente C (1997c). Simulation studies of the ca3 hippocampal subfield modelled as an attractor neural network. *Neural Networks* 10: 1559–1569.

Rolls ET, Treves A, and Tovee MJ (1997d). The representational capacity of the distributed encoding of information provided by populations of neurons in the primate temporal visual cortex. *Experimental Brain Research* 114: 177–185.

Rolls ET, Treves A, Tovee MJ, and Panzeri S (1997e). Information in the neuronal representation of individual stimuli in the primate temporal visual cortex. *Journal of Computational Neuroscience* 4: 309–333.

Rolls ET, Treves A, Robertson RG, Georges-François P, and Panzeri S (1998). Information about spatial view in an ensemble of primate hippocampal cells. *Journal of Neurophysiology* 79: 1797–1813.

Rolls ET, Critchley HD, Browning AS, Hernadi A, and Lenard L (1999a). Responses to the sensory properties of fat of neurons in the primate orbitofrontal cortex. *Journal of Neuroscience* 19: 1532–1540.

Rolls ET, Tovee MJ, and Panzeri S (1999b). The neurophysiology of backward visual masking: information analysis. *Journal of Cognitive Neuroscience* 11: 335–346.

Rolls ET, Stringer SM, and Trappenberg TP (2002). A unified model of spatial and episodic memory. *Proceedings of the Royal Society B* 269: 1087–93.

Rolls ET, Aggelopoulos NC, and Zheng F (2003a). The receptive fields of inferior temporal cortex neurons in natural scenes. *J Neurosci* 23: 339–48.

Rolls ET, Franco L, Aggelopoulos NC, and Reece S (2003b). An information theoretic approach to the contributions of the firing rates and the correlations between the firing of neurons. *J Neurophysiol* 89: 2810–22.

Rolls ET, Kringelbach ML, and De Araujo IET (2003c). Different representations of pleasant and unpleasant odors in the human brain. *European Journal of Neuroscience* 18: 695–703.

Rolls ET, O'Doherty J, Kringelbach ML, Francis S, Bowtell R, and McGlone F (2003d). Representations of pleasant and painful touch in the human orbitofrontal and cingulate cortices. *Cereb Cortex* 13: 308–17.

Rolls ET, Aggelopoulos NC, Franco L, and Treves A (2004). Information encoding in the inferior temporal cortex: contributions of the firing rates and correlations between the firing of neurons. *Biological Cybernetics* 90: 19–32.

Rolls ET, Browning AS, Inoue K, and Hernadi I (2005a). Novel visual stimuli activate a population of neurons in the primate orbitofrontal cortex. *Neurobiol Learn Mem* 84: 111–23.

Rolls ET, Franco L, and Stringer SM (2005b). The perirhinal cortex and long-term familiarity memory. *Q J Exp Psychol B* 58: 234–45.

Rolls ET, Xiang JZ, and Franco L (2005c). Object, space and object-space representations in the primate hippocampus. *Journal of Neurophysiology* 94: 833–844.

Rolls ET, Critchley HD, Browning AS, and Inoue K (2006a). Face-selective and auditory neurons in the primate orbitofrontal cortex. *Experimental Brain Research* 170: 74–87.

Rolls ET, Franco L, Aggelopoulos NC, and Perez JM (2006b). Information in the first spike, the order of spikes, and the number of spikes provided by neurons in the inferior temporal visual cortex. *Vision Research* 46: 4193–4205.

Rolls ET, Grabenhorst F, Margot C, da Silva MAAP, and Velazco MI (2008a). Selective attention to affective value alters how the brain processes olfactory stimuli. *Journal of Cognitive Neuroscience* 20: 1815–1826.

Rolls ET, Grabenhorst F, and Parris BA (2008b). Warm pleasant feelings in the brain. *Neuroimage* 41: 1504–1513.

Rolls ET, Loh M, and Deco G (2008c). An attractor hypothesis of obsessive-compulsive disorder. *European Journal of Neuroscience* 28: 782–793.

Rolls ET, Loh M, Deco G, and Winterer G (2008d). Computational models of schizophrenia and dopamine modulation in the prefrontal cortex. *Nature Reviews Neuroscience* 9: 696–709.

Rolls ET, Grabenhorst F, and Franco L (2009). Prediction of subjective affective state from brain activations. *J Neurophysiol* 101: 1294–308.

Rolls ET, Grabenhorst F, and Deco G (2010a). Choice, difficulty, and confidence in the brain. *Neuroimage* 53: 694–706.

Rolls ET, Grabenhorst F, and Deco G (2010b). Decision-making, errors, and confidence in the brain. *Journal of Neurophysiology* 104: 2359–2374.

Rolls ET, Grabenhorst F, and Parris BA (2010c). Neural systems underlying decisions about affective odors. *Journal of Cognitive Neuroscience* 22: 1069–1082.

Rolls ET, Dempere-Marco L, and Deco G (2013). Holding multiple items in short term memory: a neural mechanism. *PLoS One* 8: e61078.

Rolls ET, Joliot M, and Tzourio-Mazoyer N (2015). Implementation of a new parcellation of the orbitofrontal cortex in the automated anatomical labeling atlas. *Neuroimage* 122: 1–5.

Rolls ET, Lu W, Wan L, Yan H, Wang C, Yang F, Tan Y, Li L, Chinese Schizophrenia Collaboration G, Yu H, Liddle PF, Palaniyappan L, Zhang D, Yue W, and Feng J (2017). Individual differences in schizophrenia. *BJPsych Open* 3: 265–273.

Rolls ET, Cheng W, Gilson M, Qiu J, Hu Z, Ruan H, Li Y, Huang CC, Yang AC, Tsai SJ, Zhang X, Zhuang K, Lin CP, Deco G, Xie P, and Feng J (2018a). Effective connectivity in depression. *Biol Psychiatry Cogn Neurosci Neuroimaging* 3: 187–197.

Rolls ET, Mills T, Norton A, Lazidis A, and Norton IT (2018b). Neuronal encoding of fat using the coefficient of sliding friction in the cerebral cortex and amygdala. *Cerebral Cortex* 28: 4080–4089.

Rolls ET, Cheng W, Gong W, Qiu J, Zhou C, Zhang J, Lv W, Ruan H, Wei D, Cheng K, Meng J, Xie P, and Feng J (2019). Functional connectivity of the anterior cingulate cortex in depression and in health. *Cereb Cortex* 29: 3617–3630.

Rolls ET, Cheng W, Du J, Wei D, Qiu J, Dai D, Zhou Q, Xie P, and Feng J (2020a). Functional connectivity of the right inferior frontal gyrus and orbitofrontal cortex in depression. *Soc Cogn Affect Neurosci* 15: 75–86.

Rolls ET, Cheng W, and Feng J (2020b). The orbitofrontal cortex: reward, emotion, and depression. *Brain Communications* 2: fcaa196.

Rolls ET, Cheng W, Gilson M, Gong W, Deco G, Lo CZ, Yang AC, Tsai SJ, Liu ME, Lin CP, and Feng J (2020c). Beyond the disconnectivity hypothesis of schizophrenia. *Cerebral Cortex* 30: 1213–1233.

Rolls ET, Huang CC, Lin CP, Feng J, and Joliot M (2020d). Automated anatomical labelling atlas 3. *Neuroimage* 206: 116189.

Rolls ET, Vatansever D, Li Y, Cheng W, and Feng J (2020e). Rapid rule-based reward reversal and the lateral orbitofrontal cortex. *Cerebral Cortex Communications* 1: tgaa087 doi: 10.1093/texcom/tgaa087.

Rolls ET, Zhou Y, Cheng W, Gilson M, Deco G, and Feng J (2020f). Effective connectivity in autism. *Autism Research* 13: 32–44.

Rolls ET, Cheng W, and Feng J (2021a). Brain dynamics: the temporal variability of connectivity, and differences in schizophrenia and adhd. *Translational Psychiatry* 11: 70.

Rolls ET, Cheng W, and Feng J (2021b). Brain dynamics: Synchronous peaks, functional connectivity, and its temporal variability. *Human Brain Mapping* 42: 2790–2801.

Rolls ET, Deco G, Huang CC, and Feng J (2022a). The effective connectivity of the human hippocampal memory system. *Cereb Cortex* 32: 3706–3725.

Rolls ET, Deco G, Huang CC, and Feng J (2022b). The human orbitofrontal cortex, vmpfc, and anterior cingulate cortex effective connectome: emotion, memory, and action. *Cereb Cortex* 33: 330–356.

Rolls ET, Deco G, Huang CC, and Feng J (2022c). The human language effective connectome. *Neuroimage* 258: 119352.

Rolls ET, Wan Z, Cheng W, and Feng J (2022d). Risk-taking in humans and the medial orbitofrontal cortex reward system. *Neuroimage* 249: 118893.

Rolls ET, Deco G, Huang CC, and Feng J (2023a). The human posterior parietal cortex: effective connectome, and its relation to function. *Cereb Cortex* 33: 3142–3170.

Rolls ET, Deco G, Huang CC, and Feng J (2023b). Multiple cortical visual streams in humans. *Cerebral Cortex* 33: 3319–3349.

Rolls ET, Deco G, Huang CC, and Feng J (2023c). Prefrontal and somatosensory-motor cortex effective connectivity in humans. *Cereb Cortex* 33: 4939–4963.

Rolls ET, Deco G, Huang CC, and Feng J (2023d). Human amygdala compared to orbitofrontal cortex connectivity, and emotion. *Progress in Neurobiology* 220: 102385.

Rolls ET, Deco G, Zhang Y, and Feng J (2023e). Hierarchical organization of the human ventral visual streams revealed with magnetoencephalography. *Cereb Cortex* 33: 10686–10701.

Rolls ET, Rauschecker JP, Deco G, Huang CC, and Feng J (2023f). Auditory cortical connectivity in humans. *Cereb Cortex* 33: 6207–6227.

Rolls ET, Wirth S, Deco G, Huang CC, and Feng J (2023g). The human posterior cingulate, retrosplenial and medial parietal cortex effective connectome, and implications for memory and navigation. *Human Brain Mapping* 44: 629–655.

Rolls ET, Deco G, Huang CC, and Feng J (2024a). The connectivity of the human frontal pole cortex, and a theory of its involvement in exploit versus explore. *Cereb Cortex* 34: 1–19.

Rolls ET, Feng J, and Zhang R (2024b). Selective activations and functional connectivities to the sight of faces, scenes, body parts and tools in visual and non-visual cortical regions leading to the human hippocampus. *Brain Struct Funct* 229: 1471–1493.

Rolls ET, Yan X, Deco G, Zhang Y, Jousmaki V, and Feng J (2024c). A ventromedial visual cortical 'where' stream to the human hippocampus for spatial scenes revealed with magnetoencephalography. *Communications Biology* 7: 1047.

Rolls ET, Zhang C, and Feng J (2024d). Hippocampal storage and recall of neocortical what - where representations. *Hippocampus* 34: 608–624.

Rolls ET, Zhang R, Deco G, Vatansever D, and Feng J (2024e). Selective brain activations and connectivities related to the storage and recall of human object-location, reward-location, and word-pair episodic memories. *Hum Brain Mapp* 45: e70056.

Rolls ET, Zhang C, and Feng J (2025a). Slow semantic learning in the cerebral cortex, and its relation to the hippocampal episodic memory system. *Cerebral Cortex* 35: bhah107.

Rolls ET, Zhang C, and Feng J (2025b). Reward-specific satiety and reward-specific motivation: neural bases and significance. *Cerebral Cortex* 35: bhaf248.

Rolls ET, Feng J, and Zhang R (2026). Visual cortical lateralization in activations and functional connectivity to the sight of faces, scenes, body parts and tools. *Human Brain Mapping* 47: e70494.

Ronchi R, Park HD, and Blanke O (2018). Bodily self-consciousness and its disorders. *Handb Clin Neurol* 151: 313–330.

Rosenberg DR, MacMaster FP, Keshavan MS, Fitzgerald KD, Stewart CM, and Moore GJ (2000). Decrease in caudate glutamatergic concentrations in pediatric obsessive-compulsive disorder patients taking paroxetine. *Journal of the American Academy of Child and Adolescent Psychiatry* 39: 1096–1103.

Rosenberg DR, MacMillan SN, and Moore GJ (2001). Brain anatomy and chemistry may predict treatment response in paediatric obsessive–compulsive disorder. *International Journal of Neuropsychopharmacology* 4: 179–190.

Rosenberg DR, Mirza Y, Russell A, Tang J, Smith JM, Banerjee SP, Bhandari R, Rose M, Ivey J, Boyd C, and Moore GJ (2004). Reduced anterior cingulate glutamatergic concentrations in childhood OCD and major depression versus healthy controls. *Journal of the American Academy of Child and Adolescent Psychiatry* 43: 1146–1153.

Rosenkilde CE, Bauer RH, and Fuster JM (1981). Single unit activity in ventral prefrontal cortex in behaving monkeys. *Brain Research* 209: 375–394.

Rosenthal D (2012). Higher-order awareness, misrepresentation and function. *Philos Trans R Soc Lond B Biol Sci* 367: 1424–38.

Rubinov M and Sporns O (2010). Complex network measures of brain connectivity: uses and interpretations. *Neuroimage* 52: 1059–69.

Rudebeck PH and Murray EA (2014). The orbitofrontal oracle: cortical mechanisms for the prediction and evaluation of specific behavioral outcomes. *Neuron* 84: 1143–56.

Rudebeck PH, Saunders RC, Lundgren DA, and Murray EA (2017). Specialized representations of value in the orbital and ventrolateral prefrontal cortex: desirability versus availability of outcomes. *Neuron* 95: 1208–1220 e5.

Rumelhart DE and Zipser D (1985). Feature discovery by competitive learning. *Cognitive Science* 9: 75–112.

Rupniak N and Gaffan D (1987). Monkey hippocampus and learning about spatially directed movements. *Journal of Neuroscience* 7: 2331–2337.

Rushworth MF, Noonan MP, Boorman ED, Walton ME, and Behrens TE (2011). Frontal cortex and reward-guided learning and decision-making. *Neuron* 70: 1054–69.

Rushworth MF, Kolling N, Sallet J, and Mars RB (2012). Valuation and decision-making in frontal cortex: one or many serial or parallel systems? *Curr Opin Neurobiol* 22: 946–55.

Rushworth MFS, Behrens TE, Rudebeck PH, and Walton ME (2007a). Contrasting roles for cingulate and orbitofrontal cortex in decisions and social behaviour. *Trends in Cognitive Sciences* 11: 168–176.

Rushworth MFS, Buckley MJ, Behrens TE, Walton ME, and Bannerman DM (2007b). Functional organization of the medial frontal cortex. *Current Opinion in Neurobiology* 17: 220–227.

Russchen FT, Amaral DG, and Price JL (1985). The afferent connections of the substantia innominata in the monkey, Macaca fascicularis. *Journal of Comparative Neurology* 242: 1–27.

Rust NC and DiCarlo JJ (2010). Selectivity and tolerance ("invariance") both increase as visual information propagates from cortical area v4 to it. *J Neurosci* 30: 12978–95.

Ryom KI, Stendardi D, Ciaramelli E, and Treves A (2023). Computational constraints on the associative recall of spatial scenes. *Hippocampus* 33: 635–645.

Sah P (1996). Ca^{2+}-activated K^{+} currents in neurones: types, physiological roles and modulation. *Trends in Neuroscience* 19: 150–154.

Sah P and Faber ES (2002). Channels underlying neuronal calcium-activated potassium currents. *Progress in Neurobiology* 66: 345–353.

Saleem KS, Miller B, and Price JL (2014). Subdivisions and connectional networks of the lateral prefrontal cortex in the macaque monkey. *J Comp Neurol* 522: 1641–90.

Salinas E and Abbott LF (1997). Invariant visual responses from attentional gain fields. *J Neurophysiol* 77: 3267–72.

Salinas E and Sejnowski TJ (2001). Gain modulation in the central nervous system: where behavior, neurophysiology, and computation meet. *Neuroscientist* 7: 430–40.

Samsonovich A and McNaughton BL (1997). Path integration and cognitive mapping in a continuous attractor neural network model. *Journal of Neuroscience* 17: 5900–5920.

Sanghera MK, Rolls ET, and Roper-Hall A (1977). Visual responses of neurones in the lateral amygdala of the rhesus monkey. *Proceedings of the International Union of Physiological Sciences* III, 1950.

Sanghera MK, Rolls ET, and Roper-Hall A (1979). Visual responses of neurons in the dorsolateral amygdala of the alert monkey. *Experimental Neurology* 63: 610–626.

Sarel A, Finkelstein A, Las L, and Ulanovsky N (2017). Vectorial representation of spatial goals in the hippocampus of bats. *Science* 355: 176–180.

Sawaguchi T and Goldman-Rakic PS (1991). D1 dopamine receptors in prefrontal cortex: Involvement in working memory. *Science* 251: 947–950.

Sawaguchi T and Goldman-Rakic PS (1994). The role of D1-dopamine receptor in working memory: local injections of dopamine antagonists into the prefrontal cortex of rhesus monkeys performing an oculomotor delayed-response task. *Journal of Neurophysiology* 71: 515–528.

Scheuerecker J, Ufer S, Zipse M, Frodl T, Koutsouleris N, Zetzsche T, Wiesmann M, Albrecht J, Bruckmann H, Schmitt G, Moller HJ, and Meisenzahl EM (2008). Cerebral changes and cognitive dysfunctions in medication-free schizophrenia – An fMRI study. *Journal of Psychiatric Research* 42: 469–476.

Schliebs R and Arendt T (2006). The significance of the cholinergic system in the brain during aging and in Alzheimer's disease. *Journal of Neural Transmission* 113: 1625–1644.

Schliebs R and Arendt T (2011). The cholinergic system in aging and neuronal degeneration. *Behav Brain Res* 221: 555–63.

Schmaal L, Hibar DP, Samann PG, Hall GB, Baune BT, Jahanshad N, Cheung JW, van Erp TGM, Bos D, Ikram MA, Vernooij MW, Niessen WJ, Tiemeier H, Hofman A, Wittfeld K, Grabe HJ, Janowitz D, Bulow R, Selonke M, Volzke H, Grotegerd D, Dannlowski U, Arolt V, Opel N, Heindel W, Kugel H, Hoehn D, Czisch M, Couvy-Duchesne B, Renteria ME, Strike LT, Wright MJ, Mills NT, de Zubicaray GI, McMahon KL, Medland SE, Martin NG, Gillespie NA, Goya-Maldonado R, Gruber O, Kramer B, Hatton SN, Lagopoulos J, Hickie IB, Frodl T, Carballedo A, Frey EM, van Velzen LS, Penninx B, van Tol MJ, van der Wee NJ, Davey CG, Harrison BJ, Mwangi B, Cao B, Soares JC, Veer IM, Walter H, Schoepf D, Zurowski B, Konrad C, Schramm E, Normann C, Schnell K, Sacchet MD, Gotlib IH, MacQueen GM, Godlewska BR, Nickson T, McIntosh AM, Papmeyer M, Whalley HC, Hall J, Sussmann JE, Li M, Walter M, Aftanas L, Brack I, Bokhan NA, Thompson PM, and Veltman DJ (2017). Cortical abnormalities in adults and adolescents with major depression based on brain scans from 20 cohorts worldwide in the enigma major depressive disorder working group. *Mol Psychiatry* 22: 900–909.

Schonfeld F and Wiskott L (2015). Modeling place field activity with hierarchical slow feature analysis. *Front Comput Neurosci* 9: 51.

Schultz W (2006). Behavioral theories and the neurophysiology of reward. *Annu Rev Psychol* 57: 87–115.

Schultz W (2016a). Dopamine reward prediction-error signalling: a two-component response. *Nat Rev Neurosci* 17: 183–95.

Schultz W (2016b). Reward functions of the basal ganglia. *J Neural Transm (Vienna)* 123: 679–693.

Schultz W (2016c). Dopamine reward prediction error coding. *Dialogues Clin Neurosci* 18: 23–32.

Schultz W (2017). Reward prediction error. *Curr Biol* 27: R369–R371.

Schultz W (2024). A dopamine mechanism for reward maximization. *Proc Natl Acad Sci U S A* 121: e2316658121.

Schultz W, Tremblay L, and Hollerman JR (2000). Reward processing in primate orbitofrontal cortex and basal ganglia. *Cereb Cortex* 10: 272–84.

Schurz M, Tholen MG, Perner J, Mars RB, and Sallet J (2017). Specifying the brain anatomy underlying temporo-parietal junction activations for theory of mind: A review using probabilistic atlases from different imaging modalities. *Hum Brain Mapp* 38: 4788–4805.

Scott D and Tamminga CA (2018). Effects of genetic and environmental risk for schizophrenia on hippocampal activity and psychosis-like behavior in mice. *Behav Brain Res* 339: 114–123.

Scott TR, Yaxley S, Sienkiewicz ZJ, and Rolls ET (1986). Gustatory responses in the frontal opercular cortex of the alert cynomolgus monkey. *Journal of Neurophysiology* 56: 876–890.

Scoville WB and Milner B (1957). Loss of recent memory after bilateral hippocampal lesions. *J. Neurol. Neurosurg. Psychiatry* 20: 11–21.

Seamans JK and Yang CR (2004). The principal features and mechanisms of dopamine modulation in the prefrontal cortex. *Progress in Neurobiology* 74: 1–58.

Segerdahl AR, Mezue M, Okell TW, Farrar JT, and Tracey I (2015). The dorsal posterior insula subserves a fundamental role in human pain. *Nat Neurosci* 18: 499–500.

Seghier ML, Fagan E, and Price CJ (2010). Functional subdivisions in the left angular gyrus where the semantic system meets and diverges from the default network. *J Neurosci* 30: 16809–17.

Seigel M and Auerbach JM (1996). Neuromodulators of synaptic strength. In Fazeli MS and Collingridge GL, editors, *Cortical Plasticity*, chap. 7, 137–148. Bios, Oxford.

Sejnowski TJ (2020). The unreasonable effectiveness of deep learning in artificial intelligence. *Proc Natl Acad Sci U S A* 117: 30033–30038.

Sejnowski TJ (2023). Large language models and the reverse turing test. *Neural Comput* 35: 309–342.

Sellami A, Al Abed AS, Brayda-Bruno L, Etchamendy N, Valerio S, Oule M, Pantaleon L, Lamothe V, Potier M, Bernard K, Jabourian M, Herry C, Mons N, Piazza PV, Eichenbaum H, and Marighetto A (2017). Temporal binding function of dorsal ca1 is critical for declarative memory formation. *Proc Natl Acad Sci U S A* 114: 10262–10267.

Seltzer B and Pandya DN (1978). Afferent cortical connections and architectonics of the superior temporal sulcus and surrounding cortex in the rhesus monkey. *Brain Research* 149: 1–24.

Shallice T and Burgess P (1996). The domain of supervisory processes and temporal organization of behaviour. *Philos Trans R Soc Lond B Biol Sci* 351: 1405–11.

Shallice T and Cipolotti L (2018). The prefrontal cortex and neurological impairments of active thought. *Annu Rev Psychol* 69: 157–180.

Shannon CE (1948). A mathematical theory of communication. *AT&T Bell Laboratories Technical Journal* 27: 379–423.

Sharpe MJ, Wikenheiser AM, Niv Y, and Schoenbaum G (2015). The state of the orbitofrontal cortex. *Neuron* 88: 1075–1077.

Sheinberg DL and Logothetis NK (2001). Noticing familiar objects in real world scenes: the role of temporal cortical neurons in natural vision. *J Neurosci* 21: 1340–50.

Shergill SS, Brammer MJ, Williams SC, Murray RM, and McGuire PK (2000). Mapping auditory hallucinations in schizophrenia using functional magnetic resonance imaging. *Archives of General Psychiatry* 57: 1033–1038.

Siddiqi SH, Schaper F, Horn A, Hsu J, Padmanabhan JL, Brodtmann A, Cash RFH, Corbetta M, Choi KS, Dougherty DD, Egorova N, Fitzgerald PB, George MS, Gozzi SA, Irmen F, Kuhn AA, Johnson KA, Naidech AM, Pascual-Leone A, Phan TG, Rouhl RPW, Taylor SF, Voss JL, Zalesky A, Grafman JH, Mayberg HS, and Fox MD (2021). Brain stimulation and brain lesions converge on common causal circuits in neuropsychiatric disease. *Nat Hum Behav* 5: 1707–1716.

Sigala N and Logothetis NK (2002). Visual categorisation shapes feature selectivity in the primate temporal cortex. *Nature* 415: 318–320.

Sikström S (2007). Computational perspectives on neuromodulation of aging. *Acta Neurochirurgica, Supplement* 97: 513–518.

Skaggs WE, McNaughton BL, Gothard K, and Markus E (1993). An information theoretic approach to deciphering the hippocampal code. In Hanson S, Cowan JD, and Giles CL, editors, *Advances in Neural Information Processing Systems*, vol. 5, 1030–1037. Morgan Kaufmann, San Mateo, CA.

Smerieri A, Rolls ET, and Feng J (2010). Decision time, slow inhibition, and theta rhythm. *Journal of Neuroscience* 30: 14173–14181.

Smith ML and Milner B (1981). The role of the right hippocampus in the recall of spatial location. *Neuropsychologia* 19: 781–93.

Smith SM, Jenkinson M, Woolrich MW, Beckmann CF, Behrens TE, Johansen-Berg H, Bannister PR, De Luca M, Drobnjak I, Flitney DE, Niazy RK, Saunders J, Vickers J, Zhang Y, De Stefano N, Brady JM, and Matthews PM (2004). Advances in functional and structural mr image analysis and implementation as fsl. *Neuroimage* 23 Suppl 1: S208–19.

Smucny J, Dienel SJ, Lewis DA, and Carter CS (2022). Mechanisms underlying dorsolateral prefrontal cortex contributions to cognitive dysfunction in schizophrenia. *Neuropsychopharmacology* 47: 292–308.

Snyder LH, Grieve KL, Brotchie P, and Andersen RA (1998). Separate body- and world-referenced representations of visual space in parietal cortex. *Nature* 394: 887–91.

Sobotka S and Ringo JL (1997). Saccadic eye movements, even in darkness, generate event-related potentials recorded in medial septum and medial temporal cortex. *Brain Res* 756: 168–73.

Sobotka S, Nowicka A, and Ringo JL (1997). Activity linked to externally cued saccades in single units recorded from hippocampal, parahippocampal, and inferotemporal areas of macaques. *J Neurophysiol* 78: 2156–63.

Song S, Yao H, and Treves A (2014). A modular latching chain. *Cognitive neurodynamics* 8: 37–46.

Song Y, Lukasiewicz T, Xu Z, and Bogacz R (2020). Can the brain do backpropagation? -exact implementation of backpropagation in predictive coding networks. *Adv Neural Inf Process Syst* 33: 22566–22579.

Song Y, Millidge B, Salvatori T, Lukasiewicz T, Xu Z, and Bogacz R (2024). Inferring neural activity before plasticity as a foundation for learning beyond backpropagation. *Nat Neurosci* 27: 348–358.

Squire LR (1992). Memory and the hippocampus: a synthesis from findings with rats, monkeys, and humans. *Psychol Rev* 99: 195–231.

Squire LR and Wixted JT (2011). The cognitive neuroscience of human memory since h.m. *Annu Rev Neurosci* 34: 259–88.

Squire LR and Zola SM (1996). Structure and function of declarative and nondeclarative memory systems. *Proceedings of the National Academy of Sciences U S A* 93: 13515–22.

Stein DJ, Kogan CS, Atmaca M, Fineberg NA, Fontenelle LF, Grant JE, Matsunaga H, Reddy YC, Simpson HB, Thomsen PH, van den Heuvel OA, Veale D, Woods DW, and Reed GM (2016). The classification of obsessive-compulsive and related disorders in the icd-11. *J Affect Disord* 190: 663–74.

Steinemann N, Stine GM, Trautmann E, Zylberberg A, Wolpert DM, and Shadlen MN (2024). Direct observation of the neural computations underlying a single decision. *Elife* 12.

Sterpenich V, Vidal S, Hofmeister J, Michalopoulos G, Bancila V, Warrot D, Dayer A, Desseilles M, Aubry JM, Kosel M, Schwartz S, and Vutskits L (2019). Increased reactivity of the mesolimbic reward system after ketamine injection in patients with treatment-resistant major depressive disorder. *Anesthesiology* 130: 923–935.

Stewart SE, Fagerness JA, Platko J, Smoller JW, Scharf JM, Illmann C, Jenike E, Chabane N, Leboyer M, Delorme R, Jenike MA, and Pauls DL (2007). Association of the SLC1A1 glutamate transporter gene and obsessive-compulsive disorder. *American Journal of Medical Genetics B: Neuropsychiatric Genetics* 144: 1027–1033.

Strehler BL (1991). Where is the self? a neuroanatomical theory of consciousness. *Synapse* 7: 44–91.

Stringer SM and Rolls ET (2000). Position invariant recognition in the visual system with cluttered environments. *Neural Networks* 13: 305–315.

Stringer SM, Perry G, Rolls ET, and Proske JH (2006). Learning invariant object recognition in the visual system with continuous transformations. *Biol Cybern* 94: 128–42.

Sugiura M, Watanabe J, Maeda Y, Matsue Y, Fukuda H, and Kawashima R (2005). Cortical mechanisms of visual self-recognition. *Neuroimage* 24: 143–9.

Sulpizio V, Galati G, Fattori P, Galletti C, and Pitzalis S (2020). A common neural substrate for processing scenes and egomotion-compatible visual motion. *Brain Struct Funct* 225: 2091–2110.

Sun J, Liu Z, Rolls ET, Chen Q, Yao Y, Yang W, Wei D, Zhang Q, Zhang J, Feng J, and Qiu J (2019). Verbal creativity correlates with the temporal variability of brain networks during the resting state. *Cereb Cortex* 29: 1047–1058.

Sutton RS and Barto AG (1981). Towards a modern theory of adaptive networks: expectation and prediction. *Psychological Review* 88: 135–170.

Sutton RS and Barto AG (2018). *Reinforcement Learning: An Introduction*. MIT Press, Cambridge, MA.

Szabo M, Almeida R, Deco G, and Stetter M (2004). Cooperation and biased competition model can explain attentional filtering in the prefrontal cortex. *European Journal of Neuroscience* 19: 1969–1977.

Szabo M, Deco G, Fusi S, Del Giudice P, Mattia M, and Stetter M (2006). Learning to attend: Modeling the shaping of selectivity in infero-temporal cortex in a categorization task. *Biological Cybernetics* 94: 351–365.

Takeuchi T, Duszkiewicz AJ, and Morris RG (2014). The synaptic plasticity and memory hypothesis: encoding, storage and persistence. *Philos Trans R Soc Lond B Biol Sci* 369: 20130288.

Tanaka K, Saito C, Fukada Y, and Moriya M (1990). Integration of form, texture, and color information in the inferotemporal cortex of the macaque. In Iwai E and Mishkin M, editors, *Vision, Memory and the Temporal Lobe*, chap. 10, 101–109. Elsevier, New York.

Taschereau-Dumouchel V, Michel M, Lau H, Hofmann SG, and LeDoux JE (2022). Putting the "mental" back in "mental disorders": a perspective from research on fear and anxiety. *Mol Psychiatry* 27: 1322–1330.

Temido-Ferreira M, Coelho JE, Pousinha PA, and Lopes LV (2019). Novel players in the aging synapse: impact on cognition. *J Caffeine Adenosine Res* 9: 104–127.

Teyler TJ and DiScenna P (1986). The hippocampal memory indexing theory. *Behavioral Neuroscience* 100: 147–154.

Thapar A, Eyre O, Patel V, and Brent D (2022). Depression in young people. *Lancet* 400: 617–631.

Thorpe SJ, Maddison S, and Rolls ET (1979). Single unit activity in the orbitofrontal cortex of the behaving animal. *Neuroscience Letters* S3.

Thorpe SJ, Rolls ET, and Maddison S (1983). The orbitofrontal cortex: neuronal activity in the behaving monkey. *Exp Brain Res* 49: 93–115.

Thouless RH (1931). Phenomenal regression to the real object. *British Journal of Psychology* 21: 339.

Tian B, Reser D, Durham A, Kustov A, and Rauschecker JP (2001). Functional specialization in rhesus monkey auditory cortex. *Science* 292: 290–3.

Tonegawa S, Nakazawa K, and Wilson MA (2003). Genetic neuroscience of mammalian learning and memory. *Philosophical Transactions of the Royal Society of London B Biological Sciences* 358: 787–795.

Tosoni A, Pitzalis S, Committeri G, Fattori P, Galletti C, and Galati G (2015). Resting-state connectivity and functional specialization in human medial parieto-occipital cortex. *Brain Struct Funct* 220: 3307–21.

Tovee MJ and Rolls ET (1995). Information encoding in short firing rate epochs by single neurons in the primate temporal visual cortex. *Visual Cognition* 2: 35–58.

Tovee MJ, Rolls ET, Treves A, and Bellis RP (1993). Information encoding and the responses of single neurons in the primate temporal visual cortex. *J Neurophysiol* 70: 640–54.

Tovee MJ, Rolls ET, and Azzopardi P (1994). Translation invariance in the responses to faces of single neurons in the temporal visual cortical areas of the alert macaque. *J Neurophysiol* 72: 1049–60.

Tovee MJ, Rolls ET, and Ramachandran VS (1996). Rapid visual learning in neurones of the primate temporal visual cortex. *NeuroReport* 7: 2757–2760.

Trappenberg TP, Rolls ET, and Stringer SM (2002). Effective size of receptive fields of inferior temporal visual cortex neurons in natural scenes. In Dietterich TG, Becker S, and Gharamani Z, editors, *Advances in Neural Information Processing Systems*, vol. 14, 293–300. MIT Press, Cambridge, MA.

Tremblay L and Schultz W (1999). Relative reward preference in primate orbitofrontal cortex. *Nature* 398: 704–8.

Tremblay L and Schultz W (2000). Reward-related neuronal activity during go-nogo task performance in primate orbitofrontal cortex. *Journal of Neurophysiology* 83: 1864–76.

Treves A (1990). Graded-response neurons and information encodings in autoassociative memories. *Phys Rev A* 42: 2418–2430.

Treves A (1991a). Dilution and sparse coding in threshold-linear nets. *Journal of Physics A* 24: 327–335.

Treves A (1991b). Are spin-glass effects relevant to understanding realistic auto-associative networks. *Journal of Physics A* 24: 2645–2654.

Treves A (1993). Mean-field analysis of neuronal spike dynamics. *Network* 4: 259–284.

Treves A and Rolls ET (1991). What determines the capacity of autoassociative memories in the brain? *Network* 2: 371–397.

Treves A and Rolls ET (1992). Computational constraints suggest the need for two distinct input systems to the hippocampal ca3 network. *Hippocampus* 2: 189–99.

Treves A and Rolls ET (1994). A computational analysis of the role of the hippocampus in memory. *Hippocampus* 4: 374–391.

Treves A, Panzeri S, Rolls ET, Booth M, and Wakeman EA (1999). Firing rate distributions and efficiency of information transmission of inferior temporal cortex neurons to natural visual stimuli. *Neural Computation* 11: 601–631.

Tsao D (2014). The macaque face patch system: a window into object representation. *Cold Spring Harb Symp Quant Biol* 79: 109–14.

Tsao DY, Freiwald WA, Tootell RB, and Livingstone MS (2006). A cortical region consisting entirely of face-selective cells. *Science* 311: 670–4.

Tsao DY, Schweers N, Moeller S, and Freiwald WA (2008). Patches of face-selective cortex in the macaque frontal lobe. *Nat Neurosci* 11: 877–9.

Tsitsiklis M, Miller J, Qasim SE, Inman CS, Gross RE, Willie JT, Smith EH, Sheth SA, Schevon CA, Sperling MR, Sharan A, Stein JM, and Jacobs J (2020). Single-neuron representations of spatial targets in humans. *Curr Biol* 30: 245–253 e4.

Tsytsarev V, Volnova A, Rojas L, Sanabria P, Ignashchenkova A, Ortiz-Rivera J, Alves J, and Inyushin M (2025). Vectorial principles of sensorimotor decoding. *Front Hum Neurosci* 19: 1612626.

Underwood AG, Guynn MJ, and Cohen AL (2015). The future orientation of past memory: the role of ba 10 in prospective and retrospective retrieval modes. *Front Hum Neurosci* 9: 668.

Ungerleider LG and Haxby JV (1994). 'what' and 'where' in the human brain. *Curr Opin Neurobiol* 4: 157–65.

Uno Y and Coyle JT (2019). Glutamate hypothesis in schizophrenia. *Psychiatry Clin Neurosci* 73: 204–215.

Uran C, Peter A, Lazar A, Barnes W, Klon-Lipok J, Shapcott KA, Roese R, Fries P, Singer W, and Vinck M (2022). Predictive coding of natural images by v1 firing rates and rhythmic synchronization. *Neuron* 110: 2886–2887.

Urgen BA and Orban GA (2021). The unique role of parietal cortex in action observation: Functional organization for communicative and manipulative actions. *Neuroimage* 237: 118220.

van den Heuvel MP, de Reus MA, Feldman Barrett L, Scholtens LH, Coopmans FM, Schmidt R, Preuss TM, Rilling JK, and Li L (2015). Comparison of diffusion tractography and tract-tracing measures of connectivity strength in rhesus macaque connectome. *Hum Brain Mapp* 36: 3064–75.

van den Heuvel OA, Veltman DJ, Groenewegen HJ, Cath DC, van Balkom AJ, van Hartskamp J, Barkhof F, and van Dyck R (2005). Frontal-striatal dysfunction during planning in obsessive-compulsive disorder. *Archives of General Psychiatry* 62: 301–309.

van den Heuvel OA, van Wingen G, Soriano-Mas C, Alonso P, Chamberlain SR, Nakamae T, Denys D, Goudriaan AE, and Veltman DJ (2016). Brain circuitry of compulsivity. *Eur Neuropsychopharmacol* 26: 810–27.

van der Linden M, Berkers RMWJ, Morris RGM, and Fernandez G (2017). Angular gyrus involvement at encoding and retrieval is associated with durable but less specific memories. *J Neurosci* 37: 9474–9485.

Van Hoesen G and Pandya DN (1975). Some connections of the entorhinal (area 28) and perirhinal (area 35) cortices of the rhesus monkey. i. temporal lobe afferents. *Brain Res* 95: 1–24.

Van Hoesen GW (1982). The parahippocampal gyrus. new observations regarding its cortical connections in the monkey. *Trends in Neuroscience* 5: 345–350.

van Strien WWJ and Hollmann MW (2025). Pain perception and modulation: fundamental neurobiology and recent advances. *Eur J Neurosci* 62: e70275.

van Vugt B, Dagnino B, Vartak D, Safaai H, Panzeri S, Dehaene S, and Roelfsema PR (2018). The threshold for conscious report: Signal loss and response bias in visual and frontal cortex. *Science* 360: 537–542.

Vann SD, Aggleton JP, and Maguire EA (2009). What does the retrosplenial cortex do? *Nat Rev Neurosci* 10: 792–802.

Veale DM, Sahakian BJ, Owen AM, and Marks IM (1996). Specific cognitive deficits in tests sensitive to frontal lobe dysfunction in obsessive-compulsive disorder. *Psychological Medicine* 26: 1261–1269.

Vedder LC, Miller AMP, Harrison MB, and Smith DM (2017). Retrosplenial cortical neurons encode navigational cues, trajectories and reward locations during goal directed navigation. *Cereb Cortex* 27: 3713–3723.

Verhagen JV, Rolls ET, and Kadohisa M (2003). Neurons in the primate orbitofrontal cortex respond to fat texture independently of viscosity. *J Neurophysiol* 90: 1514–25.

Verhagen JV, Kadohisa M, and Rolls ET (2004). The primate insular/opercular taste cortex: neuronal representations of the viscosity, fat texture, grittiness, temperature and taste of foods. *Journal of Neurophysiology* 92: 1685–1699.

Vijayabaskaran S, Zeng X, Ghazinouri B, Wiskott L, and Cheng S (2025). A taxonomy of spatial navigation in mammals: Insights from computational modeling. *Neurosci Biobehav Rev* 176: 106282.

Vinckier F, Dehaene S, Jobert A, Dubus JP, Sigman M, and Cohen L (2007). Hierarchical coding of letter strings in the ventral stream: dissecting the inner organization of the visual word-form system. *Neuron* 55: 143–56.

Vogels R, Biederman I, Bar M, and Lorincz A (2001). Inferior temporal neurons show greater sensitivity to nonaccidental than to metric shape differences. *J Cogn Neurosci* 13: 444–53.

Vogt BA (2009). *Cingulate Neurobiology and Disease*. Oxford University Press, Oxford.

Vogt BA (2016). Midcingulate cortex: Structure, connections, homologies, functions and diseases. *J Chem Neuroanat* 74: 28–46.

Vogt BA (2019). *Handbook of Clinical Neurology: Cingulate Cortex*, vol. 166 of *Handbook of Clinical Neurology*. Elsevier, Oxford, 3rd edn.

Vogt BA and Pandya DN (1987). Cingulate cortex of the rhesus monkey: II. Cortical afferents. *Journal of Comparative Neurology* 262: 271–289.

Wagner IC, van Buuren M, Kroes MC, Gutteling TP, van der Linden M, Morris RG, and Fernandez G (2015). Schematic memory components converge within angular gyrus during retrieval. *Elife* 4: e09668.

Wallis G and Rolls ET (1997). Invariant face and object recognition in the visual system. *Prog Neurobiol* 51: 167–94.

Wallis JD and Kennerley SW (2011). Contrasting reward signals in the orbitofrontal cortex and anterior cingulate cortex. *Ann N Y Acad Sci* 1239: 33–42.

Wallis JD and Miller EK (2003). Neuronal activity in primate dorsolateral and orbital prefrontal cortex during performance of a reward preference task. *Eur J Neurosci* 18: 2069–81.

Wan Z, Rolls ET, Cheng W, and Feng J (2020). Sensation-seeking is related to functional connectivities of the medial orbitofrontal cortex with the anterior cingulate cortex. *Neuroimage* 215: 116845.

Wang H, Rolls ET, Du X, Du J, Yang D, Li J, Li F, Cheng W, and Feng J (2020). Severe nausea and vomiting in pregnancy: psychiatric and cognitive problems, and brain structure in children. *BMC Medicine* 18: 228.

Wang M, Datta D, Enwright J, Galvin V, Yang ST, Paspalas C, Kozak R, Gray DL, Lewis DA, and Arnsten AFT (2019). A novel dopamine d1 receptor agonist excites delay-dependent working memory-related neuronal firing in primate dorsolateral prefrontal cortex. *Neuropharmacology* 150: 46–58.

Wang XJ (1999). Synaptic basis of cortical persistent activity: the importance of NMDA receptors to working memory. *Journal of Neuroscience* 19: 9587–9603.

Wang XJ (2002). Probabilistic decision making by slow reverberation in cortical circuits. *Neuron* 36: 955–68.

Wang XJ (2008). Decision making in recurrent neuronal circuits. *Neuron* 60: 215–234.

Wang XJ (2025). *Theoretical Neuroscience: Understanding Cognition*. CRC Press, Boca Raton, Florida.

Watanabe T and Niki H (1985). Hippocampal unit activity and delayed response in the monkey. *Brain Research* 325: 241–254.

Waters SJ, Basile BM, and Murray EA (2023). Reevaluating the role of the hippocampus in memory: A meta-analysis of neurotoxic lesion studies in nonhuman primates. *Hippocampus* 33: 787–807.

Watkins LH, Sahakian BJ, Robertson MM, Veale DM, Rogers RD, Pickard KM, Aitken MR, and Robbins TW (2005). Executive function in Tourette's syndrome and obsessive-compulsive disorder. *Psychological Medicine* 35: 571–582.

Watson JF, Vargas-Barroso V, Morse-Mora RJ, Navas-Olive A, Tavakoli MR, Danzl JG, Tomschik M, Rossler K, and Jonas P (2025). Human hippocampal ca3 uses specific functional connectivity rules for efficient associative memory. *Cell* 188: 501–514 e18.

Webb TJ and Rolls ET (2014). Deformation-specific and deformation-invariant visual object recognition: pose vs identity recognition of people and deforming objects. *Frontiers in Computational Neuroscience* 8: 37.

Webb TJ, Rolls ET, Deco G, and Feng J (2011). Noise in attractor networks in the brain produced by graded firing rate representations. *PLoS One* 6: e23620.

Weeden C, Brushfield A, Smith G, and Kesner R (2009). Subregional analysis of the hippocampus in mediating sequence learning. *Society for Neuroscience Abstracts* 780.1.

Weeden CS, Hu NJ, Ho LU, and Kesner RP (2014). The role of the ventral dentate gyrus in olfactory pattern separation. *Hippocampus* 24: 553–9.

Weghenkel B and Wiskott L (2018). Slowness as a proxy for temporal predictability: an empirical comparison. *Neural Comput* 30: 1151–1179.

Wegrzyn D, Juckel G, and Faissner A (2022). Structural and functional deviations of the hippocampus in schizophrenia and schizophrenia animal models. *Int J Mol Sci* 23: 5482.

Weiskrantz L (1968). Emotion. In Weiskrantz L, editor, *Analysis of Behavioural Change*, 50–90. Harper and Row, New York and London.

Weiss AP and Heckers S (1999). Neuroimaging of hallucinations: a review of the literature. *Psychiatry Research* 92: 61–74.

Weissman MM, Bland RC, Canino GJ, Greenwald S, Hwu HG, Lee CK, Newman SC, Oakley-Browne MA, Rubio-Stipec M, Wickramaratne PJ, et al. (1994). The cross national epidemiology of obsessive compulsive disorder. The Cross National Collaborative Group. *Journal of Clinical Psychiatry* 55 Suppl: 5–10.

Whalen PJ and Phelps EA (2009). *The Human Amygdala*. Guilford, New York.

WHO (2017). World health organization: Depression and other common mental disorders: global health estimates https://www.who.int/mental_health/management/depression/prevalence_global_health_estimates/en/.

Wiener JM, Berthoz A, and Wolbers T (2011). Dissociable cognitive mechanisms underlying human path integration. *Exp Brain Res* 208: 61–71.

Wild B and Treue S (2021). Primate extrastriate cortical area mst: a gateway between sensation and cognition. *J Neurophysiol* 125: 1851–1882.

Wilkinson ST, Katz RB, Toprak M, Webler R, Ostroff RB, and Sanacora G (2018). Acute and longer-term outcomes using ketamine as a clinical treatment at the yale psychiatric hospital. *J Clin Psychiatry* 79: doi: 10.4088/JCP.17m11731.

Williams GV, Rolls ET, Leonard CM, and Stern C (1993). Neuronal responses in the ventral striatum of the behaving macaque. *Behav Brain Res* 55: 243–52.

Willshaw DJ and von der Malsburg C (1976). How patterned neural connections can be set up by self-organization. *Proceedings of the Royal Society of London B* 194: 431–445.

Willshaw DJ, Buneman OP, and Longuet-Higgins HC (1969). Non-holographic associative memory. *Nature* 222: 960–2.

Wilson FAW and Rolls ET (1990a). Neuronal responses related to the novelty and familiarity of visual stimuli in the substantia innominata, diagonal band of broca and periventricular region of the primate. *Experimental Brain Research* 80: 104–120.

Wilson FAW and Rolls ET (1990b). Neuronal responses related to reinforcement in the primate basal forebrain. *Brain Research* 509: 213–231.

Wilson FAW and Rolls ET (1990c). Learning and memory are reflected in the responses of reinforcement-related neurons in the primate basal forebrain. *Journal of Neuroscience* 10: 1254–1267.

Wilson FAW and Rolls ET (1993). The effects of stimulus novelty and familiarity on neuronal activity in the amygdala of monkeys performing recognition memory tasks. *Experimental Brain Research* 93: 367–382.

Wilson FAW and Rolls ET (2005). The primate amygdala and reinforcement: a dissociation between rule-based and associatively-mediated memory revealed in amygdala neuronal activity. *Neuroscience* 133: 1061–1072.

Wingert AM, Agnorelli C, Peill J, Reed S, Nutt DJ, and Erritzoe D (2025). Serotonergic psychedelics for depression: A comprehensive overview. *Int Rev Neurobiol* 181: 271–304.

Wirth S, Yanike M, Frank LM, Smith AC, Brown EN, and Suzuki WA (2003). Single neurons in the monkeys hippocampus and learning of new associations. *Science* 300: 1578–1581.

Wirth S, Baraduc P, Plante A, Pinede S, and Duhamel JR (2017). Gaze-informed, task-situated representation of space in primate hippocampus during virtual navigation. *PLoS Biol* 15: e2001045.

Wiskott L (2003). Slow feature analysis: A theoretical analysis of optimal free responses. *Neural Computation* 15: 2147–2177.

Wiskott L and Sejnowski TJ (2002). Slow feature analysis: unsupervised learning of invariances. *Neural Comput* 14: 715–70.

Witter MP and Amaral DG (2021). The entorhinal cortex of the monkey: Vi. organization of projections from the hippocampus, subiculum, presubiculum, and parasubiculum. *J Comp Neurol* 529: 828–852.

Wolkin A, Sanfilipo M, Wolf AP, Angrist B, Brodie JD, and Rotrosen J (1992). Negative symptoms and hypofrontality in chronic schizophrenia. *Archives of General Psychiatry* 49: 959–965.

Wood CM, Tizkar RB, Fort M, Zhang X, Mulvihill KG, Liao N, Cockcroft GJ, McIver LB, Sawiak SJ, and Roberts AC (2025). Dysfunction in primate dorsolateral prefrontal area 46 affects motivation and anxiety. *Science* 389: 840–844.

Woolnough O, Donos C, Rollo PS, Forseth KJ, Lakretz Y, Crone NE, Fischer-Baum S, Dehaene S, and Tandon N (2021). Spatiotemporal dynamics of orthographic and lexical processing in the ventral visual pathway. *Nat Hum Behav* 5: 389–398.

Wyss R, Konig P, and Verschure PF (2006). A model of the ventral visual system based on temporal stability and local memory. *PLoS Biol* 4: e120.

Xie C, Jia T, Rolls ET, and al e (2021a). Reward versus nonreward sensitivity of the medial versus lateral orbitofrontal cortex relates to the severity of depressive symptoms. *Biol Psychiatry Cogn Neurosci Neuroimaging* 6: 259–269.

Xie C, Jia T, Rolls ET, Robbins TW, Sahakian BJ, Zhang J, Liu Z, Cheng W, Luo Q, Zac Lo CY, Wang H, Banaschewski T, Barker GJ, Bokde ALW, Buchel C, Quinlan EB, Desrivieres S, Flor H, Grigis A, Garavan H, Gowland P, Heinz A, Hohmann S, Ittermann B, Martinot JL, Paillere Martinot ML, Nees F, Orfanos DP, Paus T, Poustka L, Frohner JH, Smolka MN, Walter H, Whelan R, Schumann G, Feng J, and Consortium I (2021b). Reward versus nonreward sensitivity of the medial versus lateral orbitofrontal cortex relates to the severity of depressive symptoms. *Biol Psychiatry Cogn Neurosci Neuroimaging* 6: 259–269.

Yamins DL and DiCarlo JJ (2016). Using goal-driven deep learning models to understand sensory cortex. *Nat Neurosci* 19: 356–65.

Yamins DL, Hong H, Cadieu CF, Solomon EA, Seibert D, and DiCarlo JJ (2014). Performance-optimized hierarchical models predict neural responses in higher visual cortex. *Proc Natl Acad Sci U S A* 111: 8619–24.

Yan W and Mao D (2024). Disrupted spatial selectivity in the primate hippocampal-orbitofrontal circuit during navigation in virtual reality. *Society for Neuroscience Abstracts* PSTR 140.22.

Yang C and Naya Y (2020). Hippocampal cells integrate past memory and present perception for the future. *PLoS Biol* 18: e3000876.

Yang C and Naya Y (2023). Sequential involvements of the perirhinal cortex and hippocampus in the recall of item-location associative memory in macaques. *PLoS Biol* 21: e3002145.

Yang C, Chen H, and Naya Y (2023). Allocentric information represented by self-referenced spatial coding in the primate medial temporal lobe. *Hippocampus* 33: 522–532.

Yaxley S, Rolls ET, and Sienkiewicz ZJ (1988). The responsiveness of neurons in the insular gustatory cortex of the macaque monkey is independent of hunger. *Physiology and Behavior* 42: 223–229.

Yaxley S, Rolls ET, and Sienkiewicz ZJ (1990). Gustatory responses of single neurons in the insula of the macaque monkey. *Journal of Neurophysiology* 63: 689–700.

Yeatman JD and White AL (2021). Reading: The confluence of vision and language. *Annu Rev Vis Sci* 7: 487–517.

Zaborszky L, Hoemke L, Mohlberg H, Schleicher A, Amunts K, and Zilles K (2008). Stereotaxic probabilistic maps of the magnocellular cell groups in human basal forebrain. *Neuroimage* 42: 1127–41.

Zaborszky L, Gombkoto P, Varsanyi P, Gielow MR, Poe G, Role LW, Ananth M, Rajebhosale P, Talmage DA, Hasselmo ME, Dannenberg H, Minces VH, and Chiba AA (2018). Specific basal forebrain-cortical cholinergic circuits coordinate cognitive operations. *J Neurosci* 38: 9446–9458.

Zaharia AD, Goris RLT, Movshon JA, and Simoncelli EP (2019). Compound stimuli reveal the structure of visual motion selectivity in macaque mt neurons. *eNeuro* 6: ENEURO.0258–19.2019.

Zanos P and Gould TD (2018). Mechanisms of ketamine action as an antidepressant. *Mol Psychiatry* 23: 801–811.

Zanos P, Moaddel R, Morris PJ, Georgiou P, Fischell J, Elmer GI, Alkondon M, Yuan P, Pribut HJ, Singh NS, Dossou KS, Fang Y, Huang XP, Mayo CL, Wainer IW, Albuquerque EX, Thompson SM, Thomas CJ, Zarate J C A, and Gould TD (2016). Nmdar inhibition-independent antidepressant actions of ketamine metabolites. *Nature* 533: 481–6.

Zeidman P and Maguire EA (2016). Anterior hippocampus: the anatomy of perception, imagination and episodic memory. *Nat Rev Neurosci* 17: 173–82.

Zhang B, Rolls ET, Wang X, Xie C, Cheng W, and Feng J (2024a). Roles of the medial and lateral orbitofrontal cortex in major depression and its treatment. *Mol Psychiatry* 29: 914–928.

Zhang C, Rolls ET, and Feng J (2026a). Invariant visual object and face learning in the ventral cortical visual pathway: a biologically plausible model. *PLoS Computational Biology* 22: e1013959.

Zhang M, Livi A, Carter M, Schoknecht H, Burkhalter A, Holy TE, and Padoa-Schioppa C (2024b). The representation of decision variables in orbitofrontal cortex is longitudinally stable. *Cell Rep* 43: 114772.

Zhang R, Rolls ET, and Feng J (2026b). Sex differences in activations to the sight of faces, scenes, body parts and tools in visual and non-visual cortical regions leading to the human hippocampus .

Zhu Z, Hubbard E, Guo X, Barbosa DAN, Popal AM, Cai C, Jiang H, Zheng Z, Lin J, Gao W, Zhang J, Bartas K, Macchia D, Derdeyn P, Halpern CH, Mayberg HS, Beier KT, Zhu J, and Wu H (2021). A connectomic analysis of deep brain stimulation for treatment-resistant depression. *Brain Stimul* 14: 1226–1233.

Zhuang C, Yan S, Nayebi A, Schrimpf M, Frank MC, DiCarlo JJ, and Yamins DLK (2021). Unsupervised neural network models of the ventral visual stream. *Proc Natl Acad Sci U S A* 118: e2014196118.

Zola-Morgan S, Squire L, and Amaral D (1986). Human amnesia and the medial temporal region: enduring memory impairment following a bilateral lesion limiited to field ca1 of the hippocampus. *Journal of Neuroscience* 6: 2950–2957.

Zorumski CF, Izumi Y, and Mennerick S (2016). Ketamine: Nmda receptors and beyond. *J Neurosci* 36: 11158–11164.

Index

Publisher contact:
The MIT Press
Massachusetts Institute of Technology
77 Massachusetts Avenue, Cambridge, MA 02139
mitpress.mit.edu

EU Authorised Representative:
Easy Access System Europe, Mustamäe tee 50,
10621 Tallinn, Estonia
gpsr.requests@easproject.com

Printed by Integrated Books International,
United States of America